6th International Scientific Conference on Reliability and Durability of Railway Transport Engineering Structures and Buildings (Transbud-2017)

MATEC Web of Conferences Volume 116 (2017)

Kharkiv, Ukraine
19 - 21 April 2017

Editors:

Glib Vatulia
Andrii Plugin
Oleksandr Darenskyi

ISBN: 978-1-5108-4433-9

Printed from e-media with permission by:

Curran Associates, Inc.
57 Morehouse Lane
Red Hook, NY 12571

Some format issues inherent in the e-media version may also appear in this print version.

No changes have been made to the content of these proceedings. There may be changes to pagination and minor adjustments for aesthetics.

Printed by Curran Associates, Inc. (2019)

For additional information, please contact EDP Sciences – Web of Conferences at the address below.

EDP Sciences – Web of Conferences
17, Avenue du Hoggar
Parc d'Activité de Courtabœuf
BP 112
F-91944 Les Ulis Cedex A
France

Phone: +33 (0) 1 69 18 75 75
Fax: +33 (0) 1 69 28 84 91

contact-edps@webofconferences.org

Additional copies of this publication are available from:

Curran Associates, Inc.
57 Morehouse Lane
Red Hook, NY 12571 USA
Phone: 845-758-0400
Fax: 845-758-2633
Email: curran@proceedings.com
Web: www.proceedings.com

TABLE OF CONTENTS

ISSN (electronic edition): 2261-236X

PREFACE

The 6th International Scientific Conference **"Reliability and Durability of Railway Transport Engineering Structures and Buildings"** (TransBud 2017) was held on April 19-21, 2017 in Ukrainian State University of Railway Transport at the address: Feuerbach square 7, Kharkiv, Ukraine.

This conference has been held since 2006 every two years at the Ukrainian State University of Railway Transport.

Ukrainian State University of Railway Transport, Building Academy of Ukraine, Transport Academy of Ukraine, Kharkiv National University of Urban Economy named after Oleksii Beketov, PJSK "Ukrainian Railways", LLC "MC-Bauchemie", JSC "Tines Capital Group", LLC "MVO Invest" were the organizers of the 6th conference.

The plenary session and three sections worked at the conference: section 1 - "Construction Materials, Protection and Repair of Structures and Buildings"; section 2 - "Structures, Buildings and Facilities"; section 3 - "Railways and Highways, Subways and Industrial Transport". 311 participants presented 150 reports at the conference.

The detailed information about the conference is available on the website http://conf.kart.edu.ua/en/

TRANSBUD 2017 COMMITTEE

Conference Committee Chair
Prof. Sergii Panchenko, Ukrainian State University of Railway Transport, Kharkiv, Ukraine
Conference Committee Co-Chairs
Prof. Dmytro Goncharenko, Building Academy of Ukraine, Kharkiv National University of Construction and Architecture, Ukraine
Prof. Anatolii Turenko, Transport Academy of Ukraine, Kharkiv National Automobile-Road University, Ukraine
Prof. Volodimir Babaev, Kharkiv National University of Urban named after Oleksii Beketov, Ukraine

Scientific Committee

Section "Construction Materials, Protection and Repair of Structures and Buildings"
Section Head – Prof. Andrii Plugin, Ukrainian State University of Railway Transport, Kharkiv, Ukraine
Prof. Leonid Dvorkin, National University of Water Engineering and Environmental, Rivne, Ukraine
Dr. Georhii Linnyk, PJSK "Ukrainian Railways", Kyiv, Ukraine
Prof. Arkadii Plugin, Ukrainian State University of Railway Transport, Kharkiv, Ukraine
Artem Pryimachenko, LLC "MC-Bauchemie", Bottrop, Germany
Prof. Kateryna Pushkaryova, Kyiv National University of Construction and Architecture, Ukraine
Prof. Raisa Runova, Kyiv National University of Construction and Architecture, Ukraine
Dr. Hans-Bertram Fisher, Weimar University "Bauhaus", Germany
Prof. Galyna Shabanova, National Technical University "Kharkiv Polytechnical Institute", Ukraine

Section "Structures, Buildings and Facilities"
Section Head – Prof. Valery Shmukler, Kharkiv National University of Urban Economy named after Oleksii Beketov, Ukraine
Section Co-Head – Prof. Glib Vatulia, Ukrainian State University of Railway Transport, Kharkiv, Ukraine
Prof. Viacheslav Astanin, National Aviation University, Kyiv, Ukraine
Prof. Andrii Bambura, SE "Scientific Institute of Building Constructions", Kyiv, Ukraine
Prof. Oleksandr Vainberg, PJSC "Ukrhydroproject", Kharkiv, Ukraine
Prof. Vitalii Kozhushko, Kharkiv National Automobile-Road University, Ukraine
Prof. Jozef Melcer, University of Žilina, Slovakia
Prof. Oleksandr Semko, Poltava National Technical University named after Yuri Kondratyuk
Prof. Stanislav Fomin, Kharkiv National University of Construction and Architecture, Ukraine

Section "Railways and Highways, Subways and Industrial Transport"
Section Head – Prof. Oleksandr Darenskyi, Ukrainian State University of Railway Transport, Kharkiv, Ukraine
Prof. Eduard Danilenko, State Economical and Technological University of Transport, Kyiv, Ukraine
Prof. Valeriy Zhdanyuk, Kharkiv National Automobile-Road University, Ukraine
Sergii Karpenko, ME "Kharkiv Subways"
Prof. Mykola Kurgan, Dnipropetrovsk National University of Railway Transport named after Vsevolod Lazaryan
Dr. Oleksii Skorik, Ukrainian State University of Railway Transport, Kharkiv, Ukraine
Yuzef Tulej, OJSC PJSK "Ukrainian Railways", Kyiv, Ukraine
Prof. Sergii Tolmachov, Kharkiv National Automobile-Road University, Ukraine
Dr. Volodymir Harlan, PJSK "Ukrainian Railways", Kyiv, Ukraine
TomashSzuba, JSC "Tines Capital Group", Krakov, Poland

Conference Secretariat
Dr. Olga Borziak, Ukrainian State University of Railway Transport, Kharkiv, Ukraine
Dr. Oleksii Lobiak, Ukrainian State University of Railway Transport, Kharkiv, Ukraine
Dr. Oleksii Pluhin, Ukrainian State University of Railway Transport, Kharkiv, Ukraine
Dr. Dmytro Potapov, Ukrainian State University of Railway Transport, Kharkiv, Ukraine

Statement of Peer review

In submitting conference proceedings to MATEC Web of Conferences, I certify to the Publisher that I adhere to the **Policy on Publishing Integrity** of the journal in order to safeguard good scientific practice in publishing.

1. All articles have been subjected to peer review administered by the proceedings editors.
2. Reviews have been conducted by expert referees, who have been requested to provide unbiased and constructive comments aimed, whenever possible, at improving the work.
3. Proceedings editors have taken all reasonable steps to ensure the quality of the materials they publish and their decision to accept or reject a paper for publication has been based only on the merits of the work and the relevance to the journal.

Title, date and place of the conference

6th International Scientific Conference "Reliability and Durability of Railway Transport Engineering Structures and Buildings", Transbud-2017, April 19-21, 2017, Kharkiv

Proceedings editor(s):

Prof. Glib Vatulia, Prof. Andrii Plugin, Prof. Oleksandr Darenskyi

Date and editor's signature

June 19, 2017

Effect of microfillers on the concrete structure formation

Olga Borziak[1,*], *Stanislav* Vandolovskyi[2], *Vitalij* Chajka[2], *Volodymyr* Perestiuk[3], and *Oleksandr* Romanenko[1]

[1]Ukrainian State University of Railway Transport, Building Materials and Structures Department, Feuerbach sq. 7, 61050 Kharkiv, Ukraine
[2]Kharkiv National University of Civil Engineering and Architecture, Building Materials and Products Department, Sumska st. 40, 61002 Kharkiv, Ukraine
[3]Tines Ukraine, B. Khmelnitskogo st. 16-22, 01030 Kiyv, Ukraine

Abstract. Physical and mechanical properties of concrete as a composite material are determined by structure formation processes. This paper examines the effect of dispersed mineral fillers of anthropogenic origin on concrete structure formation processes and physical and mechanical properties of fine concrete. The presence of surface charge and electrosurface potential leads to the formation of electrically heterogeneous contacts between oppositely charged surfaces which determine the strength of the material as a whole. The introduction of steel microfillers leads to the formation of strong electrically heterogeneous contacts, hardening of concrete structure and increase in its strength.

1 Introduction

Concrete is a multi-component composite material, whose physical and mechanical properties are determined by structure formation processes. The structure formed thereby is multilevel, from the microstructure to macrostructure of concrete which is a composite material [1]. Each level of the structure is determined by the dimensions of structure-forming elements and has certain structural characteristics, in particular the grain separation factor and water-cement ratio [1-4]. The concrete strength properties are defined by electrically heterogeneous contacts, i.e. the interaction between the oppositely charged surfaces with structural elements of all specified levels. The regulation of structural elements and the creation of electrically heterogeneous contacts allow us to control the properties of concrete on the whole.

Introduction of dispersed mineral admixtures, including active admixtures and fillers which have a significant impact on the structure processes, is one of such effective ways to improve the technical properties of concrete. The substances of technogenic origin and the production waste, i.e. substances with a certain mineral composition and a certain degree of dispersion are highly recommended for use as fillers.

* Corresponding author: borziak.olga@gmail.com

Mineral and polymer fibers [3], mineral powders such as silicate, carbonate powders or ash [1, 4] are most commonly used as fillers. Increase in the structure strength of a filled composite material is reached by achieving the optimum concentration of a dispersed phase; the definite particle dispersibility provides the best structural characteristics, including close packing and the maximum system consolidation [1, 4]. Empirically, it has been found that the dependence of concrete strength on the amount of admixtures of microfillers is of an extreme nature. Excessive content of high-dispersed microfillers may result in the occurrence of self-stressing sections and cracking formation [1, 4].

The aim of the study is to disclose the mechanism of the structure formation of fine concrete with silicate and metallic microfillers.

2 Development of hypothesis on structure formation processes in fine concrete with steel powder as filler

According to [5-9], the structure and strength of composite materials such as concrete depend on the electrosurface properties of all the components.

In water-dispersed systems, which include the cement stone of concrete, the particles of an inorganic dispersion phase have a surface charge and an electrical surface potential [7]. Electrically heterogeneous contacts (EHC) are formed between oppositely charged surfaces, which define the strength of a material on the whole (Fig. 1). The strength of these contacts is ensured by ion-ion interactions between the potential-determining ions (PDI) of the surfaces, and ion-dipole interactions between the potential-determining ions and the molecules of adsorption layers of water [8]:

$$R^{EHC} = \frac{\varepsilon \cdot z^{-} \cdot 2e \cdot \psi_{ES}^{E}}{4\pi \cdot \varepsilon_1 \cdot h^2 d_1} + \frac{\mu_W \cdot \psi_{ES}^{E}}{2\pi \cdot \varepsilon \cdot d_1 \cdot h_\mu^3} \tag{1}$$

Where z – the valence of potential-determining ions; e – the electronic charge; ε – the relative dielectric permittivity in the space between the potential-determining ions and the counter ions ($\varepsilon = 1$); ε_1 – the dielectric permittivity of a structured water layer; ψ_{ES}^{E} – the equilibrium electrosurface potential; d_l – the distance between the potential-determining ions and counter ions when in contact; h and h_μ – the distance between the potential-determining ions, or between the potential-determining ions and the water dipole; μ_W – the dipole moment of a water molecule.

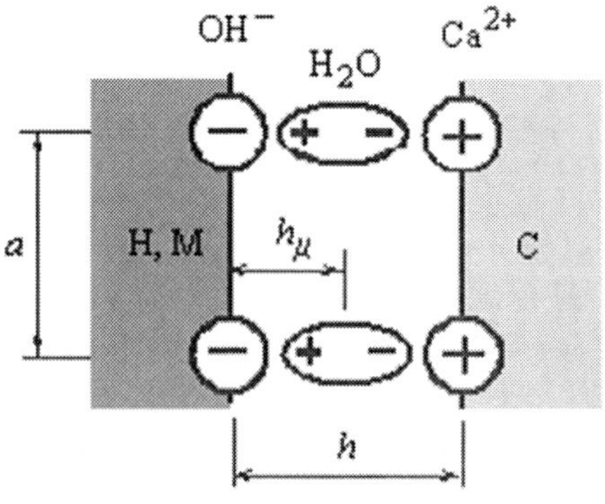

Fig. 1. Elemental electroheterogeneous contact in concrete: C – crystal hydratation products; H, M – gel hydratation products or microfiller particles; a – the distance between potential-determining ions on the surface of the substance; h – the distance between potential-determining ions in EHC; h_μ – the distance between potential-determining ions and the dipole of water.

Such contacts are formed between crystalline hydration products, which have a positive surface potential, and gel products of hydration, which have a negative surface potential. Ions Ca^{2+} are potential-determining ions of crystalline substances, while OH^- are potential-determining ions of gel substances. The first member describes ion-ion interaction between Ca^{2+} and OH^-, while the second describes ion-dipole interaction between OH^- and the adjacent dipoles of H_2O.

The concrete tensile strength R^C, Pa may be defined as the product of strength of electroheterogeneous contact R^{EHC}, Pa, by the surface portion of such contacts in the concrete section γ_s^{EHC}:

$$R^C = R^{EHC} \cdot \gamma_s^{EHC} \tag{2}$$

According to [8], γ_s^{EHC} depends on the water-cement ratio (W/C) and mineral composition of the cement clinker, as well as the composition and the dispersion of fillers.

The structuring role of mineral fillers is determined by their surface properties. A variety of silicates, which are composed of more than 90% silica, are commonly used as fillers in composite materials based on mineral cementing agents. Silicate minerals have negative surface potential, which is 1.26 V.

The metal powder consisting of steel particles used as filler has a negative electrosurface potential in the liquid phase of concrete (dispersion environment). In case of cement hydration, the particles of steel will form an EHC with portlandite $Ca(OH)_2$. Portlandite has a cation Ca^{2+}, as a potential-determining ion and an equilibrium electrosurface potential equal to + 0.53 V. According to the image method used in calculations of silicate cementing agents [8; 9], this potential-determining ion will cause an opposite charge in a steel particle equal to –0.53 V, and the corresponding EHC with a link of the e^-–H_2O–Ca^{2+} [8] type, the strength of such contacts according to (1) is:

$$R^{EHC} = \frac{1 \cdot 1 \cdot 2 \cdot 1.6 \cdot 10^{-19} \cdot 0.53}{4 \cdot 3.14 \cdot 2.16 \cdot 2.4 \cdot 10^{-10} \cdot (4.8 \cdot 10^{-10})^2} + \frac{6.1 \cdot 10^{-30} \cdot 0.53}{2 \cdot 3.14 \cdot 1 \cdot 2.4 \cdot 10^{-10} \cdot (2.4 \cdot 10^{-10})^3} = 273.3 MPa$$

Thus, the introduction of the dispersed metallic (steel) microfiller into concrete will increase the quantity of strong EHC and improve the strength of concrete.

The strength of concrete is also affected by the dispersity of fillers. In accordance with [8], the introduction of large unit area fillers will increase the γ_s^{EHC} – EHC concentration which, according to (2), will increase the strength of concrete as a whole. It should be taken into account [1] that the excessive amount of particles with a similar surface charge will cause the decrease in the quantity of EHC, the occurrence of repulsive forces, and the strength of concrete will consequently decrease. The maximum strength will be achieved if the ratio of these components equals 1.

3 Research methods and materials

Experimental studies have been conducted to test theoretical assumptions about the impact of steel microfillers on the strength of fine concrete. Sample bars made from cement-sand mixture with microfiller – steel powder in the amount of 20, 30 and 40 % of the mass of cement have been produced. The steel powder of the «PGR 3» grade has been used [10], the particle size is: >0.16 mm – 5.9%; 0.16÷0.045 mm – 75.3%; < 0.045 mm – 18.8%. For comparison, control samples (sand and cement ratio is 3:1, W/C = 0.4) have been produced.

The hardened samples were tested for bending and compression at ages of 7, 14 and 28 days. The concrete structure was explored using a scanning electron microscope.

4 Results of experimental studies

The diagrams (Fig. 2, 3) show the results of the strength measurement of control (filler content 0%) and basic samples (filler content 20, 30 and 40%) at ages of 7, 14 and 28 days.

As the results of the experimental studies show, the introduction of steel fillers significantly improves the strength of concrete in the early stages of hardening. This is particularly noticeable in the modification of the deflection strength values: the addition of 30% steel filler increases the strength of concrete at the age of 7 days by more than 4 times (8.6 MPa) compared to the control samples of concrete (1.9 MPa). Subsequently the increase in strength becomes slower and at the age of 28 days the maximum deflection strength with 30% of steel filler is 12.5 MPa, which by 2.5 times exceeds the strength of the control samples (4.9 MPa).

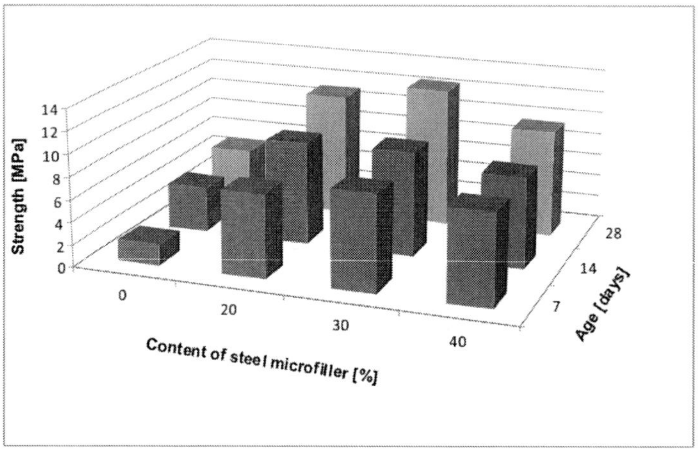

Fig. 2. Dependence of deflection strength of fine concrete on the content of steel microfiller.

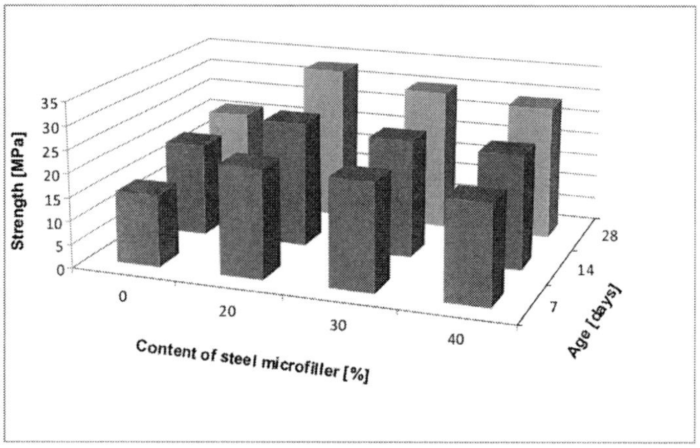

Fig. 3. Dependence of compression strength of fine concrete on the content of steel microfiller.

Fig. 4 shows the electron micrograph scanning of the shear of concrete with steel powder used as microfiller. In the center of the image there is a particle of steel microfiller having a well-defined crystal structure. The surface of the particle has been covered with prismatic crystals of hydration products. As the image shows, there is a very close contact between the steel filler particle and the cement stone.

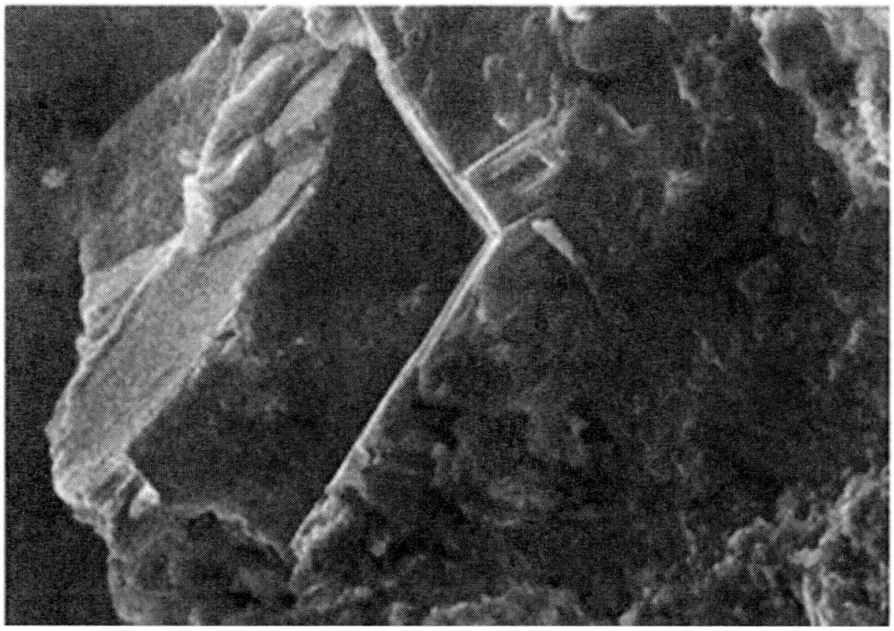

Fig. 4. The electron micrograph scanning of concrete with metal filler (×1500).

5 Discussion of results obtained in the study

The results of the experimental study confirm the hypothesis developed. The introduction of steel powder as microfiller significantly increases deflective and compression strength of fine concrete. A decrease of strength due to the introduction of steel powder in the amount of more than 30% is attributed to the deviation of the number of EHC from the optimum quantity (to the deficiency of Portlandite crystalline hydrates) and the emergence of repulsive forces between like-charged (negatively charged) particles of steel powder and calcium silicate hydrate.

6 Conclusions

The results of the theoretical and experimental studies have found that the introduction of steel powder into a concrete mixture significantly influences the structure formation of concrete. A great quantity of strong EHC is formed between steel filler and crystallhydrate products. The optimum content of steel filler, which is 30% of the mass of cement, has been substantiated. This content of steel microfiller will increase the deflection strength by 2.5 times and the compression strength by 1.4 times.

References

1. K.K. Pushkar'ova, et al. *Suchasni budivel'ni materialy i konstruktyvni systemy dlya zvedennya dostupnoho zhytla ta ob"yektiv infrastruktury* (VIK Print, Kyiv, 2015),
2. A.A.Plugin, A.N.Plugin, O.A.Kalinin, S.V.Miroshnichenko, D.A.Plugin, L.V.Trikoz, 18 Internationale Baustofftagung, Tagungsbericht, **1**, 0874–0881 (2012)
3. A.A. Plugin, V.A. Arutyunov, T.A. Kostyuk, D.A. Bondarenko, O.V. Starkova, Sovremennyiy nauchnyiy vestnik, **32(171)**, 109-116 (2013)

4. E.A. Belichenko, A.N. Plugin, A.A. Plugin, S.N. Tolmachev, EEJET, **6/5(60)**, 13-17 (2012)
5. A.V.Romanenko, A.A.Plugin, A.N. Plugin, O.A. Kalinin, Al.A.Plugin, EEJET, **5/10(59)**, 28-32 (2012)
6. A.N. Plugin, A.A. Plugin, Stroitelnyie materialyi, **7 (631)**, 68-71 (2007)
7. A.N. Plugin, A.A. Plugin, O.A. Plugin, A.S. Kaganovskii, O.V. Gradoboev, 18 Internationale Baustofftagung, Tagungsbericht, **1**, 0656-0663 (2012)
8. A.N. Plugin, A.A. Plugin et al. *Osnovy teorii tverdeniya, prochnosti, razrusheniya i dolgovechnosti portlandtsementa, betona i konstruktsiy iz nikh* (Naukova dumka, Kyiv, 2012)
9. V.V. Kapranov Tverdenie vyazhuschih veschestv i izdeliy na ih osnove (South Ural Publishing House, Chelyabinsk, 1976)
10. Derzhavnyy standart GOST 9849 – 86

Regulation of stress-strain state and calculation of deformations of continuous foundations under increased multiple loads

Anatoliy Butenko[1,*], and *Oleg* Gerasimenko[2]

[1]TOV Trans Stroj Komplex (LLC under the laws of Ukraine), Simferopolskiy off st. 6, 61052 Kharkiv, Ukraine
[2]Ukrainian State University of Railway Transport, Building Materials and Structures Department, Feuerbach sq. 7, 61050 Kharkiv, Ukraine

Abstract. The analysis of design and operation experience of grain elevators is executed. Metal silos of high capacity are representatives of structures with continuous foundations that perceive increased and multiple loads. Methods of regulating the stress-strain state with the help of creation of artificial bases are considered in detail. A technique for estimating the stress-strain state and calculating the deformations of continuous foundations is proposed.

The development of agricultural industry in Ukraine has caused the need for developing infrastructure. Currently, the construction of railway transfer platforms for crops and other loose products and materials is a vital area. Such products are stored in silo-type elevators using high-capacity metal silo towers.

The dimensions of such structures determine the dimensions of foundations. Generally, flat bottom silos, which are the most loaded, are constructed with the following types of foundations:

– natural foundations;
– piled foundations;
– artificial foundations.

Natural base is used for continuous foundations in rare cases, due to the fact that the design resistance of soil is extremely limited. For instance, with an approximate height of the silo of 25 meters and with the corresponding volume of product, the load will be approximately 20 tons / m^2 (0.2 MPa). After taking into account the dead load of the foundation, the silo and its technological equipment as well as other loads (snow, etc.), the total load on the foundation is over 30 tons / m^2 (0.3 MPa). Only certain types of soils can carry such loads.

Piled foundations are extremely expensive and they increase the construction time. Besides that, the construction of piled foundations is not always efficient and justified. While choosing the foundation type, one of the criteria is availability and position of support ground, which determines the length of the poles and influences both economic

* Corresponding author: butenkoanatoliy@gmail.com

rates and reliability of the foundation. The distance between poles is impossible to decrease, even if the bearing capacity of poles is high enough to carry the load of the foundation, since the stiffness and strength parameters of the grill are limited.

The most interesting foundations are artificial, including foundations with ground (sand or crushed stone) beds, and also foundations on stabilized soil, reinforced soil and strengthened soil with the help of jet grouting. ***Soil stabilization*** is increasing bearing capacity of soil with several methods: mechanical (tamping), dewatering [1, 2], injection, thermal (calcinating, freezing). ***Injection methods*** of soil stabilization assume the injection of hardening materials into the soil (including chemical methods – cementation, silicification, tarring).

Large part of the territory of Ukraine is in difficult soil conditions for construction and maintenance. One of the most important tasks is providing reliability and durability of foundations. The difficulty of soil conditions is illustrated by loose unstable soils, covering major territories. Loose or unstable soils are the ones that drastically change their structure due to additional external factors and cause significant subsidence. Structural breakdowns of these soils happen under physical or mechanical stresses. Structurally unstable soils are clayey, collapsible forest soils, the ones that constrict heavily while soaking. Such soils deform easily, which leads to large subsidence and cause additional expenses for maintenance and strengthening.

Limits for use of various stabilization methods are determined according to the soil permeability and the properties given to the stabilized soil (table 1).

Table 1. Limits for use of various stabilization methods

Soil type	**Gravel**	**Sand**					**Loamy sand**		**Aeolian soil**		**Loam**	**Clay**
Soil impermeability, meters per day	**500**	**100**	**80**	**50**	**20**	**5**	**1**	**0.3**	**2**	**0.2**	**0,05**	**0.005**
Efficient stabilization methods												
Cementation	stabilization with strength from 0 to 50 MPa	+	+									
Clay grouting		consolidation with strength from 2 to 5 MPa										
Silicification:												
two-solution			stabilization with strength from 0 to 5 MPa									
one-solution						consolidation with strength from 2 to 5 MPa			+	+		
gas			stabilization with strength from 0 to 5 MPa						stabilization with strength from 0 to 5 MPa			
tarring					stabilization with strength from 0 to 5 MPa							
electrochemical stabilization											consolidation with strength from 2 to 5 MPa	

Taking this into account, the methods are divided into the following types:

1)rapidly increase mechanical strength and impermeability of soil;

2)provide stabilized soils with impermeability only;

3)increase water resistance and thickness of soil (dewater clayey soil and muck).

Table 1 shows that cementation should be used to provide soil stabilization for gravel and sandy soils with soil permeability index from 80 to 500 meters per day. Lower limit is characterized by coarse sands, which can be penetrated by modern cement.

In order to give impermeability to sandy soils, where we cannot use cement, it is replaced with silicate and clay-silicate solutions, which can provide impermeability for sandy soils with permeability index from 20 to 100 meters per day. Sandy soil is stabilized by two-solution silicification. This method is viable for soils with permeability index from 2 to 80 meters per day.

A number of one-solution silicification methods (with phosphoric acid, sulfuric acid and aluminum sulfate, sodium aluminate, etc.) can be used in sandy soils with permeability index from 0.5 to 5 meters per day. For the most part these methods provide soils with impermeability. The method of one-solution silicification using hexafluorosilicic acid provides sandy soils with strength and impermeability and can be used for soils with permeability index from 0.2 to 20 meters per day.

This method allows for the improved formulas with increased penetration capability and strength to be used [3-6]. The method of one-solution silicification, which involves chemically active substances from the soil itself, is used to stabilize collapsible forest soils. This method is used for collapsible soils with impermeability index from 0.2 to 2.0 meters per day, where the soil dampness is below 17%, which means that the forest has to be located above ground water level. With higher dampness, approx. 20-22%, gas silicification is used. First, the soil is injected with carbon dioxide, then silicate, and again carbon dioxide. These processes can be only performed by experienced professionals. As a result of gas silicification, the soil gets strength and water resistance.

Stabilization of sandy soils combined with providing impermeability is achieved with the help of tarring, if the soil has impermeability index from 0.5 to 20 meters per day.

Method of electrochemical stabilization allows us to give water resistance to clay soils saturated with water, and, as a result, eliminate soaking and bulging in water. Sphere of use for this method is limited by the following soil impermeability indexes: using two-solution electro-silicification - from 0.05 to 0.2 meters per day, using one-solution electro-silicification - from 0.005 to 0.2 meters per day. Between these limits electric current multiplies the penetration of solution into the soil by 1.5…2.

Stress-strain properties of soils change dramatically during stabilization and the modulus of deformation multiplies by 5-10. So, while choosing the type of soil stabilization and calculating the soil settlement, calculation of stabilized soil footing is performed as if for linearly deformed environment. Preset parameters of stabilized soil are used in the "footing-foundation" design diagram. It is worth mentioning that in case of reconstruction the reverse problem is solved – the adjustment of parameters is made for existing foundation.

Extraordinary complexity of silo foundations consists in operating mode under repeatable load-unload rotations in a relatively short time span [7].

Loose soils under the foundations can lead to the breakdowns in stability of footing. Careful attention should be given to the soils, which are in watered condition, because the soil can change its structure, consistency and lose stability as a result of destruction of bonds. This process is possible for the soils, which are in the zone of footing and foundation or in case of preconsolidation of surface soil levels of footing with further water restraint in soil mass. In a similar situation the most optimal decision will be to strengthen loose or destroyed soils of the footing.

While strengthening footing soils through injections or with the help of jet grouting, we should consider stress-strain behavior of foundation. There are structural elements with different stiffness properties and different tension under the bottom of foundation, so the head pressure or the depth of jet grouting differs under various structural elements of the foundation as well. It is necessary to take into account the fact that the head pressure value must not exceed the value of design resistance of soils.

Extended foundation and the outlined in the plan foundations deform during stabilization of soils and under work, which determines the overall deformation diagram of the foundation and its stress-strain behavior (SSB) in the future.

The problem of determining SSB comes down to solving complex configuration plate with stiffness characteristics, which differ significantly from one another, on quasi-elastic basis. Stiffness parameters of a large continuous foundation depend on geometrical dimensions (Fig. 1). The value of soil resistance under the bottom of foundation depends on stiffness of the foundation itself and the applied loads, which are different as a result of distribution across the footprint of the structure. There are numerous loads across the foundation: the pressure, transferred from the upper body of the silo (snow load); the dead weight of the upper body of the silo; the load of loose materials on the walls of the silo resulting from friction; the load of the weight of the product applies to the other parts of the silo. There are zones on the footprint of the foundation, which are in quasilinear deformed environment and zones of soil behavior beyond the elasticity limits.

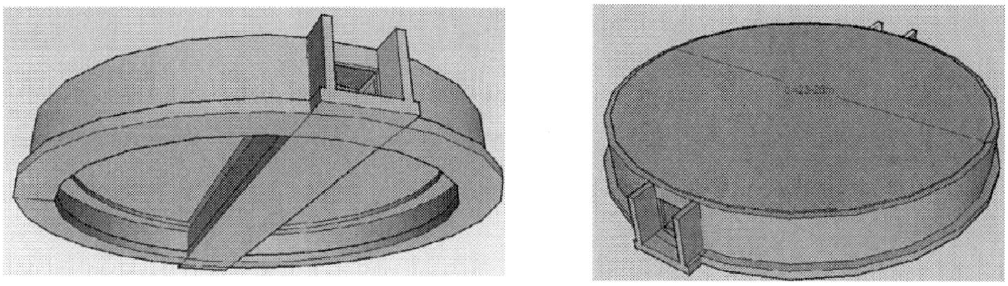

Fig. 1. Exterior design of continuous silo foundation.

The experience in operating existing elevator complexes, and also the results of geodesic monitoring of silo bodies and the experience in observing settlement of foundations [7] show that the modern calculation methods, used in designing, are inefficient. The data, received from calculations, do not match the data of geodesic observations of the settlement of foundations and their structural parts. The usage of finite element models while predicting the settlement of continuous foundations is not enough. For the most part, the reason for this is the approximate estimation of parameters as a result of possible negative processes in soils during operation of foundations or geological properties of soils. Multiple repeatable loads of silos cause additional settlement of the foundations and worsen the condition of soils.

The method of equivalent layer, which takes into account transverse extension of soil unlike the method of gradual summation, is the most reliable method for evaluation and prediction of foundation settlement. With the width of the foundation bottom $b \geq 10$ m and the deformation modulus of soils $E \geq 10$MPa (also allowed to use for soil layers with the deformation modulus $E<10$ MPa in compressed soils, if their combined width does not exceed 0.2 from the compressed width) the method of linearly deformed layer is applied. However, increasing the foundation dimensions, which is typical for continuous foundations, decreases the accuracy of settlement calculations.

In order to estimate the settlement and stress-strain behavior of continuous foundations the engineering method based on design principles of DBN (National building code) B.2.1-

10-2009 [8] is suggested. This method takes into account structural features of large silo foundation, mutual influence of neighbor parts of the foundation in proportion to tension under their footing and stiffness of the foundation, calculations of settlement under specific areas of foundations. In order to estimate SSB we should also consider stiffness of foundation basis and prestress caused by the injection of solutions for soil strengthening.

References

1. A.M. Plugin, O.S. Gerasimenko, S.V. Miroshnichenko, Plugin A.A., Collected scientific works of Ukrainian State University of Railway Transport, **63**, 63-69 (2004)
2. A.M. Plugin, A.A. Plugin, O.S. Gerasimenko, Collected scientific works of Ukrainian State University of Railway Transport, **77**, 162-175 (2006)
3. A.N. Plugin, A.A. Plugin, L.V. Trikoz, Collected scientific works of Ukrainian State University of Railway Transport, **87**, 108-120 (2007)
4. A.N. Plugin, O.S. Gerasimenko, L.V. Trikoz, A.A. Plugin, Naukovii visnyk budivnitstva, **48**, 154-162 (2008)
5. O.S. Gerasimenko, *Increasing the penetration capability and strength of liquid glass* (Kharkiv, UkrGAZHT, 2008)
6. O.S. Gerasimenko, A.A. Plugin, A.M. Plugin, L.V. Trikoz, *Composition of liquid glass*, patent UA87795C04B28/26 (2009)
7. A.A. Butenko, O.V. Kichaeva, Suchasni tehnologii ta metodi rozrahunkiv u budivnytstvi. Collection of scientific articles, **5**, 308-315 (2016)
8. *Footings and foundations of buildings. Basis principles of structural design.* DBN (National building code of Ukraine) B.2.1-10-2009, (Kyiv, 2009)

Mathematical model of consolidation of fine concrete mixtures with different mobility, casted by vacuumizing and axial pressing in layers

Elena Dedeneva[1,*], *Ibrahim* Kazimagomedov[1], *Saad* Salem[1], *Tatiana* Kostyuk[1], *Younis* Basheer N.[1], and *Marina* Yakimenko[1]

[1]Kharkiv National University of Civil Engineering and Architecture, 61002 Kharkiv, Ukraine

Abstract. A mathematical model allowing establishing regularities in the consolidation processes of fine-grained concrete mixtures with different mobility and compaction methods has been worked out. This study is based on two-phase systems and nonlinear character of their consolidation. It resolves the question of the choice of vacuumizing optimal parameters and axial pressing in layers for molding of thin-walled products such as concrete roof tiles and concrete pipe products. Finally, we can get products without heat treatment by the materials and energy-saving technologies.

1 Relevance of the work

Development and implementation of a new generation of concrete are considered very important for modern construction that will allow realizing the most interesting architectural projects. Concrete production is based on a thorough study of the processes that reveals new patterns of structure at different levels: micro-, meso- and macro-levels. The concrete structure is the main determining factor of its basic physical and mechanical properties. Therefore, the study of methods of forming a homogeneous structure, achieving high levels of strength at early stages of hardening, and producing highly functional concrete is extremely important. Dependencies between concrete structure and its properties are common regularities for all kinds of concrete, which consider the structural components of the composite theory. By installing these regularities, we can impact concrete structure, form its properties in order to obtain concretes of specified technical parameters [1, 2, 3]. Modern technology intensive construction methods require removing mold forming in short time having a sufficient quality of concrete. Therefore, High Performance Concrete (HPC) characterized by high operating ability is more frequently used [3, 4]. The main branches of the HPC use are the high-rise construction, nuclear power plants, waterworks, pipelines, bridges and engineering structures, road surfaces of monolithic and precast-monolithic structures, runways of airfields, launching complexes for space systems and other special units [2].

Recently, as a result of sharp increase in intensity of traffic flows, environmental degradation and impact of aggressive environment, all these factors have led to increasing

* Corresponding author: kstuca.bmiv@gmail.com

the load on reinforced concrete structures of transport objects, bridges, overpasses, tunnels, and deformation of these structures, thus requiring constant repair. It is known [3] that after a long operation local defects in the concrete cover appear in the elements of reinforced concrete bridges. It assists to protect the steel bars. The whole evolution of the improvement of the concrete structure to increase its strength is associated with a decrease in the size of coarse aggregate from 20-40 to 3-10 mm [4]. Thus, fine-grained concrete is increasingly in demand. Fine-grained concrete is used for production of load-bearing elements of concrete and reinforced concrete structures. As it is known, the structure of concrete is the main and determining factor of its physical-mechanical properties. It is compacted both during the molding and curing processes. The period of formation of the concrete structure and its properties depend on the composition of the initial materials and the method of molding [4 - 7].

For the selection of optimal parameters of effective methods for compacting concrete and mortar mixtures, a mathematical model with the establishment of appropriate regularities in the course of these complex processes is important for theoretical justification.

The object of the authors' research is thin-walled products in the form of concrete pipes (circular and trough configuration) and flat concrete tiles. In Ukraine, as well as in the world, there is a great demand for roofing materials and materials for water disposal and land reclamation. Concrete materials of this functional purpose allow replacing expensive ceramic tiles and pipes, as well as environmentally unsafe asbestos-cement materials.

By analogy with the study of consolidation processes in soils, this paper is based on the biphasic of concrete systems: solid and gaseous one in rigid mixtures and solid and liquid one in ductile [3]. At the first stage, consolidation, the particles of the "skeleton" are mechanically approached and the material acquires new properties, in particular the structural strength R_{str} without chemical transformations of the clinker minerals that make up the molding mixture. At the hardening stage, active crystallization processes take place, in the result of which the material of building products acquires specified properties (strength, density, etc.). The problems of accelerating the molding process, both at the first and second stages, have great importance for the theory and practice of obtaining building materials and products.

The solution of the nonlinear equation of consolidation allows us to link process parameters, load, time of its application, thickness of the layer of the material compacted, with the physical and mechanical characteristics of the molding mixture obtained on the basis of an experimental study of samples of the mixture at a compression test. The processing of results allows us to establish the dependence of the relative porosity ψ on the effective pressure σ_{eff}, the filtration coefficient K_f on the relative porosity, and on the basis of the obtained data to calculate the constants for each mixture composition. It is essential that the description of the consolidation process of molding mixtures is carried out within the framework of a mathematical model for mixtures of various mobility both tough and low-plastic ones.

2 Research and discussion of the results

The process of molding mixture compaction is similar to soil compaction in the approach of the particles of the composite (crushed stone, sand, cement and bound water) with filtration of the medium filling the gaps in the skeleton. If the mixture is rigid, is air (axial pressing by layers) if it is plastic (vacuumizing) the filtered medium is the water-cement mixture. In the end of the consolidation process, when the particles will be closer to the contact of the bound water shells (the end of the filtration process) the material acquires new properties and, most importantly, structural strength, which allows removing mold forms immediately.

Further increase in the applied load (vacuum) does not lead to changes in mechanical properties, since even a high-pressure fluid application cannot delete associated fails. The consolidation process is very fleeting, thus, the more rigid moldable mixture, the faster it compacts Thus, in the case of axial pressing in layers, the compaction of a layer up to one-centimeter-thick takes less than a second, and in the case of vacuumizing, it takes 5-7 seconds. This makes it possible to disregard processes of dissolution and hydrolysis of the initial particles of the binder and, especially, crystallization of calcium hydro silicate and calcium hydroxide in the pore electrolyte, but the binder is water, and there are processes of dissolution and hydrolysis of the initial, especially small particles of the binder, and formation of contacts between original particles and ones of a new formation [5]. The material constants can be determines as follows

$$\xi = \frac{yh}{\sqrt{2\theta_i C_k t}} \tag{1}$$

where h is the thickness of a compacted layer; C_k is the coefficient of consolidation in compression; t is the loading duration; $\theta = ce^{-0.5\,(T-1)\,q}$; q is the compression pressure; a, n, α are dimensionless parameters determined by compression testing and incoming depending on

$$\psi = ce^{-\alpha q} \tag{2}$$

$$K_f = K_f^0 \psi^n \tag{3}$$

Relating the degree of compaction of the concrete mix with the process parameters compression tests of fine mixtures were held at different values of w/c and humidity of a mixture as shown in Table 1 below. There are specified values of structural strength R_{str} determined experimentally.

Table 1. The dependence of structural strength of concrete mixtures on moisture and water-cement ratio.

Number of composition	Water-cement ratio, W/C	Concrete moisture, W, %	Structural strength R_{str}, kg/cm^2
1	0.33	5.25	0.42
2	0.42	5.25	0.41
3	0.42	6.8	0.31
4	0.39	18.0	0.42
5	0.41	18.9	0.40

Compression tests were carried out in a cylinder of a diameter of 51 mm with a pressure from 0.1 to 0.5 MPa, stepwise, with a uniform step of 0.1 MPa, which completely covers the range of pressures used for axial pressing in layers. A vacuum mold form with dimensions of 20×20×5 cm was also used for tests of fine-grained mixtures compacted by vacuumizing. The range of generated vacuum pressures is from 0.02 MPa to 0.1 MPa with an interval of 0.02 MPa, which corresponds to the values of generated pressures in the molding of concrete roof tiles. As shown by compression studies of concrete mixtures in a wide range of composition and humidity changes [4], the coefficient of initial filtration K_f^o values from 1.8×10^{-7} to 7.4×10^{-7}. The graphical dependence of the relative porosity on the effective pressure for mixtures 1-5 is shown in Fig. 1 in semi logarithmic coordinates.

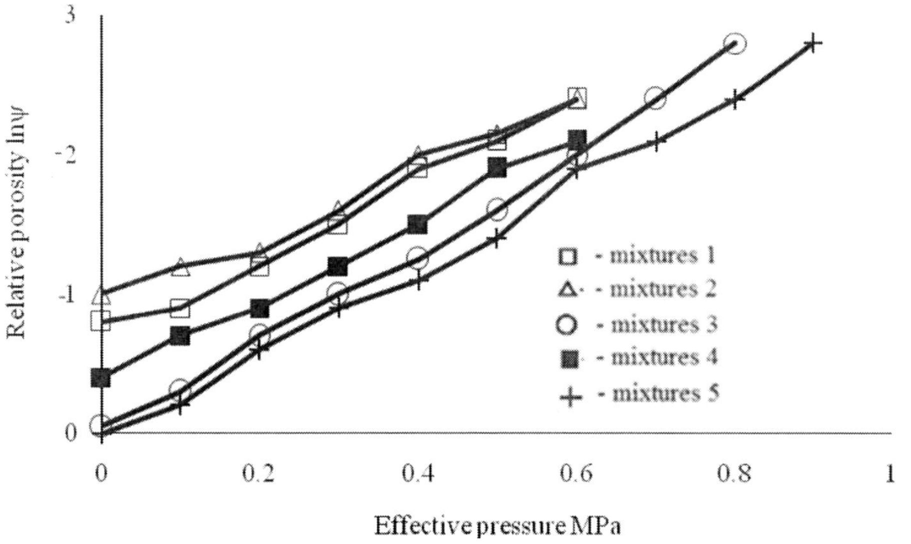

Fig. 1. Dependence of the relative porosity on the effective pressure.

As the filtration characteristics of concrete mixtures to the value $ln\psi$=-1.0 showed, the $\ln\frac{K_f}{K_f^o}$ dependence is linear and the determination of the coefficient n in expression (3) does not run into difficulties. So for the considered compositions, n = 1.75 and the parameters of the compression curves c = 0.905-1.0, α = 3.5-3.33, that is, with sufficient accuracy for practical application, can be set **c** = 1.0, α= 3.4, n = 1.75.

Expression (1) includes all the basic technological parameters of the compaction process under axial compression by layers, and vacuumizing.

3 Conclusions

It has been established that, when molding concrete roof tiles, the application of the linear compaction theory leads to an increase of the vacuumizing time necessary to obtain a given structural strength. Thus, when the shingles are formed with a specified structural strength of 0.42 kg/cm^2 and a thickness of 12 mm at the vacuum pressure 0.05 MPa, the time required for the nonlinearity process is 9 seconds, and according to the linear theory it is 15 seconds. Thus, the nonlinear model for the consolidation processes of solution and fine-grained mixtures, compacted by vacuuming and axial pressing in layers, proposed by the authors, makes it possible to obtain thin-walled products with a sufficiently high structural strength without thermal treatment in terms of material and energy-saving technologies.

References

1. R. Runova, V. Gotz., M. Sanitsky, *Konstruktsiyni materiali new pokolinnya that tehnologii ih vprovadzhennya in budivnitstvo* (2008)
2. Z. Jamrozy, *Beton i jego technologie* (Warzawa, Wydawnictwo naukowe Pwn, 2000)
3. N. Zaichenko, *High-strength fine-grained concrete with a complex microstructure of the modified* (2009).

4. O. Vandolovsky, I. Kazimagomedov E., O. Dedenova, News ODABA **35,** 50-55 (2009)
5. A.Vandolovsky, E. Dedeneva, HDTUBA HOTV OMB, **41,** 127 (2007)
6. A. Vandolovsky, E. Dedeneva, HDTUBA HOTV OMB, **47,** 192 (2008)
7. B. Younis, E. Dedeneva, T. Kostiuk, Modern scientific achievements: experience exchange, Morrisville, Lulu Press, 28-31 (2017)

The definition of basic parameters of the set of small-sized equipment for preparation of dry mortar for various applications

Inga Emelyanova[1], *Vladimir* Blazhko[1,*], *Svetlana* Dansheva[1], and *Natalia* Shyshko[1]

[1]Kharkiv National University of Civil Engineering and Architecture, Department of Mechanization of Construction Processes, Sumska st. 40, 61002 Kharkiv, Ukraine

Abstract. Based on the conducted information retrieval and review of the scientific literature, unsolved issues have been identified in the process of preparation of dry construction mixtures in the conditions of a construction site. The constructions of existing technological complexes for the production of dry construction mixtures are considered and their main drawbacks are identified in terms of application in the conditions of the construction site. On the basis of the conducted research, the designs of technological sets of small-sized equipment for the preparation of dry construction mixtures in the construction site are proposed. It is found out that the basis for creating the proposed technological kits are new designs of concrete mixers operating in cascade mode. A technique for calculating the main parameters of technological sets of equipment is proposed, depending on the use of the base machine of the kit.

1 Introduction

Modern dry building mixtures are complex compositions of components, different in their properties. Mixing these components together and creating a homogeneous mixture is a difficult task in a technical sense. Creation of effective equipment for these purposes is an urgent problem of modern construction. With its solution, special attention should be paid to the development of technological kits.

At present, in the process of preparation of dry construction mixtures, there is a number of unresolved issues: increased wear of the working bodies and mixer bodies, long time for the preparation of complex compositions, the presence of segregation during the mixture preparation and the complexity of preparing the mixtures directly at the construction site [1-4].

At present, technological complexes, which have the ability to produce mixtures of different types, are used for the preparation of dry construction mixtures [5-7].

The complexes are equipped with high-quality dosing devices and forced-action mixers with one or two shafts, on which various stirring devices are fixed [3, 9]. However, modern concrete mixers do not always allow to produce homogeneous mixtures with a small filling factor of the working volume of the machine. In addition, existing mixers do not allow to

* Corresponding author: blagko-2008@ukr.net

create a single organized work cycle in one place. From these positions, technological kits that allow to combine all operations in time are more attractive.

In addition, the variety of existing equipment, as well as technological lines and kits, is not aimed at solving the problem of making dry construction mixtures directly on the construction site.

Creation of effective small-sized equipment for the preparation of dry mixtures is an urgent task. The results of laboratory and industrial tests of new design mixers operating in a cascade mode, confirm their effectiveness in the preparation of dry building mixtures, which makes it possible to equip these machines with technological kits and complexes [8].

Main material and results. It is proposed to develop and introduce small-size sets of equipment for the new mixers operating in a cascade mode in the production of small amounts of work (with individual low-rise construction, construction in the agricultural sector, repair and reconstruction of existing buildings). Technological equipment kits are shown in Fig. 1. All equipment of these technological kits is based on a common base and is selected according to the performance of the base machine-mixer.

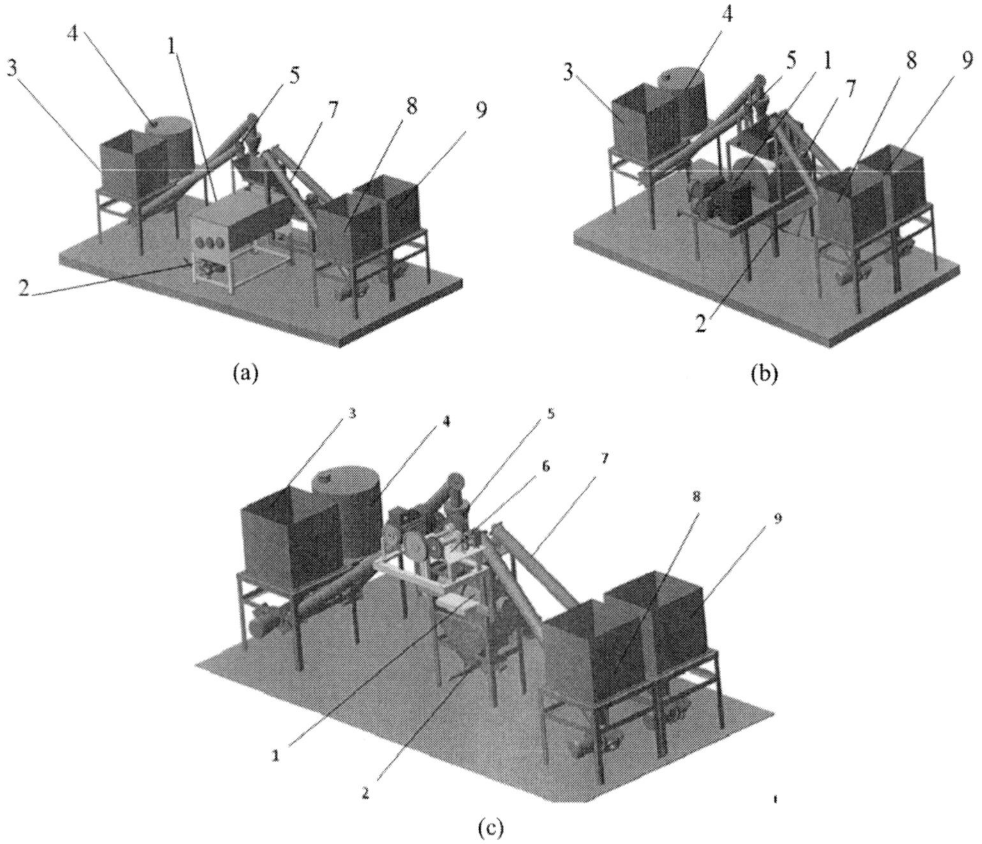

Fig. 1. Technological sets of equipment for the preparation of dry building mixtures: (a) the base machine is a three-shaft mixer; (b) the base machine is a mixer of gravitational-forced action; (c) base machine - two-rotor turbulent mixer

The technological set of equipment, presented in Fig. 1, includes the following components: 1 - mixer (basic machine); 2 - filling machine 3 - sand bunker; 4 - silo of cement; 5 - weighing batcher; 6 - fiber cutter; 7 - auger feeder; 8 - bunker of granite grains; 9 - filler hopper.

The basic machines of the technological set (Fig. 1) can be concrete mixers of a new constructive solution: a three-shaft concrete mixer, gravity-forced concrete mixer, a two-side turbulent mixer [5]. A feature of the work of these machines is the creation of a complex multi-path motion of the particles of the mixture in the body of the machine [9]. The adjacent equipment of the technological set is unified and has a traditional design solution.

2 Theoretical part

The productivity of a technological equipment set is determined on the basis of the technical performance of the base machine used in the composition of the technological set.

The productivity of the technological set with the use of a three-shaft concrete mixer is found according to the dependence [10]

$$P_{tech} = 3600 \cdot \frac{\pi}{4} \cdot (D^2 - d^2) \cdot b \cdot n \cdot z_b \cdot Sin\alpha \cdot k_l^{aver} \cdot k_r^{II}, \quad (1)$$

where D – is the diameter of the shaft along the end of the blade, m;
d – is the diameter of the average shaft, m;
b – blade width, m;
Z_b – there is number of blades of the middle shaft;
α – is the angle of installation of the blades, deg;
k_l^{av} – load factor of the mixer relative to the average shaft, $k_l = 0.75$;
n – frequency of rotation of the shaft of the working element;
k_r^{II} – coefficient of return of the concrete mixture of the second zone.
The technical performance of a set of equipment can be defined as

$$P_{tech} = V_{ov} \cdot Z_l = \frac{1}{2}[\pi \cdot L_b(R_b^2 \cdot k - r_{sh}^2) - \pi \cdot r_n^2 \cdot l_b \cdot z_l - z_1 \cdot b_1 \cdot h_1 \cdot c_1 - z_2 \cdot b_2 \cdot h_2 \cdot c_2] \cdot Z_c, \quad (2)$$

where V_{ov} – is the total volume of the mixture in the mixer, m^3;

$Z_c = 3600 / t_c$ – there is number of cycles of machine operation per hour (t_c – is the duration of one cycle, which consists of the sum of time for loading components t_1, mixing them t_2 and unloading the finished mixture t_3 ($t_c = t_1 + t_2 + t_3$), s);

L_b – length of the mixer body, m;
R_b – is the radius of the mixer body, m;
k – is a coefficient that takes into account the position of the mixture in the body;
R_b, l_b, z_b, r_{sh} – radius, length, number of legs of shaft blades and shaft radius, m;
b_1, b_2, h_1, h_2, c_1, c_2 – length, height and thickness of blades of the body and blades of the mixer shaft, m.

The productivity of the technological set when using a two-rotor turbulent mixer is found according to the dependence

$$P_{tech} = V_{ov} \cdot Z_b \cdot k_m, \quad (3)$$

where is V_{ov} – the total volume of the mixture in the mixer, m^3;

$Z_b = 3600 / t_b$ – there is number of cycles of machine operation per hour (t_b is the duration of one cycle, which consists of the sum of the time for loading components t_1, mixing them t_2 and unloading the finished mixture t_3 ($t_b = t_1 + t_2 + t_3$), s);

k_m – is the machine utilization factor (0.85).

In this case, the feeding of the components of the concrete mixture into the mixer will be as follows:

$$M = \int_0^t (Qx_{ent} - Qx_{ex})dt, \quad (4)$$

where M – total mass of components;

t – the time interval for loading all components;

Qx_{ent} – the flow of the components of the mixture upon leaving the storage bin;

Qx_{ex} – the flow of the components of the mixture as it leaves the feeder.

The ratio of the total mass of M components to the entire working volume of mixer V is as follows:

$$\frac{M}{V} = \Delta x = \frac{Q}{V}\int_0^t (x_{ent} - x_{ex})dt , \tag{5}$$

where x_{ent}, x_{ex}, x – respectively, the content of the components at the inlet, outlet and inside the mixer;

Q – the capacity of the mixer under steady operating conditions.

Among the main performance indicators of a small-sized technological set of equipment for the preparation of a building mixture, it is worth mentioning the power costs.

The power costs for the operation of a technological equipment set intended for the preparation of fiber-concrete mixtures consist of the sum of the power costs of the individual types of equipment that are included in the kit (Fig. 1c): the machine-tool of the fiber, the base machine (concrete mixer), feeders and dispensers. Thus, the dependence for determining the power of the technological set of equipment has the form [7]

$$N_{t.c} = N_{p.c} + N_p + N_f + N_d , \tag{6}$$

where $N_{t.c}$ – Total costs of power for the operation of the technological set of equipment, kW;

$N_{p.c}$ – power costs of the base machine, kW;

N_p – power consumption of a fiber cutter, kW;

N_f – power consumption of feeders, kW;

N_d – power consumption of dispensers, kW.

Taking into account the design features of the base machine of the three-shaft concrete mixer (Fig. 1), when working on dry building mixtures, $N_{t.c}$ should be defined as for a three-shaft mixer $N_{b.m}$, taking into account the operation of the machine on dry building mixtures

$$N_{t.c} = \frac{N_1 + N_2 + N_3}{1000 \cdot \eta} , \tag{7}$$

where N_1 – the power used to mix the components of the mixture with the upper shaft;

N_2 – the power used to mix the components of the concrete mix and transport it to the discharge port by the middle shaft;

N_3 – the power used to mix the components of the mixture by the lower shaft;

η – the efficiency of the mixer drive.

$$N_1 = N_1^{/} + N_1^{//} , \tag{8}$$

where $N_1^{/}$ – the power expended on the mixing process in the first zone,

$N_1^{//}$ – power costs for the mixing process in the second zone

$$N_1^{/} = \omega \cdot F_b \cdot C_1 \cdot f \cdot (R^2 - r^2) \cdot Z^1{}_b \cdot k_l^{up} \cdot \rho_1 , \tag{9}$$

where ω – angular velocity of the shaft, s^{-1};

F_b – blade area, m^2;

C_1 – coefficient of resistance to movement of particles of dry concrete mixture along the blades in the first zone, $C_1 = 9$;

f – coefficient of friction of the dry mixture over the surface of the blades, $f = 0{,}4$;

R – outer radius of rotation of the blade, m;
r – internal radius of rotation of the blade, m;
Z^I_b – there is number of blades on the upper shaft in the first zone;
k_l^{up} – load factor of the concrete mixer relative to the upper shaft;
ρ_l – average density of dry mix, kg/m^3.

$$N_1^{II} = \omega \cdot F_b \cdot C_2 \cdot f^I \cdot (R^2 - r^2) \cdot Z^{II}{}_b \cdot k_l^{up} \cdot \rho_2, \tag{10}$$

where C_2 – coefficient of resistance to movement of particles of the dry mixture along the blades in the second zone, $C_2 = 7$;

f^I – coefficient of friction of the mixture along the blade surface, $f^I = 0.1...0.2$;

$Z^{II}{}_b$ – there is number of blades on the upper shaft in the first zone.

After the transformations, the dependence for determining the power costs by the upper shaft of the mixer has the form

$$N_1 = \omega \cdot F_b \cdot \rho \cdot (R^2 - r^2) \cdot k_3 \cdot C_1 \cdot f \cdot Z_b + C_2 \cdot f^I \cdot Z_b^I, \tag{11}$$

The power expended on mixing the components of a concrete mix and its transportation to the discharge port by an average shaft

$$N_3 = N_2^I + N_2^{II}, \tag{12}$$

where N_2^I – the power expended on the process of mixing the components of the dry building mixture in the first zone and its transportation to the second zone;

N_2^{II} – power expended in the second zone of the mixer.

$$N_2^I = m_c \cdot g \cdot W \cdot \omega \cdot R_s, \tag{13}$$

where m_c – mass of the components of the mixture in the auger zone, kg;
W – Drag coefficient (when the body moves in hostile environments $W = 4$);
ω – Angular speed of rotation of the shaft, s;
R_s – Radius of screw, m;
g – acceleration of gravity

$$N_2^{II} = \omega \cdot F_b \cdot C_2 \cdot f^I \cdot (R^2 - r^2) \cdot Z_b^{II} \cdot k_l^{up} \cdot \rho_2, \tag{14}$$

where Z_b^{II} – there is number of blades on the middle shaft in the second zone.

The power required to mix the components of the mixture with the lower shaft

$$N_3 = N_3^I + N_3^{II}, \tag{15}$$

where N_3^I – power expended on the process of mixing in the first zone;
N_3^{II} – the power expended on the mixing process in the second zone of the mixer;
N_3^I and N_3^{II} – are determined by analogy with the upper shaft.

For a gravity-forced-action mixer [5] $N_{t.c}$, spent on the process of preparing a dry construction mix, consists of the power required to rotate the mixer body and the power required to rotate the blade shaft

$$N_{t.c} = N_b + N_s, \tag{16}$$

where N_b – power required to rotate the body of the mixer,

$$N_b = N_s^{(1)} + N_s^{(2)} = \frac{0.85 \cdot G_{c.m} \cdot h \cdot Z \cdot \omega_b}{\eta_d \cdot 1000} + \frac{F_{fr} \cdot V_{a.v} \cdot z_1}{\eta_d \cdot 1000}, \tag{17}$$

where $G_{c.m}$ – weight of the components of the mixture, N;
h – vertical coordinate of mass mixture displacement in the body, m;

Z – there is number of mixes circulating in the body of the machine;
ω_κ – angular speed of rotation of the mixer body, s^{-1};
η_d – efficiency of the machine drive;
F_{fr} – the frictional force arising when the particles of the mixture move along the surface of the blade of the mixer body, N;
$V_{a.v}$ – absolute velocity of the mixture particles along the body blades, m/s;
z_2 – number of blades on the inner surface of the body.
The power expended on actuating the working member of the machine

$$N_{sh} = N_b^{(1)} + N_b^{(2)} = \frac{\omega_{sh} \cdot M_{sh}}{\eta_d \cdot 1000} + \frac{F_{fr} \cdot V_{a.v} \cdot z_2}{1000 \cdot \eta_d}, \quad (18)$$

where ω_{sh} – angular speed of rotation of the mixer's blade shaft, s^{-1};
M_{sh} – torque of the blade shaft, N·m;
η_d – efficiency of shaft drive;
F_{fr} – the friction force that occurs when a particle of a concrete mix moves along the blade surface, N;
$V_{a.v}$ – absolute velocity of the mixture particle along the shaft blade, m/s;
z_2 – there is number of blades on the shaft.
When installing a two-rotor turbulent mixer

$$N_{set} = \omega_{b.s} \cdot F_b \cdot q_{mix.} \cdot f \cdot (R - r) \cdot Cos\beta \cdot Z_b \cdot k_l \cdot \rho_1, \quad (19)$$

where $\omega_{b.s}$ – angular speed of rotation of the mixer's blade shaft, s^{-1};
F_b – the area of the blade, m^2;
q_{mix} – pressure on the blade from the mixture side, Pa;
f – coefficient of friction of the components of the mixture along the surface of the blades of the working element of the mixer of construction mixtures f= 0.4;
R – outer radius of blade rotation, m;
r – internal radius of blade rotation, m;
Z_b – there is number of blades on the upper shaft in the first zone;
k_l – load factor of the concrete mixer;
ρ_1 – average density of dry mix, kg/m^3;
β – blade angle.
Thus, the proposed technological kits of small-sized equipment can be successfully used to work directly on the construction site.

3 Conclusions

1. Technological sets of small-sized equipment of a new configuration and a new design solution are proposed to perform small volumes of work in the conditions of a construction site.
2. Dependencies are given allowing to determine the performance indicators of new technological equipment sets for the preparation of dry building mixtures on the construction site.

References

1. I.A. Emelyanova, A.N. Baranov, V.V. Blazhko, Materials of the International Scientific and Technical Conference "Interstroymech-2005", **1**, 38-43 (2005)

2. I.A. Emelyanova, V.V. Blazhko, A.I. Anishchenko, Naukovii visnik budivnitsva, **52**, 317-325 (2009)
3. A. Kudrolli, Reports on progress in physics, **67** (**3**), 209-247 (2004)
4. C.S. Campbell, Annual Revue of Fluid Mechanics, **22**, 57-92 (1990)
5. I.A. Emelyanova, V.V. Blazhko, O.V. Dobrokhodova, A.I. Anishchenko, Materials of the International Scientific and Technical Conference "Interstroymeh-2009", 398-403 (2009)
6. I.A. Emelyanova, V.V. Blazhko, O.V. Dobrokhodova, A.I. Anishchenko, Proceedings «HEAVY MACHINERY – HV 2011», 11-14 (2011)
7. I.A. Emelyanova, V.V. Blazhko, O.V. Dobrokhodova, A.I. Anishchenko, Proceedings «HEAVY MACHINERY – HV 2014», p. 35 – 38 (2014)
8. I. Emel'yanova, A. Anishchenko, S. Evel, *Concrete-mixers working in cascade mode: monograph* (Tim Public Library, Kharkov, 2012)
9. I.A. Emelyanova, A.I. Anishchenko, F.A. Stoyanov, New materials and technologies in engineering: a collection of scientific papers on the results of the International Science and Technology Conference, **16**, 137-140 (2012)
10. I.A. Emelyanova, V.V. Blazhko, A.S. Neporozhnev, V. Yu. Shevchenko, New materials and technologies in machine building: a collection of scientific Works on the results of the International Scientific and Technical Conference, **13**, 155-157 (2011)

Influence of complex chemical additives on the water resistant silicate materials

Olena Hryhorenko[1,*], *Anastasia* Lobanova[1], *Ibrahim* Kazimagomedov[1], *Kyrylo* Plakhotnikov[1], and *Olexandr* Lulyko[1]

[1]Kharkiv National University of Civil Engineering and Architecture, Building Materials and Products Department, Sumska st. 40, 61002 Kharkiv, Ukraine

Abstract. In this paper, it has been established that the use of silica fume, low-base slag and calcium hydroxide to modify clays in constrained contact conditions leads to an increase in the number of contacts between the particle components and provides the conditions for the formation of strong contacts and the formation of dense microstructures, and therefore reduces its water absorption and increases water resistance of unburned building materials based on local aluminum silicate raw materials.

1 Introduction

The increasing water resistance of unburned building materials based on local aluminum silicate raw materials is very difficult and complex task which depends on many factors and can be solved using the methodology of system analysis. The main problem in production of building materials by unburned material technology from local clays is their complex and heterogeneous polymineral and granulometric composition. This problem can be efficiently solved by carrying out experimental studies with the construction of a model for substantiation of the solution.

2 Theoretical part

The loam soil with a plasticity number of 14.35 from the Kharkovskoye-2 brick raw material field was used for the study. Visually, the initial mixture consists of loose red-brown rock with the carbonate content to 3%. The granulometric and chemical composition of the clay used is shown in Tables 1 and 2.

Table 1. Size distribution of the clay used in the study, %

Particle size, mm			
more than 0.1	0.1 - 0.05	0.05 - 0.005	less than 0.005
12.32	21.83	44.65	21.20

* Corresponding author: elenkagrigorenko@meta.ua

Table 2. Chemical composition of the clay used in the study

Content of components,%									
SiO_2	Al_2O_3	Fe_2O_3	TiO_2	MgO	CaO	Na_2O / K_2O	SO_3	H_2O^+	H_2O^-
77.20	9.14	3.16	0.61	1.01	2.10	1.31	0.31	3.22	1.94

Table 1 shows that the loam silt used has a high content of fraction (44.65%). By the number of its plasticity the sample is classified as heavy loam. The significant number of loam particles of the size less than 0.005 mm (21.2%) in the clay determines the interaction between colloidal particles which means the consequence of molecular attraction and electrostatic forces directly between the particles themselves and between particles and water molecules contained in the loam. The presence of highly dispersed clay particles in the loam (more than 65%, Table 1) contributes to the increase in strength of clay materials, especially when they are firmed.

The chemical composition (Tab. 2) shows a high content of silica fume in the rock that characterizes the studied loam as acidulous.

The mineralogical composition of the clay was investigated by X-ray analysis method. The mineralogical composition was studied by means of X-ray diffraction analysis using a DRON-1.5 diffractometer. The results are shown in Fig. 1.

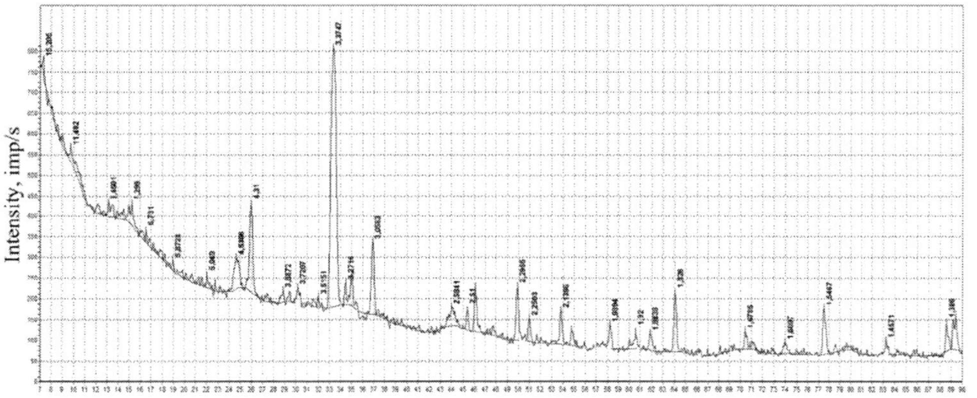

Fig. 1. X-ray diffraction pattern of the clay raw material before the modification.

The diffraction reflects the peaks characteristic for quartz stacking distances - 4.31, 3.37, 2.47, 2.29, 2.14, 1.83, 1.55, 1.39 Å), dickite – β-form $Al_2O_3 \cdot 2SiO_2 \cdot 2H_2O$ (stacking distances – 7.299, 4.54, 3.72, 2.58, 2.51, 2.25, 1.68, 1.46 Å), halloysite (stacking distances – 11.37, 5.87, 4.51, 3.51, 1.68 Å) and calcite (stacking distances - 3.88, 306, 2.09, 1.92, 1, 88 Å).

There are also marks of micaceous minerals, feldspars and iron oxides in the material, which are accessory clay minerals. The presence of the background on the X-ray allows us to assume the presence of an X-ray amorphous phase in the sample.

Thermal-oxidative breakdown was studied by differential thermal (DTA) method of analysis on the derivatograph of F. Paulik and L. Erdei (Hungary) in the temperature range 20-1000°C at a heating rate of 10 deg/min in open atmosphere. The sample weighted 100 mg. Three curves for the heating process of the samples were recorded during the study: T-curve for the temperature change, TG-curve for sample weight change during heating and DTA curve for the change in enthalpy.

Thermographic studies showed typical effects in the DTA curve (Fig. 2) for the samples of the Kharkivskoye-2 loam field which is typical for various clay minerals.

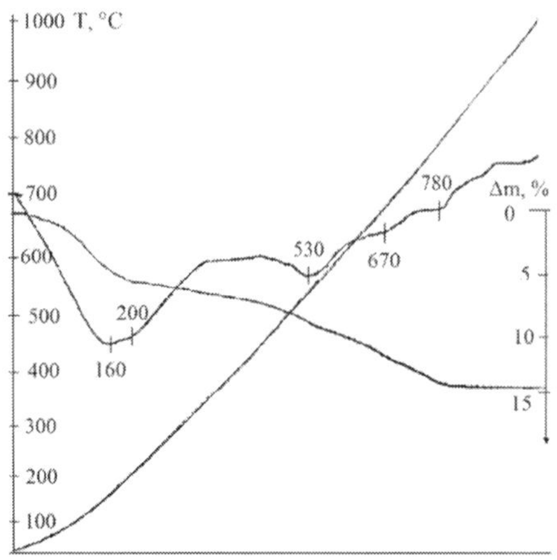

Fig. 2. Thermogram of loam samples in the study.

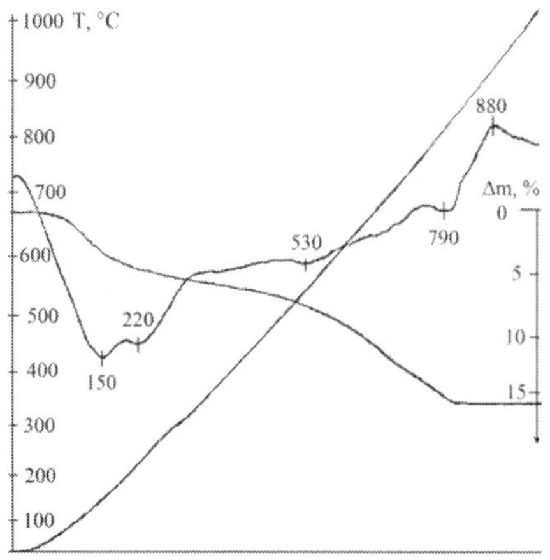

Fig. 3. Thermograms of the samples of the water-resistant clay-lime-slag material.

On DTA thermograms of loam samples (Fig. 2), the intense endothermic effect with a maximum at temperature of 160°C is caused by the removal of adsorbed water. The endoeffect in the area of 530°C may indicate the constitutional allocation of water with the destruction of the dickite crystal lattice. The third exothermic effect at 920°C is concerned apparently with the crystallization of the amorphous silica.

The quantitative mineralogical composition of brick raw materials at the Kharkovskoye-2 field is shown in Tab. 3.

Table 3. The mineralogical composition of the clay used

Content of components,%				
quartz	dickite	halloysite	calcite	mixed-formation and X-ray amorphous phase
47	14	3	2	34

The particle dispersion of the raw material investigated contributes to the increase in strength of clay raw materials, especially if they are intensively densified.

The thermodynamic instability and polymineral composition of the rock presented indicate the possibility of stabilizing the structure of the aluminosilicate raw material presented by interacting with modifying binder additives containing the exchange cations Mg^{2+}, Ca^{2+}, Sr^{2+}, Ba^{2+}, Al^{3+}, Fe^{3+} in their composition, which interaction with hydroxyl OH- clay minerals at elevated values of pH leads to exchange of a hydrogen by a cation of this metal with formation of more stable structures.

As a result of the studies [1-8], it has been established that the production of waterproof materials is based on unburned clays which can be obtained by addition of complex modifying additives with slag-alkali materials (silica fume, slags and hydrated lime) in their composition. Thus, a water-resistant composition with a compressive strength of 10.7 MPa, a flexural strength of 8.4 MPa, the softening coefficient K_p = 0.83, the water absorption Wm = 8 and the freeze resistances F = 50 cycles was developed.

The formation of waterproof unburned building materials based on aluminosilicate raw materials depends on the formation of sparingly soluble hydrate compounds of clay minerals. In order to analyze the structure, phase and mineralogical composition of waterproof construction materials based on unburned clays obtained as a result of experimental studies, a number of studies were carried out with microscopy, X-ray and thermogravimetric methods of analysis.

The aim of the study was to establish the restructuring of clay materials in the samples with the optimum composition of slag, silica fume and lime, after heat and moisture treatment, in a dry and moist state, and to compare the data obtained with ordinary clay (Fig. 1). The X-ray diffraction pattern of the water-resistant clay-lime-slag material studied is shown in Fig. 4.

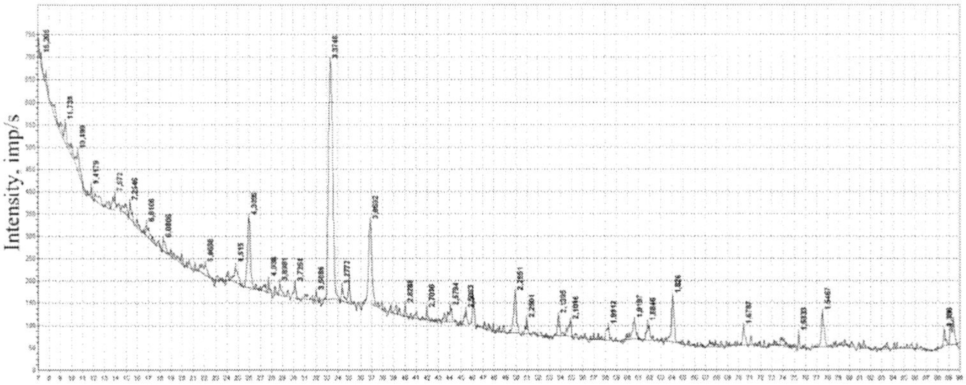

Fig. 4. X-ray phase analysis of the samples of studied water-resistant clay-lime-slag material.

The most intense lines in Fig. 4 refer to quartz contained in clay. Besides, the X-ray diffraction pattern in Fig. 4 contains lines of diffraction reflection of the same mineral phases as in the clay sample in Fig. 1. However, the X-ray diffraction pattern shows the presence of diffraction reflection lines of the following compounds:

- $2CaO{\cdot}Al_2O_3{\cdot}8H_2O$ – hexagonal eight-aqueous calcium hydroaluminate with the interplanar distances d = 10,499; 2.83; 2.7 Å;
- $4CaO{\cdot}Al_2O_3{\cdot}19H_2O$ – four-calcareous nineteen-aqueous hydroaluminate with the interplanar distances d = 10.6; 4.04; 2.83; 2.71 Å;
- $Ca_3Si_6O_{15}{\cdot}7H_2O$ – hydrosilicate of nekoite-type with the interplanar distances d = 9.42; 7.97; 2.82 Å.

The thermogram of DTA samples of the clay-lime-slag composition investigated is shown in Fig. 3.

On the DTA thermograms of the test samples (Fig. 3), the intense endothermic effect with the maximum at a temperature of 150°C is caused by the removal of adsorbed water and the crystallization water of calcium hydroaluminates. The endo-effect at 530°C may indicate the dehydration of $Ca(OH)_2$ and dickite. The third endothermic effect at a temperature of 790°C is obviously associated with the dissociation of calcite ($CaCO_3 \rightarrow CaO + CO_2$). In Fig. 3 the thermogram displacement and subsidence of the endothermic peak associated with the dissociation of calcite (780°C) into the region of higher temperatures (up to 790°C) are observed. This subsidence, as well as the presence of a clearly expressed exoeffects in the area 880°C, indicates the presence of nekoite-type neo-formations in the hardened clay-lime-slag material, and is explained by its dehydration ((-) 700-750°C) and further crystallization of dehydration products ((+) 750-800°C).

To analyze the structure of samples of waterproof building materials, microstructural studies were carried out using an optical microscope with a 600-fold magnification. A microstructural analysis of modified samples after heat and moisture treatment was carried out, as standard samples of unmodified clay were taken (Fig. 5).

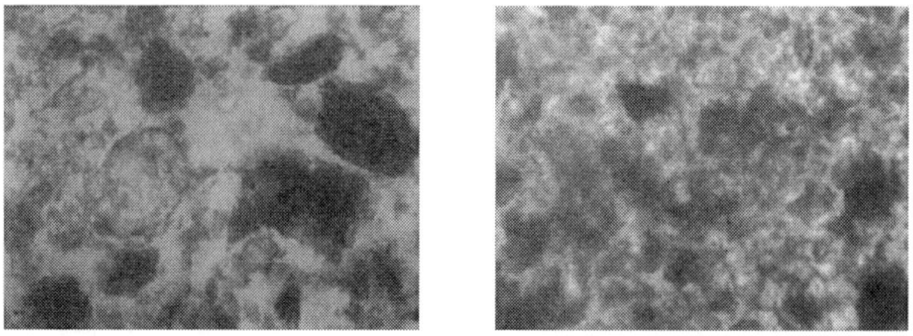

(a) (b)

Fig. 5. Microstructure of clay samples: (a) unmodified clay; (b) modified clay.

In a sample of unmodified clay (Fig. 5a), a disordered lumpy micro texture is observed, being typical for alluvial clays.

Addition of a modifier and implementation of the process in constrained contact conditions (Fig. 5b) lead to formation of the coagulation-crystallization structure. The structure has a higher density than unmodified clay, although the preparation conditions in both cases are similar (pressing pressure 20 MPa). In the sample in Fig. 5b formation of the structure of newgrowths is observed in the form of low-basic calcium hydrosilicates hypothetically. Crystalloptical studies were carried out using POLAM-L211 and MIN-8 microscopes with the use of immersion liquids (Fig. 6).

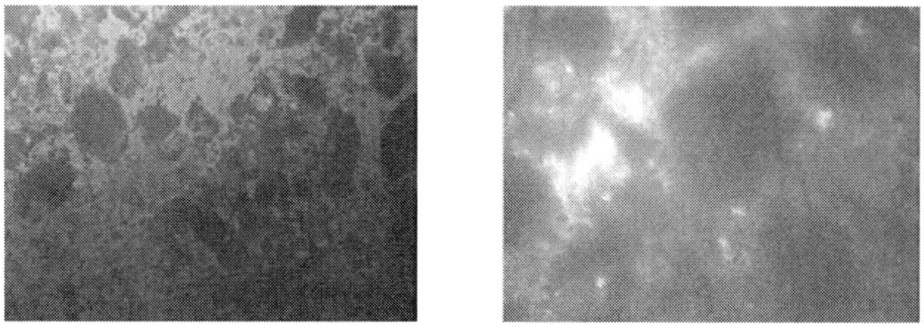

(a) (b)

Fig. 6. Micrograph of the immersion specimen of a clay-lime-slag material sample: (a) in transmitted light without analyzer; (b) in polarized light with an analyser.

Microscopically, immersion preparations are represented by three phases (Fig. 6):

1) an amorphous isotropic bunched siliceous mass (clot size 20-50 μm), the content of which is ~ 60 vol. %;

2) fine crystalline quartz (grain size 3-7 μm) ~ 25 vol. %;

3) the transparent grains of the carbonate phase are presented in the form of spherulites (from 10 to 25 μm), whose content is ~15 vol. %.

In Fig. 6b, the anisotropic crystalline phase is clearly visible in the form of separate luminous grains against the background of an isotropic amorphous basic siliceous mass.

The composition and structure of crystalline newgrowths and their phase composition were studied by IR spectroscopy (Fig. 7).

The spectrum shows bands of crystalline quartz: 460, 480, 700, 790, 810, 1080 and 1190 cm^{-1} clearly. Wide blurred bands in the region of 1460-1540 and 2900-2940 refer to the carbonate phase, probably represented by polymorphic modification of calcium carbonate.

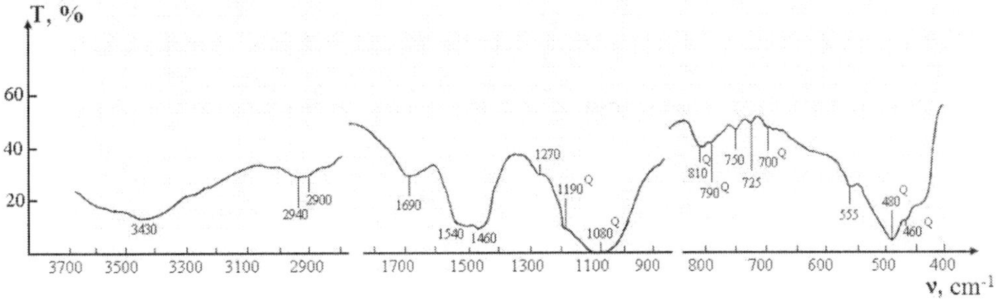

Fig. 7. Infrared absorption spectrum of a sample of clay-lime slag material: Q – bands of crystalline quartz (SiO_2).

The presence of a small but clearly defined peak of 555 cm^{-1}, small bands of 725 and 750 cm^{-1}, as well as the characteristic position and shape of the bands in the region of valence and deformation vibrations of molecular water (H_2O) indicate large amount of hydrated amorphous silica. It is confirmed by the data of crystal-optical analysis.

The data obtained with the help of IR spectroscopy make it possible to conclude that this sample consists of three main phases:

1) fine crystalline quartz (25 vol.%);
2) hydrated amorphous silica (60 vol.%); and
3) calcium carbonate (presumably, vaterite) (15% vol.%).

The results of electron microscopic studies of the structure of the clay raw material used, the composition of the developed composition and the composition of the developed composition after water storage are shown in Fig. 8 to 10.

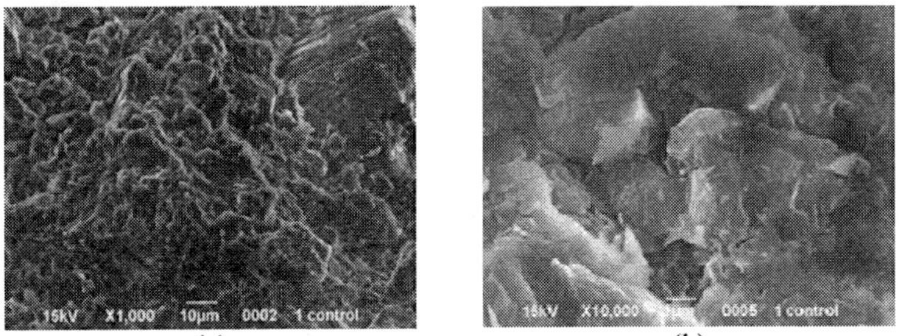

(a) (b)

Fig. 8. The results of electron microscopic analysis of the clay raw materials structure used in the study: (a) × 1000; (b) × 10000.

(a) (b)

Fig. 9. The results of electron microscopic analysis of the structure of waterproof building materials based on the raw clay of the modified composition: (a) × 1000; (b) × 9500.

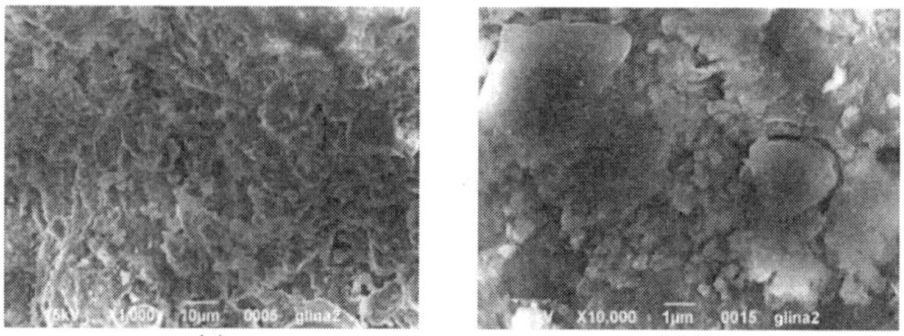

(a) (b)

Fig. 10. The results of electron microscopic study of the structure of waterproof building materials based on the raw clay of the modified composition after storage: (a) × 1000; (b) × 10000.

Comparative characteristics of the samples show that the modification of the clay raw material with a calc-siliceous modifier in alkaline medium leads to the destruction of the crystalline structure of clay minerals (Fig. 8a) and to formation of an amorphous phase in the gel form (Fig. 9a), which fills the volume between particles of clay raw materials, leads to compaction of the modified material structure, and therefore reduces its water absorption and increases water resistance.

The microscopic and EMC research conducted allows us to refine the model of structure and structural elements of the unburned clay-lime-slag composition (Fig. 11).

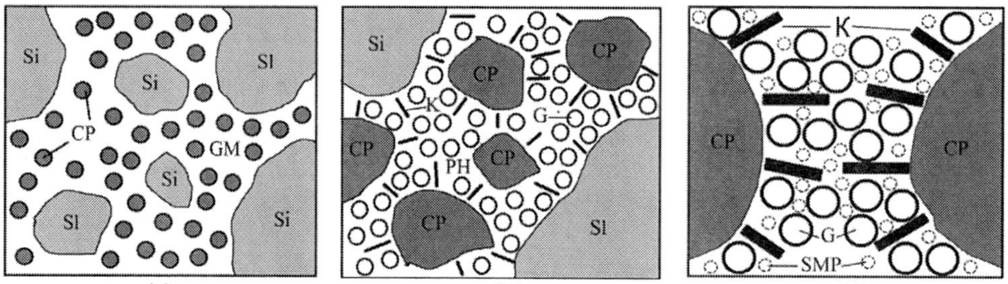

(a) (b) (c)

Fig. 11. Structure model of the unburned clay-slag composition elements on: (a) meso-, (b) micro-, (c) submicro-level; Si – quartz sand (0,2 - 1,25 mm); Sl – large grains of unreacted slag (0,2 - 0,63 mm); GM – a gelatinous mass; CP – not completely reacted clay particles (50 - 100 μm); K – sub microcrystalline splices of hydrated newgrowths (0.5 - 10 μm); G - particles of hydro silicate gel; PH – the products of hydration of the slag-alkali component; SMP - sub micropores (3 - 8 nm).

The material used in this study is the multilevel matrix filled composite. At the macro level, the composition is homogeneous, since there is no large filler. The granulometric composition used for the study of clay (Table 1), and the dispersion additives used for the modification is no more than 1.25 mm. Therefore, at the meso-level (Fig. 11a), the filler will be quartz sand contained in clay rock, and large grains of unreacted slag (mainly its crystalline phase), and the matrix is gel-like mass consisting of hydrated newgrowths. At the micro-level (Fig. 11b), the matrix is hydration products of the slag-alkali component, and the filler is fine-dispersed clay particles. At the sub micro-level (Fig. 11c), the matrix is hydrosilicate gel, and the filler is sub microcrystalline splices of hydrated newgrowths of diffuse structure and crystal sizes. At the sub micro-level, there is an increased content of sub micropores measuring 3 ± 8 nm (Fig. 11c).

3 Conclusions

This study shows that the use of silica fume, low-base slag and calcium hydroxide to modify clays in constrained contact conditions leads to increase in the number of contacts between the particle components and provides the conditions for the formation of strong contacts and dense microstructures due to the synthesis of newgrowths in the form of calcium hydroaluminates, calcium carbonate in the form of vaterite and significant amount of hydrated amorphous silica.

References

1. A. Wandolovskiy, O. Hryhorenko, Sc. bull. constr., **72**, 228-223 (2013)
2. O. Hryhorenko, Sc. bull. constr., **75**, 205-208 (2014)
3. A. Wandolovskiy, O. Hryhorenko, Sc. bull. constr., **78**, 143-145 (2014)
4. A. Wandolovskiy, O. Hryhorenko, T. Kostyk, Bull. of the Odessa St. Ac.y of Constr. and Architect., **60**, 36-40 (2015)
5. A. Wandolovskiy, O. Hryhorenko, Sc. bull. Ukr. St. Ac. of Railw. Transp., **152**, 199-204 (2015)
6. A. Wandolovskiy, O. Hryhorenko, O. Dedeneva, Sc. bull. Inform. proces. sys., **3(140)**, 199-202 (2016)
7. A. Wandolovsky, N. Younis Basheer, Yousef Riyed AL-hawari, IJESIT, **6**, 1-8 (2017)
8. O. Hryhorenko, A. Lobanova, The lat. techn., equip. and manag. sys. in const., 29-33 (2016)

Investigation of multiphase equilibria in the subsolidus of BaO–CoO–Fe_2O_3–Al_2O_3 system

Oleg Kostyrkin[1], *Galina* Shabanova[2], *Sergei* Logvinkov[3], *Natalia* Tsapko[3], and *Maryna* Ivashchenko[1,*]

[1]Ukrainian State University of Railway Transport, Department of Occupational Safety and Environmental Protection, Feyerbakh sq. 7, 61050 Kharkiv, Ukraine
[2]National Technical University "Kharkov Polytechnic Institute", Department of Ceramics Technology, Refractories, Glass and Enamels, Kyrpychova str. 2, 61002 Kharkiv, Ukraine
[3]Semon Kuznets Kharkiv National University of Economics, Department of Technology, Ecology and Security of Vital Activities, Nauky avenue 9-A, 61166 Kharkiv, Ukraine

Abstract. One of the most important problems related to the development of new nonmetal materials and their performance characteristics is to predict the phase composition. The most comprehensive information on phase interactions and the thermodynamic stability of phase combinations is given by the state diagrams. The materials synthesized in the system subsolidus domain can be predicted the most accurately, because their sintering occurs without participation of the melt. Due to the above fact, the studies of the subsolidus structure of BaO – CoO – Fe_2O_3 – Al_2O_3 system are of great interest, because on the basis of this system we can obtain a huge amount of nonmetal materials with prescribed properties, for example ferrimagnetic materials to protect from electromagnetic radiation, because the system compounds have cementing, refractory and ferrimagnetic properties. To study the structure of BaO – CoO – Fe_2O_3 – Al_2O_3 system in detail the authors summed up already known data on the thermodynamic constants of system compounds. This allowed us to do the thermodynamic analysis of multiphase equilibrium processes that occur in the subsolidus of BaO – CoO – Fe_2O_3 – Al_2O_3 system that was used as a basis for the plotting of the state diagram for the subsolidus domain of the system. A promising field for the application of obtained data is the cement production technology. The produced cement can be used independently and as a binding material to produce special cements and materials that retain their properties when exposed to the action of high-frequency electromagnetic radiation.

1 Introduction

Prediction of the phase composition is one of the most important problems for the development of new types of refractory nonmetal materials and the conditions for their use. The most complete information about phase interrelations and the thermodynamic stability of phase combinations is given in the state diagrams in which thermodynamically

* Corresponding author: marina.sh.225@gmail.com

equilibrium compositions are interconnected with the temperature. The $BaO - CoO - Fe_2O_3 - Al_2O_3$ system is of great interest for the researchers from the standpoint of the possibility of the production of ferrimagnetic materials that protect from the electromagnetic radiation, because its compounds have cementing, refractory and ferromagnetic properties.

However, the unavailability of data on the structure of the above system hampers the development of special materials with prescribed properties. Therefore, this scientific paper delves into the studies of the structure of $BaO - CoO - Fe_2O_3 - Al_2O_3$ system and solid-phase reactions in it.

2 Analysis of the latest research papers and publications

The $BaO - CoO - Fe_2O_3 - Al_2O_3$ system has a rather complicated structure that has not been properly studied yet.

The data on the structure of binary systems of refractory oxides were summarized in the references based on computational and experimental data [1].

The $Fe_2O_3 - Al_2O_3$ system was studied by the authors [2]. The obtained data show that it contains the hematite Fe_2O_3, corundum Al_2O_3 and $Fe_2Al_2O_6$. The last compound is stable in the air right up to the temperature of 1410°C. At a higher temperature it is decomposed into corundum and the spinel-like solid solution.

The $CoO - Al_2O_3$ system was studied in the subsolidus domain [3] at a temperature of 1400°C. $CoAl_2O_4$ is the only stable compound in it.

According to the data obtained by the author [4] the $CoO - Fe_2O_3$ is thermally stable right up to the temperature of 1200°C and they established the existence of one compound in it, in particular cobalt ferrite $CoFe_2O_4$.

The authors of scientific papers [5-8] gave consideration to $CoO - BaO - Al_2O_3$, $CoO - BaO - Fe_2O_3$, $CoO - Al_2O_3 - Fe_2O_3$, and $BaO - Al_2O_3 - Fe_2O_3$ systems. Their thermodynamic and geometro-topological characteristics were given and that allows us to proceed to more detailed studies of $CoO - BaO - Fe_2O_3 - Al_2O_3$ system.

3 Defining research objectives and problems

The data of tetrahedration of these systems give important information on the prediction of the direction (vector) of chemical interactions and phase combinations for four-component physical and chemical systems synthesized in the products of solid phase reaction. In the tetrahedrated system connodas combine coexisting equilibrium phases by pairs; compound composition points located at the apexes of elementary triangles consolidate triple equilibrium combinations of the phases; the apexes of elementary tetrahedrons correspond to the points of compound compositions that can coexist in the four-phase equilibrium combinations. Therefore, the main problem on the determination of the subsolidus structure for four-component systems is related to the fragmentation of the concentrated tetrahedron of the system into elementary triangles, i.e. tetrahedration.

Based on the above, we arrived at a conclusion that it is topical to study the problem on multiphase equilibrium processes that occur in the subsolidus of $BaO - CoO - Fe_2O_3 - Al_2O_3$ system. The set goal can be achieved based on the thermodynamic analysis done for all possible combinations of the phases, peculiar for the above system.

4 Core of the Research

To do a theoretical research for the $BaO - CoO - Fe_2O_3 - Al_2O_3$ system it is reasonable to carry out the thermodynamic analysis of the processes that occur in it. This can be done, if initial thermodynamic constants are available. Table 1 gives the initial data for the thermodynamic analysis.

Table 1. Initial thermodynamic data

Compound formula	$-\Delta H^0_{298}$, kJ/mole	S^0_{298}, J/mole·K	$C_p=a+b\cdot T+c\cdot T^{-2}$, J/mole·K		
			a	$b\cdot 10^3$	$-c\cdot 10^{-5}$
CoO	238,9	52,97	48,28	8,535	1,67
BaO	558,15	70,29	53,30	4,35	8,3
Fe_2O_3	821,36	89,96	98,28	9,04	14,85
α-Al_2O_3	1675,61	50,92	114,77	12,08	35,44
$CoAl_2O_4$	1948,9	101,7	161,17	32,97	30,32
$CoFe_2O_4$	1087,4	134,7	131,8	141,42	–
$Fe_2Al_2O_6$	1262,73	71,55	50,84	6,54	21,93
$BaFe_{12}O_{19}$	5886,09	609,19	348,60	1168,82	–
$BaFe_2O_4$	1282,81	158,99	172,38	20,88	5,94
$Ba_2Fe_2O_5$	1763,97	229,07	295,03	–	51,09
$Ba_3Fe_2O_6$	2620,61	312,63	219,03	90,80	14,48
$Ba_5Fe_2O_8$	3635,56	809,76	288,11	150,90	6,05
$Ba_7Fe_4O_{13}$	5541,71	687,26	506,88	112,10	37,99
$Ba_2Fe_6O_{11}$	3752,80	437,4	398,94	119,50	59,91
$BaCoO_2$	837,13	123,08	78,99	38,34	4,49
$BaAl_2O_4$	2334,17	123,43	148,32	35,44	29,25
$Ba_3Al_2O_6$	3537,91	267,78	247,86	48,53	17,41
$BaAl_{12}O_{19}$	10740,33	376,56	738,22	70,50	221,75
$Ba_4Al_2O_7$	4014,49	329,99	275,85	56,89	23,33
$Ba_5Al_2O_8$	4569,03	427,82	298,73	123,87	15,51
$Ba_7Al_2O_{10}$	5689,5	567,66	374,18	164,90	14,03
$Ba_8Al_2O_{11}$	6238,10	611,68	441,99	96,23	25,31
$Ba_{10}Al_2O_{13}$	7350,52	778,85	478,25	226,40	12,24
$Ba_2Co_2Fe_{12}O_{22}$	6848,23	785,98	812,71	203,71	140,65
$BaCo_2Fe_{16}O_{27}$	7987,63	895,36	1029,37	151,06	214,45
$Ba_3Co_2Fe_{24}O_{41}$	12609,45	1396,22	1566,51	255,75	314,43

The combination of triple systems triangulated earlier [5-8] into the concentrated tetrahedron of $BaO - CoO - Fe_2O_3 - Al_2O_3$ system closes automatically elementary tetrahedrons:

1) $BaO - Ba_5Fe_2O_8 - BaCoO_2 - Ba_{10}Al_2O_{13}$;
2) $Ba_5Fe_2O_8 - BaCoO_2 - Ba_{10}Al_2O_{13} - Ba_7Fe_4O_{13}$;
3) $BaCoO_2 - Ba_{10}Al_2O_{13} - Ba_7Fe_4O_{13} - Ba_3Fe_2O_6$;
4) $BaCoO_2 - Ba_{10}Al_2O_{13} - Ba_3Fe_2O_6 - Ba_8Al_2O_{11}$;
5) $BaCoO_2 - Ba_8Al_2O_{11} - Ba_3Fe_2O_6 - Ba_7Al_2O_{10}$;
6) $BaCoO_2 - Ba_7Al_2O_{10} - Ba_3Fe_2O_6 - Ba_5Al_2O_8$;
7) $BaCoO_2 - Ba_5Al_2O_8 - Ba_3Fe_2O_6 - Ba_4Al_2O_7$;
8) $BaCoO_2 - Ba_4Al_2O_7 - Ba_3Fe_2O_6 - Ba_3Al_2O_6$;
9) $BaCoO_2 - Ba_3Al_2O_6 - Ba_3Fe_2O_6 - BaAl_2O_4$;
10) $BaCoO_2 - BaAl_2O_4 - Ba_3Fe_2O_6 - Ba_2Fe_6O_{11}$;
11) $BaCoO_2 - BaAl_2O_4 - Ba_2Fe_6O_{11} - BaFe_2O_4$;
12) $BaCoO_2 - BaAl_2O_4 - BaFe_2O_4 - Ba_2Fe_6O_{11}$;
13) $BaCoO_2 - BaAl_2O_4 - Ba_2Fe_6O_{11} - BaFe_{12}O_{19}$.

The $BaCoO_2 - BaAl_2O_4 - BaFe_{12}O_{19}$ face (the elementary tetrahedron No13) is simultaneously coincidental with the faces of the closed tetrahedron domain with the apexes of $BaCoO_2$, $BaBaFe_{12}O_{19}$ and CoO. In addition, the points of the compositions of triple compounds $Ba_3Co_2Fe_{24}O_{41}$ and $Ba_2Co_2Fe_{12}O_{22}$ belong to one of the faces of this domain and the points of the compound compositions that allow for the partition of the domain into elementary domains are unavailable on other faces and ribs. This circumstance is indicative of the fact that the tetrahedration of specified domain is only possible due to the stability of "internal" connodas that pass through the concentration space of studied system: $BaAl_2O_4 - Ba_3Co_2Fe_{24}O_{41}$, and $BaAl_2O_4 - Ba_2Co_2Fe_{12}O_{22}$. These connodas define the existence of elementary tetrahedrons:

14) $BaCoO_2 - BaAl_2O_4 - Ba_2Co_2Fe_{12}O_{22} - CoO$;
15) $BaCoO_2 - BaAl_2O_4 - Ba_2Co_2Fe_{12}O_{22} - Ba_3Co_2Fe_{24}O_{41}$;
16) $BaCoO_2 - BaAl_2O_4 - Ba_3Co_2Fe_{24}O_{41} - BaFe_{12}O_{19}$;
17) $BaAl_2O_4 - Ba_3Co_2Fe_{24}O_{41} - BaFe_{12}O_{19} - CoO$;
18) $BaAl_2O_4 - Ba_3Co_2Fe_{24}O_{41} - Ba_2Co_2Fe_{12}O_{22} - CoO$.

The binary compositions $BaFe_{12}O_{19}$, $Fe_2Al_2O_6$, and $CoFe_2O_4$ proximal to the apex of Fe_2O_3 that are located on the sides of concentrated tetrahedron of the system specify the tetrahedral domain circumscribed by appropriate points of the compositions of these compounds. In addition, the connoda $Ba_2Fe_{12}O_{19} - CoFe_2O_4$ also has a point of the composition of triple compound $Ba_2Co_2Fe_{16}O_{27}$ that coexists with Fe_2O_3 due to the availability of $Fe_2O_3 - Ba_2Co_2Fe_{16}O_{27}$. The only option for the partition of tetrahedral domain (circumscribed by the points of Fe_2O_3, $BaBaFe_{12}O_{19}$, $Fe_2Al_2O_6$, and $CoFe_2O_4$) into elementary tetrahedrons is realized if the connoda $Ba_2Co_2Fe_{16}O_{27} - Fe_2Al_2O_6$, is available:

19) $Fe_2O_3 - BaFe_{12}O_{19} - Fe_2Al_2O_6 - BaCo_2Fe_{16}O_{27}$;
20) $Fe_2O_3 - Fe_2Al_2O_6 - CoFe_2O_4 - BaCo_2Fe_{16}O_{27}$.

The analogous situation is peculiar for the tetrahedral domain circumscribed by the points of the compositions of compounds $BaFe_{12}O_{19}$, $Fe_2Al_2O_6$, $CoFe_2O_4$, Al_2O_3 that has the connoda $Fe_2Al_2O_6 - Ba_2Co_2Fe_{16}O_{27}$. The partition of this domain into elementary tetrahedrons is possible using the $Al_2O_3 - Ba_2Co_2Fe_{16}O_{27}$ connoda:

21) $BaFe_{12}O_{19} - Fe_2Al_2O_6 - BaCo_2Fe_{16}O_{27} - Al_2O_3$;
22) $Fe_2Al_2O_6 - BaCo_2Fe_{16}O_{27} - CoFe_2O_4 - Al_2O_3$.

The further tetrahedration requires the analysis of phase equilibriums for solid-phase exchange reactions:

$$BaAl_{12}O_{19} + 3\ BaCo_2Fe_{16}O_{27} = 4\ BaFe_{12}O_{19} + 6\ CoAl_2O_4 \quad (1)$$

$$BaAl_{12}O_{19} + 6\ CoFe_2O_4 = BaFe_{12}O_{19} + 6\ CoAl_2O_4 \quad (2)$$

$$BaAl_{12}O_{19} + 8\ CoFe_2O_4 = BaCo_2Fe_{16}O_{27} + 6\ CoAl_2O_4 \quad (3)$$

The computation data $\Delta G = f(T)$ for the reactions (1)–(3) are given in the Table 2 and show that it is preferable for the reaction to progress in the direct direction except for the reaction (3) that is reversible at 1565 K, and in the high-temperature range the $BaAl_{12}O_{19}$, and $CoFe_2O_4$ phase combination is more stable thermodynamically in comparison with $Ba_2Co_2Fe_{16}O_{27}$ and 6 $CoAl_2O_4$.

Table 2. Computation data $\Delta G = f(T)$

Reaction No	Gibbs energy value, kJ, at T, K	
	1000	1500
1	–656	–1493
2	–340	–408
3	–232	–39
4	–888	–1532
5	–572	–447

The phase variability due to the reaction reversibility (3) can be of importance only for particular cases, because the reactions (1) and (2) are more advantageous thermodynamically. In the indicated thermodynamic range the changes in the signs ΔG of summative reactions (4) = (1) + (3), (5) = (2) + (3), are not observed (Table 2):

$$BaAl_{12}O_{19} + BaCo_2Fe_{16}O_{27} + 4\ CoFe_2O_4 = 2\ BaFe_{12}O_{19} + 6\ CoAl_2O_4 \quad (4)$$

$$2\ BaAl_{12}O_{19} + 14\ CoFe_2O_4 = BaFe_{12}O_{19} + BaCo_2Fe_{16}O_{27} + 12\ CoAl_2O_4 \quad (5)$$

A negative value of ΔG for the reactions (4) and (5) is indicative of the coexistence of $BaAl_{12}O_{19}$ with $CoAl_2O_4$ and $BaCo_2Fe_{16}O_2$ with $CoAl_2O_4$. The elementary tetrahedrons are closed by the connoda $BaAl_{12}O_{19} – CoAl_2O_4$, respectively.

23) $CoO – BaFe_{12}O_{19} – CoAl_2O_4 – BaAl_2O_4$;
24) $BaFe_{12}O_{19} – CoAl_2O_4 – BaAl_2O_4 – BaAl_{12}O_{19}$;
25) $BaFe_{12}O_{19} – CoAl_2O_4 – BaAl_{12}O_{19} – Al_2O_3$.

The connoda $BaCo_2Fe_{16}O_{27} – CoAl_2O_4$ completes the partition of analyzed system into elementary tetrahedrons:

26) $CoO – BaFe_{12}O_{19} – BaCo_2Fe_{16}O_{27} – CoAl_2O_4$;
27) $CoO – BaFe_{12}O_{19} – BaCo_2Fe_{16}O_{27} – Al_2O_3$;
28) $CoO – BaCo_2Fe_{16}O_{27} – CoAl_2O_4 – CoFe_2O_4$;
29) $BaCo_2Fe_{16}O_{27} – CoAl_2O_4 – CoFe_2O_4 – Al_2O_3$.

Therefore, the subsolidus structure of $BaO – CoO – Fe_2O_3 – Al_2O_3$ system was defined only by 6 "internal" connodas that distinguish 29 elementary tetrahedrons that define all the 2, 3 and 4-phase combinations of coexistent compounds of the system (Fig.1).

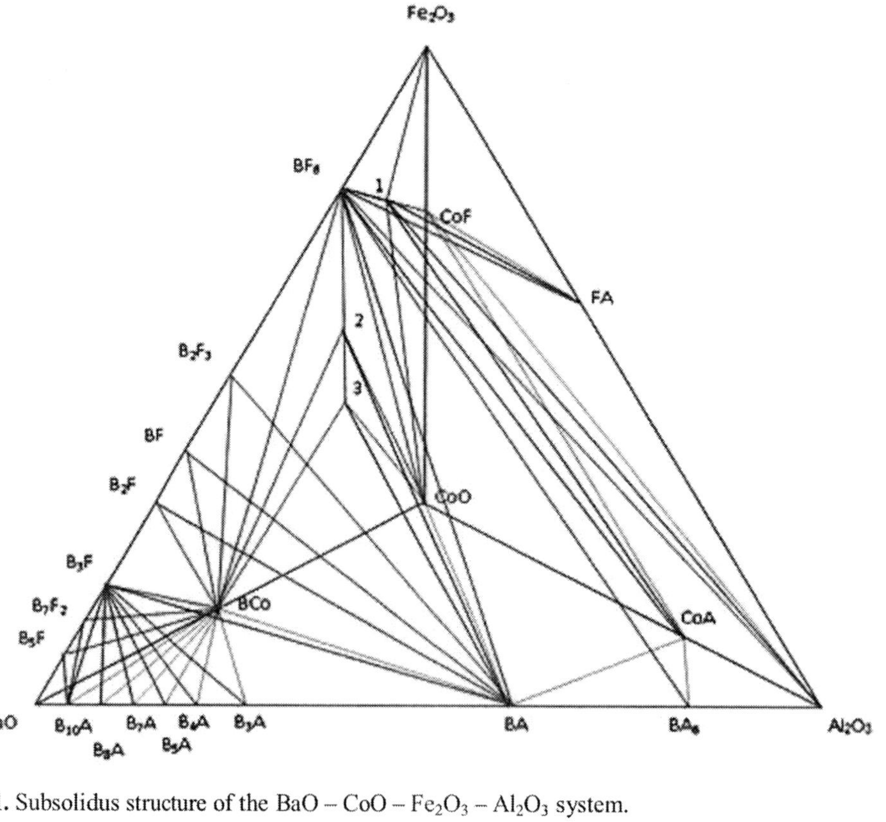

Fig. 1. Subsolidus structure of the $BaO – CoO – Fe_2O_3 – Al_2O_3$ system.

5 Research-related conclusions, outlooks and further progress in this area

Theoretical research done for the subsolidus of $BaO - CoO - Fe_2O_3 - Al_2O_3$ system is of interest for advanced applications of this system. For example, the obtained data can be used for the production of ferrimagnetic cementing materials using the principle of reaction-formed structures that have high values of the refractoriness combined with good protective properties to prevent the impact of electromagnetic radiation of different wave lengths. A promising field for the application of obtained data is the cement production technology. The produced cement can be used independently and as a binding agent to produce special concretes and materials that retain their properties under the action of high-frequency electromagnetic radiation.

References

1. N.A. Toropov, V.P. Barzakovskyi, V.V. Lapin, N.N. Kurtseva, *Silicate System State Diagrams: manual Issue 1: Binary systems* (Monograph, Leningrad Dept., 1969).
2. A. Muan, C.L. Gee, J. of the Amer. Ceram. Soc. **39**, 6 (1956).
3. J. Sticher, H. Schmalzried, Chem., **35** (1961).
4. N.S. Kurnakov, P.H. of the Acad. of Scien. of the USSR (1940).
5. G.N. Shabanova, M.Y. Ivashchenko, M.I. Vorozhbian, O.V. Kostyrkin, N.S. Tsapko. Chem. and Chem. Techn. Problems, **2** (2014).
6. G.N. Shabanova, S.M. Logvinkov, O.V. Kostyrkin, N.S. Tsapko. Refractories and techn. ceramics, **9** (2014).
7. G.N. Shabanova, S.M. Logvinkov, O.V. Kostyrkin, N.S. Tsapko, Refractories and techn. ceramics, 7-8 (2015).
8. M.Y. Ivashchenko, G.N. Shabanova, M.I. Vorozhbian, O.V. Kostyrkin, N.S. Tsapko. *Proc. of the Nat. Techn. Univ. "KhPI"* **51 (1093)**, (2014).

Increase of brick masonry durability for external walls of buildings and structures

Tetiana Kropyvnytska[1,*], *Roksolana* Semeniv[1], and *Hanna* Ivashchyshyn[1]

[1]Lviv Polytechnic National University, Department of Building Production, Karpinskyy str. 6, 79013, Lviv, Ukraine

Abstract. The current article is focused at efflorescence formation and durability of brick masonry. The chemical and phase composition, microstructure of efflorescence on brick and masonry mortar on exterior walls of buildings are studied. Improved properties of brick masonry for external walls of buildings are achieved by the use of masonry mortar based on multi-component cement, which contains zeolite. It is shown that the use of multi-component cement with low clinker content in the mortar reduces the content of calcium hydroxide and SO_3. It is determined that synergistic combination of mineral additives of different groups with substantial reduction of cement clinker in cements containing zeolite, improves durability of brick masonry without efflorescence formation. Physic-chemical modification of mortar by complex air-entraining admixture allows obtaining high-quality modified mortars with improved quality parameters. The surface modification of ceramic bricks by silicon and acrylic compounds provides protection and durability of brick masonry of buildings and structure walls.

1 Introduction

Buildings and structures are exposed to the environmental factors (rainfall, temperature changes) during the exploitation. This has a negative impact on the physical and technical properties of brick masonry for external walls of buildings and salts formation on the surface of the masonry. One of the reasons of brick masonry destruction is readily soluble salts of sulphuric salts in the alkaline and alkaline earth metals from ceramic brick. They can connect the water of crystallisation forming crystal hydrates. The porous structure of ceramic bricks causes the diffusion of alkalis from cement mortars, reinforcing a destructive effect on the masonry. However, the formation of salt efflorescence on the brick masonry surface is also connected with the salt solution penetration in the brick. The salt efflorescence appear on the masonry, when we use the mortar, based on the Portland cement with the high content of clinker and the addition of lime. Herewith, calcium hydroxide, which is a product of Portland cement hydration, washed on the surface of the material and during the interaction with CO_2 of the air forms efflorescence [1, 2].

On the other hand, increased content of $Ca(OH)_2$ in cement and lime mortar leads to the more intensive leaching of lime and subsequent carbonation on the masonry surface to form

* Corresponding author: tkropyvnytska@ukr.net

insoluble calcium carbonate efflorescence. Also one of the reasons of efflorescence forming is the use of well soluble frost-resistance additives and accelerators of hardening. In the process of installation, a high capacity to water absorption of ceramic bricks in masonry seams leads to dehydration of mortar mixture with forming a highly porous layer of the mortar. Free water from the masonry mortar, penetrating in the brick, dissolves the salts followed by crystallization of salt efflorescence not only on the surface but also in the pores. It results in the appearance of cracks and destruction of the masonry integrity during exploitation [3].

One of the effective ways to improve the durability of masonry and prevent the formation of salt efflorescence is the application of mortars, based on the cements with a low content of Portland cement clinker [4-6]. Herewith, an active mineral additive of natural origin, zeolite tuffs, plays the main role in building mortars. A high activity of zeolite tuffs (a large group of minerals - water aluminosilicate sodium and calcium from subclass of frame silicates) on the binding of lime and gypsum in the hydrated phases indicates the possibility of efficient use of the natural zeolites in cements for building mortars. At the same time, the presence of channels in the minerals structure of the zeolite group and high porosity of zeolite tuffs allow active absorption of SO_3. The building mortar modified with natural zeolites is characterized by good sorption properties due to their porous structure and can accumulate a sufficient amount of salt [7-10].

Currently special attention is paid to mortars, modified with air entraining admixtures of new generation that improves the technological properties of mortar mixtures and mortar performance properties. The use of air entraining admixtures in building mortar excludes the addition of lime in their composition that provides the increased frost resistance of masonry without forming of salt efflorescence. The right choice of mortar composition determines the durability of masonry in different conditions of buildings exploitation [10, 11]. It is widely used the method of surface impregnation by repel compounds, resulting the reduction of porosity and permeability of the material, improve the performance properties and increase the durability of brick masonry. The main types of impregnating products used for the treatment of porous building materials mostly have repellent action that does not provide enough resistance to the action of gases, vapors and liquids. To the greatest extent, acrylic and organosilicon compounds are used to achieve these requirements. These compounds, modified by oxide and mineral substances of different chemical origin, due to the high permeability are mudding the pores and cracks on the surface and providing the necessary hydrophilicity and stability of the modified matrix [12]. Therefore, the aim is to study brick masonry durability for external walls of buildings using mortars based on cement which contains zeolite, and modified by air-entraining admixtures, determine properties of masonry surface treated with water repellent compounds.

2 Materials and methods

Portland cement (OPC) CEM I 42,5 JSC "Ivano-Frankivskcement" based on Portland cement clinker with normal mineral composition (mass.%: C_3S – 62.42; C_2S – 13.62; C_3A – 7.06; C_4AF – 12.32) was used in the investigations. Zeolite tuff were used for production of masonry cement MC 22,5X (in the amount of SiO_2 - 70-75 %) as an active mineral additive from Sokyrnytsky quarry, and limestone (with content of $CaCO_3$ - 95 mass.%) as microfiller.

The mortar prisms (40×40×160 mm) with cement–sand–water ratio of 1/3/0.5 were prepared. The compressive and flexural strength were determined on two mortar prisms for each testing age according to EN 196-1. Physical and mechanical tests of cements and cement-sand mortars were carried out according to usual procedures. Air-entraining admixture MasterAir 81 (BASF) was used to improve the quality parameters of plasters.

Organosilicon and acrylic compounds were used for bricks covering as hydrophobic additives.

Study of fractional composition and grinding fineness of cements were carried out by sieve analysis and determination of the specific surface area by Blaine. The chemical composition of cements and salt efflorescence was defined using X-ray spectrometer ARL 9800 XP. The phase composition of the resulting products was determined by X-ray powder diffraction: patterns were recorded at room temperature with Cu Kα radiation on a PANalytical X'pert Pro diffractometer equipped with the X'celerator detector in the 2θ range from 5° to 70° (step 0.033°, time/step 50s). The powder was sieved with 63 μm mesh to prepare the samples. Scanning electron microscope was used for studying the crystals of salt efflorescence on the samples and morphology of the cement stone surface.

3 Results of investigation

Investigation of salt efflorescence forming on the exterior walls surface of brick masonry for different building structures is made. As shown in Fig. 1a, salt efflorescence (from brick) is characterized by a high content of SO_3 (51.2 мас.%), alkaline oxides Na_2O (36.7 mass.%) and K_2O (12.4 mass.%). Therewith, are fixed the negligible inclusion CaO (2.6 mass.%) and SiO_2 (1.8 mass.%). The chemical composition of salt efflorescence from masonry mortar (Fig. 1b) is presented by a high content of CaO (29.9 mass.%) and the negligible quantity of SiO_2 (5.8 mass.%). Thus the content SO_3 is 7.52 mass.%. Alkaline oxides (Na_2O and K_2O) are used in much smaller quantities.

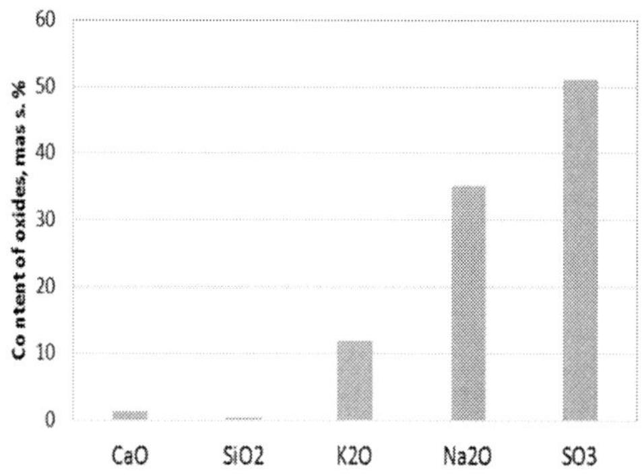

(a)

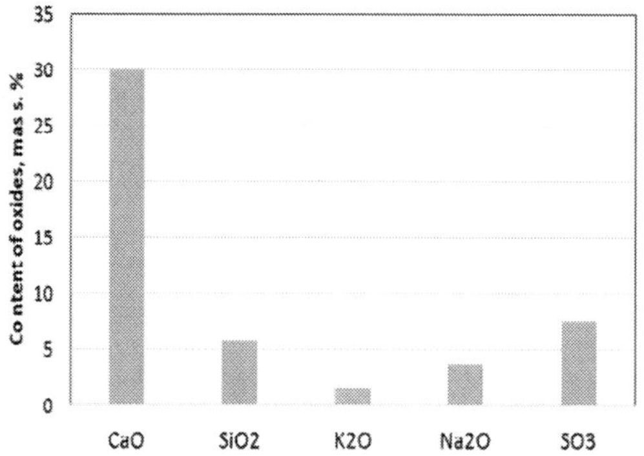

(b)

Fig. 1. The chemical composition of salt efflorescence: (a) from ceramic brick masonry; (b) from mortar masonry

According to X-ray phase analysis (Fig. 2a), crystalline phase of salt efflorescence from brick is represented mainly by lines of Na_2SO_4 (d/n=0.467; 0.384; 0.318; 0.278; 0.264; 0.232 nm), arcanite K_2SO_4 (d/n=0.288; 0.221; 0.208 nm) and singenite $K_2Ca(SO)_4 \cdot H_2O$ (d/n=0.951; 0.285 nm). Microstructure of salt efflorescence (Fig. 2b) is represented by different phases Na_2SO_4 and K_2SO_4. As seen in microphotography (Fig. 2c), salts Na_2SO_4 are characterized by loose, finely crystalline structure. At the same time, salts of K_2SO_4 are crystallized by aggregates similar to plates (10-20 μm) in separate groups (Fig. 2d). The presence of Na_2SO_4 and K_2SO_4 phases is confirmed by EDX (Fig. 2e, g). Sodium sulphate accumulates and is bound sulphate anion to form the mineral singenite due to the increased content K_2O in salt efflorescence. This can be the reason of destructive effects in the matrix of ceramic bricks and masonry. The salt crystals, filling the pores of the bricks outer layers, create the pressure and destroy the surface layers of ceramic material, which leads to the peeling of the bricks front surface and the crushing of brick masonry.

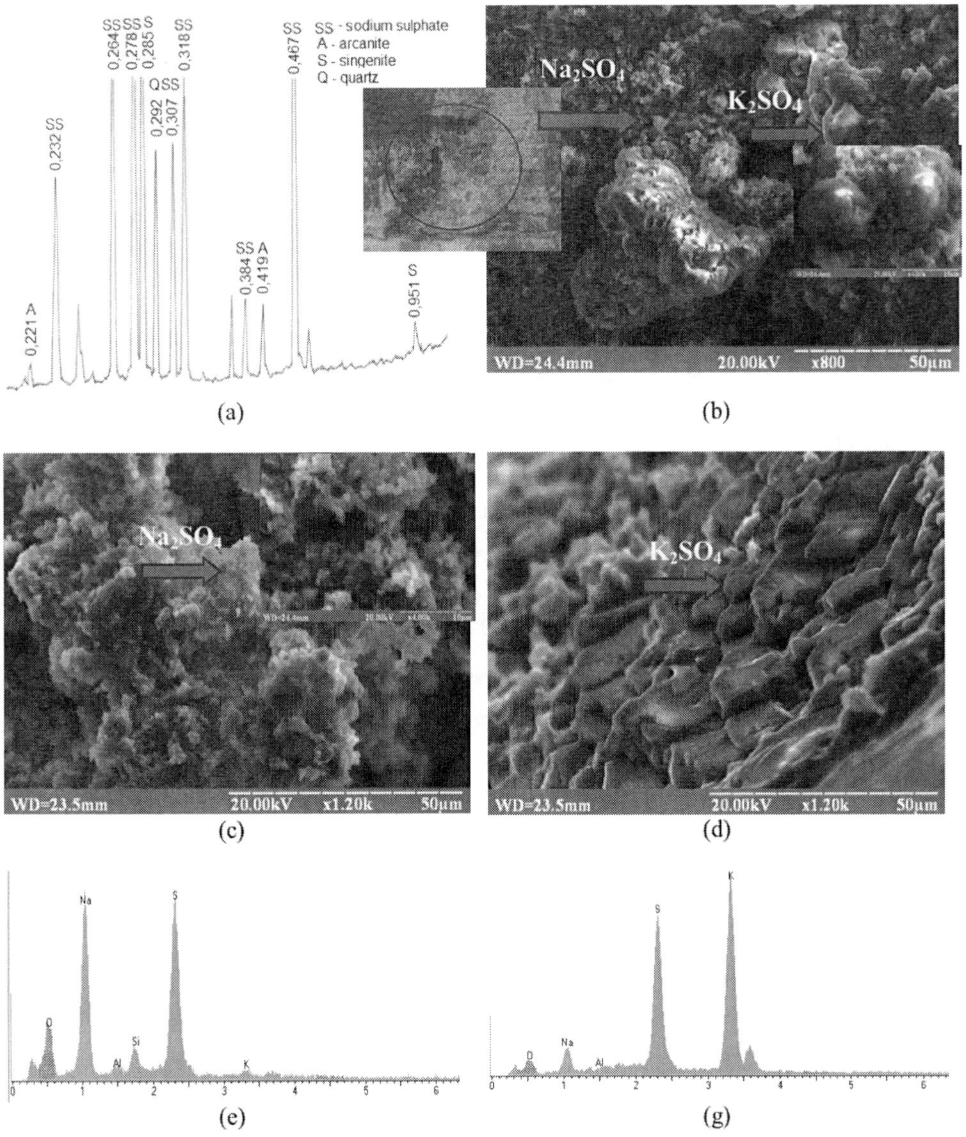

Fig. 2. X-ray powder diffraction (a), SEM images ((b, c, d) and the EDX spectrum (e, g) of salt efflorescence from ceramic bricks

The lines of calcite $CaCO_3$ (d/n=0.303; 0.249; 0.228 nm), quartz (d/n=0.334 nm), Na_2SO_4 (d/n=0.467; 0.384; 0.278; 0.264; 0.232 nm), calcium sulphate dihydrate $CaSO_4{\cdot}2H_2O$ (d/n=0.756; 0.427 nm) characterize the salt efflorescence from mortar masonry. The negligible reflexes of singenite (d/n=0.951; 0.285 nm) are used. The main mineral phase of salt efflorescence from mortar masonry is calcite with addition of anhydrous sodium sulphate (tenardite) (Fig. 3b). Moreover, the layer of the mineral rashes, adjacent directly to the brick, consists mainly of calcite which increases the tenardite (Fig. 3c). It should be noted the prevailing growth of ions concentration of Ca^{2+} and SO_4^{2-} due to the formation of gypsum dihydrate $CaSO_4{\cdot}2H_2O$.

DOI: 10.1051/matecconf/201711601007

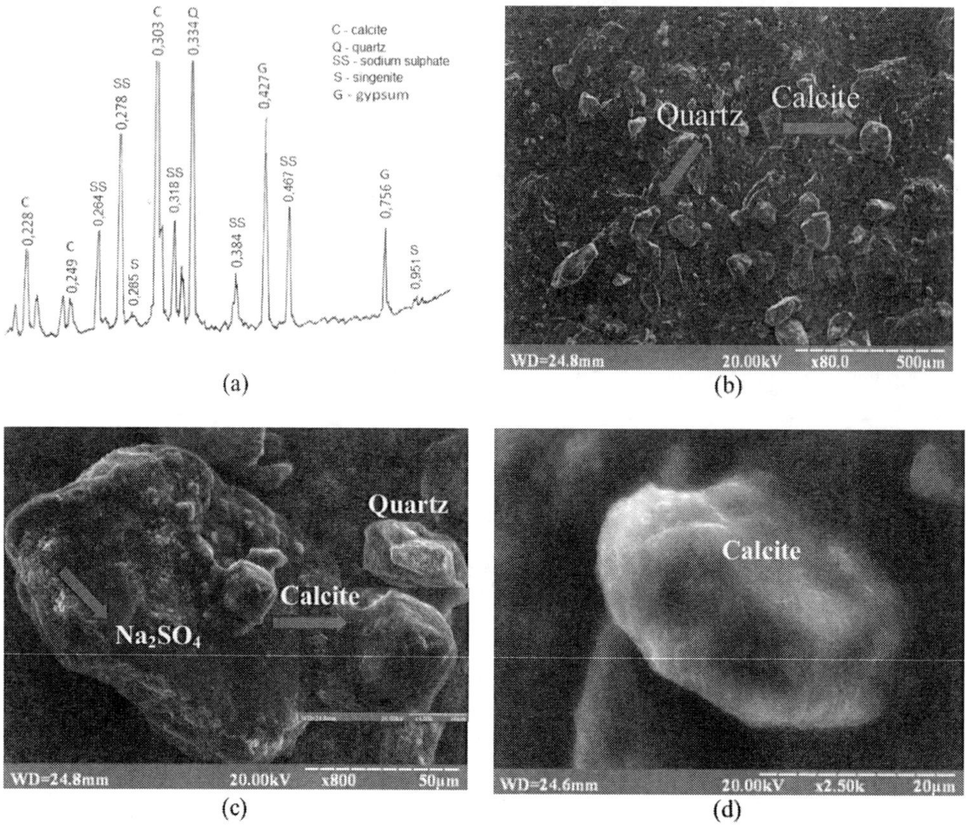

Fig. 3. X-ray powder diffraction (a) and SEM images (b, c, d) of salt efflorescence from mortar masonry

The special masonry cement complied with class MC 22,5 X DSTU B EN 413-1:2015 with low-clinker content (40 mass.%) is used to prevent the formation of salt efflorescence on the masonry surface of exterior walls in the mortar. Water demand of multi-component cement (SSA=7600 cm^2/g) is 31.8%, initial setting time – 160 min, final – 210 min, volume coefficient of water separation is 4.8%. Standard cone flow for multi-component cement is 162 mm (Fig. 4a). Compressive strength after 2, 7 and 28 days of hardening is 7.6, 16.1 and 34.7 MPa, respectively, according to EN 196-1 (Fig. 4b).

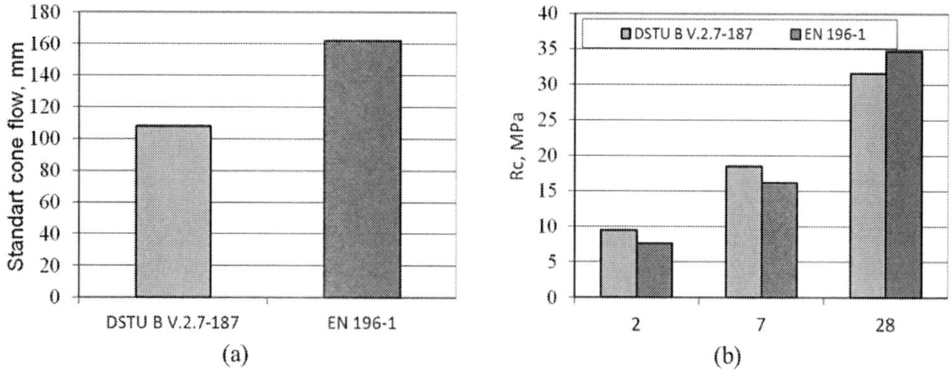

Fig. 4. Standart cone flow (a) and compressive strength (b) of cement MC 22,5 X according to DSTU B. V.2.7-187 and EN 196-1

The lines of calcite (d/n=0.303; 0.249 nm) and crystalline hydrate phases: calcium hydroxide (d/n=0.490; 0.263; 0.192 nm), ettringite (d/n=0.973; 0.388 nm) and a certain amount of calcium hydroaluminate C_4AH_{13} (d/n=0.81 nm) are observed on diffractograms of cement stone based on MC 22,5 X (W/C=0.34) after 28 days of hydratation by the X-ray analysis method (Fig. 5). It is observed a decrease of calcite lines $CaCO_3$ and presence of intensive lines of hydrocarboaluminate $C_3A{\cdot}CaCO_3{\cdot}H_2O$ (d/n=0.76; 0.380 nm), which indicate partial transformation of hexagonal hydroaluminate into hydrocarboaluminate after 28 days of hydration.

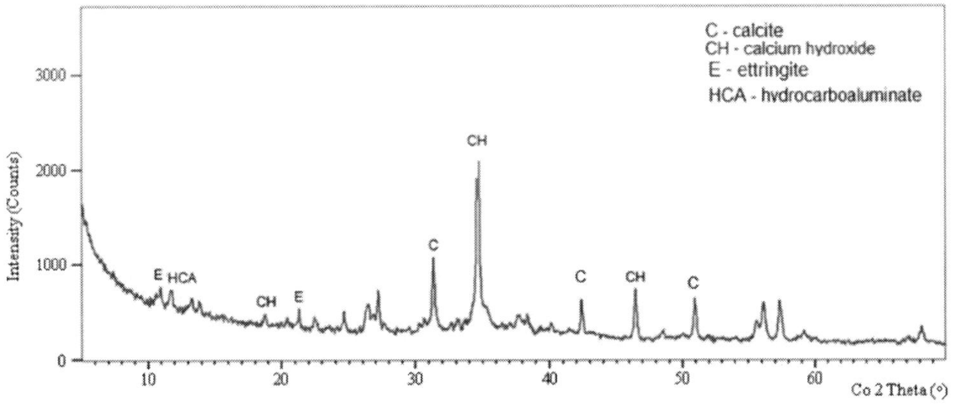

Fig. 5. X-ray powder diffraction of cement MC 22,5 X after 28 days of hardening

The addition of highly active mineral additives to cement MC 22,5 X allow reducing the calcium hydroxide in 2.2 times compared with a stone based on CEM I after 28 days of hydration. The dense structure of the hydrated solid phase is provided by AF_m- i AF_t-phases in gel-like C-S-H phase. Active forms of SiO_2 and Al_2O_3 in the composition of aluminum-containing pozzolana promote better binding of calcium hydroxide into low-basic calcium hydrosilicates, which indicates the acceleration of puzzolanic reaction. Fine limestone stabilizes hydration products of tricalcium aluminate with formation of hexagonal hydrocarboaluminate $C_4A{\cdot}CO_2{\cdot}12H_2O$ during hydration of the clinker component.

The quantity of air-entraining admixtures largely determines the porosity of mortar. For mortar with the addition of lime the total porosity is 23.17% (open – 16.68%, closed – 6.49%). At the same time, the total porosity for modified mortar (ρ=1860 kg/m^3) increases to 26.77%, but the ratio between the open and closed porosity changes (7.61 and 19.16% respectively). The addition of 0.03...0.05% of air-entraining admixtures by mass leads to the reduction of the average pores size to 2.77...1.95. The application of multi-component cement, which contains zeolite, provides for formation of a denser mortar matrix which reduces the motion of pore hydrates and thus reduces efflorescence formation.

The method of scanning electronic microscopy proves that structure of cement stone is condensed by colmatation of open pores of Portlandite and AF_m-phases (Fig. 6a). Hydrophobic plastifying admixtures cause the formation of numerous elastic air bubbles which spread on the whole cement paste uniformly. Fig. 6b displays the closed air pore of 5 mm in diameter. In the freshly-mixed mortar, spherical bubbles reduce the inner friction, plasticize and prevent its components from segregation. Due to the aeration, the building mortar is characterized by lower absorption ability, it is more resistant to atmospheric influences and more resistant to the multiple cycles of freezing and thawing, as water that freezes in the capillaries has the ability to penetrate into empty bubbles which prevent the cracking, and thus play the role of "buffers" – stress dampers [10].

(a) (b)

Fig. 6. The microstructure of the cement stone after the 28-day hydration with the addition of compound air entraining agent: (a) open pore, (b) closed air pore

According to the thermogravimetry data, in modified mortars comprising after 3 months of hardening in the air-dry conditions LOI=7.38 mass.%, the content of bound water equals to 3.10 mass.%, or in relation to cement – 14.39 mass.%, the content of $Ca(OH)_2$ and $CaCO_3$ – 2.80 and 7.27 mass.% accordingly. This shows that over 50% of water evaporates during the hardening. For the mortar with the addition of 20 mass.% of lime paste LOI=9.16% (including the amount of bound water – 2.50%), and the content of $Ca(OH)_2$ and $CaCO_3$ equals to 2.61 and 12.15 mass.% accordingly, thus in a mortar containing lime paste the loss of water is more intense and through the open capillary pores is also accompanied by substantial carbonization.

The necessary complex of the quality parameters of high-performance mortar had been analyzed and the results showed that the workability of modified mortar mixtures was characterized by the ability to water-retention 96-97% and by segregattion - 7-8%. Modified mortars (f_c = 10.4 MPa), possess minor shrinkage deformations (0.31 mm/m), 75 cycles of freezing and thawing and are characterized by reduced efflorescence.

The mortar is characterized by smaller water absorption and higher resistance against the atmospheric influences and repeated cycles of freezing due to aeration, because the water that freezes in the capillaries can compress into empty bubbles that prevent cracking, acting as "shock absorbers" – the stresses dampers. Modified masonry mortars based on MC 22,5X provide the high quality of masonry (completeness and uniformity of the seam filling, observance of their rational thickness), an adhesion of the mortar and also strength, deformation and stability to salt efflorescence on the surface of the walls. The compressive strength of masonry is 3.5 MPa, when modified mortar and ceramic bricks are used. So the compression strength of masonry due to the emergence of a complex stress state is 2.8 times less compared with the strength of brick and mortar.

One of the effective ways to protect the ceramic brick from the destructive action of moisture and salts is its processing impregnation by ecologically safe chemicals compound of multifunctional action. It was done the physical and chemical surface modification of ceramic bricks (depth impregnation - 2.1 - 3.1 mm) by organosilicon and acrylic compounds. Experimental investigations have established that ordinary ceramic bricks (compression strength – 10 MPa, average density - 1647 kg/m^3, porosity - 24.3%) and modified ceramic bricks with protective coating according to water absorption index reduced 8.2 - 8.4 times and water resistance and frost resistance increased, respectively, by 6.2 - 17.5% and 55 - 64%. The use of organosilicon compounds provides a compaction of the structure of ceramic brick surface, the high resistance to water permeability and salt solutions and durability.

4 Conclusions

The local white spots of salt efflorescence lead to destructive phenomena in the brick masonry. The use of multimodal cements provides compacting mortar matrix structure, reduces shrinkage and water absorption, increases water resistance and fracture toughness of mortars and allows obtaining plastered surfaces without efflorescence. Synergistic combination of mineral additives of different groups with substantial reduction of high-energy-consumption clinker component content in multi-component cement MC 22,5X allows improving the technological characteristics of mortar mixes and quality of brick masonry. Physical and chemical modification of mortars with highly effective complex aeration additives becomes one of the main solutions to the problem of salt efflorescence and masonry durability for exterior walls of buildings and structures.

This research was supported by Prof., Dr.Sc., Myroslav Sanytsky

References

1. H. Brocken, T. G. Nijland. Constr. and Build. Materials, **18**, 315–323 (2004)
2. C. Dow, F.P Glasser. Cem. Concr. Res., **33**, 147–154 (2003)
3. N. Sghaier, M. Prat. Transport in Porous Media, **80,** 441 – 454 (2009)
4. A. Mobili, A. Belli, C. Giosuè, T. Bellezze, F. Tittarelli. Cem. Concr. Res., **88**, 198–210 (2016)
5. M. Muller, H-M. Ludvig, M. Ben Haha, M. Zajac. *19th Internationale Baustofftagung*, **1**, 449-456 (2015)
6. M. Sanytsky, T. Kruts, T. Kropyvnytska, B. Rusyn. *14th International Congress on the Chemistry of Cement*, **1**, 265 (Beijing, China, 2015)
7. C. Bilim. Constr. and Build. Materials, **25**, 3175-3180 (2011)
8. J. Shibata, N. Murayama, T. Masui, S. Yao, T. Toyofuku, **41**, 67-71 (2015)
9. B. Klimek, D. Barnat–Hunek, M. Franus. *Proceedings of Lviv Polytechnic National University*, **781**, 72-78 (2014)
10. M. Sanytsky, T. Kropyvnytska, R. Kotiv. Advanced Materials Research, **923**, 42–47 (Switzerland, 2014)
11. A. Izaguirrea, J. Lanasb, J. Álvareza. Cem. Concr. Res., **39**, 1095–1104 (2009)
12. H. Zhu, P. Wang, R. Wang, G. Zhang. Materials science, **20**, 345-350 (2014)

Effects of elevated temperatures on the properties of nanomodified rapid hardening concretes

Uliana Marushchak[1,*], *Myroslav* Sanytsky[1], and *Yuriy* Olevych[1]

[1]Lviv Polytechnic National University, Building Production Department, S. Bandera str. 12, 79013, Lviv, Ukraine

Abstract. Effects of elevated temperature exposure and hardening age on mechanical properties of nanomodified concrete are presented. The mass loss, compressive strength and shrinkage of the exposed to the elevated temperatures and cooled specimens were determined. Nanomodification by ultrafine mineral additives and polycarboxylate eter superplasticizer provides direct early structure formation, higher strength at early and later ages, and improved thermal resistance in a range of 200-350 °C of cement-base materials. High specific surface area of ultrafine mineral additive particles and their "excess surface energy" determine interaction degree with powder part components of concrete and allow realizing dense packing, filler, high water-reducing, early pozzolanic reaction, high early strength effects to increasing the rigidity of the cement matrix to resist deformation at impact of elevated temperature.

1 Introduction

The improving of the reliability of buildings and structures including traffic, tunnel structures and fortifications, which exposed to fire and high temperatures, ensures their safe exploitation. Exposure to elevated temperatures significantly affects the performance of concrete structures, since concrete is not stable at high temperatures. Subjecting concrete to a higher temperature leads to severe deterioration and it undergoes a number of transformations and reactions, thereby causing progressive breakdown of cement gel structure, reduced durability, increased tendency of drying shrinkage, and structural cracking [1, 2]. Strict requirements regarding the resistance of structures to fire and elevated temperatures, have forced to develop new solutions to improve existing cement-based materials due to the designing of thermal stable high strength concrete. The use of such concrete for repairing works on intensive traffic zones, roads, facilities in airports and railways sets high requirements concerning hardening time.

One of the innovative ways to obtain rapid-hardening concretes with high performance is the use of nanotechnology techniques related to the directed process of structure formation by modification with nanoscale elements [3, 4]. The increasing of thermal stability of paste matrix can be achieved by adding ultrafine supplementary cementitious

* Corresponding author: ulmarushchak@ukr.net

materials. The ultrafine particles less than 1 μm (nanostructure scale), which have defined a supply of free surface energy, increase the interfacial surface that can accelerate chemical reactions, detect catalytic activity and cause more substantial influence of superficial atoms on synthesis of cementitious system strength. The ultrafine particles become an integrated part of the hydrated cement paste and improve homogeneity of the interfacial zone.

Cement pastes modified with nanosilica showed that the nanosilica exhibited increased thermal stability. This effect can be attributed to improvement of thermal stability of calcium silicate hydrate gel (C-S-H) and to the filler effect [5]. Due to the reactivity of nanosilica more high-density C-S-H was produced, and samples containing nanosilica had still higher residual compressive strength. Presence of nanosilica contributes to reduction of micro-cracks due to its pozzolanic activity [6]. One of the widely used methods of concrete nanotechnology is the modification with the surface active substances and ultrafine additives. Highly efficient polycarboxylate superplasticizers with the nano-designed molecular chains make it possible to increase strength and decrease surface spalling due to significant reduction of the water-cement ratio [7]. The development of rapid hardening concretes with high thermal resistance requires complex nanomodification of cement matrix by ultrafine supplementary material and effective polycarboxylate superplasticizers.

2 Materials and methods

2.1. Materials

Ordinary Portland cement CEM I 42.5 (according to EN 197-1 standard) was used in the investigations. Natural sand of Zhovkva quarry (MF=2.1) and coarse aggregate fractions 2-5 mm were used for fine-grained concrete production. Low calcium fly ash, silica fume and methakaoline were used as ultrafine supplementary cementitious materials. The dosages of polycarboxylate eter superplasticizer GLENIUM ACE 430 (PCE) which incorporated into nanomodified mixes was 1.5% by weight of cement. The nanomodified cement containing ultrafine mineral additives, polycarboxylate superplasticizer are characterized by the gap-graded particle size distribution which leads to the best packing density of particles.

2.2. Preparation of the specimens and testing

Consistency of fresh concrete was determined by flow-table method conforming to EN 1015-3 standard. The fine-grained concrete components were mixed according to EN 196-1 procedure. The samples 40 mm x 40 mm x 160 mm were cured in normal conditions for the hardening of concrete (90-100% RH at 20±2°C). After 1, 2 and 7 days the samples were dried in a chamber at a constant temperature of 105 °C. Then some specimens were kept in medium-temperature furnace and subjected to elevated temperatures of 200°C and 350°C for 2 hours according to DSTU B.V.2.7-249:2011. After cooling, samples underwent the compressive strength tests according to EN 196-1. Weight loss and shrinkage of concrete were determined after exposure to elevated temperatures.

3 Result and discussion

The consistencies of reference fresh mixture (RC) and nanomodified fresh mixtures (NMC) determined by the flow table method are 180–190 mm. However nanomodified fine-grained mixture characterizes significant water-reduction effect ΔW/C=32%.

The early strength of nanomodified fine-grained concrete after 1 day is 33.3 MPa and after 2 days – 58.4 MPa that is, respectively, 1.9 and 2.3 3 time higher than the strength of reference concrete (Fig. 1). The compressive strength of nanomodified fine-grained concrete after 28 days of hardening at normal conditions increases up to 104 MPa. The parameter R_{c2}/R_{c28} of nanomodified concrete is 66.9%, which allows attributing these concrete to rapid hardening high strength concretes.

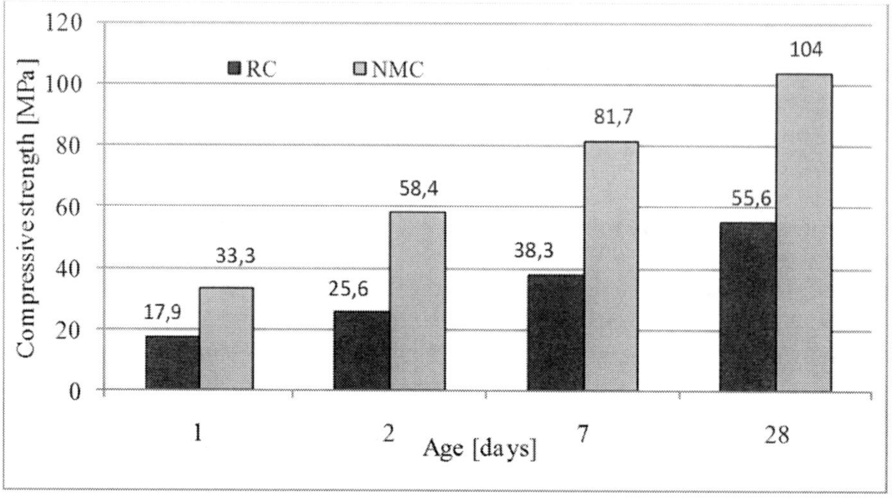

Fig. 1. Compressive strength of fine-grained concrete in normal condition.

Up to 350°C the strength of heat-treated samples was still higher than those of the control samples (Fig. 2). Thus, the strength of nanomodified fine-grained concrete after 1 day hardening in normal conditions and exposure to 200°C increases 2.6 times compared to reference concrete. The strength increasing of nanomodified concrete after 2 days when exposed to 200°C is 47.6%, and after 7 days – 13.5%.

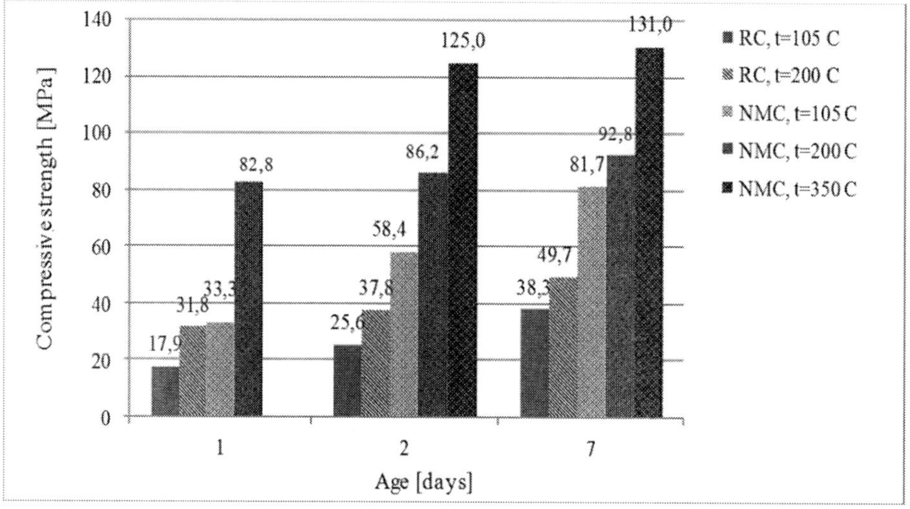

Fig. 2. Residual compressive strength of fine-grained concrete subjected to elevated temperature.

After exposure to 350°C and subsequent cooling to room temperature, there was a significant improvement in strength for samples of nanomodified fine-grained concrete. Thus, strength of nanomodified concrete after 2 and 7 days of hardening at normal condition is, respectively, 125.0 and 131.0 MPa. The reason is that among a certain higher temperature range, the pore pressure in the hardened cement mortar is increased by the evaporation of internal water, forms a high-temperature autoclaving environment, so as to promote the unhydrated cement particles forming more hydration product, provides fuller pozzolanic reaction and enhances the compressive strength of specimens [2].

The effect of increasing temperature on hydrated cement paste depends on the degree of hydration and moisture state. The presence of large quantities of evaporable water (free, capillary and adsorbed water as well as water from the cement hydration products) can cause shrinkage and damage to concrete in the form of surface spalling. The changes in the mass loss of fine-grained concrete are highly related to age of hardening (Fig. 3). Rapid water binding into hydration products in case of nanomodified concrete causes significant decrease in mass loss. It can be seen that changes in the mass loss of nanomodified concrete exhibited a mass loss 1.6-times lower after 1 day and 2-times lower after 7 days as compared to the samples of reference concrete.

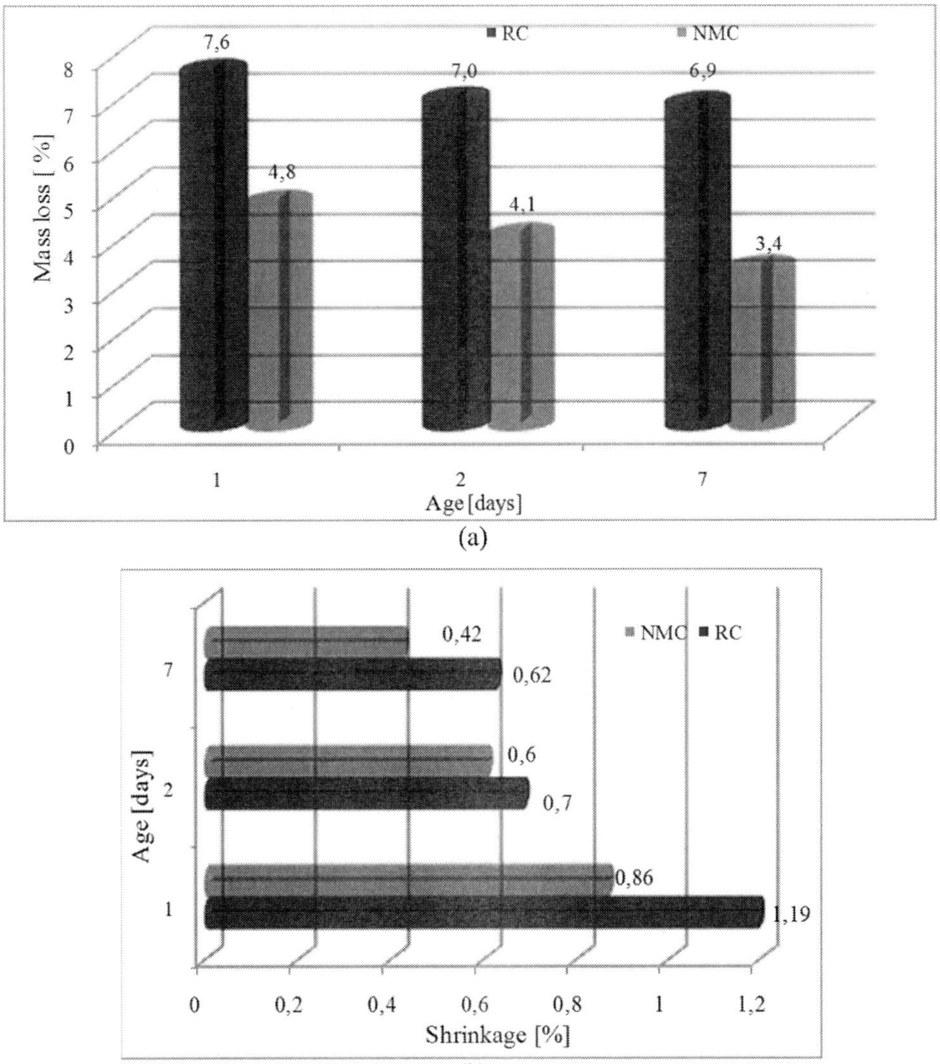

(a)

(b)

Fig. 3. Mass loss (a) and shrinkage (b) of fine-grained concrete exposure to 200 °C.

A higher w/c ratio increases the total shrinkage and the rate of shrinkage of the cement paste by providing more space for free water diffusion and reducing the rigidity of the solid matrix to resist deformation. Ultrafine additives densify the hydrated cementitious paste, thereby slowing down the rate of water evaporation and, hence, shrinkage. Thus, thermal shrinkage of nanomodified concrete after 1 day of hardening and exposure to 200°C is 0.86%, which is 27.7% lower than the shrinkage of reference concrete.

Conclusion

The nanomodified concretes with ultrafine mineral additives are characterized by high strength at early and later age, exhibit enhanced stability of mechanical properties when exposed to temperatures in a range of 105 to 350°C. The possibility of obtaining rapid hardening concretes with high thermal resistance is achieved by complex nanomodification with chemical admixture and ultrafine mineral additives, which is provided by water demand reducing, system particle packing optimization, increasing cement matrix density, stimulating nucleation processes in the intergranular space, acceleration of hydration process and pozzolanic reaction.

References

1. G. Wang, Ch. Zhang, B. Zhang, Q. Li, Z. Shui, Fire Mater., **39**, 741 (2015)
2. C. Alonso, L. Fernandez, J. Mater. Sci. **39**, 3015 (2004)
3. H.R Ashani, S.P. Parikh, J.H. Markna, Int. J. Nanosi. Ser., **2(5)**, 32 (2015)
4. U. Marushchak, M. Sanytsky, T. Mazurak, Yu. Olevych, Eastern Eur. J. Enterprise Technologies, **6 (84)**, 50 (2016)
5. P. Aggarwal, R.P. Singh, Y. Aggarwal, Cogent. Eng., **2** 1078018 (2015)
6. E. Horszczaruk, P. Sikora, K. Cendrowski, E. Mijowska, Constr. Build. Mater., **137**, 420 (2017)
7. J. Plank, C. Schroefl, M. Gruber, N. Lesti, R. Sieber, J. Adv. Concr. Technol., **7-1**, 5 (2009)

Improvement in strength calculation for pavement structure of rail station areas

Vladimir Mozgovyi[1,*], *Sergei* Baran[1], and *Alexander* Kutsman[1]

[1]National Transport University, Road Construction Materials and Chemistry Department, Boychuka st. 42M, 01103 Kyiv, Ukraine

Abstract. The paper discusses the issue of how to increase the durability of road pavements at stations. At the same time, the behavior of asphalt-concrete layers is considered taking into account thermoreological properties and their materials. An increased duration of traffic loads on station areas influences changes in strength and rigidity characteristics of the asphalt concrete. Mathematical models to increase the accuracy of strength calculation of pavement designs on station areas have been developed.

Nowadays the most common material used as the pavement of city streets and squares is the asphalt concrete. However, strength and durability of such a coating on station areas are considerably lower in comparison with these characteristics for sections of streets and roads of intensive traffic. The research conducted by the authors showed that an increase in continuous loads on the asphalt concrete pavement causes its more intensive destruction.

In accordance with the normative design procedure for asphalt concrete pavement under Branch Construction Norms VBN B.2.3-218-186 [1], 0.1 sec is the design loading duration which corresponds to the average speed of movement for design vehicles on general-purpose roads. At the same time, the traffic on station areas significantly differs from the traffic on public roads. The traffic mode on station areas is characterised by lower speeds, longer stops and parking, which causes an increased loading duration, as well as a change in the properties of thermo-rheological sensitive materials.

An assessment of an effect of design characteristics of asphalt concretes on the stress-strain state and durability of the pavement covering under multiple static and short-term loads was made on the basis of work [2]. Consequently, it was proved that longer loads resulted in lower coefficient of elasticity of asphalt concretes (Table 1) and other materials, which led to a decrease in the maximum horizontal tensile stresses in the upper layer of the base of asphalt concrete. And values of the tensile strength in bending R_{ben} of the material in which σ_{max} occur increase (Table 1).

It should also be taken into account that the tensile strength at bending R_{ben} with an increase in the loading duration decreases much faster than the maximum horizontal tensile stresses σ_{max}. It is exactly the reason why fatigue cracks appear prematurely on the sections where heavy vehicles slow down very frequently.

* Corresponding author: mozgoviy@gmail.com

 DOI: 10.1051/matecconf/201711601009

Table 1. Deformation and strength characteristics of the highly porous asphalt concrete on Road Bitumen 60/90 under various loading duration.

Design characteristics	Loading duration, sec			
	0.1	1.0	10.0	600.0
Modulus of elasticity, MPa	2100	1510	1050	510
Tensile strength at bending, $R_{ben.}$, MPa	5.65	3.18	1.79	0.64

For the analysis the comparative calculations of soft pavement with black rubble as the base and the pavement of the same strength with cemented crushed stone as foundation was conducted.

The results of the calculation regarding the applied design load before the pavement deterioration (relative durability) are given in Table 2.

Table 2. Relative durability of road pavement.

Type of road pavement	Loading duration, t, sec	Relative durability, $N(t)/N(t=0.1\text{sec})$
Not reinforced	0.1	1
	1	0.305
	10	0.170
	100	0.108
Reinforced	0.1	1
	1	0.314
	10	0.175
	100	0.138

Table 2 shows that the number of the design load application before the pavement destruction decreases with an increase in the loading duration. Besides, it can be concluded that strengthened base slightly increases its durability, and to a greater extent with a longer duration of the load.

Thus, the study conducted shows that durability of the asphalt-concrete coating of non-rigid road pavement significantly decreases with an increased loading duration. And the durability of asphalt-concrete coating considerably depends on the thermoreological sensitivity of the base materials.

Let us consider in more detail the influence of these factors on durability of asphalt-concrete coating for non-rigid road pavements with bases of various thermoreological sensitivity by examples of loading regimes typical for the areas characterised by slow traffic and stops of vehicles.

Taking into account the well-known fact that loading duration, temperature and humidity influence the deformation and strength characteristics of road pavement materials according to their composition and structure, the authors proposed a classification of materials by their thermoreological sensitivities. In this case, the ratio of instantaneous and long moduli of elasticity was taken as an indicator of thermoreological sensitivity. Thus, road construction materials were divided into four groups (Table 3).

By the method described in [6, 7] the research into the durability of non-rigid road pavements with asphalt-concrete coatings of different base models according to their thermoreological sensitivity and the same strength criteria by method [1] (Fig. 1) was conducted.

The obtained results show that these road pavements have different durability when a loading duration increases (Fig. 1).

The research conducted shows that for the roads of slow traffic and stops of vehicles, the use of materials of lower thermoreological sensitivity for the base of road pavement contributes to a significant increase in the pavement durability (2-5 times or more).

Table 3. Groups of road building materials by the thermoreological sensitivity.

Types of road construction materials by groups			
I (Highly sensitive)	II (Sensitive)	III (Moderately sensitive)	IV (Poorly sensitivity)
Asphalt-concrete of all types, organomineral mixtures, etc.	Clay and silty roadbeds, crushed stone and gravel, treated with organic binders, etc.	Crushed stone, gravel, sand, untreated or treated with mineral binders of less than 5%, sandy and sandy loam soils and gravel-sand mixtures treated with mineral binders, etc.	Crushed stone, gravel, sand, coarse-grained soils and gravel-sand mixtures treated with mineral binders in an amount of more than 5%, lean concrete, etc.

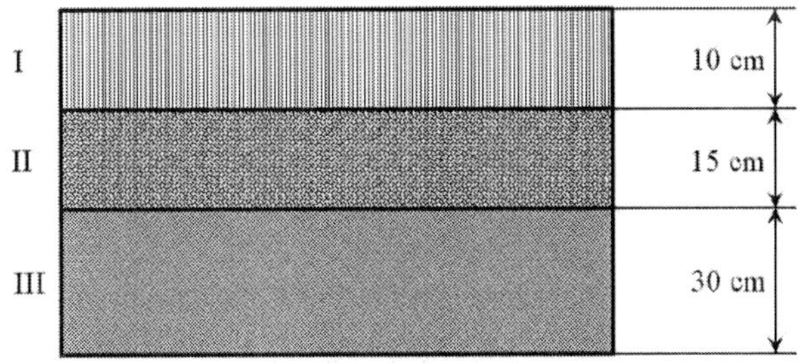

(a)

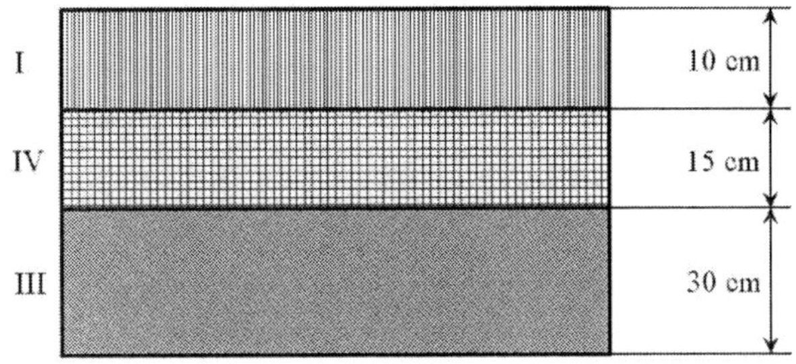

(b)

Fig. 1. Road pavements equivalent to the traditional criteria of the limiting state: (a) with a base of materials of increased thermoreological sensitivity (1st variant); (b) the same of reduced thermoreological sensitivity (2nd variant).

Thus, the conducted researches allow offering more perfect models of asphalt-concrete coatings taking into account the different loading duration on various base models containing materials of various thermoreological sensitivity.

In connection with the above mentioned, the following is suggested.

For calculation of asphalt concrete pavements on station areas it is advisable to produce the complex multiple impact as short-term loading (t_l = 1.0 sec) and static loading (t_l =600 sec).

For such a combination of binary loading we use the fundamental propositions of the kinetic theory of strength of solids [2, 3-5] when considering the issue of the asphalt concrete pavement durability.

In this case using Miner's rule changes in relative durability expressed as the measure of damageability M from each binary loading cycle can be written as the following dependence:

$$M = M_1(\sigma_1(t_l = 1.0\ s) + M_2(\sigma_2(t_l = 600\ s) < 1, \quad (1)$$

where $$M_1 = \frac{n\,(t_l=1.0\ s)}{\sum N_{p1}(t_l = 1{,}0s)};\ M_2 = \frac{n\,(t_l=600\ s)}{\sum N_{p2}(t_l = 600s)}. \quad (2)$$

Under the combined influence of the binary cycle on fatigue destruction of asphalt concrete pavements at the known values $\sum N_p$ of loading units and the share δ_2 in the total number of loads $\sum N_p$ from vehicles creating the loading duration $t_н > 1.0\ sec$ and determined on the basis of the feasibility studies the following expressions will be true:

$$N_1 + N_2 = \Sigma N; \qquad N_1 = (1 - \delta_2) \cdot \Sigma N; \qquad \delta_2 = \frac{N_2}{\Sigma N}. \quad (3)$$

Therefore, the above calculation testify that for the corresponding loading regimes the limiting quantity of loads on the non-rigid asphalt concrete pavement for the binary cycle of loading at t_{l1}=1.0 sec and t_{l2}=600 sec can be established.

With the help of this technique, the calculations of road pavement structure with asphalt concrete pavement were conducted.

The structure includes the following layers:

- dense asphalt concrete on Road Bitumen 60/90, a thickness of 40 mm;
- porous asphalt concrete on Road Bitumen 60/90, a thickness of 80 mm;
- highly porous asphalt concrete on Road Bitumen 60/90, a thickness of 220 mm;
- crushed-gravel-sandy mixture, treated with cement M20, a thickness of 260 mm;
- soil road bed – dusty sandy loam.

Preliminary strength calculation in accordance with [1] for a design loading duration of 0.1 sec showed that the selected design met all strength criteria. However, with an increase in the loading duration on pavements from 0.1 sec to 1 sec, in accordance with the proposed procedure, the calculation results showed that the chosen design did not satisfy the criterion of tensile strength at bending, which requires the introduction of appropriate adjustments. Due to correction, it turned out that in order to compensate for the decrease in the strength coefficient with an increased loading duration from 0.1 sec to 1 sec it is efficient to replace the highly porous asphalt concrete with dense asphalt concrete of increased strength and resistance to repeated loads. Similar results were obtained when calculating the structure for 600-sec duration.

Therefore, in order to ensure the necessary strength and increased durability of asphalt-concrete layers for pavement on rail station areas it is reasonable to replace the bending material in a layer with a material of increased strength and resistance to repeated loadings.

References

1 Transport facilities. Flexible type of pavement.VBNV.2.3-218-186-2004

2 V.V. Mozgovyi, *Scientific foundations for ensuring the temperature crack resistance of asphalt concrete coatings*: dis. for the degree of Doctor of Technical Sciences. Kiev, (1996)
3 A.A. Ilyushin, On a theory of long-term strength Engineering Journal MTT,**3**(1967).
4 A.A. Ilyushin, B.E. Pobedrya, Fundamentals of the mathematical theory of thermoviscoelasticity (Science, Kiev, 1970)
5 V.V. Moskvitin, Resistance of viscoelastic materials (Science, Kiev, 1972)
6 V. Smolyanets, *Improving the design of non-rigid asphalt pavement in a city*: dis. for the degree of Ph.D. (Kiev, 2005)
7 L.A. Mozgovaya, A.V. Mozgovoy, V.V. Mozgovoy and others, Road engineering, 126-139 (2007)

Formation of structure of high-strength composites with account of interactions between liquid phase and disperse particles

Andrii Plugin[1,*], *Elena* Dedeneva[2], *Tatiana* Kostyuk[2], *Dmitro* Bondarenko[2], and *Olga* Demina[2]

[1]Ukrainian State University of Railway Transport, Building Materials and Structures Department, Feuerbach sq. 7, 61050 Kharkiv, Ukraine
[2]Kharkiv National University of Civil Engineering and Architecture, Building Materials and Products Department, Sumska st. 40, 61002 Kharkiv, Ukraine

Abstract. The ways of composites' directed structuring are covered in this work. The qualitative and quantitative assessment of the role of liquid phase is made. The regime of the effective vibro vacuumizing processof cement composite mixture, which makes it possible to extract free water is developed. Thin-walled articles obtained in this case possess high density, strength with the bend, lowered water absorption and porosity.

1 Introduction

In the contemporary construction practice the need for thin-walled articles made of fine-grained concrete increases. These are the decorative sheets, which imitate natural stone, concrete tile, landscaping architecture of complex configuration, etc.

Improvement of the mechanism of the composite structure formation gives possibility of obtaining the high-quality articles made of fine-grained concrete. It is known that the fine-grained concrete mixtures possess increased waterdemand. Quantity of liquid phase, which ensures the necessary packability of the fine-grained concrete mixture W_{lph}= 53-65% (from the mass of cement) [1]. Quantity of free waterwhich is able to form bug holes, is W_{fw} = 34%.

Forobtainingthearticlesofdensestructureandhighstrengththeapplicationofintensivemethod sofconcretemixturepackingwiththepossibilityof free (technological) water removal is required[2]. One of the effective methods from this point of view is vibro vacuumizing [3]. It was established earlier that vibro vacuumizing process gives the possibility to move away from the concrete mixture 29-30% of water, i.e., all the free liquid [3,4]. The residual water content is more than a quantity of bound water by 4-5%. For the comparison, in the concrete mixtures packed by vibration, the residual water content is 20-23%. Thus, the method appears to be sufficiently effective looking at the quantity of dehydrated water. The study of different vibro vacuumizing regimes gives the possibility of further increase of the structural and physical and physical and mechanical characteristics of thin-walled concrete articles.

* Corresponding author: aaplugin@gmail.com

 DOI: 10.1051/matecconf/201711601010

With the formation of concrete structure three phases interact: gaseous, liquid and solid. The liquid phase is water, in which the additional chemical substances which role requires separate examination can be present. From different forms of interactions of liquid and solid phases it is possible to isolate three basic ones. 1) The chemical interaction, caused by hydration and crystallization of the solution of new formations, and formation of the gel of mineral binding. 2) The physicochemical interaction caused by the adsorption of water molecules in the action zone of the surface force field of solid phase. 3) The technological interaction, which makes it possible to lower the viscosity of stock mixture causing the necessary packability of concrete mixture. As it is evident, directed structuring is impossible, first of all, without the qualitative and quantitative assessment of the role of liquid phase.

The goal of the work is obtaining cement composite for the thin-walled articles of the increased density and strength.

For achievement of the stated goal it was necessary to solve the following tasks: to estimate the role of liquid phase in the process of the structuring of cement composite; to develop the effective regime of its packing.

2 Materials and methods

Cement composite (fine-grained concrete) in the work was obtained from the following materials. For the laboratory models the Portland cement PC II/A-S-400 (CEM II/A-S-32.5) was used, which corresponds to Ukrainian national standard DSTU B V.2.7-46:2010 requirements "Building materials. Cements of general purpose. Specifications". To achieve high density and waterproofing capacity of concrete, and also more uniform surface of articles, the quartz small sand with the module of the coarseness 1.5 was used. Concrete mixtures of composition C: S=1:2 had initial water-cement ratio W/C = 0.65.

For conducting the studies of the influence of different packing regimes on the structural and physical and physical and mechanical properties of fine-grained concrete the models with the size of 350×350×12 mm were prepared. For determining the strength with the bend within the periods 0, 1, 3, 7, 14, 28 of twenty-four hours there were 18 models for each of the regimes. For determining density and water absorption there were 3 models for each regime.

Average density, water absorption and porosity were in accordance with the DSTU B V.2.7-170:2008 requirements "Building materials. Concretes. The methods of determination of average density, moisture, water absorption, porosity and waterproofness".

The determination of strength with the bend of models was conducted in accordance with the DSTU B V. 2.7-43:96 requirements "Heavy concretes. Specifications". The models were experienced on the hydraulic PSU-50 press. The visual inspection of models showed that they do not have cracks. Their structure is uniform.

3 Theoretical justification for forming high-strength composites structures

It is known that during mixing concrete ingredients with water their chemical interaction occurs. It is the hydration which means the dissolution of clinker minerals, formation of the supersaturated aqueous solution relative to hydrates, and the formation of hydration products. The part of the water is connected chemically, especially by the crystal hydrates of calcium hydrosulfoaluminates (CHSA). A quantity of water, connected by CHSA, gradually increases from 1-2% during the first minutes of their formation to 13% toward the end the gripping.

Together with the chemical one the physicochemical interaction occurs, which includes a change in the polar properties of water molecule (Fig. 1). Falling into the power surface field of solid phase, which can have electro-surface potential (0 and electro-kinetic ξ-potential surface different in sign and value (for example, SiO_2 is negative, $Ca(OH)_2$, CaO, CHSA is positive), the water dipoles are attracted to the active centers by pole with the opposite sign [1, 5].

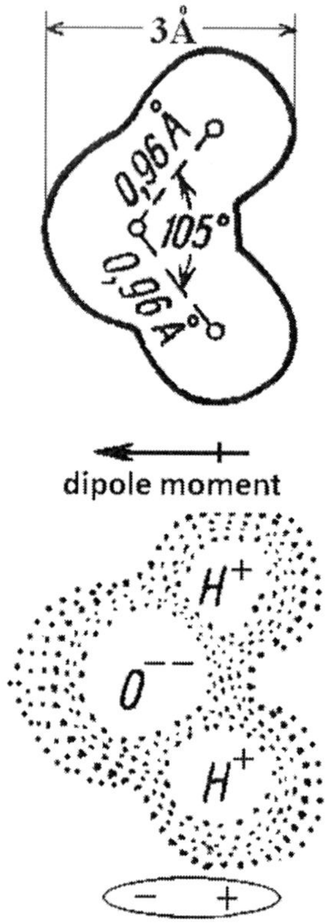

Fig.1. Change in the configuration of water molecule with change in the dipole moment.

On the rigid surface they form a film in the form of the salvation sheath, which consists of several adsorptive and diffusion layers. The thickness of films depends on mineralogical composition and mass of the particles of solid phase.

With its properties the adsorbed water approaches a pseudocrystalline body with the density 1.2-2.0 of g/cm^3, the significant elasticity, the strength with the shift at about 1×10^4 Pa, the low freezing point being lower than 195 K [3]. Such anomalous boundary layers of water depend on the presence of the dual electrical layer (DEL) on the border of the division of solid and liquid phases.

DEL is formed in the thin capillaries on the surfaces of colloidal and highly dispersed particles due to the adsorption of ions from the solution on the rigid surface. As a result of interaction with the solution the colloidal new formation micelle appears, in which the solid phase (nucleus) is connected with the solution through stabilizing DEL, and sometimes it happens through the salvation sheath (Fig. 2a, b).

On the border of adsorptive and diffuse DEL zones ξ-potential appears during the motion of liquid relative to the surface of the nucleus of micelle, (Fig. 2b), which characterizes the intensity of electro-kinetic phenomena and corresponding physical chemistry processes, which lead to the appearance of the crystal hydrate structure of the concrete block [3].

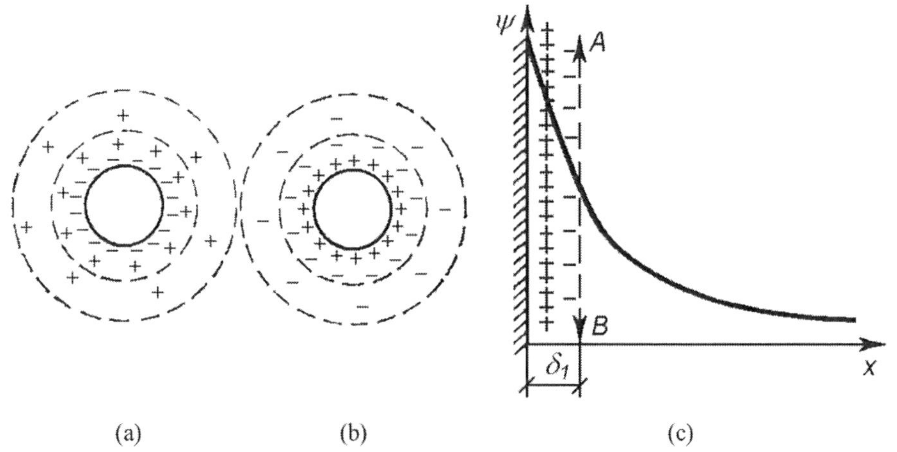

Fig.2. Forms of connection and the value of ψ potential on the division border of "solid phase-liquid": (a) with the negatively charged particle; (b) with the positively charged particle; (c) charge and potential distribution in the micelle.

The structural strength of concrete to a considerable extent depends on the electro-surface properties of the components [5-7, 9-12]. During the first minutes the positively charged particles of CHSA of colloidal sizes are actively formed in the freshly prepared concrete mixture. SiO_2 sand particles are minus charged. Since the hydrogenous bonds are oriented equally probably in all directions in the volumetric free water, and strengthening the dipole in parallel oriented moments the excessive orientation of hydrogenous bonds in the direction of the surface of solid phase appears in the boundary layer, the orienting action of electrical surface field is strengthened. This causes the appearance of durable electro-heterogeneous contacts between the particles of cement and sand (Fig. 3). Water demand for hydrosilicate gel is $K_g = 0.18$ C.

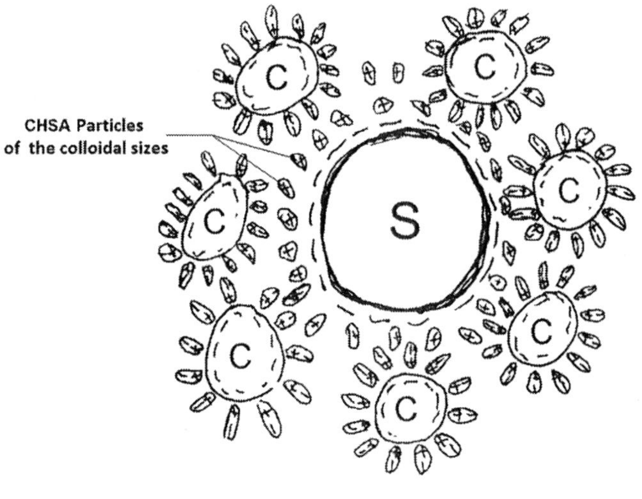

Fig. 3. Scheme of the formation of electroheterogeneous contacts between the particles of cement and sand in the initial period of the formation of fine-grained concrete structure

The water, which did not enter the crystal-hydrates, is structured having different effect on the physical and mechanical properties of concrete block and after that of concrete and solution.

A quantity of physically bound water depends on the summary specific area of the particles of solid phase, i.e., the granulometric composition of the fillers of concrete mixture. For the fine-grained concrete mixtures a quantity of physically bound water is more than for the concretes on the coarse filler. The role of the structured liquid phase especially grows for the fine-grained concretes, the water demand for fine sands is K_s = 0.12 and for coarse sands is K_s = 0.08 [1].

With respect to the value of the specific surface area of filler a quantity of physico chemical bound water for the concrete mixture varies in the limits of K_{ph-ch} = 0.26-0.30. Technological interaction between the liquid and solid phases is specified by the need of guaranteeing the packability of concrete mixture, which to a considerable degree depends on the raw-material base of fillers. Taking into account that at the present time most regions have got fine-grained sands, the industry of building materials must use the technologies, which give the possibility to obtain concretes of dense structure on the fine and very fine sands.

Due to the fact that physicochemical bound water near the solid surface has the special properties similar in the indices to the properties of solid body, it is necessary to insert the physically free liquid for guaranteeing the packability of concrete mixture on the fine fillers Its quantity can be determined according to the value of the specific area of the particle of solid phase (table 1). The indirect index, which considers the composition of cement and its physical properties, is the normal thickness of the cement test $K_{n.th}$.

Table 1. Dependence of thickness and volume of solvate film on the grains of quartz sand on average size of its particles and their specific surface area

Average size of quartz sand grains D_{av}, mm	The specific surface area of quartz sand grains S_s, sm^2/g	Thickness of solvate film, $d_{s.f}$, μm	Volume of solvate film, $V_{s.f}$, sm^3
1.60	18.5	4.1	0.07585
0.90	33.0	4.0	0.13200
0.63	52.8	4.0	0.21120
0.45	66.0	4.0	0.26400
0.23	129.0	3.9	0.50310

In the opinion of I.N. Akhverdov [8] taking into account the recommendations of V.S. Fadeeva molding humidity and necessary quantity of liquid phase must exceed $K_{n.th}$, also, in the technology of concretes with the contemporary methods of packing (vibration, centrifugation) on the average comprises:

$$K_{l.ph} = 1.65K_{n.th}. \quad (1)$$

where 1.65 is the coefficient, depending on the specific area of the particles of solid phase with their average coarseness.

In the dependence (1) possible deviations in the compositions of concretes are not considered, in particular, the granulometry of fine filler. On the basis of the analysis of the experimental data about the water-requirement for concrete mixtures on the fine-grained fillers, sands of different coarseness, [1, 4, 10-12] the coefficient value, which considers the increased water-need of fine sands, is obtained. Dependence (1) for the fine-grained concretes on the fine sands takes the form:

$$K_{l.ph} = 1.98K_{n.th}. \quad (2)$$

For example, for the fine-grained concrete on PC II/A-S-400 cement with $K_{n.th}$ = 0.27 the total quantity of liquid phase, necessary for the preparation of the fine-grained concrete mixture with $K_{l.ph}$ = 1.98×0.27 = 0.53 (taking into account technological interactions we assume $K_{l.ph}$ = 0.60); K_{cr} = 0.01 enters into crystal-hydrates; it is physic-chemically connected in the salvation sheaths - $K_{ph\text{-}ch}$ = 0.30. Then a quantity of free water, capable to form the bug holes during the evaporation is K_{fr} = 0.60–0.01–0.30 = 0.29.

With the introduced quantity of liquid phase for obtaining the concrete of dense structure it is necessary to develop the effective regime of vibro air free process, which provides the total removal of free water.

It is known [2, 4] that molding articles made of fine-grained concretes is divided into two stages: 1) the vibratory compaction of concrete mixture packed in the form during 10 s; 2) the vacuum processing of the provibrated concrete mixture with the negative pressure 0.07 MPa during 10 s.

In view of the fact that liquid with this regime is moved away from the condensed concrete mixture, in the space, earlier occupied by water, the directed capillaries, by micro- and ultra micro pores are formed. It is difficult to assume how they subsequently will be filled up with the products of new formations. For obtaining a maximally dense concrete structure it is necessary to bring a quantity of free water moved away from the mixture to 34%, and to also ensure liquidation of the directed capillaries. Repeated short-term vibration must be effective additional technological operation for this, it is desirable with tightening weight.

4 Research results and their discussion

Studies regarding the stages of the technological regimes of packing for the structural and physical and physico-mechanical properties of the fine-grained concretes have been carried out. Three series of the technological regimes of packing have been accepted: I - control room is the vibratory compaction of concrete mixture during 10 s; II - vibratory compaction during 10 s with the subsequent air free process during 9 s; III - vibratory compaction during 10 s with the subsequent air free process during 9 s and the repeated vibration during 10 s with tightening weight of 5 kg/sm^2.

The duration of vibration and air free process are assigned employing the procedure [4]. Models hardened under the normal hydrothermal conditions in the course of 28 days.

The influence of the regimes of concrete mix compaction on the bending strength of concrete R_b is illustrated by the graph, represented in Fig. 4

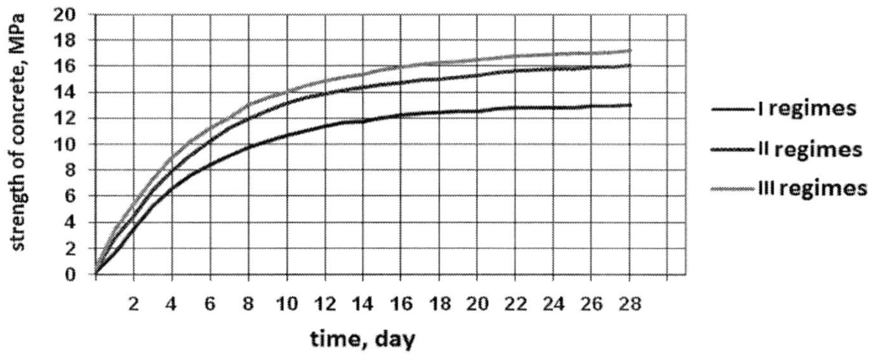

Fig. 4. The influence of the compaction regimes on the strength of concrete.

It is evident from the graph that the best regime for the extraction of free water during packing of thin-walled articles is the regime III. It combines vibratory compaction, vacuumizing with the subsequent vibratory compaction of mixture with tightening weight. This technological solution contributes to the destruction of the directed capillary holes formed in the air free process. As a result, the structure of cement composite is packed which contributes to an increase in strength (Fig. 4) and in the waterproofing of thin-walled articles made of the cement composites.

In the course of experiment, a quantity of water extracted from the mixture was controlled, as well as density, porosity, water absorption and strength of concrete. Results are given in the table 2.

Table 2. The structural and physical and physic-mechanical indices of fine-grained concretes with different technological regimes of the packing

Regime	Quantity of water extracted from the mixture, %	Density, kg/m^3	Water absorption, %	Porosity,%	Bending strength R_b, MPa, at the age of:					
					1 hour	days				
						1	3	7	14	28
I	-	2355	6	11.0	0.20	1.60	5.20	9.10	11.70	13.00
II	29.3	2430	5	10.1	0.36	2.75	6.44	11.27	14.40	16.10
III	33.6	2500	3	7.2	0.42	3.35	7.30	12.04	15.35	17.20

5 Conclusions

It was determined, that for obtaining the required mobility of mixtures a quantity of water in the volume sufficient for creating the layers, removed from the action of the potential of rigid surface is necessary. As a result of studies the obtained value of the coefficient, which considers the increased water demand of small fractions, has given the opportunity to determine the total quantity of liquid phase necessary for the preparation of the fine-grained cement mixture - $K_{l.ph}$ = 0.6 (from the cement), and also a quantity of free water, capable during the evaporation to form the bug holes–K_{fr} = 0.29 (from the cement).

For obtaining the cement composites with the high density it is necessary to provide the conditions of creating thin structured layers of the liquid between the particles of the solid phase of mixture. It is determined, that the most rational method of the extraction of free water is the air free process.

Experimental studies established that the developed technological regime III (vibratory compaction during 10s with the subsequent air free process during 9s and the repeated vibration with tightening weight of 5 kg/cm^2 during 10s) ensures the complete removal of the free liquid phase from the packed mixture.

Fine-grained vacuum concrete, packed under the developed regime, has the increased structural and physical and physical and mechanical characteristics in comparison with the concrete packed by vibration. Thus, increase in concrete density comprises 6%; reduction in water absorption – 52%, of the porosity –65%. The structural strength of concrete increases 2 times and composes 0.42MPa, which gives the possibility to achieve the immediate removal of the molds of articles. The strength of concrete at the age of 28 twenty-four hours increases by 32%.

References

1. Yu.M. Bazhenov, *Vysokoprochnyy melkozernistyy beton dlya armotsementnykh konstruktsiy* (Gosstroyizdat, Moscow, 1963).
2. A.G. Vandolovskiy, Ye.B. Dedeneva, Naukoviy visnik budivnitstva, **41**, 127-132 (2007)
3. A.G.. Vandolovskiy, Ye.B. Dedeneva, Naukoviy visnik budivnitstva, **47**, 192-195 (2008)

4. V.S. Babushkin, Ye.B. Dedeneva, Ye.V. Yemets, D.A. Bondarenko, Naukoviy visnik budivnitstva, **31**, 123-127 (2005)
5. A.N. Plugin, A.A. Plugin, Collected scientific works of Ukrainian State University of Railway Transport, **77**, 5-70 (2006)
6. A.N.Plugin, A.A.Plugin, O.A.Pluhin, A.S.Kaganovskii, O.V.Gradoboev, 18 Ibausil, Weimar, **1**, 0656-0663 (2012)
7. A.N.Plugin, A.A.Plugin, O.A. Kalinin et al. *Osnovy teorii tverdeniya, prochnosti, razrusheniya i dolgo-vechnosti portlandtsementa, betona i konstruktsiy iz nikh* (Naukova dumka, Kyiv, 2012)
8. I.N. Akhverdov, *Osnovy fiziki betona* (Stroyizdat, Moscow, 1981)
9. A.N. Plugin, *Elektrogeterogennyye vzaimodeystviya pri tverdenii tsementnykh vyazhushchikh*, DSc(Chem) thesis (Dumanskiy Colloid Chemical and Water Chemical Instite of NAS of Ukraine, Kyiv, 1989)
10. A.N. Plugin, A.A. Plugin, O.A. Kalinin, Cement, **2**, 28-32 (1997)
11. A.A.Plugin, T.A.Kostyuk, M.G.Saliya, D.A.Bondarenko, Sbornik nauchnykh trudov MGSU posvyashchennyy 90-letiyu fakulteta PGS, 224-227 (2011)
12. A.A.Plugin, M.G.Saliya, T.A.Kostyuk, D.A.Bondarenko, O.V.Starkova. VIII Mezhdunarodna nauchno-prakticheska konferentsia «Nauchnoye prostranstvo Yevropy», **29**, 62-67 (2012)

Electric potentials and charges on buildings and areas around deep water wells

Arkadiy Plugin[1,*], *Oliksii* Pluhin[2], *Olena* Palant[1], *Oleksandr* Konev[1], and *Andrii* Nikitinskii[1]

[1] Ukrainian State University of Railway Transport, Building Materials and Structures Department, Feuerbach sq. 7, 61050 Kharkiv, Ukraine
[2]Ukrainian State University of Railway Transport, Structural Mechanics and Hydraulics Department, Feuerbach sq. 7, 61050 Kharkiv, Ukraine

Abstract. Consideration was given to the mechanisms of propagation of electric potentials in the locality of water wells and their effects on damages in the structure of buildings. Electric potentials were measured using the digital multimeter. It was established that exploitable water wells determine the division of electric charges and potentials in the adjacent territory due to the origination of macroelectrokinetic potential and the evacuation of cations from the ground. These charges and potentials can result in nonuniform strains of bases and foundations, cracks and damages in the structure of buildings and also electrocorrosion intensification nearby DC rail tracks. The measures to protect the structures from indicated destructive factors have been recommended.

1 Introduction

Railways have a branched water-supply infrastructure that dates back to steam-engine times. Water supply wells are situated very often near railway tracks. Railway workers state that damages done to buildings and structures situated above the wells are more vigorous in comparison with ones located in remote districts. One of the most aggressive factors that condition electrocorrosion damages in railway structures is leakage currents [1, 2] and electric potentials induced by them [3]. However, the influence of water supply wells on electric potentials in adjacent territories and the electrocorrosion damage to structures has not been studied yet. This research was done by the authors under the leadership of Prof. Andrey Plugin with the participation of engineer Vladimir Kasianov.

The objective of the research was to establish the mechanisms of propagation of electric potentials over the territory adjacent to a water-supply well, and also reveal their impact on damages to the structure of buildings and work out recommendations on prevention of such impacts.

* Corresponding author: pluginarkadiy@gmail.com

2 Analytical Review of the Reference Data

It is known that structures near DC railways and other sources are exposed to electrocorrosion. It was earlier assumed that metal structures and the iron concrete reinforcement are exposed to such impacts [1, 2]. A.A. Staroselsky was the first to notice corrosion changes of the cement stone near the reinforcement [3]. The concrete electrocorrosion was studied in detail in [4-6]. It was established that the concrete electrocorrosion is manifested by the intensive electromigration leaching of the cement stone exposed to the pulsed unidirectional electric potential. This potential is created by trains running on DC rail sections. Papers [7] and [8] state that the alternating current and potential also condition the iron concrete and concrete electrocorrosion, though it is less intensive.

It was established in [9-12] that electric potentials can affect the concrete strength [9] and result in internal stresses [10] and cracks [11-12] in concrete structures. Papers [13-15] note the influence of potentials on stability of railway slopes, side slopes and roadbeds and also the building position on the whole [16].

Due to the above, the investigation of reasons for formation of electric potentials and charges in the locality and assessment of a risk level of their influence on buildings and structures is of great importance. Papers [4, 6, 14] describe the mechanisms of propagation of potentials from rail tracks in the locality and structures, including their configuration on the plane, and [14-17] describe propagation mechanisms of the potentials created by the polarization of ground massifs, including transfer of ions, mainly cations, water streams, filtration, diffusion and electromigration of flows and vertical polarization of structures and buildings [16, 17]. However, the influence that vertically extended structures, in particular water-supply wells, have on electric potentials in the territory adjacent to them and the electrocorrosion damages to structures were not studied.

3 Research methods

The damages to the buildings of water-supply well were studied by the method of visual inspection and measurements by taking pictures and mapping these damages.

Electric potentials (the potential difference between the measurement points) were measured with a digital PC510 multimeter *Sanva* according to the diagram given in Fig.1. The contact with measurement points was provided by: 0 (the well casing) immediately by the multimeter probe; 1.3 (building structures) using special copper-sulfate electrode (the contact of brass measurement plate and the structure is provided by the sponge impregnated with the saturated copper sulfate solution); 2, 4 (ground) using steel rods with a diameter of 12 mm driven into the ground at a depth of 0.5 m. A value of the potential was measured at each point 10 times with a 1-minute interval and thus an average value was taken.

4 Research data

The water well at the Osnova station of the South railway (Kharkov, Ukraine) was studied. The well has the following parameters: the depth is 766m; the debit is 984m^3/24hrs; the well casing is made of steel; the compound water intake tube is made of steel; a 380V submersible pump was used. The nearest DC rail track ran at a distance of 100m from the well.

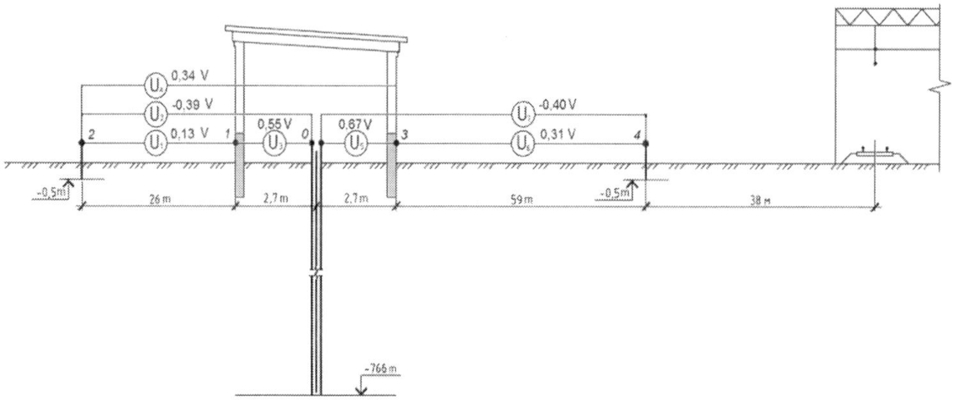

(a)

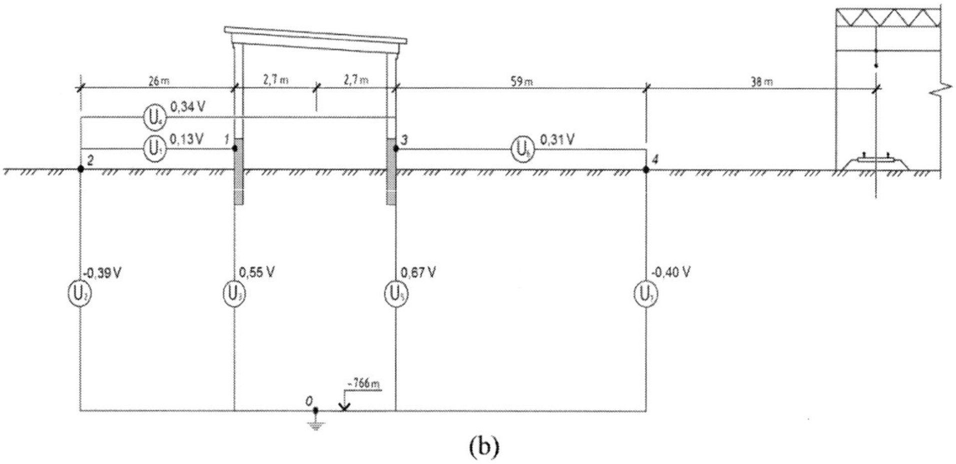

(b)

(c)

Fig.1. Measurement of electric potentials induced on water well structures: (a) measurement circuit; (b) reduced scheme; (c) Prof. Andrey Plugin and Dr. Aleksandr Konev take measurements

Fig. 2 gives the pictures of facades and appropriate maps of damages. The picture shows that the main damages are represented by opening upwards cracks in the walls and the surface fracture of the wall facing and masonry.

Fig.1 gives the measured values of electric potentials. The potential values on the building base surface are equal to 0.55–0.67V and on the ground surface these are equal to minus 0.39–0.4 at a distance of 29–62 m from the well. The availability of influence that passing rail trains with electric traction have on the value of potentials and that of the switching on/off of submersible pump was not revealed.

5 Discussing the research data

A possible reason for absence of influence on the value of potentials that passing rail trains with electric traction can have is the current leakage from the rail track into the ground to the lower part of a casing pipe.

A change of electric potential in the locality ranging from positive values of 0.55–0.67V near the well to negative of 0.39–0.4V when moving away from it can be explained by the charge separation by the analogy with [16,17]. It is known that the Earth has an electric charge that is negative excess charge near its surface with the appropriate electric field. The field intensity near the Earth surface is equal on the average to minus 130V/m, but it ranges from minus 1000 to plus 200 V/m. Therefore, the excess charge (and the field intensity) is changed considerably at the bottom of rivers and along their length due to the wash-out of cations from the bottom ground entrained by the water flow. This can explain repeated abnormally high strains and damages to river bridges up to their collapse (the collapse of Takomsky bridge is the most known event) [16, 17].

The data obtained allowed us to assume that analogous macroelectrokinetic phenomena also occur around the water-supply wells analyzed. During the water intake in the water-saturated ground surrounding the well the the flow macropotential occurs [18]. This flow washes the counterions of binary electric layers of its particles, i.e. the cations, mainly Na^+, K^+ and Ca^{2+}, out of the ground; they are further entrained by the pumped water. The ground particles carry an excess negative charge resulting from their potential-defining ions OH^-, and the ground surrounding the well acquires a negative excess charge whose propagation range is extended with time. Not very high values of measured potentials result from the origination of counter-field due to the polarization of ground that has a high dielectric permittivity.

The excess charge presupposes the ground swelling on the foundation of the pump station building. It can result in considerable inhomogeneous strains of the foundation and, consequently, formation of cracks in the walls. The pattern of the cracks observed (Fig.2) going upwards confirms the ground swelling in the middle of the foundation.

According to Fig.1 the upper portion of foundations has a positive potential against the ground in the depth. It results in the entrainment of cations Ca^{2+} from the concrete and the foundation masonry slurry into the ground and their drainage from the upper portion of walls to the bottom (Fig.3) and appearance of the potential difference between the upper and lower parts of the wall. In its turn, it results in the macroelectroosmosis of water to the upper part of walls and their moistening. When a rainy weather changes to hot, it results in the alternate moistening and drying of the facing and masonry and their damage as can be seen in Fig.2. To prevent such damages to buildings we need to provide a qualitative horizontal waterproofing not only at a level of the base top, but also at a level of the riprap. The authors recommend investigations into the possibility of bypassing the pipe casing and building structures.

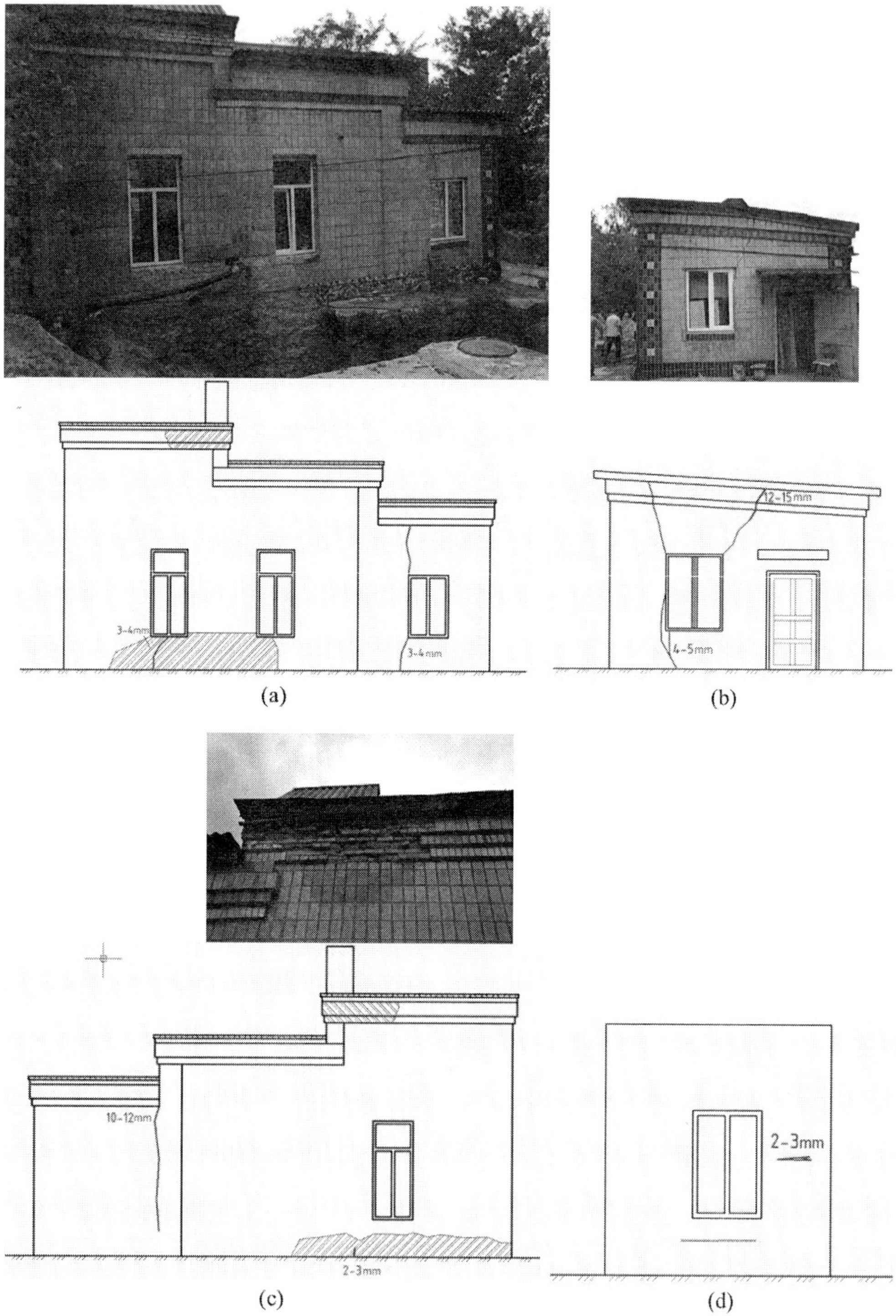

Fig.2. The pictures and maps of damages of the building facades of the water station: (a) eastern; (b) northern; (c) western and (d) southern.

In addition, when DC rail lines are located near the water well, the negative excess charge of the ground around it presupposes the intensification of the electrocorrosion of rail fastening, reinforced concrete ties, foundations, support butts of overhead contact systems, etc. To prevent this, we recommend using the superstructure design for the railways and tram roads with the isolated rail or on isolated blocks [19].

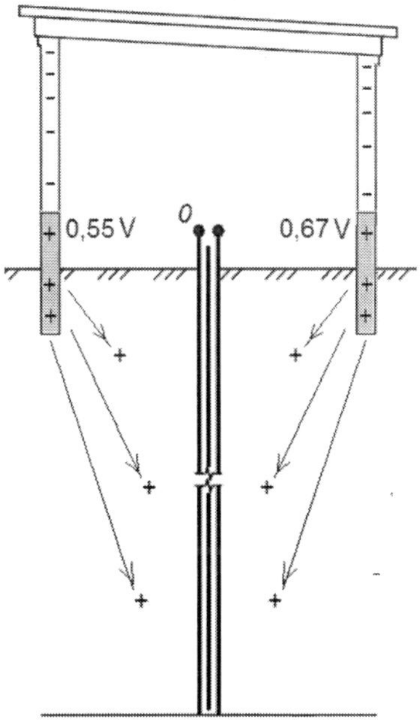

Fig. 3. The diagram of entrainment of cations Ca^{2+} from the foundation and walls of buildings.

6 Conclusions and recommendations

It has been established that exploitable water wells condition the separation of electric charges and potentials in the locality due to the origination of macroelectrokinetic potential and the entrainment of cations from the ground. The potentials around the explored well are changed from positive values of 0.55–0.67V near the well to negative 0.39–0.4V as the distance to the well increases. These charges and potentials can result in the nonuniform strain of bases and foundations, formation of cracks and damages to the structure of buildings situated above the wells and near them and also in intensive electrocorrosion of structures situated near DC rail lines.

To prevent such damages to buildings a qualitative horizontal waterproofing should be provided not only at a level of the base top, but also at a level of the riprap. We recommend investigating the possibility of bypassing the pipe casing and building structures.

To prevent the electrocorrosion of rail fastening, reinforced concrete ties, foundations, and support butts for overhead contact systems near the water wells, we recommend using the superstructure design for the railways and tram roads with the isolated rail.

References

1. I.A. Kornfel'd, V.A. Pritula, *Zashchita zhelezobetonnykh konstruktsiy ot elektrokorrozii, vyzyvayemoy bluzhdayushchimi tokami* (Stroyizdat, Moscow, 1964)
2. A.V. Kotel'nikov, *Bluzhdayushchiye toki elektrifitsirovannogo transporta* (Transport, Moscow, 1986)
3. A.A. Starosel'skiy, *Elektrokorroziya zhelezobetona* (Budivel'nik, Kyiv, 1978)

4. O.A. Pluhin, O.S. Borzyak, V.B. Martinova, O.K. Khalyushev, *Yelektrichni vplivi na beton*, ed. A.A. Plugin and M.M. Zaychenko (Fort, Kharkiv, 2013)
5. A.N. Plugin, A.A. Plugin, O.A. Pluhin, O.A. Dudin, O.S. Borzyak, 17 Internationale Baustofftagung, Weimar, **2** (2009).
6. O.A. Pluhin, Zbirnik naukovih prac' Ukrainian State Academy of Railway Transport, **115** (2010)
7. A.N. Ostapenko, Transport of Ural, **4**, 31 (2011)
8. A.N. Plugin, A.A. Plugin, O.A. Dudin, O.A. Pluhin, O.S. Borzyak, A.A. Konev, Visnyk Odes'koyi derzhavnoyi akademiyi budivnytstva i arkhitektury, **43** (2010)
9. O.A. Pluhin, A.A. Zabiyaka, V.V. Perestyuk, S.G. Nesterenko, A.A. Plugin, Zbirnik naukovih prac' Ukrainian State Academy of Railway Transport, **130** (2012)
10. A.N. Plugin, A.A. Plugin, O.A. Pluhin, A.A. Konev, A.A. Zabiyaka, Zbirnik naukovih prac' Ukrainian State Academy of Railway Transport, **138** (2013)
11. O.A. Pluhin, A.N. Plugin, A.A. Plugin, A.A. Konev, A.A. Zabiyaka, Zbirnik naukovih prac' Ukrainian State Academy of Railway Transport, **142** (2013)
12. A.A. Plugin, O.A. Zabiyaka, G.O. Linnik, Zbirnik naukovih prac' Ukrainian State Academy of Railway Transport, **115** (2010)
13. A.N. Plugin, L.V. Trykoz, O.S. Herasimenko, A.A. Plugin, D.A. Plugin, O.A. Pluhin, Visnyk Dnipropetrovs'koho natsional'noho universytetu zaliznychnoho transportu im. V. Lazaryana, **39** (2011)
14. A.A. Plugin, A.M. Plugin, L.V. Trykoz, O.S. Herasimenko, Railway Transport of Ukraine, **5** (2010).
15. L.V. Trykoz, O.S. Herasimenko, Eastern European Journal of Enterprise Technology, **4/5**, 70 (2014)
16. A.N. Plugin, L.V. Trykoz, A.A. Plugin, D.A. Plugin, A.A. Konev, O.S. Borzyak, Zbirnik naukovih prac' Ukrainian State Academy of Railway Transport, **148**, 2 (2014)
17. A.N. Plugin, A.A. Plugin, L.V. Trykoz, D.A. Plugin, O.A. Pluhin, Railway Transport of Ukraine, **2** (2014)
18. Ye.D. Shchukin, F.V. Pertsov, Ye.A. Amelina, *Kolloidnaya khimiya*, ed. 4 (Vysshaya shkola, Moscow, 2006)
19. T. Shuba, V. Chystyak, V. Perestyuk, O. Yelyakina (Palant), O.A. Zabiyaka, A.A. Plugin, Zbirnik naukovih prac' Ukrainian State Academy of Railway Transport, **122** (2011)

Research into the effectiveness of grounded screens of electroconductive silicate compositions for electrocorrosion protection

Dmytro Plugin[1*], *Vladimir* Kasyanov[1], *Vitalii* Konev[1], *Sergii* Nesterenko[2], and *Aleksander* Afanasiev[1]

[1]Ukrainian State University of Railway Transport, Building Materials and Structures Department, Feuerbach sq. 7, 61050 Kharkiv, Ukraine
[2]O.M. Beketov National University of Urban Economy in Kharkiv, Land Administration and GIS Department, Kulikovska descent 12, 61002 Kharkiv, Ukraine

Abstract. The effectiveness of coatings based on Portland cement and sodium silicate with graphite filler as an electrocorrosion protection of grounded screens is investigated. An original technique is developed, consisting in comparing the values of the current *I* through the reinforcement and concrete of the structure model without protection and with protection by grounded screens-coatings. As a result of experimental studies it is found that, in comparison with the model without protection, the *I* value of the coated model of the Portland cement-based composition is less: by 49-66%, and for the model coated by sodium silicate-based composition - by 34-60%.

1 Introduction

According to [1, 2], structures close to railways and other sources electrified by direct current are subjected to electrocorrosion. Mainly metal structures and reinforcement of reinforced concrete are corroded. It was noted in [3] that concrete is also corroded near the reinforcement. It was established in [4-6] that the electrocorrosion of concrete proceeds not only near the reinforcement, but even in unreinforced structures and masonry mortar, and consists in the intensive electromigration leaching of the cement stone under the action of a pulsating unidirectional electric potential. This potential is created when trains run along a rail section electrified by a direct current.

In [4, 7] it was proposed to apply the diode grounding of metal-injection jacket for electrocorrosion protection of culvert structures. Such jacketing is used to repair damaged pipe structures (Fig. 1 a, b). The metal sheathing of the jacket protects concrete and mortar from carbonization and leaching, cement stone of concrete and inside covering – from electrochemical corrosion. However, under the influence of leakage currents, the metal sheathing of the jacket is subjected to electrocorrosion in the anode zones (Fig. 1 c). Its diode grounding makes it possible to protect both the sheathing itself from electrocorrosion by transferring the anode zones to the ground, and the concrete and the solution of the pipe

* Corresponding author: plugin.da@gmail.com

structures from the current flow (Fig.1 d). However, the metal-injection jacket is an expensive construction which is useful when it is necessary to restore the bearing capacity of a structure with significant damage. Replacement of metal-injection jacketing by coating, a screen made of an electrically conductive composition based on mineral binders with its grounding, will make it possible to reduce the cost of repair and protection of the structure. However, the effectiveness of such electrocorrosion protection has not previously been investigated. Such studies were carried out by the authors under the guidance of Prof. Andrii Plugin with the participation of Dr. Oleksii Pluhin.

The aim of the research is to determine the magnitude of the reduction in current strength through the concrete structure model as a result of its protection by grounded coating with compositions based on mineral binders.

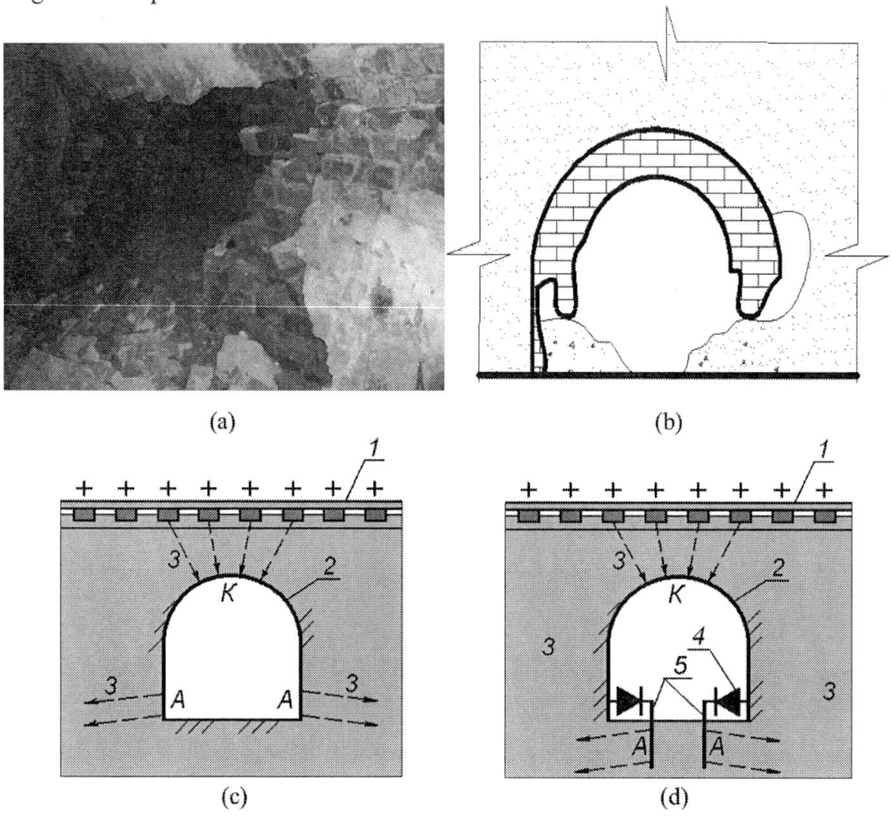

Fig. 1. Repair and corrosion protection of the culvert pipe in the railway embankment: (a, b) damage in the form of piles; (c, d) leakage current flow from the rails through the unprotected (c) and diode-protected grounding (d) metal-injection jacket: 1 – rail track; 2 – metal-injection jacket; 3 – leakage currents; 4 – valve (diode) of the protective device; 5 – grounding rod; K – cathodic area; A – anodic area (corrosive)

2 Material and methods

The effectiveness of grounded coatings, screens made from Portland cement-based compositions as well as sodium silicate hardened with sodium silicofluoride with graphite powder filler, is investigated. The composition formulas are developed at Ukrainian State University of Railway Transport [8].

An original technique for experimental research of the effectiveness of grounded screens made of electrically conductive compositions for electrocorrosion protection of structures is developed. The technique compares the electrical current I magnitude flowing through the reinforcement and concrete of the model of a structure without a protective screen (I_0) and the model of a structure with a protective screen (I_1). The reduction in current intensity through concrete and reinforcement of the model with protection I_1 in comparison with that without protection I_0 was taken as the efficiency index of the screen:

$$\Delta I = I_1/I_0 \times 100\ \%.$$

The models of the structure are developed on the basis of analysis of leakage currents from the rail track through sections of passenger platforms on sections with a positive potential on rails (Fig. 2) without protection (Fig. 2 a, b) and with grounded screens (Fig. 2 c, d). As can be seen, in rainy weather, the leakage currents flow through concrete, reinforcement and grounding for the platform without protection, and through the screen and grounding bypassing concrete and reinforcement for the platform with the screen.

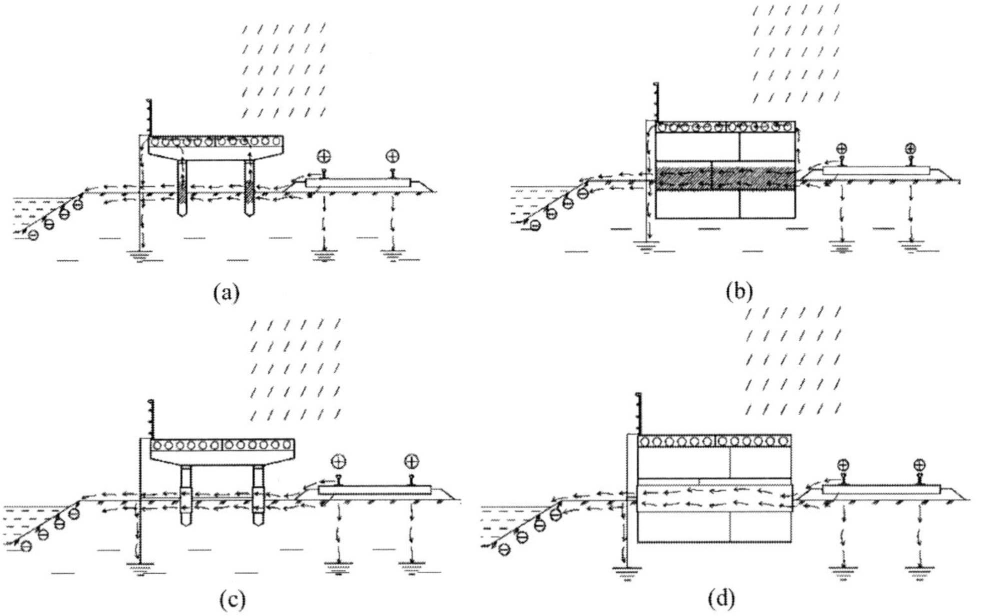

Fig. 2. The leakage current flows from the rail track in areas with a positive potential on rails through the construction of high passenger platforms: (a, c) on reinforced concrete racks and crossbars; (b, d) on concrete block supports; (a, b) without protection; (c, d) with grounded screen coatings.

The studies are carried out using a special experimental setup (Fig. 3). The setup consists of a model of the structure – the prism of heavy concrete of a compressive strength of 20 MPa *1* with an upward built-in rod of reinforced steel *2*. The concrete of the model corresponds to the concrete of the structure between its faces (Fig. 2a) or the concrete of a protective layer between grounded reinforcement and face of the structure (Fig. 2b). This model prism is installed in a model of moistened soil, very fine sand *4*, put into a metal container *3*. The current flowing into the structure from direct current sources, e.g., an electrified rail track, is created (simulated) by application of a potential difference of 40 V between container *3* and reinforcement rod *2* in prism *1*, and the current strength is measured with the ammeters included in the chain (digital multimeters Sanwa PC510) *8* and *9*. The measurements begin at a dry state of the soil, and then it is water-saturated; they continue during 240 hours with a periodicity of 1 hour first, then 6, 12 and 24 hours. The results are presented as the graphs of dependence of current I on time t and the value of

 DOI: 10.1051/matecconf/201711601012

current reduction through the concrete and reinforcement of the model with protection in comparison with the model without protection ΔI.

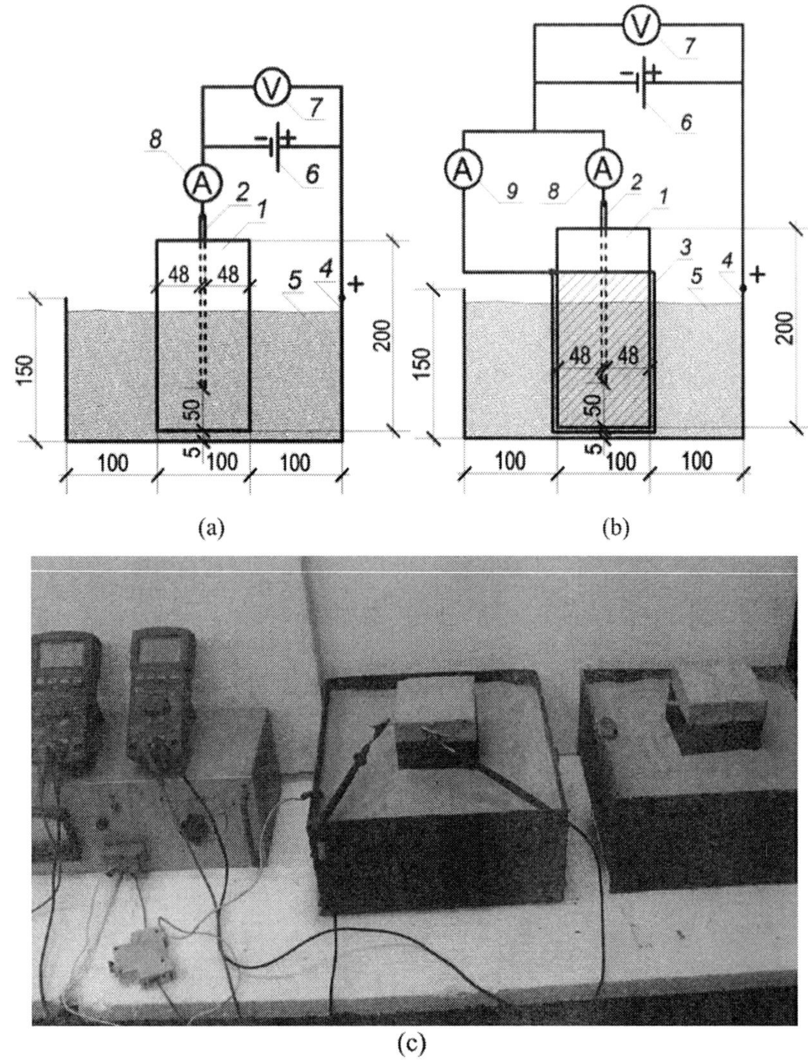

Fig. 3. Setups for the experimental study of effectiveness of screens of electroconductive compositions for electrocorrosion protection of structures: (a, b) a schematic diagram of the setup with a control sample (model of a structure) without a screen (a) and with a test sample (structure model) with a screen of electroconductive composition (b); (c) view of experimental setups; *1* – concrete of the model; *2* – reinforced rod of the model; *3* – protective screen (coating); *4* – metal container; *5* – moistened sand; *6* – current source; *7* – voltmeter; *8* – milliammeter for measuring the current intensity through reinforcement and concrete I_{rc}; *9* – milliammeter for measuring the current intensity through the screen and the soil I_{sg}.

3 Research results and their discussion

The results of the research into effectiveness of grounded screens made of electroconductive compositions based on mineral binders for electrocorrosion protection of structures are given in Fig. 4.

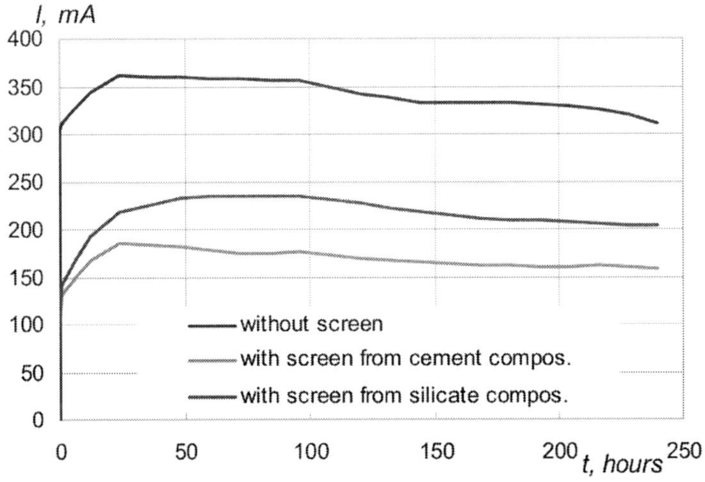

(a)

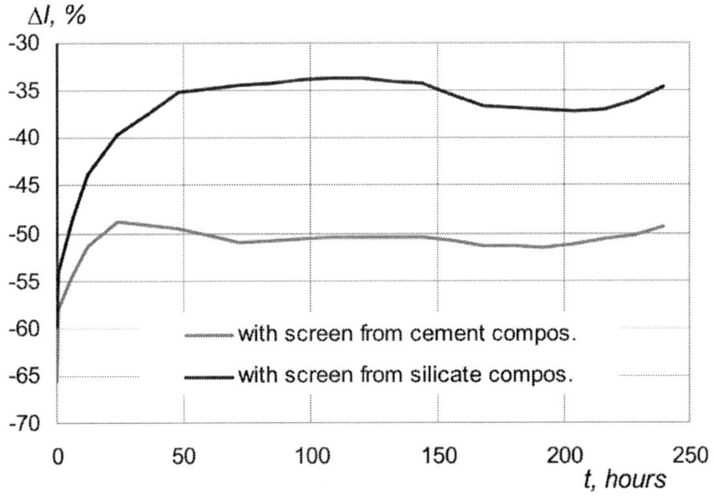

(b)

Fig. 4 Results of the research into effectiveness of screens made of electroconductive compositions based on Portland cement and sodium silicate: (a) dependence of current I through reinforcement and concrete of the model without a protective screen (control) and with protective screens on the time since the soil is water saturated t: (b) dependence of change in current strength through reinforcement and concrete of the model with protective screens in comparison with the model without protection ΔI on time t.

As can be seen from Fig. 4, in the case of a dry state of the soil (at the beginning of the experiment at t = 0), the current intensity through reinforcement and concrete I is minimal for both the model of the structure without protection and models with protective screens. When the soil is water saturated the current strength increases by three orders of magnitude, and in comparison with the model without protection, the current strength I in the model with a screen made of Portland cement composition is less by 49-66 %, and in the model with a screen made of sodium silicates and calcium composition it is less by 34-60 %. This confirms the effectiveness of grounded screens for electrocorrosion protection of structures, e.g., passenger platforms, from leakage currents.

However, during the study, the effect of the contact area of the screen grounding on the measured current strength is noted. The study of this influence is underway.

4 Conclusions

A technique for the experimental study into the effectiveness of screens made of electroconductive silicate compositions for electrocorrosion protection of structures is developed; it consists in comparing the magnitude of the electric current I flowing through reinforcement and concrete of the model of a structure without a protective screen, and the model of a structure with a protective screen, and defining the reduction in the current strength through concrete and reinforcement of the model with protection in comparison with the model without protection ΔI.

The effectiveness of grounded screens for electrocorrosion protection of structures, e.g., passenger platforms from leakage currents is experimentally confirmed. Thus, it is established that in the case of dry soil conditions, the current strength through reinforcement and concrete I_{rc} is minimal in both the model of a structure without protection and models with protective screens. When the soil is water saturated, the current strength increases by three orders of magnitude, and in comparison with the model without protection, the value I in the model with the screen of Portland cement composition is less by 49-66 %, and in the model with the screen of sodium silicates and calcium composition it is less by 34-60 %.

References

1. I.A. Kornfeld, V.A. Pritula, *Zashchita zhelezobetonnykh konstruktsiy ot elektrokorrozii, vyzyvayemoy bluzhdayushchimi tokami* (Stroyizdat, Moscow, 1964)
2. A.V. Kotelnikov, *Bluzhdayushchiye toki elektrifitsirovannogo transporta* (Transport, Moscow, 1986)
3. A.A. Staroselskiy, *Elektrokorroziya zhelezobetona* (Budivelnik, Kyiv, 1978)
4. O.A. Pluhin, O.S. Borzyak, V.B. Martinova, O.K. Khalyushev, *Yelektrichni vplivi na beton*, ed. A.A. Plugin and M.M. Zaychenko (Fort, Kharkiv, 2013)
5. A.N. Plugin, A.A. Plugin, O.A. Pluhin, O.A. Dudin, O.S. Borzyak, 17 Internationale Baustofftagung, Weimar, **2**, 1151-1156 (2009)
6. O.A. Pluhin, Collected scientific works of Ukrainian State University of Railway Transport, **115**, 56-67 (2010)
7. O.A. Pluhin, O.S. Borzyak, O.A. Dudin, J.A. Konev, A.A. Plugin, Collected scientific works of Ukrainian State University of Railway Transport, **127**, 204-212 (2011)
8. A.A. Plugin, T.O. Kostiuk, O.A. Pluhin, D.O. Bondarenko, N.M. Partala, Yu.A. Sukhanova. Kompozytsiia pronyknoi dii dlia hidroizoliatsii ta zakhystu vid elektrokorozii: UA113600[P]. (2017-02-10)

The effect of structural characteristics on electrical and physical properties of electrically conductive compositions based on mineral binders

Oleksiy Pluhin[1], *Andrii* Plugin[2], *Dmytro* Plugin[2,*], *Olga* Borziak[2], and *Oleksiy* Dudin[3]

[1]Ukrainian State University of Railway Transport, Structural Mechanics and Hydraulics Departmen, Feuerbach sq. 7, 61050 Kharkiv, Ukraine
[2]Ukrainian State University of Railway Transport, Building Materials and Structures Department, Feuerbach sq. 7, 61050 Kharkiv, Ukraine
[3]Ukrainian State University of Railway Transport, Track and Track Sector Department, Feuerbach sq. 7, 61050 Kharkiv, Ukraine

Abstract. The present-day situation is such that industrial production, municipal economy, traffic, etc consume a continuously increasing amount of energy. As a result, the range of propagation of electric stray currents, including their appropriate potentials and leakage currents that affect building structures and facilities is expanded and this is of importance. It requires stricter demands regarding protection of building structures and facilities exposed to electric currents and potentials. This study gives the research data on the influence of structure on electrophysical properties, in particular specific electric resistance of the silicate compositions obtained for electrocorrosion protection. The theoretical research done allowed us to plot the relationships to define the specific electric resistance and the specific conductivity of a composite material using the data on its composition and the specific electric resistance of the filler and matrix. The algorithm to define the specific electric resistance of the composite material was developed for research and/or technological purposes by using the data on specific electric resistance of the filler, matrix and composition.

1 Introduction

Special requirements are set to the specific electric resistance and/or specific electric conductivity of electrical concretes and mortars that are considered to be special materials. It has nothing to do with high or low requirements; they are limited by a narrow range of values. Heavy-weight concretes of rail constructions and rail supports have an increased electric resistance [1, 2] to provide the electrocorrosion protection and reliable operation of rail chains. The authors of this research paper developed the National Standard DSTU B V.2.6-2009:2016 "Preliminary Stressed Reinforced Concrete Railway Ties for the

* Corresponding author: plugin.da@gmail.com

Railroads of the 1520 and 1435-mm Tracks. Technical Specifications" to provide the electrocorrosion protection by normalization of specific electric resistance of the concrete making it equal to at least 110 Ohm·m (for a certain humidity). An increased electrical resistance is also required to provide the electrocorrosion protection during the structure repair, using for this purpose the polymer-cement slurry [3]. Electroconductive concretes are used for heating elements [4-6] and the ionizing radiation protection [7]. The electric conductivity of concrete is controlled with electroconductive fillers. These are usually carbon-graphite [5, 6] and metal containing [7] fillers. The composition (the component ratio) [8] and the composite structure [9, 10], especially the microstructure of it [7, 10, 11] also affect the electrical conduction, including technological impacts that arise during the composite production [8, 12, 13]. The electric conductivity of concrete greatly depends on humidity [14] that can create problems for operation of special structures. The empiric method is used for the control of electric conductivity in composite material production; its dependence on receptor-process factors is experimentally studied. This is a rather laborious and time-consuming approach. Therefore, the problem of prediction of specific electric resistance of the composite material using the data on electric resistance of filler and matrix and the summarized characteristics of its composition is rather vital.

The objective of the research was to develop the algorithm for the prediction of specific electric resistance of the composite material using the data on specific electric resistance of its components and summarized characteristics of its composition.

2 Research Methods and Materials

This is a theoretical investigation. Based on the review of reference data, the physical (three-dimensional geometric) model of the structure of composite material was developed and it was converted into the mathematical model, using the computation method. The mathematical model served as the basis for the development of an algorithm to predict the specific electric resistance of the composite material on the basis of data on the specific electric resistance of filler, matrix and composition.

The model and the algorithm were developed for composite materials whose specific electric conductivity and resistance should meet special requirements ranging from electroconductive to electric insulating properties. The matrix of these materials is represented by hardened cementing materials, for example Portland cement stone, solidified liquid glass, etc. Fillers with high electric conductivity are graphite powder, ground coke, metal powders, etc. Fillers with high electric resistance are quartz, granite, etc. To realize the algorithm we need to have reference or experimental data on specific electric conductivity and/or specific electric resistance of matrix and filler.

3 Research data

To provide the electrocorrosion protection we will study the influence of the structure on electrophysical properties, in particular, specific electric resistance of silicate compositions. The composite material (the matrix composite [15]) consists of structure-forming particles (filler) connected together by the matrix. To study and describe quantitative mechanisms in [16] the structure is conditionally represented by the spatial geometric model with the regular distribution of ball-shaped structure forming particles of the same diameter. Fig. 1a gives the diagram of the structure of such a composition and Fig. 1b gives the diagram of its conditional spatial elementary cell, which consists of one particle of the filler of the average size (diameter) d, m, surrounded at the three sides (in space) by the matrix of an average thickness in places of a maximum proximity of the filler particles δ, m, Fig. 1b.

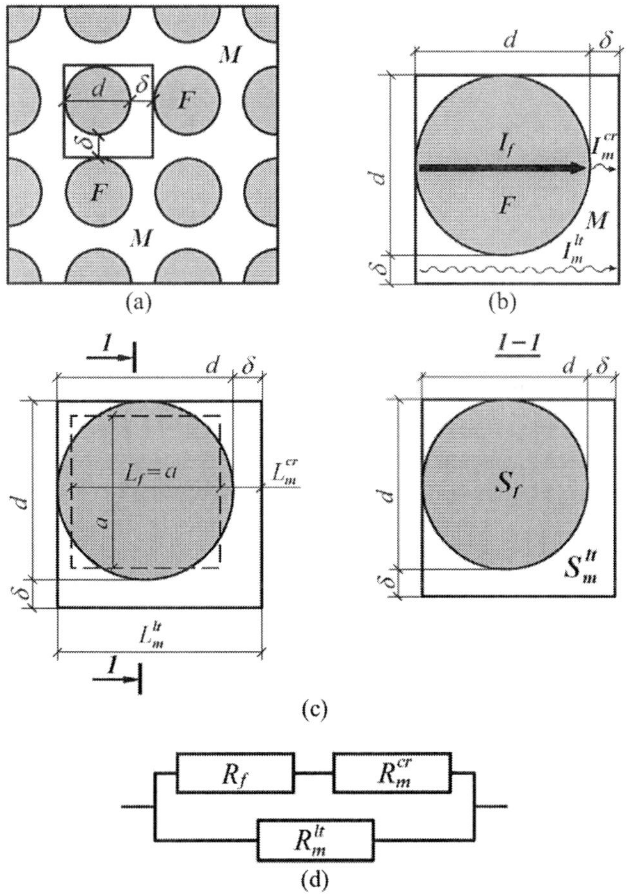

Fig. 1. The diagram of formation of the electric resistance of composite material: (a) composite material structure diagram; (b) the diagram of conditional elementary cell of the composite material and the current flowing through it; (c) the equivalent circuit of conditional elementary cell of the composite material and the electric current flowing through it; (d) the equivalent circuit of electric resistance of the elementary cell of composite material; (F) is the filler and (M) is the matrix.

Fig. 1d gives the equivalent circuit of the conditional elementary cell of composite material and the electric current flowing through it. Fig. 1b, c, shows that the total electric current I that flows through the conventional elementary cell is subdivided into the currents that flow through the filler particle (I_f) and the interlayers between the filler particles in transverse and longitudinal directions (I_m^{lt}) of the current flow. Since the length of the filler particle is a variable value that varies in the range of 0–d the length of the transverse section of the matrix is also a variable value that varies in the range δ–$\delta + d/2$, then we will conditionally change the form of the ball-shaped filler particle given in the diagram in Fig. 1a to the cube of the same volume (Fig. 1c):

$$V_{ball} = \frac{\pi d^3}{6} = V_{cube} = a^3 \tag{1}$$

The length of the cube edge is equal to:

$$\frac{\pi d^3}{6} = a^3, \quad a = d\sqrt[3]{\frac{\pi}{6}} \tag{2}$$

According to the diagrams given in Fig. 1b, c the specific electric resistance of the composite material is equal to the specific electric resistance of its conditional elementary cell:

$$\rho_c = \frac{RS}{L}, \tag{3}$$

where R is the electric resistance, Ohm; S is the cross-section area, m^2; L is the length, m.

$$S = (d+\delta)^2;\ L = d+\delta. \tag{4}$$

Proceeding from such a current flow diagram, the equivalent circuit of the electric resistance of conditional elementary cell of the composite material can be presented in the form given in Fig. 1d by analogy with [14], where the electric supports (conductors) of the filler R_f and the matrix interlayer have tandem connection in the transverse direction of the current flow R_m^{cr} and the electric resistance of matrix interlayer in the longitudinal direction of the current flow R_m^{lt} is connected in parallel to the first two (R_f and R_m^{cr}).

According to the diagram in Fig. 1c, d the electric conductivity $1/R_c$ (an inverse value of the electric resistance R_c) of conditional elementary cell of the composite material will be equal to:

$$\frac{1}{R_c} = \frac{1}{R_f + R_m^{cr}} + \frac{1}{R_m^{lt}}, \tag{5}$$

where R_f is the electric resistance of a filler particle, Ohm; R_m^{cr}; R_m^{lt} are the electric resistances of matrix interlayers in the transverse and longitudinal directions of the current flow, respectively, Ohm:

$$R_f = \frac{\rho_f L_f}{S_f} = \frac{\rho_f d\sqrt[3]{\frac{\pi}{6}}}{d^2\sqrt[3]{\left(\frac{\pi}{6}\right)^2}} = \frac{\rho_f\sqrt{\frac{6}{\pi}}}{d} = \frac{\rho_f \cdot 1.38}{d} \tag{6}$$

$$R_m^{cr} = \frac{\rho_m L_m^{cr}}{S_f} = \frac{\rho_m \cdot d\left(1-\sqrt[3]{\frac{\pi}{6}}\right)+\delta}{d^2\sqrt[3]{\left(\frac{\pi}{6}\right)^2}} = \frac{2\delta}{\pi d^2} + \frac{2\rho_m}{\pi d} - \frac{\rho_m\sqrt{\frac{6}{\pi}}}{d} = \frac{2\delta}{\pi d^2} + \frac{2\rho_m}{\pi d} - \frac{\rho_m \cdot 1.38}{d} \tag{7}$$

$$R_m^{lt} = \frac{\rho_m L_m^{lt}}{S_m^{lt}} = \frac{\rho_m (d+\delta)}{(d+\delta)^2 - d^2\sqrt[3]{\left(\frac{\pi}{6}\right)^2}} = \frac{\rho_m d+\delta}{(d+\delta)^2 - \frac{\pi d^2}{2}} = \frac{\rho_m d+\delta}{(d+\delta)^2 - 1.57d^2} \tag{8}$$

where L_f is the length of a filler particle, which is equal to the cube of the edge length a, m;

$$L_f = a = d\sqrt[3]{\frac{\pi}{6}} \tag{9}$$

L_m^{cr} is the size of a matrix interlayer in the transverse direction of the current flow, which is equal to the thickness of it δ, m;

$$L_m^{cr} = (d+\delta) - a = d + \delta - d\sqrt[3]{\frac{\pi}{6}} = d\left(1 - \sqrt[3]{\frac{\pi}{6}}\right) + \delta \tag{10}$$

L_m^{lt} is the size of a matrix interlayer in the longitudinal direction of the current flow, which is equal to $(d+\delta)$, m; S_f is the cross-section area of a filler particle, m^2; S_m^{cr}, S_m^{lt} are the cross-section areas of matrix interlayers for the transverse and longitudinal directions of the current flow, respectively m^2:

$$S_f = S_m^{cr} = a^2 = \left(d\sqrt[3]{\frac{\pi}{6}}\right)^2 = d^2\sqrt[3]{\left(\frac{\pi}{6}\right)^2};\ S_m^{lt} = S - S_f = (d+\delta)^2 - d^2\sqrt[3]{\left(\frac{\pi}{6}\right)^2} \tag{11}$$

By substituting (3; 4; 6; 7; 8) into (5) we obtain the equations to determine the specific electric conductivity for the composition $1/\rho_c$, the inverse value of its specific electric resistance ρ_c:

$$\frac{1}{\rho_c(d+\delta)} = \frac{1}{\frac{\rho_f\sqrt{\pi/6}}{d} + \frac{2\delta}{\pi d^2} + \frac{2\rho_m}{\pi d} - \frac{\rho_m\sqrt{\pi/6}}{d}} + \frac{1}{\frac{\rho_m d + \delta}{(d+\delta)^2 - \pi d^2/2}} =$$

$$= \frac{d + \pi d^2 + \pi d - d}{\rho_f\sqrt{\frac{\pi}{6}} + 2\delta + 2\rho_m - \rho_m\sqrt{\frac{\pi}{6}}} + \frac{(d+\delta)^2 - \frac{\pi d^2}{2}}{\rho_m(d+\delta)} =$$

$$= \frac{\pi d^2 + \pi d}{0.72\rho_f + 2\delta + 1.28\rho_m} + \frac{(d+\delta)^2 - 1.57d^2}{\rho_m(d+\delta)};$$

$$\frac{1}{\rho_c} = \frac{\pi d^2 + \pi d\cdot(d+\delta)}{0.72\rho_f + 2\delta + 1.28\rho_m} + \frac{(d+\delta)^2 - 1.57d^2}{\rho_m}. \tag{12}$$

We can see that the specific electric conductivity of the composition $1/\rho_c$ and its specific electric resistance depend on the specific electric resistance of the filler ρ_f and the matrix ρ_m and such characteristics of the composition structure as d, δ and their ratio δ/d. We will study the influence of the composition of composite material on the structure characteristics.

By analogy with [16, 17] we will use for this purpose the structural characteristic of the composite material, in particular a factor of separation of the filler by the matrix η. Using the approach described in [18] we obtain the formula for the separation factor η. Let us assume that the spatial model of the composite material given in Fig. 1a is formed by ball-shaped filler particles separated by the matrix with regard to the initial state. In their initial state, the balls or spheres have a contact type arrangement and a simple cubic packing (Fig. 2a). In comparison to other dense packing (Fig. 2) the simple cubic packing is less dense [19]. Nevertheless, many researchers believe [20; 21] that the concrete structure and the structure of analogous composite materials correspond to a great extent to it. In the separated state, the order of relative position of spheres is retained.

It was assumed earlier that the simplest regular element of the packing that corresponds to the conditional spatial elementary cell of the structure is the cube with the sphere inscribed into it. The cube side is equal to the sphere diameter (Fig. 3a). In this case the volume of voids V_v is equal to the difference between the cube volume and the sphere volume:

$$V_v = d^3 - \frac{\pi d^3}{6} = \left(1 - \frac{\pi}{6}\right)d^3. \tag{13}$$

The hollowness of this packing of filler particles, in particular the spheres is equal to:

$$E_f = \frac{\left(1 - \frac{\pi}{6}\right)d^3}{d^3} = 1 - \frac{\pi}{6} = 0.4764. \tag{14}$$

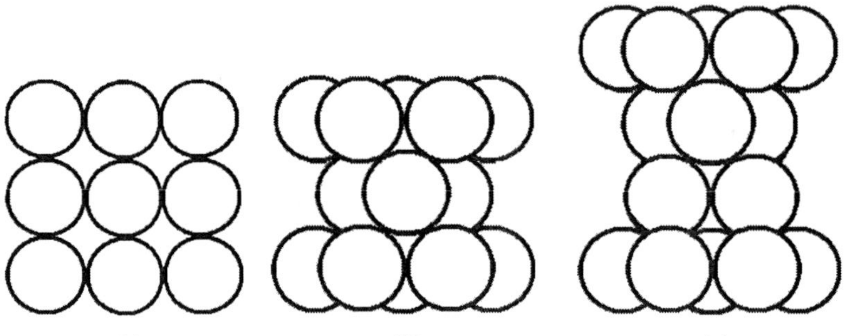

(a) (b) (c)

Fig. 2. Diagrams of the packing of the spheres of the same diameter, (a) the simple cubic packing; (b) the densest face-centered cubic packing; (c) the densest hexagonal packing.

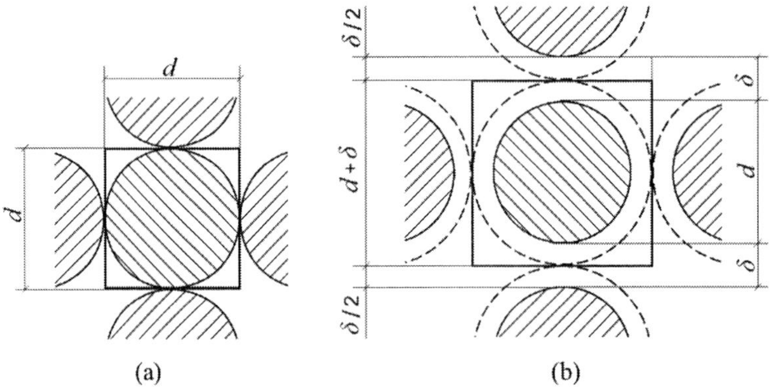

(a) (b)

Fig. 3. The diagram of conditional elementary cell of the structure of composite material that corresponds to a simple cubic packing of the spheres: (a) without separation, (b) with separation.

In the case of contact-type arrangement of the spheres the matrix volume is equal to the volume of voids. The separation of the spheres by the matrix results in the formation of an interlayer between them with the thickness δ (Fig. 3b) and the matrix volume consists of the volume of voids between the spheres with the diameter $(d+\delta)$ and the difference in the volumes of spheres with diameters $(d+\delta)$ and d:

$$V_m = \left(1 - \frac{\pi}{6}\right)(d+\delta)^3 + \left[\frac{\pi}{6}(d+\delta)^3 - \frac{\pi}{6}d^3\right] = (d+\delta)^3 - \frac{\pi}{6}d^3. \tag{15}$$

The factor of separation of filler particles by the matrix η is the ratio of the matrix volume V_m and the volume of voids V_v between the filler particles in the initial state for their contact arrangement without the separation:

$$\eta = \frac{V_m}{V_v} = \frac{(d+\delta)^3 - \frac{\pi}{6}d^3}{\left(1-\frac{\pi}{6}\right)d^3} = 2.10\left(1+\frac{\delta}{d}\right)^3 - 1.10. \tag{16}$$

The derived equation corresponds to the analogous formula in [17].

If denser packing is formed each sphere occupies lesser total volume of the system in comparison with the simple cubic packing. Let us assume that a portion of the cube volume is equal to k. In this case the volume of voids V_v and the hollowness E_f will be equal to:

$$V_v = kd^3 - \frac{\pi d^3}{6} = \left(k - \frac{\pi}{6}\right)d^3. \tag{17}$$

$$E_f = \frac{\left(k-\frac{\pi}{6}\right)d^3}{kd^3} = 1 - \frac{\pi}{6k}. \tag{18}$$

The value k can be expressed via the hollowness by transforming (18):

$$k = \frac{\pi}{6\left(1-E_f\right)}. \tag{19}$$

If the particles are drawn apart at the distance δ, the matrix volume and the separation factor are equal, respectively, to

$$V_m = \left(k-\frac{\pi}{6}\right)(d+\delta)^3 + \left[\frac{\pi}{6}(d+\delta)^3 - \frac{\pi}{6}d^3\right] = k(d+\delta)^3 - \frac{\pi}{6}d^3. \tag{20}$$

$$\eta = \frac{k(d+\delta)^3 - \frac{\pi}{6}}{\left(1-\frac{\pi}{6}\right)d^3} = \frac{k}{k-\frac{\pi}{6}}\left(1+\frac{\delta}{d}\right)^3 - \frac{\frac{\pi}{6}}{k-\frac{\pi}{6}}. \tag{21}$$

By substituting (19) into (21) we obtain

$$\eta = \frac{1}{E_f}\left(1+\frac{\delta}{d}\right)^3 - \frac{1}{E_f} + 1. \tag{22}$$

The relationship δ as a function of d, E_f and η will be obtained by properly transforming (22):

$$\delta = d\left[\sqrt[3]{E_f\left((\eta-1)+1\right)} - 1\right] \tag{23}$$

for the calcium silicate and sodium silicate-based composition:

$$\eta = \frac{V_m}{V_v} = \frac{\frac{Sl}{\rho_{sl}} + \frac{LG}{\rho_{lg}} + \frac{FS}{\rho_{fs}}}{\frac{F}{\rho_f^a} E_f} = \frac{\frac{Sl}{LG} \times \frac{1}{\rho_{sl}} + \frac{1}{\rho_{lg}} + \frac{FS}{LG} \times \frac{1}{\rho_{fs}}}{\frac{F}{LG} \times \frac{E_f}{\rho_f^a}}, \quad (24)$$

where *Sl, LG, FS, F* are the consumption of cinder, liquid glass, silicon fluoride, sodium and filler, kg/1 m^3 of the composition, respectively; $E_f = 1 - \rho_f^a / \rho_f$ is the hollowness of the package of filler particles in the compacted state (analogous to the hollowness of concrete fillers); ρ_{sl}, ρ_{lg}, ρ_{fs}, ρ_f are the true hollowness of cinder, silicon fluoride, sodium and filler, kg/m^3, respectively; ρ_f^a is the average filler density in the compacted state (analogous to the bulk density of concrete fillers) kg/m^3; *Sl/LG, FS/LG, F/LG* are the quantitative mass relationship of appropriate composition components;
for Portland cement-based compositions:

$$\eta = \frac{V_m}{V_v} = \frac{\frac{S}{\rho_s} + \frac{C}{\rho_c} + \frac{W}{\rho_w}}{\frac{F}{\rho_f^a} E_f} = \frac{\frac{S}{C} \times \frac{1}{\rho_s} + \frac{1}{\rho_c} + \frac{W}{C} \times \frac{1}{\rho_w}}{\frac{F}{C} \times \frac{E_f}{\rho_f^a}}, \quad (25)$$

where *S, C, W, F* are the consumption of the sand, cement, water and the filler, kg/1 m^3 of the composition, respectively; $E_f = 1 - \rho_f^a / \rho_f$ is the hollowness of the packing of filler particles in the compacted state (analogous to the hollowness of concrete fillers); ρ_s, ρ_c, ρ_w, ρ_f are the true densities of the sand, cement, water and filler, kg/m^3 , respectively; ρ_f^a is the average density of the filler in the compacted state (analogous to the bulk density of concrete fillers), kg/m^3; *S/C, W/C, F/C* are the quantitative mass relationship of appropriate composition components.

4 Conclusions

Thus, the algorithm used for the definition of electric resistance of a composite material using data on the specific electric resistance of the filler and matrix and the composition of composite material for research or technological purposes can be described as:

1) the separation factor of filler particles by the matrix η is determined using data on composition and physical properties of the components of composite materials and formulae (24) and (25) or analogous formulae;

2) the matrix interlayer thickness between the filler particles δ is determined by formula (23);

3) the specific electric conductivity and specific electric resistance of composite materials as the inverse value of it are derived from formula (12), using data on the specific electric resistance of the filler ρ_f and the matrix ρ_m, the average size of filler grains d and the matrix interlayer thickness between them δ;

4) for the composite material with the multilevel structure the computation is done using the following procedure; first for a finer level (for example, the micro-level represented by the cement stone and the filler), then for the next level using the value of specific electric resistance obtained as that of the specific electric resistance of the matrix for the next coarser level, in particular the mezzo-level with the sand filler, and the macro-level with the crushed stone filler.

The authors express their gratitude and sincere appreciation to Dr. Oleg Kalinin and eng. Vladimir Kasyanov for their assistance in the research work.

References

1. A.A. Starosel'skij, Korrozija i zashhita zhelezobetonnyh konstrukcij v sooruzhenijah jelektrificirovannyh zheleznyh dorog (KhIIT, Kharkov, 1988)
2. A.N. Plugin, A.A. Plugin, Al.A. Pluhin, A.A. Dudin, Collected scientific works of Ukrainian State University of Railway Transport, **111**, 245-261 (2009)
3. V.V. Palij, A.N. Pshin'ko, A.N. Plugin, A.A. Plugin, S.G. Nesterenko, A.A. Konev, Collected scientific works of Ukrainian State University of Railway Transport, **134**, 235-241 (2012)
4. L.E. Vrublevskij, Vozmozhnosti ispol'zovanija jelektroprovodnogo betona (betjela) v grazhdanskom stroitel'stve (CNTI, Moscow, 1971)
5. E.K. Pushkareva, S.H. Huziy, Budivnytstvo, **9**, 54-57 (2001)
6. A.N. Lopanov, E.A. Fanina, O.N. Guzeeva, ARPN Journal of Engineering and Applied Sciences, **9 (11)**, 2250-2253 (2014)
7. O.V. Khrystych, M.S. Lemeshev, Visnyk of Vinnytsia Politechnical Institute, **2**, 18-23 (1998)
8. A.N. Lopanov, E.A. Fanina, O.N. Tomarovshhenko, Bulletin of BSTU named after V.G. Shukhov, **1**, 130-133 (2017)
9. V.R. Serdyuk, M.S. Lemeshev, O.V. Khrystych, Visnyk of Vinnytsia Politechnical Institute, **2**, 5-9 (1997)
10. K.K. Pushkar'ova, S.H. Huziy, Resursoekonomni materialy, konstruktsiyi, budivli ta sporudy, **5**, 82-85 (2000)
11. E.K. Pushkareva, S.H. Huzyy, Naukovo-praktychni problemy modelyuvannya ta prohnozuvannya nadzvychaynykh sytuatsiy, **5**, 23-29 (2002)
12. A.N. Lopanov, E.A. Fanina, O.N. Guzeeva, VESTNIK MGSU, **8**, 78–87 (2016)
13. A.N. Lopanov, O.N. Guzeeva, I.V. Prushkovsky, K.V. Tikhomirova, ARPN Journal of Engineering and Applied Sciences, **9 (11)**, 2275–2278 (2014)
14. A.F. Bernackij, Ju.V. Tselebrovskij, V.A. Chunchin, *Elektricheskie svojstva betona* (Energiya, Moscow, 1980)
15. L. Brautman, R. Krok, B. Noton, *Kompozitsionnyy materialy. T. 3: Primenenie kompozitsionnykh materialov v tekhnike* (Mashinostroenie, Moscow, 1978)
16. A.A. Plugin, *Sovershenstvovanie sostava i struktury betona s uchetom ehlektro poverhnostnyh svojstv ego sostavlyayushchih dlya povysheniya prochnosti i stojkosti izdelij kolcevogo secheniya* (Thesis of Cand. Tech. Sciences, Kharkov, 1994)
17. A.A. Plugin, O.A. Plugin, H.-B. Fisher, G.N. Shabanova, 1 Weimarer Gipstagung, Weimar, 435-443 (2011)
18. A.N. Plugin, *Ehlektrogeterogennye vzaimodejstviya pri tverdenii cementnyh vyazhushchih* (Thesis of DSc (Chem), Kiev, 1989)
19. M.P. Shaskolskaya, *Kristallografiya* (Vysshaya Shkola, Moscow, 1976)
20. A.E. Sheikin, Yu.V. Chekhovsky, M.I. Brusser, *Struktura i svojstva cementnyh betonov* (Strojizdat, Moscow, 1979)
21. I.A. Rybev, *Stroitelnye materialy na osnove vyazhushchih veshchestv* (Vysshaya Shkola, Moscow, 1978)

Phase composition and strength of cement stone with a complex additive

Galina Shabanova[1], *Alla* Korohodska[1], and *Vita* Shumejko[1,*]

[1]National Technical University "Kharkov Polytechnic Institute" (NTU "KhPI"), Technology of Ceramics, Refractory Materials, Glass and Enamels Department, Kyrpychova str. 2, 61002, Kharkiv, Ukraine

Abstract. Ecologically friendly construction is impossible without the use of materials that are characterized by a low emission of dangerous substances from the surface and have other appropriate properties. A solution of the problem related to the elaboration of special measures taken to reduce the emission of ammonia from construction materials and the structures manufactured using the Portland cement has been offered. An optimized composite dope was developed and optimized for the Portland cement composition that contains the nitric acid calcium, which is one of the ingredients that can considerably reduce the concentration of nitrogen-containing compounds, making thus the construction materials less dangerous to the human health. Strength characteristics of the cement stone with the composite dope were studied and it was established that a level of the intensification of a gain in the strength is rather significant and by the seventh day the hardening exceeds the values peculiar for the dope-free compositions at an age of 28 days by more than 45 %.

1 Introduction

Technological progress in the field of construction material science requires an increase in the efficient action of the dopes of a different type. The use of chemical dopes is one of the most universal and available methods of the control of concrete technology and adjustment of its properties. In the past some chemical products were widely used as the dopes for construction materials and at the present time the dopes specially prepared for cement solutions and concretes are more preferable. Many dopes that improve certain characteristics of the concrete mixture or the concrete sometimes fail to improve other characteristics and even worsen them. To overcome these side effects we use composite dopes that consist of several independent components. In addition to evident advantages, the chemicalization of concrete production technologies poses new problems related to the quality of concrete from the standpoint of ecology and hygiene [1, 2].

At the present time, a concept of the "green" (or ecological) construction, in particular the construction of such objects that have a minimum negative effect on the environment becomes increasingly popular. The buildings and installations erected according to the ecological concept will provide comfortable and safe living conditions and also preserve

* Corresponding author: shum-vita@ukr.net

and improve the environment. Meanwhile, the experts [3] believe that building and finishing materials and structures, including concrete ones that release toxic substances are one of the sources of chemical pollution of the air inside the living quarters. The most important aspect is the problem of ammonia emission from concrete structures in the living quarters of residential and public buildings. The emission of ammonia by the concrete is a widely spread phenomenon peculiar for newly built houses and it has a dreadful effect on human beings and the environment.

2 Analysis of the latest research papers and publications

The analysis of scientific sources and patent information allows us to draw a conclusion that the problem of ammonia emission by the concrete is also peculiar for the foreign practice of construction [4-7]. It was noted, that ammonia release sources are Portland cement-based solutions and concretes. The paper [8] gives consideration to one more possible reason for the accumulation of nitrogen-containing compounds in the Portland cement; in particular this is the formation of Portland cement clinker under reduction conditions due to the annealing and cooling. The research done enabled the theoretical substantiation of the possibility of formation of iron nitrides during the formation of Portland cement clinker that can be a source of ammonia emission from construction materials and the structures. A solution was found for the elaboration of special measures taken to reduce the ammonia emission from construction materials and the structures fabricated using the Portland cement. The authors [8] assume that ammonia emission sources accumulated in the hardened cement stone have a continuous harmful effect due to a permanent emission of ammonia, and the research done by them showed a positive influence of the dopes that are capable of intensifying the ammonia emission during the cement mixing and at an early stage of the hardening of cement stone. It was shown that nitric acid calcium added in low concentrations facilitates an early ammonia emission from the cement stone.

3 Defining research goals and problems

Ecologically friendly construction is impossible without the use of materials characterized by a low emission of dangerous substances from the surface and that have other appropriate properties. In this connection, a creation of the composite dope for the Portland cement composition that includes $Ca(NO_3)_2$ is of great interest. The latter is one of the gradients that can considerably reduce the concentration of nitrogen-containing compounds making the construction materials less dangerous to the human health.

Since the Portland cement is an active component of the concrete, the characteristics of this artificial material will be defined to a great extent by the properties of hardened Portland cement stone. The use of nitric acid calcium in substantial concentrations (more than 1mass %) is efficient for the control of time required for the hardening of cement stone [9]. It is clear that these factors are not efficient at very low concentrations of nitric acid calcium. However, the research showed that $Ca(NO_3)_2$ added even in low concentration has a substantial influence on physical and mechanical properties of the cement stone.

The composition of composite dopes should be "designed" to allow the components to intensify the effects provided by each component individually.

An efficient use of the dopes for contemporary construction technologies is defined by the choice of their rational composition taking into consideration the chemical phenomena of additivity and synergism, i.e. a compatibility of the dopes of a different nature during their interaction, which enables the intensification of obtained effects. For cement systems, the efficiency is defined by the hydration activity that characterizes an early strength and the grade strength.

The purpose of this scientific paper was to optimize cement-containing compositions with modifying dopes and analyze the phase composition of hydration products that provides for a decrease in the emission of ammonia from construction materials and the structures fabricated using the Portland cement.

4 Major Research

When selecting a dope and defining a dose we must clearly see the role of each component in the polyfunctional modifier. All the composite dopes can conventionally be subdivided into the three groups: the mixture of electrolytes; the mixture of surfactants and the mixtures of electrolytes and surfactants. The introduction of electrolytes enables a control of the hardening rate and structural and mechanical properties and the surfactants control the mobility, imparting some special properties, for example rheological to the cement solution and the concrete mixture.

To do research we used the dope-free Portland cement of a PC I-500-H grade (DSTU BV.2.7-46:2010) produced by PJSC "Eurocement-Ukraine". The nitric acid calcium (NC), sodium hexametaphosphate (SHMP) and sodium tripolyphosphate (STPP) were used as modifying dopes.

Sodium salts of phosphoric acids are the electrolytes and perform as the plasticizing additive. The sodium hexametaphosphate has a polymer nature and it is known as the Graham's salt. Therefore the mixture of sodium hexametaphosphate and sodium tripolyphosphate contributes to the properties not only of electrolyte. This mixture forms duel electric layers on hydrated cement particles and contributes to the rarefaction of cement paste due to the electrostatic repulsion. Beside, having the surfactant properties, the mixture adsorbs on cement particles and amplifies the rarefaction-plasticization effect due to their steric repulsion during the cycle rupture and the orientation of linear polymer chains being forming.

Experimental verification of the individual compatibility of tested dopes with Portland cement enabled the establishment of their optimal quantity to be used to provide an improvement in the strength characteristics of cement stone in comparison with those of the produced Portland cement.

To define an optimal composition for the offered composite dope consisting of the SHMP, STPP and NC mixture in different proportions and carry out the comparative tests we fabricated the following cement compositions and analyzed their compression strength:

-1)- PC I-500-H;

-2)- PC I-500-H + 0.075 % (33.33 % NC : 33.33 % STPP : 33.33 % SHMP);

-3)- PC I-500-H + 0.075 % (33.33 % NC : 44 % STPP : 22.67 % SHMP);

-4)- PC I-500-H +0.075 % (33.33 % NC : 22.67 % STPP : 44 % SHMP).

To define the strength, the specimens in the form of cubes 20x20x20 mm were made of the cement paste, A compression strength of the specimens of hydrated dope-free Portland cement and the doped cement were tested using the hydraulic press during their hardening for 2, 7 and 28 days. The comparative characteristics of strength properties are given in Fig. 1, in which the curves are numbered in compliance with the above cement compositions.

The data given in Fig. 1 show that the Portland cement, modified with the dope based on the composite PC I-500-H + 0.075 % (33.33 % NC : 22.67 % STPP : 44 % SHMP) provides a more substantial increase in the strength of cement stone. A level of the intensification of the gaining in strength is significant and by the seventh day of hardening it exceeded that of the strength values peculiar for the dope-free composition of 28 days old by more than 45 %.

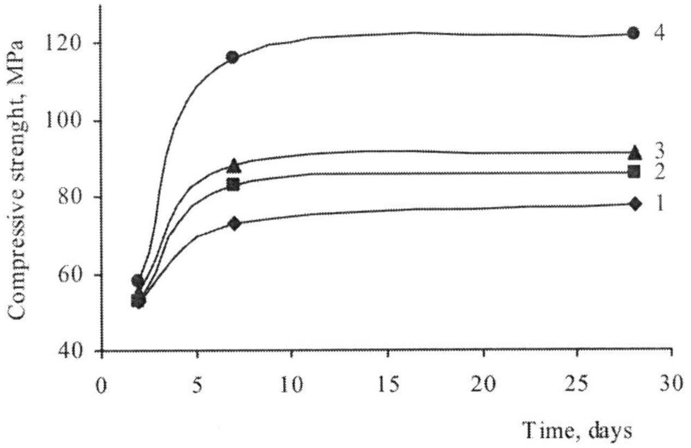

Fig. 1. Strength characteristics of tested cement compositions.

To define the phase composition of hydration products the specimens of hydrated dope-free Portland cement and those with an optimal composition of the composite dope were subjected to X-ray phase analysis. The X-ray phase analysis of tested specimens was done using the diffraction meter DRON-3M. The research data are given in the form of bar - roentgenograms in Fig. 2.

The bar - roentgenograms given in Fig. 2 show the intensity of diffraction maxima for crystal compounds in hydration products and confirm the key conceptions of the mechanisms of a change in physical and mechanical characteristics of the tested objects. We believe that an increase in the strength of cement stone containing the introduced composite dope is caused by structural and phase changes during the process of hydration hardening.

The effect of strengthening in the presence of dope is provided due to a change in the character of the formation of crystal hydrate clusters and also due to qualitative and quantitative changes in the crystallization conditions of key phases. The dope contributes to a more complete hydration of basic clinker phases and an additional formation of tobermorite-like phases (interplane distances are 1.41, 1.268, 0.3076, 0.3001 nm) of a different crystallinity and basicity and also to the formation of AF_m-phase (interplane distances correspond to those of ettringite that are identified by its strong peaks of 0.973, 0.561 and 0.388 nm).

A close connection exists between the structure strength and the character of it. However, these both properties are defined by the type and the volume of hydration products of the cementing material and different hydration phases possess different binding capacity. Tobermorite phases manifest the highest binding capacity in the cement paste. A.A. Govorov showed [10] that the stone strength is provided by new formations that are related to tobermorite and gyrolite groups and the strength of specimens based on the aluminum-substituted tobermorite is comparable with that of pure "xonotlite" structures. The formation of low-base calcium hydrosilicates with high specific surface results in a decrease of the efficient radius of pores and facilitates the formation of close pores that are not participating in the process of moisture transfer. All these changes decrease a degree of shrinkage and the gas and water permeability and result in an increase of the strength, deformability, crack resistance and the longevity that is provided due to a high content of calcium hydrosilicates of a reduced basicity that are capable of protecting hydroaluminate phases.

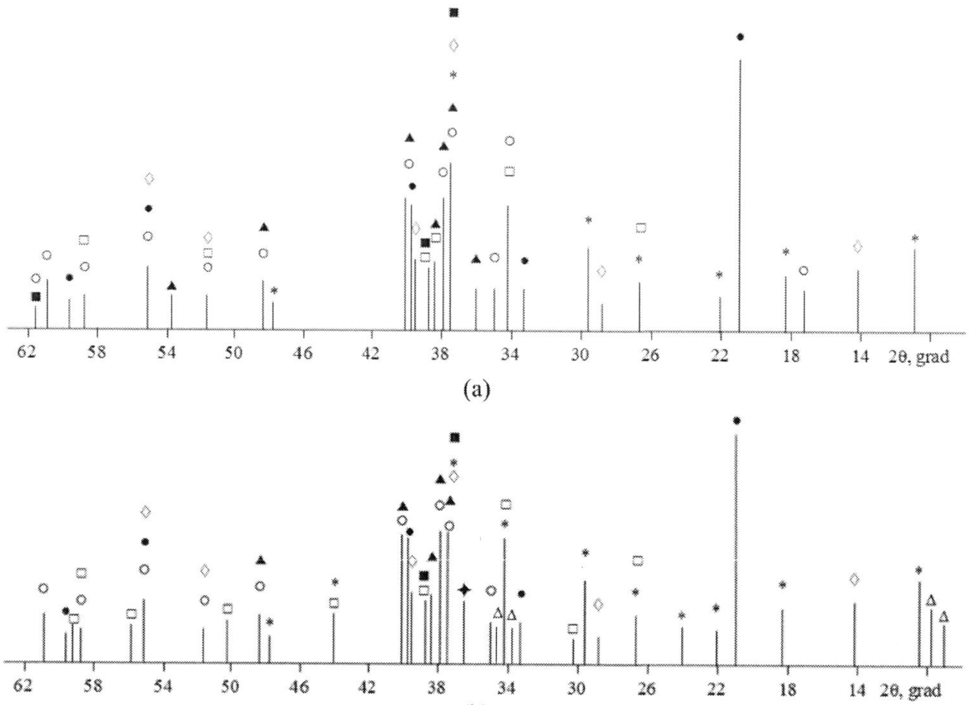

(b)

* – $Ca_6Al_2(SO_4)_3(OH)_{12}$·$26H_2O$, • – $Ca(OH)_2$, O – Ca_3SiO_5, ▲ – Ca_2SiO_4,
■ – $Ca_3Al_2O_6$, ◇ – $Ca_2(Al,Fe)_2O_5$, □ – $CaCO_3$, Δ – tobermorite-like phases, ✦ – AF_m-phase

Fig. 2. The bar - roentgenograms of hydrated cement compositions of 28 days old: (a) dope-free Portland-cement; (b) Portland cement with an optimal composition of the composite dope.

According to V.V. Timashev's data [10], the theoretical strength of low-base calcium hydrosilicates defined by their chemical composition and the structure of crystal lattice is actually two times higher than that of high-base calcium hydrosilicates. According to Y. Yambor data [10], the pore size distribution is an important factor that has a substantial influence on the material strength. In particular, it was shown that the binding compositions that contain hydration products of a similar type are characterized by such pore size distribution. Tobermorite and tobermorite phases contribute to the formation of the pores of an average size in the cement stone; the smallest of them have a size of 10 to 40 nm in contrast to other hydration products [10].

5 Conclusions made using the research data and the prospects for further development in this area

The main functional purpose of electrolyte dopes is considered to be an intensification of the gain in the concrete strength and the facilitation of cement hydration processes, producing a positive influence on the phase composition character of new hydrate formations and the pore space. An optimal relation of the dope ingredients that make up the cement-containing composition provides a considerable increase in the strength by the seventh day of hardening, more than 45 % in comparison with the dope-free Portland cement at an age of 28 days.

 DOI: 10.1051/matecconf/201711601014

Application of the developed composite dope seems to be rather promising due to the modifying effect it has on the hydration hardening processes and a gain in the strength of hardening compositions and it can be also used for the production of materials that are characterized by a low emission of hazardous substances from the surface and have, as a result, an increased ecologic compatibility.

References

1. S.P. Sivkov, I.C. Kudysheva, *Ammonia emission from cement concretes* (19 Internationale Baustofftagung, Weimar, 2015)
2. V.N. Shimanov, *Problem of emission of ammonia from concrete constructions* (Access mode: http://www.science-education.ru/105-7025, 2015)
3. L.V. Predelskiy, O.E. Prihodchenko, *Building ecology* (Train aid, Rostov on Don, 2003)
4. Higuchi Takayuki, *Method for electrochemical treatment of ready mixed concrete* (Patent JP2004122620, 2004)
5. Pollak Vladimir, Chodak Ivan, *Vtthod for the treatment of concrete* (Patent SK151099, 2001)
6. Robert F. Rathbone, Thomas L. Robl. University of Kentucky Center for Applied Energy Research. Final Report. 63 (2001)
7. Z. Bai. Environment International **32**, 303 (2006)
8. G.N. Shabanova, S.M. logvinkov, T.V. Shkolnikova, V.N. Shumejko. Collected scientific works of Ukrainian State University of Railway **162**, 189 (2016)
9. V.S. Izotov, U.A. Sokolova, *Chemical additions for modification of concrete* (Monograph, Moscow, 2006)
10. *Intercommunication of structure and properties of concrete* (Access mode: http://stroi-archive.ru/shlakoschelochnoy-beton/1077-vzaimosvyaz-struktury-i-svoystv-betona.html, 2016)

Nanotechnological and energy-saving methods of production of building composites

Olena Shynkevych[1,*], *Yevgen* Lutskin[1], *Oleksandr* Koishev[1], *Georgiy* Bondarenko[1], *Andriy* Tertychnyi[1], and *Igor* Myronenko[2]

[1]Odessa State Academy of Civil Engineering and Architecture, Didrikhson st. 4, 65029 Odessa, Ukraine
[2]Odessa National Maritime University, Mechnikov st. 34, 65029 Odessa, Ukraine

Abstract. The bases of silicate materials' manufacture of non-autoclave hardening are developed. The realisation opportunity of technological decision complex of using of mineral substances' structure in silicate materials' manufacture of non-autoclave hardening and wall's products on their basis is theoretically proved and practically confirmed. The complex activation of mixture with water materials, as slip, defined the transition from autoclave treatment to thermo-moisture one of silicate materials. The possibility of the practical realization of structure mineral substances' reserve for the energy consumption of silicate material production are proved experimentally. The optimization of the composition and the hardening conditions in the conditions of thermo-moisture treatment on the basis of experimentally-statistical modelling ensured the receipt of materials with the required properties. The analysis connection between hardening conditions and contents has been fulfilled on experimental-statistic models. The changing of silicate materials' properties under the influence of inorganic modifier' surface, of hardening conditions and content of gypsum addition have been estimated. Correlation analysis allows to receive new information about the influence of the contents and the hardening conditions on a degree of interrelation between the structure and the properties of building materials. Optimal compositions and the regimes of hardening are recommended for the receipt of wall's articles of a different purpose.

1 Introduction

Development and introduction of the energy conservation and competitive building materials are the actual question of building branch. In these conditions the normative requirements to heat-protection of constructions increase considerably. Substantial increase of production volume of heat-effective building materials is planned.

In total amount of wall material productions the significant volume is allocated with silicate materials, as effective and ecologically pure materials. Production of silicate

* Corresponding author: elena_shinkevich@ukr.net

concrete has reserves on decrease of bulk density, heat conductivity and charge of material and energy.

Within these problem the production with using of modern technology process equipment the wall's silicate non-autoclave hardening materials and their optimization on a complex of quality parameters, first of all the heat conductivity, is a perspective direction which will allow to reduce expenses of energy as at a stages of materials and articles' production, and at a stage of construction operation.

The production's opportunity of silicate materials non-autoclave hardening due to use active silica has been proved with positions of thermodynamics in works [1, 2].

The production's opportunity of silicate materials at normal temperature and pressure (T=85±2°C, d=0.1 MPa) allows to speak about essential potential reserves of production of this kind of articles [3, 4].

2 Technological bases of reception of silicate materials of non-autoclave hardening

On the basis of experimental-theoretical researches the opportunity of realization of the non-autoclave technologies of silicate materials' reception is established. Hydrated lime-silica binding substances at temperature T=85°C and atmospheric pressure is realized due to high index pH with using as components binding not slaked to exhaust, which is grounded together with quartz sand, and chemically active amorphous silica with the optimum specific surface, which is entered instead of the ground quartz sand's part.

The necessity of calcium hydroxide's linkage in autoclave processing conditions is caused by chemical thermodynamics of lime and quartz: with rise in temperature the solubility to exhaust falls, though the speed of dissolution increases, and the solubility of silica increases (fig.1.) [5]. Under certain conditions synthesis of hydrosilicates of autoclave hardening is possible at temperature below 100°C and atmospheric pressure [1, 2, 6].

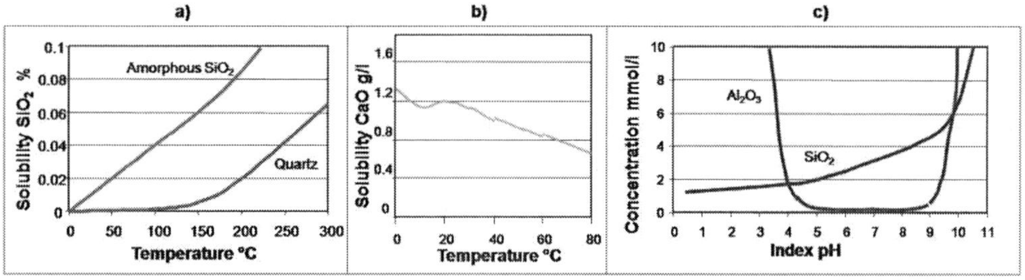

Fig. 1. Influence of temperature midst on solubility of silicate concrete mixture components: SiO_2 - (a) (J.Shtark), CaO - (b) (V.Babushkin, O.Mchedlov-Petrosyan) and an index pH on concentration of Al_2O_3 and SiO_2 - (c) (J.Shtark).

With regulation of lime speed and silica dissolution it is possible to displace balance in system in this or that side and to change kinetic of hydrations and properties of silicate materials.

The technology provides complex activation of a silicate concrete mixture, including a fine-grained filler, warming up a mix in forms at a stage of preliminary ageing and thermo-moisture treatment (TMT) at T=85°C. The preliminary moderate warming up a mix in forms is provided by application of not slaked ground lime together with quartz sand. Speed regulation of lime's thermal emission is provided by the optimum water-firm relation, the content of gypsum additives and plasticizer [7].

Thermal activation of components of a silicate concrete mixture is traditionally carried out in autoclaves where under the conditions of the raised temperature and pressure there is a hydrothermal synthesis of calcium hydrosilicates (V.Mihaelis). Transition from hydrothermal synthesis of hydrosilicates of calcium in autoclaves to thermo-moisture processing is carried out due to association of the following ways of activation:

– mechano-chemical activation of crystal quartz in an aquatic midst of the mixer-activator,

– chemical activation – due to introduction of the inorganic modifiers, containing amorphous silica, and the raised values pH midst,

– thermal activation – due to application of not slaked ground lime.

3 Nanotechnology receptions of producing the complex activated composites on silicate matrix

Developed by the authors the complex activation includes the continuous loop of the different types and methods of activation: mechanical, chemical and thermal [6, 8, 9, 10]. Each type of activation is accompanied by effects which create the conditions for the possibility of subsequent type of activation.

The result of the complex activation is the formation of linear defects, dislocations and point positions and substitutions. Furthermore, there may be angle changing between the bonds and the appearance of dangling bonds, which leads to the formation of free radicals in the crystals with covalent bonds and the amorphization for molecular crystals.

Each type of activation will be caused by the prevalence of any type of deformations of the solid phase structure. The differences will be responsible for the nature and kind of dislocations with allowance for the extent and duration of external and internal influences.

3.1. Mechanical activation.

Mechanical effects in the dispersion medium speed mixer activator provide mechanochemical activation of crystalline quartz surface. The term "mechanochemical reaction" was introduced by W.Ostwald in 1891.

Mechanochemical activation is carried out to the special properties of the newly formed surfaces, especially changing the local chemical and phase content of solids, as well as their aggregate state under the influence of mechanical effects of high intensity.

As nanotechnology reception the mechanochemical activation reduces the viscosity of 3 or more dispersed system containing lime [6, 8, 9, 10, 11]. This effect of viscosity reducing was used to compensate the increased water demand of the mixture introducing the composite porous opal-cristobalite rocks and using the activation of the binder together with the fine aggregate.

3.2. Thermal activation.

Using as a binder quick lime promotes the "inner activation" of dispersed system under an elevated temperature $T= 40\div60°C$, the formation of multiple point contacts in the dislocation fields are appeared and the conditions for the hydration hardening unrelated in hydrous calcium lime are created.

External thermal activation occurs in conditions of thermal-moisture treatment. In such conditions at $T=85°C$ contradiction is canceled, which is connected with the increasing the quartz solubility and decreasing lime solubility with increasing its dissolution [2, 4]. In addition, the increase of pH system causes to create favorable conditions for longevity growths GSK during the operational phase when $11.5 \leq pH \leq 12.5$.

3.3. Chemical activation.

According to E. Avvakumov's works [12] with the presence of water in the inorganic solid-phase system "method of soft mechanochemical synthesis" is happened. To obtain composite oxides from the simple oxide hydroxides are used as the starting components one of which is characterized by acidic, the other one is characterized by basic properties. By means of the neutralization reaction the intensification of the processes takes place. In addition, in such system the conditions for the hydrothermal processes are created.

The acid activation occurs by introducing the amorphous-crystalline silica, alkaline activation – the introduction of increased amounts of lime; it raises pH. The replacement of ground quartz sand in the binder by opal-cristobalite rocks enhances to increase frost and water resistance of the silicate matrix while reducing its density matrix to 20-25%.

Moreover, the presence of opal-cristobalite rock in the dispersion causes formation of nanoscale in the GSK pores of these rocks; in this case they are "nanoreactor"; which walls restrict the growth of neoplasm [13]. Also, the presence of particles of porous rocks with different dispersion binder allows adjusting the speed and reaction kinetics of hydration [6, 8, 9].

The use of quick lime determines the possibility of using of high modulus liquid glass as one of the alkali addition, so the temperature rise on the forming stage allows regulating rapid setting of such mixture.

Alkali and alkali-containing additives increase the thermodynamic instability of systems by shifting the equilibrium caused by the formation of additional defects on the surface of silica-containing components. Alkali-containing additives are capable to aeration of mixtures under the certain conditions [14].

In this work, the low-temperature aeration of concrete during introduction of the activation to the mixture for the silicate matrix by liquid glass additives $Na_2O{\cdot}nSiO_2{+}mH_2O$ and sodium hydroxide NaOH was provided.

4 Experiment

The select of the plan and factors for experiment is carried out in the view of following positions. The inorganic modifier with the set specific surface carry out a role of modifiers of structure and allow to regulate a kinetics of hydration and structurization processes [10, 11, 12, 13]. Depending on a specific surface of inorganic modifier, reaction can go in conditions of lime's excess, deficiency or at its full linkage.

The basic natural experiment is realized on sated to 24 punctiform six factors plan of type MTQ "mixture-technology-quantity" [15, 16, 17]. The specific surface of the inorganic modifier was fixed as three dependent mixture factors at levels: $\upsilon_1=S_{sp1}=350m^2/kg$, $\upsilon_2=S_{sp2}=425m^2/kg$, $\upsilon_3=S_{sp3}=500m^2/kg$. The factors are connected by linear dependence: $\upsilon_1+\upsilon_2+\upsilon_3=1$. The duration of a preliminary ageing ($\tau_{p.a.}=0\div12$ hours), duration of TMT ($\tau_{TMT}=10\div18$ hours) at T=85°C and contents of the additive of gypsum ($C_g=0\div5\%$) were changed as three independent factors. In each point of the plan there were generated on 15 identical structures. By the plan it is calculated two identical complexes of six-factorial experimentally-statistical models by type (1) [15, 16, 17]. The models describe dependence "content-technology-properties" and "content-technology-structure". It has allowed to study dependence "characteristics of structure- properties".

The results of physical and chemical researches have shown a presence in structure of hydrosilicates of calcium of different sorts, morphology and quantities depending on a specific surface of the inorganic modifier, the contents of the additive of gypsum and hardening conditions. By the results of electronic microscopy, DTA and X-ray hydrosilicates of calcium of are identified: hillebrandite B, its version hillebrandite C, foshahite and partially crystal CSH (II) [6, 8, 9].

The frost resistance of monomineral samples of hillebrandite (F100) is on the order above, than frost resistance CSH (B) (F10) and to bermorite (F15) which form mainly firm phase of autoclave materials [1, 2].

5 Results and discussion

By results of experiment experimentally-statistical models of influence of contents and conditions of hardening on properties and characteristics of structure of silicate materials of non-autoclave hardening are calculated.

The models of change of a porousness of the general, opened and closed, and also the quantitative contents of hydrosilicates of calcium under influence of six factors are calculated. The changing of the general porousness on 10 % the relation of opened porousness to closed porousness increased in three times. Character of a capillary porousness was estimated by the relative average size of capillaries (d_k) which changes in 4.4 times and by the coefficient of uniformity of distribution of sizes of capillary pores (α_k) which changes in 2.5 times.

The contents and hardening conditions which provide a maximum of compressing strength do not coincide with values of these factors which provide a minimum of heat conductivity or a maximum of frost resistance. So, under influence of all six factors the increase compressive strength calculated on a full ES model, makes 6.4 times. Joint influence on compressive strength of three groups of factors: grain-growing content (S_{sp1}, S_{sp2}, S_{sp3} – the block D), hardening conditions (τ_{TMT} and $\tau_{p.a.}$ – the block B) and additives of gypsum (C_g) in quantitative expression is equivalent. Each group of factors is able to provide a gain of compressive strength more than twice (fig. 2.).

Table 1. The relative modification of the properties under the influence of the different groups of factors technology and content

	Groups of factors	R_b	F	λ
1.	The various specific surface area S_{sp} of the inorganic modifier	2.1	2.5	1.6
2.	The hardening conditions $\tau_{p.a.}$, τ_{TMT}	2.1	1.0	1.5
3.	The additive of gypsum C_g	2.2	2.0	1.0
4.	All six factors S_{sp}, $\tau_{p.a.}$, τ_{TMT}, C_g	6.4	4.8	3.0

Table 2. The value of the specific surface of the inorganic modifier S_{sp} m^2/kg, wich provide the optimum properties

	Without gypsum	With gypsum
R_b max	S_{sp2}=425	S_{sp1}=350
F max	$0.5S_{sp1}+0.5S_{sp3}$	
λ min	S_{sp1}=350	S_{sp3}=500

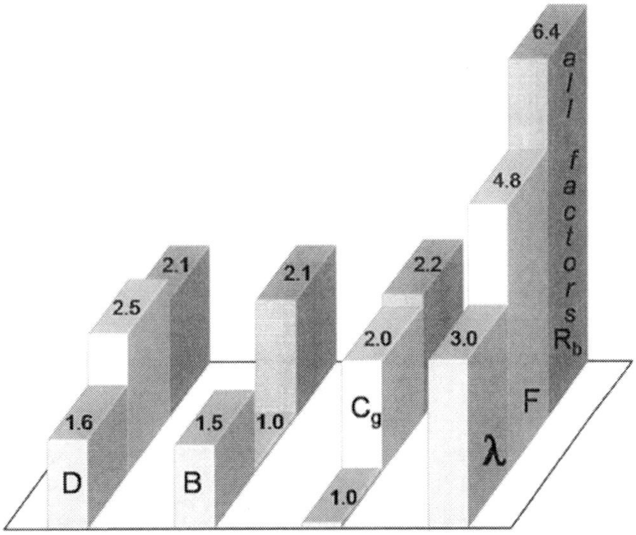

Fig. 2. The analysis of relative influence of various groups of factors: grain-growing content S_{sp1}, S_{sp2}, S_{sp3} (the block D), the hardening conditions τ_{TMT} and $\tau_{p.a.}$ (the block B), additives of gypsum (C_g) and their interactions (a full model) on a coefficient of heat conductivity λ, a frost resistance F and a compressive strength R_b.

The coefficient of heat conductivity λ changes in three times depending on a specific surface of the inorganic modifier and hardening conditions: 0.43 - 1.3 Wt/m·K. So, under influence of a specific surface of a inorganic modifier the relative change of a coefficient of heat conductivity equal 1.6 (the block D), under influence of hardening conditions - 1.5 (the block B). what is connected to the relative average size of capillary pores d_k changes in 6 times (a correlation coefficient r $\{\lambda;d_k\}$ = 0.68÷0.89) and coefficient of uniformity of distribution of sizes of capillary pores α_k - in 2 times (a correlation coefficient r $\{\lambda;\alpha_k\}$ = -0.75÷ - 0.91).

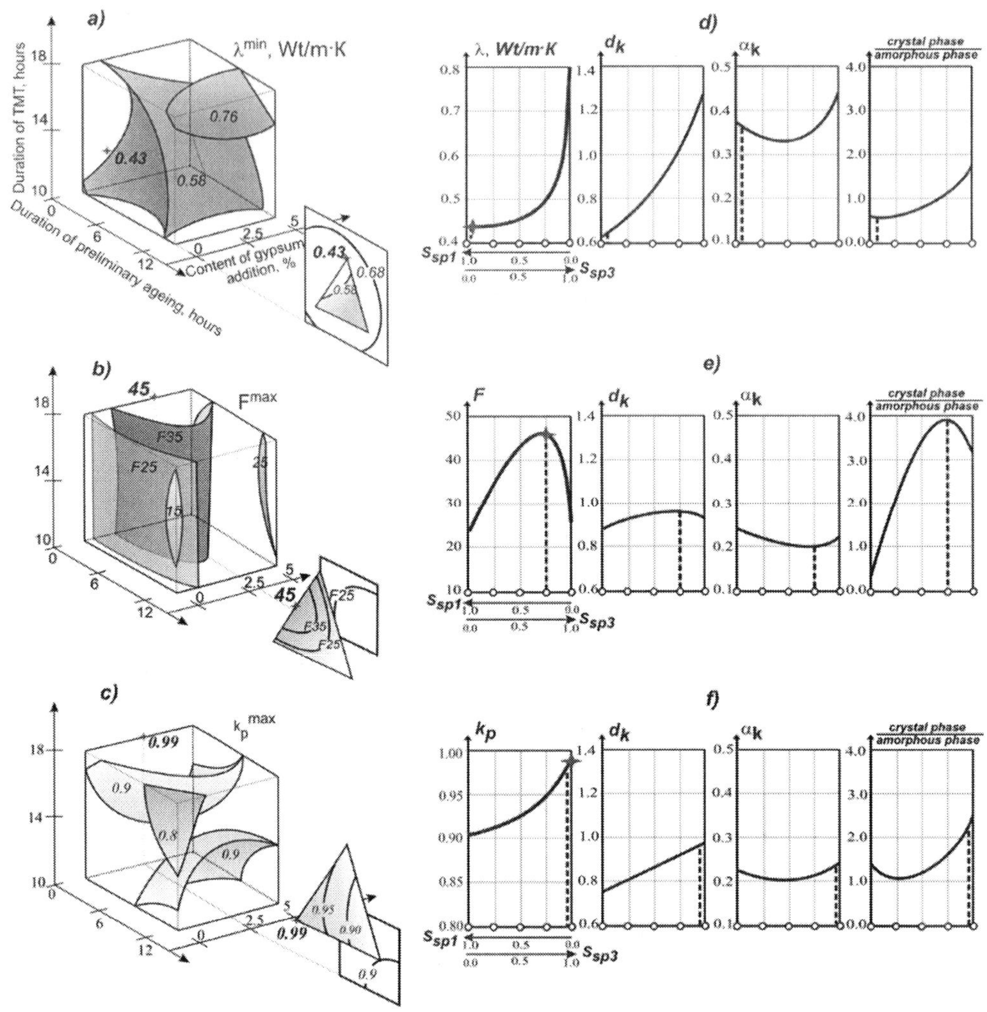

Fig. 3. The change of a coefficient of heat conductivity (a), of a frost resistance (b), of a coefficient of water resistance (c) under influence of hardening conditions and contents of the additive of gypsum at optimum values of size of a specific surface of the inorganic modifier (λ^{min}, F^{max}, k_p^{max} accordingly) (isosurfaces are inside the cube - a full model); hardening conditions at the fixed contents of gypsum (isolines on the squared diagrams - the block B); a specific surface of the inorganic modifier at the fixed values of hardening conditions and the contents of gypsum (isolines on the three-cornered diagrams - the block D). Influence of content of inorganic modifier S_{sp1} and S_{sp3} on the relative average size of capillary pores d_k, coefficient of uniformity of distribution of sizes of capillary pores α_k and a ratio of crystal phase and amorphous one for the optimum hardening conditions and the contents of the additive of gypsum (λ^{min} - (d), F^{max} - (e) and k_p^{max} - (f)).

The great value for a coefficient of heat conductivity has also a ratio between amorphous and crystal phases in cementing substance. This ratio changes more than ten time. The coefficient of heat conductivity of crystal quartz two times more than of amorphous silica. Downturn of density on 20-25 % as a result of increase in a share of the closed porosity of the inorganic modifier also causes decrease in coefficient of heat conductivity.

Due to the additive of plaster at constant size of coefficient of heat conductivity λ the compression strength increases more than twice.

It enables receptions of wide scale of composites with a complex of differing properties. The specific surface of the inorganic modifier (the block D) influences on a frost resistance 2.5 times more, than hardening conditions (the block B). Character of a porousness for a maximum of frost resistance F and a minimum of a coefficient of heat conductivity λ is various: the relative average size of capillaries $d_k(F)$ =0.95> $d_k(\lambda)$ =0.85 and the coefficient of uniformity of distribution of sizes of capillary pores $\alpha_k(F)$ =0.2 <$\alpha_k(\lambda)$ =0.38 (fig.3). Increase of frost resistance is provided due to formation in structure of additional contacts from the particles of secondary quartz.

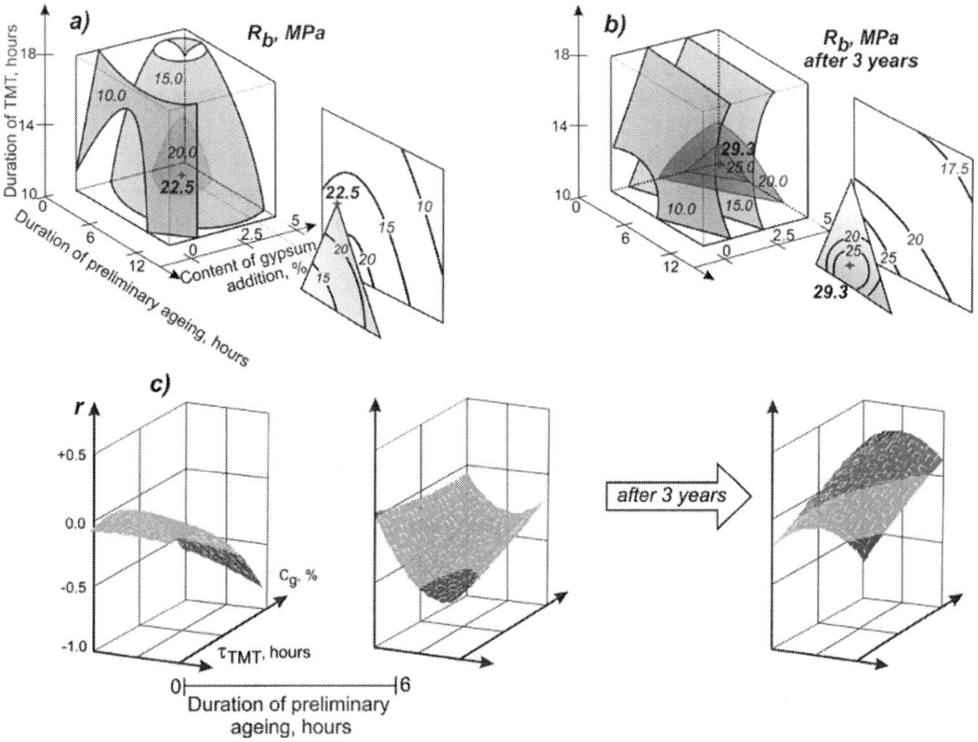

Fig. 4. The change of compressive strength after TMT (a) and age of 3d years (b) under influence of hardening conditions and contents of the additive of gypsum at optimum values of size of a specific surface of the inorganic modifier (R_b^{max}) (isosurfaces are inside the cube - a full model); hardening conditions at the fixed contents of gypsum (isolines on the squared diagrams - the block B); a specific surface of the inorganic modifier at the fixed values of hardening conditions and the contents of gypsum (isolines on the three-cornered diagrams - the block D). The coefficient of correlation ***r*** between R_b and the ratio of amorphous and crystal phases at various stages of processes of hydration and structurization (c).

Increase of frost resistance is promoted by formation in structure of additional contacts from particles of secondary quartz. Crystallization of amorphous silica blocks capillaries on their length. The additive of gypsum changes a coefficient of uniformity of distribution of sizes of capillary pores α_k with 0.6 till 0.85. The estimation of effects of interactions between parameters of structure and properties on the blocks of experimental-statistic models allows to define the most sensitive to regulation of the characteristic of structure. The value of interrelation between characteristics of structure and of properties is changing in the time as a result of spatial changes of parameters of structure. It is shown the various contribution of a characteristics of structure in formation of properties at different stages of process of structurization, for example on the compressive strength (fig. 4).

6 Conclusion

The bases of manufacture of lime-silica composites (non-autoclave hardening) are developed.

The complex activation of high-mobility silicate concrete mixture, including fine-grained filler, provides transition from autoclave processing to thermo-moisture treatment at T=85°C.

Technical and economic calculation has shown, that transition from autoclave processing of the products received by a method of pressing to thermo-moisture treatment processing of products, reduces expenses for manufacture more than 25%; and introduction of wall's products allows to reduce a heat consumption through non-load-bearing designs to 30%.

The contribution of various characteristics of structure to formation process of properties changes in the time.

The kinetics of hydration processes and structurization can be adjusted by a specific surface of the inorganic modifier, defining individual levels of properties. Influence of content, in particular of specific surface of the inorganic modifier, and hardening conditions on various properties in quantitative expression is ambiguous.

References

1. O.P. Mtschedlov-Petrosyan, V.I. Babuschkin, Poc. 4th Int. Symp. Chem. Cem, **1**, 533-544 (1960)
2. V.I. Babushkin, O.P. Mchedlov-Petrosyan, *Silicate Waterproof Products* (Kiev, 1962)
3. V.I. Bolshakov, A.I. Golovko, A.V. Koval, Yu.I. Mustafin, S.A. Sherbak, *Structurization of the silicate systems* (Dnepropetrovsk, 2000)
4. J. Shtark, B. Viht, *Durability of Concrete* (Kiev, Transl. from German, 2004)
5. H. Taylor, *Chemistry of Cement* (Moscow, the World, 1996)
6. E. Shinkevich, N. Sidorova, E. Lutskin, V. Sidorov, S. Politkin. *Raw Mix for Obtain Modified Silicate Materials and Method of Its Prepare*. Declared patent 64603 A, 7 C04B28/20, Ukraine (2004)
7. P. Krivenko, Proceeding of International Conference "Alkali Activated Materials – Research, Production and Utilization", 313-332 (2007)
8. E. Shinkevich, E. Lutskin, O. Gnyp, A. Koichev, J. Dotsenko, Proceeding of the 8th Int. Symp. Brittle Matrix Composites 8, 517-525 (2006)
9. E. Shinkevich, Proceeding of 13th Int. Congress on the Chemistry of Cement, 351-358 (2011)
10. Y. Lutskin, E. Shinkevich. Proceeding of 14th Int. Congress on the Chemistry of Cement, **2**, 632 (2015)
11. E. Shinkevich, Y. Zaytsev, E. Lutskin, G. Bondarenko, A. Tymnyak, Proceeding of 2nd International conference on Microstructural-related Durability of Cementitious Composites, 1462-1468 (2013)
12. E.G. Avvakumov, M. Senna, N. Kosova, W. Soft, *Mechanochemical Synthesis a Basis for New Chemical Technologies* (Netherlands, Kluwer Academic Publishers, 2001)
13. C. Poole, F. Owens, *Nanotechnology* (Moscow, Technosphere, 2005)
14. V.D. Gluhovskij, R.F. Runova, S.E. Maksunov, *Binders and composite materials of contact hardening* (Kiev, Budivelnik, 1991)

15. V. Voznesenskiy, T. Lyashenko at al., *Computer and optimization of composite materials* (Kiev, Budyvelnik, 1989)
16. E. Lutskin, E. Shinkevich, Proceeding of 13th International Congress on the Chemistry of Cement, 359 (2011)
17. E. Lutskin, E. Shinkevich, Technical Journal, **9** (**1-2015**), 27-33 (2015)

MATEC Web of Conferences 116, 01016 (2017) DOI: 10.1051/matecconf/201711601016
Transbud-2017

Fracture properties of high-strength concrete obtained by direct modification of structure

Serhiy Solodkyy[1], *Taras* Markiv[2], *Khrystyna* Sobol[1], and *Oleksii* Hunyak[1,*]

[1]Lviv Polytechnic National University, Department of Highways and Bridges, Bandery st. 12 S., 79013 Lviv, Ukraine
[2]Lviv Polytechnic National University, Department of Building Production, Bandery st. 12 S., 79013 Lviv, Ukraine

Abstract. High-strength concrete is effectively used worldwide in the last three decades, but it is more brittle in comparison with normal strength concretes. Partial substitution of cement in concrete by active mineral additives and usage of chemical admixture of plasticizing and air-entraining action can considerably change their fracture properties. The obtained results show that the increase of the fracture properties is observed in concretes modified with chemical admixtures incorporating mineral additives such as zeolite and limestone due to consolidation of the concrete microstructure. Densification takes place as a result of limiting the amount of calcium hydroxide (CH) due to its reaction with active silica included in the zeolite and the formation of larger amounts of hydrated calcium silicates of tobermorite type as well as calcium hydroaluminate and hydrocarboaluminate with the simultaneous adsorption modification of hydrated products by chemical admixtures.

1 Introduction

High-strength concrete (HSC) is used in most cases for the construction of high-rise buildings, long-span bridges, etc. According to the American Concrete Institute the concrete with the compressive strength higher than 41 MPa is usually defined as high-strength concrete. In the last three decades it has been effectively applied worldwide. However, HSC tends to be more brittle in comparison with normal strength concretes (NSC) [1, 2]. Concrete failure is a process of crack formation and growth. Unlike other widespread building materials concrete is a quasi-brittle material, which means that it acts both ways: being brittle (like glass) and elastic-plastic (like steel) simultaneously. The application of linear elastic fracture mechanics (LEFM) and fracture energy approaches are proved to be good mechanisms to predict and prevent failure of cement-based man-made structures. The main fracture properties of concrete can be determined using simple three-point bend notched beam specimens: critical stress intensity factor K_{Ic}, fracture energy G_F, and characteristic length χ_{tb} [3].

Recent researches have shown [4, 5] that partial substitution of cement in concrete by active mineral additives can considerably change their fracture properties. Usage of these

* Corresponding author: oleksii.m.guniak@lpnu.ua

additives does not only allow lowering the environmental harm associated with cement production, but also receiving an economic effect and obtaining a durability improvement in received cement-based composites [6].

Natural zeolite is a promising mineral additive with the relatively high pozzolanic activity [7]. The previous researches [8, 9] of concretes have shown, that namely 10 mass.% of natural zeolite in combination with chemical admixtures of plasticizing and air-entraining action result in the improvement of concrete properties and durability.

A small amount of limestone is proved to be a good additive due to its high grindability and ability to decrease the temperature of cement hydration [10]. Grains of limestone in combination with zeolite act as a centre of crystallization of hydrated phases [9].

Researches have shown that modification of HSC structure with both mineral additives and chemical admixture of plasticizing and air-entraining action improves the strength and fracture characteristics of hardened concretes made with fresh concrete mixes of stiff consistency [11]. However, such important properties as fracture toughness and fracture energy of zeolite containing high-strength concretes have not been thoroughly researched.

Therefore, the objective of this study is to determine the fracture properties of modified concretes made of fresh concrete mixes of semi-fluid consistency.

2 Experimental Program

2.1 Materials and mix-proportions

Locally available commercial Portland cement CEM I 42.5R was used in this study as a main binding material. Physical and mechanical properties of the cement were tested according to [12], and the results summarized in Table 1.

Table 1. Physical and mechanical properties of the cement.

Specific surface, m^2/kg	Residue on sieve 008, %	Water demand, %	Setting time, min		Compressive strength, MPa	
			initial	final	2 days	28 days
340	1.6	29.5	150	210	25.4	52.6

Chemical compositions of Portland cement, natural zeolite derived from Sokyrnytsia (Ukraine), limestone coming from Mykolaiv quarry were determined by X-ray spectrometer ARL 9800 XP and presented in Table 2. Clinoptilolite content in zeolite was 65%.

Table 2. Chemical composition of cement, zeolite and limestone.

Material		CEM I 42,5R	Natural zeolite	Limestone
Chemical composition, %	SiO_2	21.49	77.64	1.95
	Al_2O_3	4.85	13.04	0.51
	Fe_2O_3	3.59	2.90	0.36
	CaO	64.54	4.77	54.66
	MgO	1.44	1.60	0.52
	SO_3	1.60	0.05	0.53
	LOI	2.49	-	41.47

The river sand with fineness modulus of 1.8 was used as a fine aggregate. The coarse aggregate of irregular shape was used for the research. Bulk density of coarse and fine aggregate was 1420 and 1440 kg/m^3, respectively, voidage was 45% for both aggregates.

Commercially available polycarboxylate-based superplasticizer with a specific gravity of 1.05 and solid content of 20% and an air-entraining agent, based on vinsol resin, with specific gravity of 1.05 and solid content of 4.4%, were used in this study.

The morphology of the investigated materials was examined with the scanning electron microscope SEM FEI Quanta 250 FEG, equipped with EDS.

Four mix-proportions were designed: control mix C and three mixes (ZL, PA, ZLPA) containing mineral additives and chemical admixtures. Mix-proportions are presented in Table 3.

Table 3. Mix-proportions and fresh concrete mix properties.

Mix identification	Portland cement (kg/m^3)	Natural zeolite (kg/m^3)	Limestone (kg/m^3)	Sand (kg/m^3)	Gravel (kg/m^3)	Superplasticizer (% of cement)	Air-entraining agent (% of cement)	Slump (mm)	w/cm ratio	Air content (%)
C	350.0	-	-	600	1230	-	-	140	0.57	2.0
ZL	297.5	35	17.5	600	1230	-	-	140	0.59	2.0
PA	350.0	-	-	600	1230	1	0.015	140	0.50	4.2
ZLPA	297.5	35	17.5	600	1230	1	0.020	140	0.52	3.9

2.2 Sample Preparations and Test Methods

Initially, the materials were thoroughly mixed in the required proportions using the laboratory mixer. Water with admixtures was then added and the concrete was mixed up to a uniform state. Subsequently, the fresh concrete was poured into the beam and cubes were moulded. A plate vibrator was used to compact the concrete in the moulds in one layer. Twenty-four hours after casting, the specimens were demoulded and cured for 28 days with the ambient climatic conditions such as temperature of 20±2^{0}C and relative humidity of 95%. The specimens were taken out from the curing tanks after 28 days and then central notches were made using a diamond saw. The size of the prism specimens was 100*100*400 mm with an effective span of 380 mm and a notch-to-depth ratio of 0.40. The size of cube specimens was 100*100*100 mm.

An estimate of fracture energy was obtained from the L–d diagram according to the RILEM method (work-of-fracture) [13], the load-displacement history up to failure was recorded by a computer-aided data acquisition system. Fracture parameters were calculated according to the Ukrainian standard DSTU B V.2.7-227 [14].

3 Results and discussion

It is well known that during the loading process different defects can be initiated and developed inside the material [15]. Damage, in most cases, begins in the concrete cement matrix or in the interfacial transition zone, namely between cement paste and aggregate. In these studies the effect of concrete structure on the fracture properties of high-strength concrete was studied. As can be seen from Fig. 1, the typical load-deflection curves obtained from fracture test show that, in general, the addition of mineral (zeolite and limestone) additives and chemical (plasticizer and air entraining agent) admixtures in spite of the approximately equal compressive strength (Fig. 2) leads to decrease in maximum load bearing capacity from 4190 N for the mixture C to 3910 N and 3380 N for mixtures PA and ZL respectively. Since the tensile bending strength is a more sensitive parameter of concrete than the compressive strength the reason may be due to presence of mineral particles that act as flaws of microstructure, which can initiate the failure earlier than that concrete without mineral additives. The decrease of area under the load-deflection plot shows that the fracture energy G_F is reduced from 468.59 J/m^2 for control concrete C to 449.32 J/m^2 for concrete ZL.

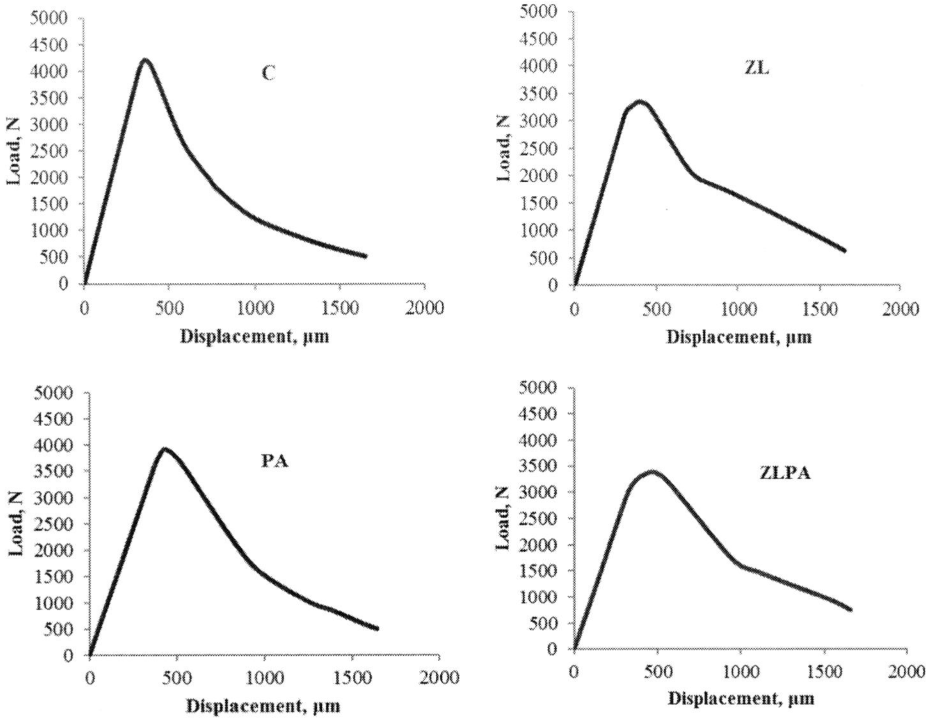

Fig. 1. Typical load-deflection curves for studied concretes.

The air-entraining agent together with superplasticizer in concrete PA has a positive effect on its fracture energy both in subcritical (G_i) and post-peak (W_L) stages and, as a result, on fracture energy G_F. The modified concrete incorporating zeolite and limestone ZLPA characterizes the highest fracture energy in subcritical (G_i) and post-peak (W_L) stages and fracture energy G_F.

Table 4. Fracture and mechanical properties of concretes.

Mix identification	G_i, J/m^2	$W_L \cdot 10^{-2}$, N·m	G_F, J/m^2	K_c, MPa·m$^{1/2}$	Tensile strength in bending, R_{tb}, MPa	Bending modulus of elasticity, E_{tb}*10^{3}MPa	Characteristic length, χ_{tb}*10^{-4},m
C	128.61	210.57	468.59	0.55	8.86	7.03	206
ZL	129.20	214.38	449.32	0.42	7.08	8.36	252
PA	148.06	217.20	497.27	0.40	8.27	8.11	190
ZLPA	163.56	219.81	521.42	0.49	7.15	9.98	238

As seen from the Fig. 2, the compressive strength of concrete ZL incorporating zeolite and limestone is lower than the control concrete C at its early age of hardening. Concrete PA (without mineral additives) containing plasticizer and air-entraining agent is characterized by higher compressive strength both on 3 and 7 days of hardening in comparison with control concrete C. However, the strength of tested concretes is almost leveled with increasing the age and in such a case the fracture energy of designed concretes can be reasonably compared.

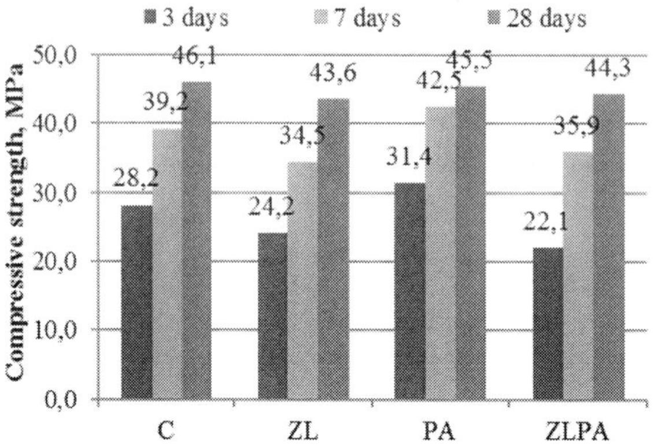

Fig. 2. Compressive strength of concretes.

The physical and mechanical properties of concrete are a direct result of a series of complex processes taking place in microstructure of concrete during hydration. The used electron microscopy method complements the data on the structure formation of cement paste. Microstructure analysis of the cement paste without additives after 28 days of hardening (Fig. 2) shows that microstructure is presented by a large number of hydration products. Some crystals have a fine needle-like or plate-like shape. In some places, dense crystalline blocks of lamellar crystals of Portlandite are observed (Fig. 3a), that is confirmed by the microprobe analysis (Fig. 3b).

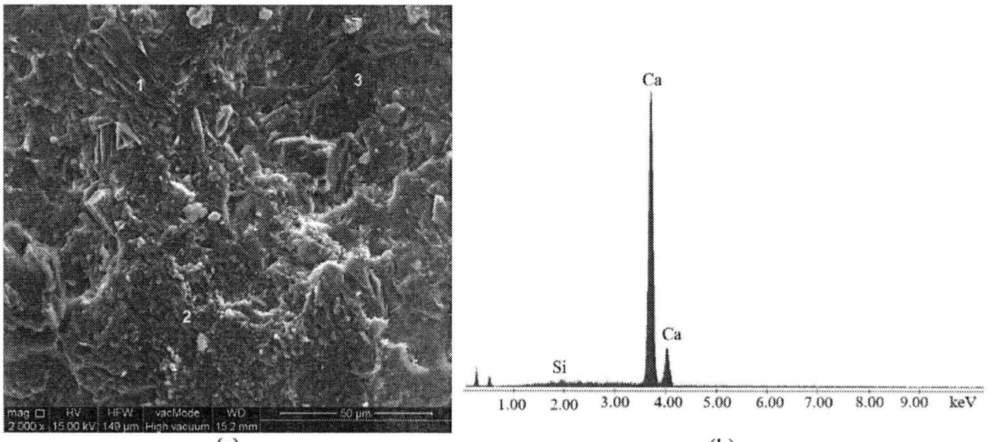

(a) (b)

Fig. 3. Microstructure (a) and EDS spectra (b) of Portland cement paste without mineral additives after 28 days of hardening.

Information regarding the phase arrangement inside the material structure of modified cement matrix of concrete, containing zeolite and limestone, is essential. As seen from Fig. 4a, such a phase, as C-S-H which is the most important component of the cement matrix strengthens the structure and improves the fracture toughness of concrete. In contradistinction to C-S-H phase, large lamellar portlandite crystals can cause a significant weakening and greater susceptibility to the brittle destruction.

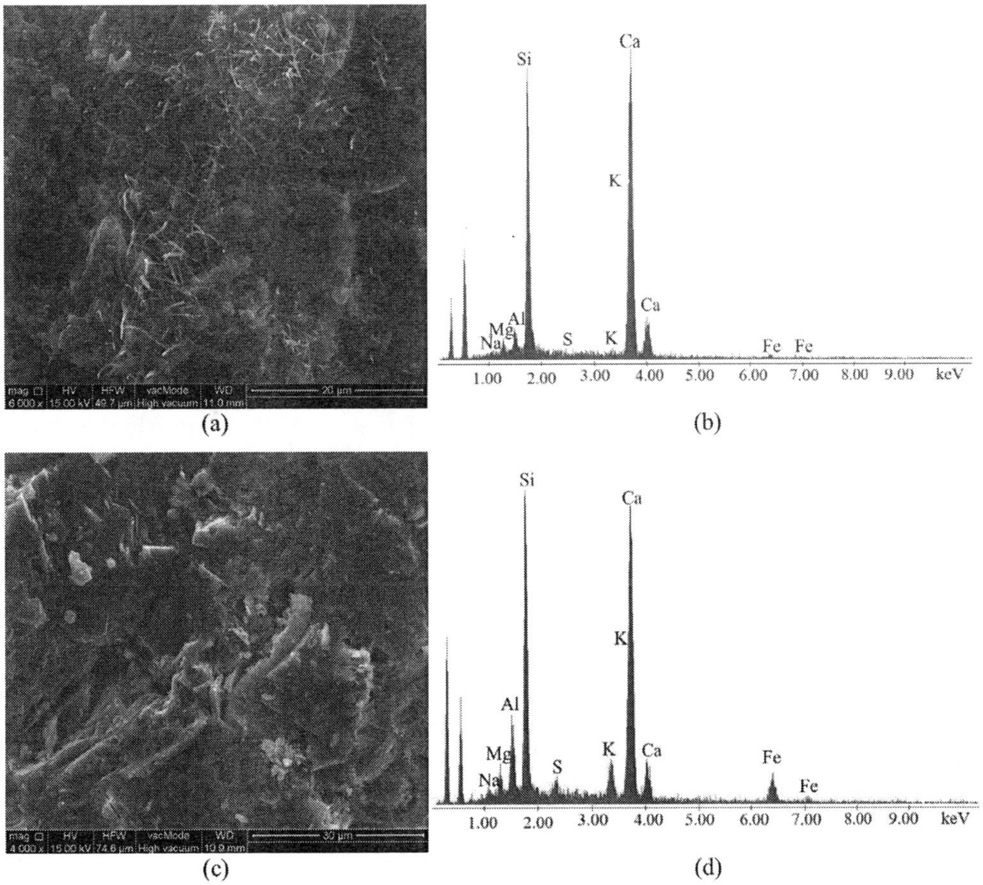

Fig. 4. Microstructure (a, c) and EDS spectra (b, d) of modified Portland cement paste containing zeolite and limestone after 28 days of hardening.

Fig. 4 shows, that zeolite and limestone cause a distinct replacement of weak calcium hydroxide (CH) phase for a compact and durable phase C-S-H (Fig. 4a) as well as calcium hydroaluminate and hydrocarboaluminate (Fig. 4a,c). The concrete containing zeolite and limestone as well as superplasticizer and an air-entraining agent have lower amount of CH and due to adsorption modification of hydration products is characterized by denser structure and smaller pore size in comparison with the control concrete. It is also confirmed by the lowest value of water absorption of concrete ZLPA containing both mineral additives and chemical admixtures. It reduces from 3.94 mass.% for control concrete C to 2.97 mass.% for modified concrete ZLPA. This compact concrete structure with pores of small diameter has a positive influence on both the strength and on the fracture properties of high-strength concrete.

4 Conclusion

The conducted tests revealed that the increase of the fracture properties is observed in concretes incorporating mineral additives such as zeolite and limestone modified with chemical admixtures. It is shown that consolidation of the concrete microstructure takes place as a result of limiting the amount of CH due to its reaction with active silica included in the zeolite and the formation of larger amounts of hydrated calcium silicates of the tobermorite type with the simultaneous absorption modification of hydrated products with chemical admixtures. As a result, the modified concretes containing zeolite and limestone can reach higher strength and are characterized by the higher fracture properties.

References

1. R.A. Einsfeld, M.S.L. Velasco, Cement Concrete Res. **36**, 576 (2006)
2. G.A. Rao, B.R. Prasad, Cement Concrete Res. **32**, 247 (2002)
3. A. Hillerborg, Mater. Struct. **18**, 291 (1985)
4. B.H. Bharatkumar, B.K. Raghuprasad,. D.S. Ramachandramurthy, R. Narayanan, S. Gopalakrishnan, Mat. Struct. **38**, 63 (2005)
5. L. Lam, Y.L. Wong, C.S. Poon, Cement Concrete Res. **28**, 271 (1998)
6. E. Gartner, Cement Concrete Res. **34**, 1489 (2004)
7. M. Valipour, F. Pargar, M. Shekarchi, S. Khani, Constr. Build. Mater. **41**, 879 (2013)
8. T. Markiv, K. Sobol, M. Franus, W. Franus, Arch. Civ. Mech. Eng. **16**, 554 (2016)
9. O. Hunyak , K. Sobol, T. Markiv, Mistobuduvannya ta terytorial'ne planuvannya, **54**, 139 (2014)
10. P. Hawkins, P. Tennis, R. Detwiler, *The use of limestone in Portland cement: a state-of-the-art review* (Skokie, Portland Cement Association, 2003)
11. S. Solodkyy, O. Hunyak, T. Markiv, Avtomobil'ni dorohy i dorozhnye budivnytstvo, **98**, 247 (2016)
12. BS EN 196, *Method of testing cement.* (British Standards Institution, London, 2016)
13. Rilem TC-50 FMC (Draft Recommendation). Mater. Struct. **18**, 258 (1985)
14. DSTU B V.2.7-227. *Metody vyznachennya kharakterystyky trishchynostiykosti (v`yazkosti ruynuvannya) pry statychnomu navantazhenni* (Ukrarhbudinform, Kyiv, 2010)
15. S. Solodkyy, *Trishchynostiykist' betoniv na modyfikovanykh tsementakh* (LPNU Publishing House, Lviv, 2008)

The properties of concrete mixes, pumped over long distances

Viktor Sopov[1,*], and *Viktor* Dolgiy[1]

[1]Kharkiv National University of Civil Engineering and Architecture, Department of Physical and Chemical Mechanics and Technologies of Building Materials and Products, Sumska st. 40, 61002 Kharkiv, Ukraine

Abstract. The influence of technological factors, chemical and mineral admixtures on pumping and thixotropic properties of concrete mixtures was studied. The causes of the loss of pressure when pumping concrete mixtures through channels of a complex section are established. Existing methods for calculating pressure loss during pumping of concrete mixtures are considered. Some results of pumping tests of modern concrete mixtures are given. It is shown that the use of powdered concrete with high mobility for pumping over long distances during the filling of long-span metal or bridge structures, concreting in tunnels where injection pressure is limited, can be the only solution in the implementation of these tasks.

1 Relevance of the problem

In the last decade, great progress has been made in pumping concrete mixes with pump equipment [1-3]. Practical recommendations for concrete pumps have been developed, basic principles of pumping have been defined, etc. But all these knowledge and rules focus on vibrating heavy concrete (CVC). Pumping of other modern types of concrete mixtures has not been sufficiently studied today. Recommendations for pumping vibrating heavy concrete will not meet the necessary conditions when pumping self-compacting concrete mixtures (SCC), mortars, fine grained concrete mixtures and powdered concrete (RPC).

In practice, the study of pumping concrete mixtures is associated with the determination of pressure losses in the system using certain pumping equipment. The causes of pressure loss can be:

- Technological (configuration, material, length and diameter of the concrete, the pressure of the concrete pump, the type of the concrete pump, the speed of the concrete mix in the concrete, the temperature of the mixture and the ambient air);
- Rheological (shear stress, yield strength, viscosity, structural strength of the mixture, thixotropy);
- Tribological (friction and lubricating layer).

Technological reasons for the loss of pressure in the concrete guide are associated with pumping equipment. In a number of cases, the calculation of the required pressure reduces to the calculation of only technological parameters without taking into account the

*Corresponding author: vsopov1959@hotmail.com

rheological properties of the concrete mixture [4]. For vibrational concretes this may be the case.However, the properties of modern concrete mixtures and concretes can be very different from those of CVC. Therefore, for more accurate prediction of pressure losses when pumping modern concrete, it is necessary to take into account the rheological and tribological characteristics of it.

2 Theoretical basis

Modern studies of pumping concrete are devoted to rheological and tribological (tribology - the science of the processes of friction and lubrication) characteristics of the flow of concrete mixture [5]. In this case, the main role is assigned to the lubricating layer (Fig. 1).

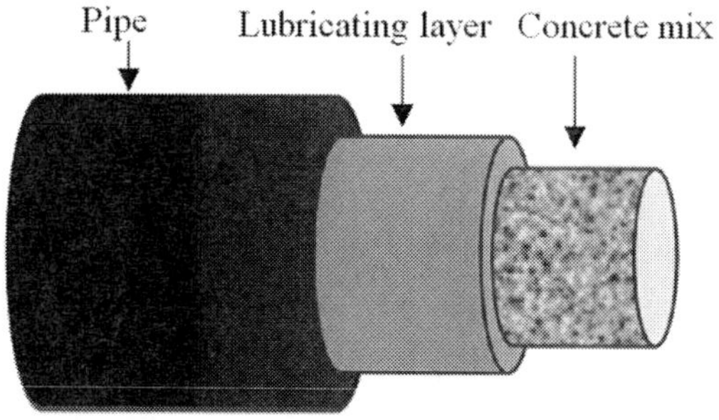

Fig.1. Schematic representation of a concrete mix in a pipe.

The most accepted approach in the quantitative evaluation of the rheological properties of a concrete mixture is the measurement of the shear stress as a function of the shear rate, since the readings of the sediment or the spreading of the cone do not give a complete picture of the processes occurring in the mixture during pumping. The rheological model of concrete mixtures obeys the Bingham law, where the shear stress depends on the shear rate [6], i.e. non-Newtonian fluid (Fig. 2):

$$\tau = \tau_0 + \frac{\mu}{\gamma}, \tag{1}$$

where γ is the velocity gradient, μ and τ_0 are the plastic viscosity and yield stress, respectively.

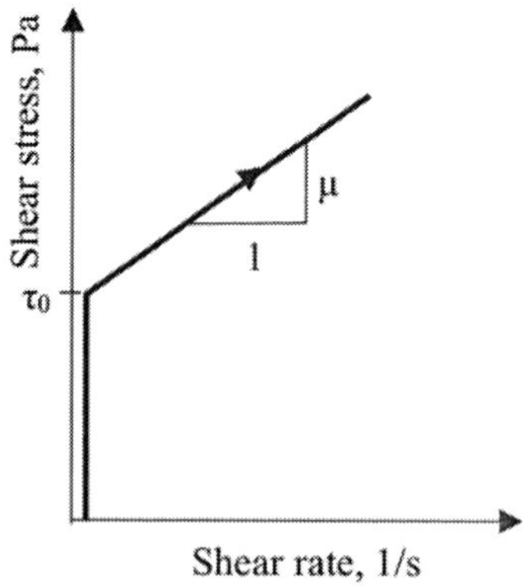

Fig. 2. Dependence of shear stress on shear rate for cement systems.

The non-Newtonian behavior of concrete mixtures is determined by the strength of the spatial structure of the disperse system, which is directly related to the nature of its behavior during flow, as well as its speed of destruction and recovery. The flow of cement systems occurs together with the destruction of the spatial structure and, the more the flow velocity, the more the system, the viscosity of which decreases, decreases. At the same time, as a result of the chemical reaction, the limit of temperature and plastic viscosity vary with time. The presence of the structure gives the system properties of ductility, fluidity, and viscosity.

Thixotropyis defined as a decrease in viscosity when applying shear, and then gradual recovery when shearing stresses are removed. An interesting feature is the cement system in determining the viscosity with increasing the shear rate and then reducing it, forming, so-called, hysteresis loops [7]. Those. The yield strength decreases as the total amount of shear energy increases. And important is the dependence of the yield point on the history of shear stresses. The shape of the loop varies depending on the cycle time, when the shear stress is increased due to hydration during a long cycle (Fig. 3) [7].

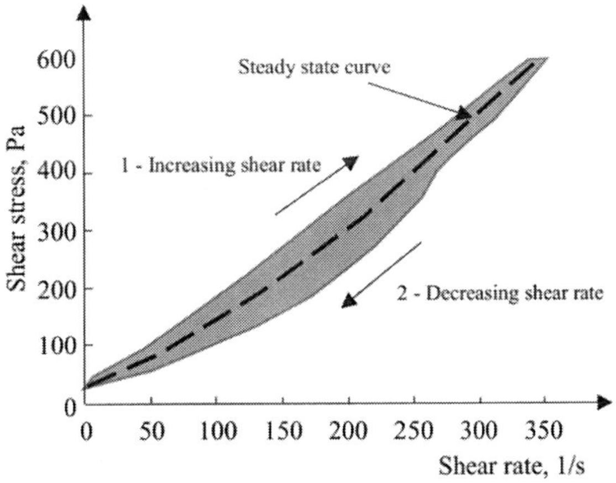

Fig. 3. Example of thixotropic loop obtained with a cement paste [7].

The components of the concrete mix differently affect its rheological properties. The influence of organic superplasticizers is caused by a decrease in the yield strength of the concrete mix without a significant effect on its viscosity. Similarly, an increase in the air content of the mixture affects the decrease in its plastic viscosity (Fig. 4). The increase in mixing water leads to a mutual decrease in rheological parameters. An interesting feature is the additive of microsilica - with increasing its content, viscosity decreases, and at a flow rate of more than 25% of cement mass, the yield strength of the cement system rises sharply. The addition of fly ash largely affects the yield strength and has a slight effect on the viscosity change.

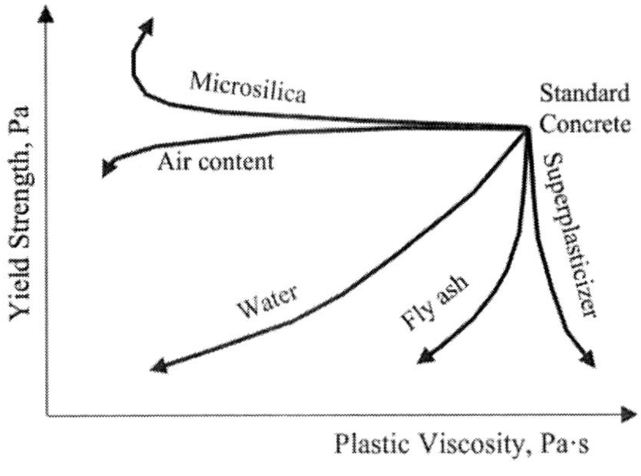

Fig. 4. Effect of components of a concrete mixture on its rheological characteristics.

The development of new-generation concrete technology is associated with a reduction in aggregate fractions [8]. So, for example, for ultrahigh-strength concrete (UHPC) (over 100 MPa), the particle size of the aggregate is commensurate with the cement particles. Such concrete is devoid of shortcomings of the contact zone with inert aggregates, where all components enter into chemical reactions, forming a more solid structure.

Replacement of a large aggregate with finely divided mineral additives gives a great advantage in pumping such mixtures due to the practical exclusion of sedimentation of large grains and obtaining a more homogeneous structure of the pumped material. Based on the properties of reactive powdered concrete, their use makes it possible to produce concrete mixtures with mobility that goes far beyond self-compacting concrete (EN 206-9) with high final characteristics, corresponding to new-generation concrete. This opens up a wide range of possibilities associated with pumping concrete over long distances, filling in elements of complex configuration, obtaining high-quality surfaces, etc.

3 Experimental part

The experiment consisted in determining the pressure loss in the pipe section 99 m when the concrete pump was pumped by a highly mobile concrete mix by measuring the internal pressure in the concrete guide at the beginning and middle of the concrete pipe (Fig. 5). Auto-concrete pump hydraulic with the length of the stroke of the piston 1500 mm and the developed pressure on the concrete mixture to 4 MPa. The choice of the composition of concrete was carried out from the conditions of the greatest distance of pumping of the concrete mixture through a concrete guide of rectangular cross-section 60 × 80 mm. The maximum aggregate size was 0.2 mm.

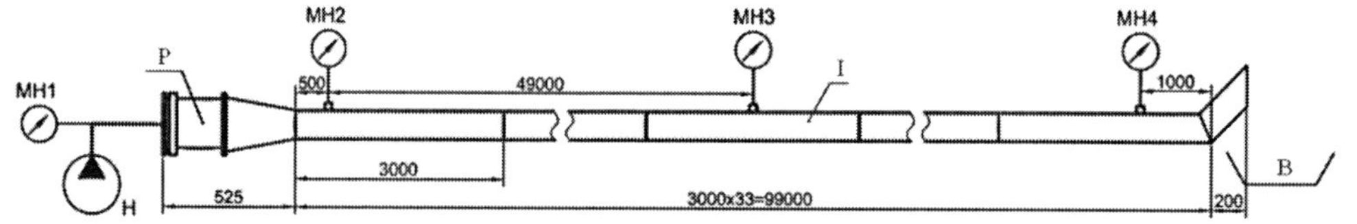

Fig. 5. General scheme of concrete pipe: I – Concrete pipe of rectangular cross-section 60 × 80mm; H – concrete pump; B – a container for draining a concrete mixture; P – adapter to the concrete pump; MH1-MH4 – manometers with membrane media separators.

The mobility of the concrete mixture was determined by the spreading of the Abrams cone and the Suttardviscosimeter, and its rheological characteristics by means of rotary viscosimeter of a Fungilab Expert L. The diffusion of the Abrams cone was 104 cm, the expansion of the Suttard viscometer cylinder was 20 cm and the dynamic viscosity on the rotational viscometer was 1274 - 5317 cP, depending on the shear rate (Fig. 6 and Fig. 7).

The speed of the concrete mix was 0.9 m/s at the minimum speed of the concrete pump.

DOI: 10.1051/matecconf/201711601017

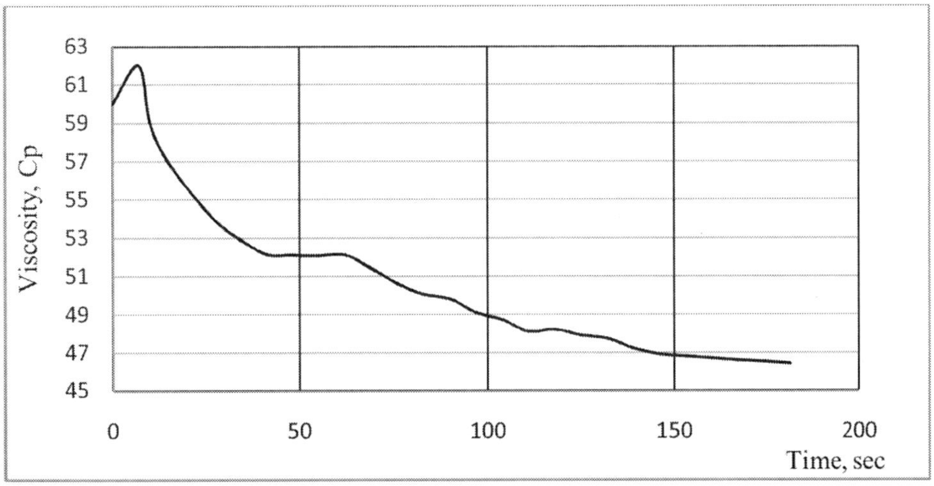

Fig. 6. The viscosity of the concrete mix at a spindle rotation speed of the viscometer 20 rpm.

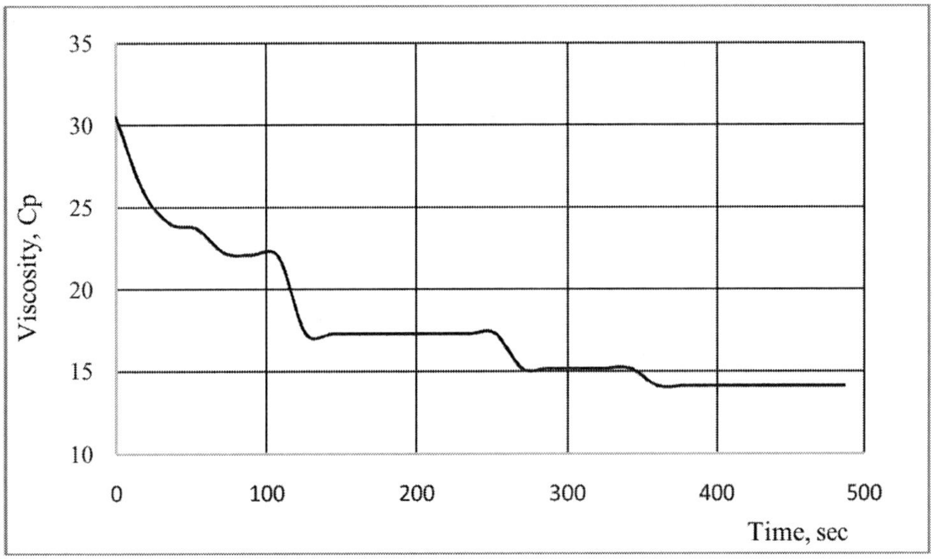

Fig. 7. The viscosity of a concrete mix at a speed of rotation of a spindle of a viscosimeter 40 – 100 rpm.

The readings of the manometer MH2 are unstable in the range of 0.0 - 0.4 MPa, the MH3 readings are more stable and are in the range of 0.0-0.1 MPa. Starting from 100 seconds the steady flow is observed according to the readings of the MH3 manometer, which coincides with the time of the mixture exit from the concrete guide. The presence of the MH2 manometer in close proximity to the concrete pump, which generates alternate translational movements of 2 pistons with a length of 1500 mm, shows the actual instability of pressure in the pipe at the very beginning (Fig. 8).

DOI: 10.1051/matecconf/201711601017

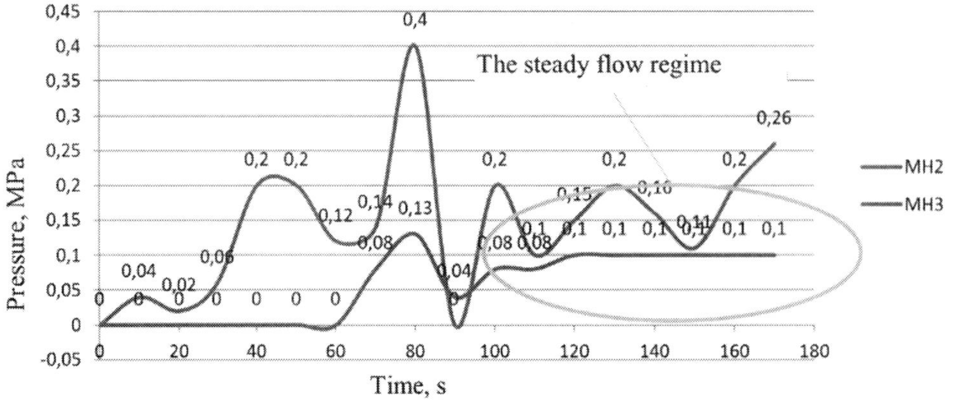

Fig. 8. Changing the pressure in the pipe in time when pouring the concrete mixture.

4 Discussion of results

In the process of injection at some point in time, the concrete pump begins to apply a constant force (Fig. 9). All the liquid in the pipe immediately come into motion can not. Just before the piston, there will be a compression region, which begins to propagate through the pipe, moving more and more distant areas of liquid inside the pipe.

The velocity of propagation of the compression region coincides with the speed of sound in this liquid. Finally, after a short period of time, a pressure distribution is established in the tube, which ensures the constancy of velocity along the tube [9]. Those. The curves MH2 and MH3 of the readings of the pressure gauges on the graph (Fig. 8) after 100 s show the already established flow regime, which determines this phenomenon.

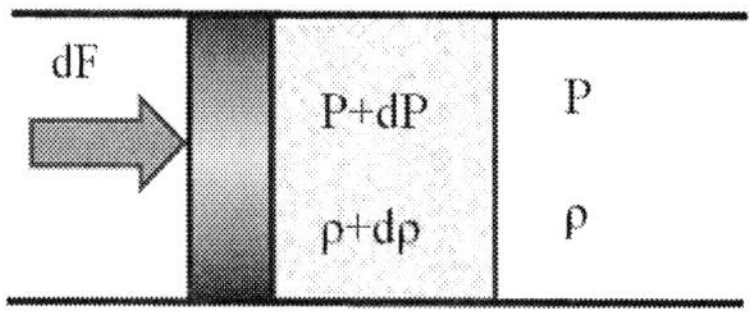

Fig. 9. Transfer of pressure from the piston.

Anyway, the pumping of concrete mixtures over long distances is associated with the use of powerful pumping equipment. However, if technological pressure losses depend on the choice of pumping equipment, then rheological and tribological depend on the composition of the mixture and, taking into account the specific features of the influence of individual components of the concrete mixture on its rheological properties, it is possible to significantly influence the pumping distance by choosing the optimal composition. Increase pumping capacity of concrete can be due to the modification of their compositions and the obtaining of viscosity and spreadability, which is beyond the scope of the classification by standards. Reducing the viscosity of mixtures should be accompanied by the preservation of uniformity.

5 Conclusions

Exclusion of a large aggregate will solve a number of problems associated with segregation of the mixture under pressure, and the introduction of organic and finely dispersed mineral additives will ensure uniformity of the mixture during pumping and high final characteristics of the concrete. The use of powdered concrete with high mobility for pumping over long distances when filling long-span metal or bridge structures, concreting in tunnels where injection pressure is limited, can be the only solution in the implementation of these tasks.

References

1. *Concrete Society. Pumping concrete*, Good Concrete Guide 2, Surrey,14 (2005)
2. P.J.M. Bartos, M. Sonebi, A.K. Tamimi, Workability and rheology of fresh concrete: compendium of tests, Report of Rilem TC 145-WSM, *Workability of special concrete mixtures*, 127 (2002)
3. D. Kaplan, *Pompage des bétons*, Ph.D.-thesis, Laboratoire Central des Pontset Chaussées, (In French), 228(2001)
4. N. Roussel, RILEM *Mater. Struct.* **39** (1), 75–83 (2006)
5. G.Tattersall, *Workability and Quality Control of Concrete* Tattersall, G London, E & FNSPON, 262 (1991)
6. O.Wallevik, *Proc 3rd IntSymp on Self-Compacting Concrete*, Reykjavik, 23-31 (2003)
7. N. Roussel, Cem. Concr. Res., **36**, 1797-1806 (2006)
8. P. F. G. Banfill, *Rheology Reviews*, 61-130 (2006)
9. Jae Hong Kima, SeungHee Kwon, Shiho Kawashima, Hong Jae Yim, Cem. Concr. Res., **77**, 60-67 (2017)

The role of chemical admixtures in the formation of the structure of cement stone

Viktor Sopov[1,*], *Lidiya* Pershina[1], *Larisa* Butskaya[1], *Ekaterina* Latorets[1], and *Olga* Makarenko

[1]Kharkiv National University of Civil Engineering and Architecture, Department of Physical and Chemical Mechanics and Technologies of Building Materials and Products, Sumska st. 40, 61002 Kharkiv, Ukraine

Abstract. The influence of sulfates and carbonates of potassium and sodium on the character of the formation of the microstructure of cement stone was studied. The role of cations in the structure formation of cement stone is shown. The efficiency of chemical additives, hardening accelerators, was estimated from the ratio of the volumes of gel and capillary micropores. The ratio of gel and capillary pores allows to determine the efficiency coefficient of the action of chemical additives. It is shown that the potassium carbonate for Portland cement is the most effective additive for hardening in terms of microstructure modification, and potassium sulfate for slag Portland cement.

1 Introduction

In predicting the durability, durability of concrete, the pore structure takes the lead. Its characteristics determine the process of destruction of concrete in the structure. In the theory of calculating the durability of concrete structures, along with other factors, the leading role of the differential pore structure has been established. The determination of these characteristics is carried out experimentally or on the basis ofsemiempirical relationships. Estimating the total porosity, these dependences do not provide answers to questions related to the shape of pores, the quantitative and probabilistic distribution of pores by size.

It is known that the most numerous and responsible for the properties of concrete is the proportion of pores - capillaries. They are permeable to water and are the reason for its penetration inside the concrete of structure, contribute to the accumulation and development of cracks. The main properties of concrete depend on the characteristics of the capillary pore structure, the formation of which begins at the early stage of cement hardening.

The process of pore formation of cement stone is much more complicated than in monomineral knitting in connection with the presence of two developing and interacting structures - hydrosulfoaluminate and hydrosilicate. The main structure-forming role is played by hydrosilicates and calcium hydroxide, which constitute the bulk of the new

* Corresponding author: vsopov1959@hotmail.com

formations of cement stone. Despite the huge morphological diversity of other hydrate neoplasms in the hardened cement stone, their effect on certain physical and mechanical properties is deterministic and can be predicted in advance. An ensemble of micropores with a corresponding distribution of their for size determines the main properties of cement stone - strength, permeability, frost resistance, etc. Changes in the pore size in the range from 2 to 100 nm can dramatically change the properties of the cement stone and thus allow to control the character of the structure formation at the hardening stage. One of the levers controlling the structure of cement stone is the use of chemical and mineral admixtures.

2 Materials and methods of research

The effect of chemical admixtures on the character of pore formation of two types of cement was studied: Portland cement PC CEM I 42.5R and slag Portland Cement II CEM II B-S 32.5 R, differing in mineral composition, slag content and, as a result, the rate of hydration and character of structure formation (Table 1).

Table 1. Cement characteristics

Showing		**CEM I 42,5R**	**CEM III/B-S 32,5 R**
Actual Activity, MPa		52.0 (28 days)	23.7 (7 days)
Activity with the steaming, MPa		32.0	25.2
Fineness of grinding cement (passed through a sieve 008), %		91.0	91.0
Setting time, hr-min: - start - end		3-20 4-20	1-50 7-00
In tests on river sand: normal density		28.0	28.1
Signs of false setting of cement		no	no
Mineral composition, %	C_3S C_2S C_3A C_4AF	57.48 21.36 7.57 12.35	50.0 30.0 3.8 16.2

To estimate the parameters of the microstructure of the cement stone (dimensions, volume and pore size distribution), the thermoporometry method was used [1]. The experimental basis of the method is differential scanning microcalorimetry. To determine the amount of free and adsorbed water in the cement stone, it is frozen in a calorimeter and phase transitions of water are recorded, the presence and intensity of which are determined by the pore size and free water content in them [2]. The relationship between the pore radius r and the crystallization temperature of the pore water T is expressed by an equation of the type:

$$r = \frac{A}{T_0 - T} + B, \quad (1)$$

where A and B are constants, depending on the experimental procedure, $T_0 = 273$ K. In this case, the thickness of the layer of adsorbed water is taken into account, which is 2.5 monolayers (8Å) according to [3].

The pore volume is determined from the equation of the pore size distribution curve

$$\frac{\Delta V}{\Delta r} = (a \cdot r^2 \exp^{-br^2})^m, \quad (2)$$

where *a, b* – the coordinates of the maximum propagation curve, *m* – width of the distribution curve at 10% of its maximum amplitude. The volume of pore space calculated according to formula (2) corresponds to the volume of the pore liquid, including free and adsorbed water.

3 The results of research

The hardening of all types of cement for 1 day is accompanied by the formation of a pore structure with two areas of distribution of micropores in size, characteristic for gel and capillary micropores (Fig. 1). For Portland cement, they are located in the range 2.2…22 nm with pore distribution maxima at 2.36, 2.45 and 14.6 nm. For slag Portland cement, the range of distribution of micropores narrows to a range of 3.1 ... 6.6 nm with maxima at 3.4, 3.7 and 6.3 nm (Fig. 1a). The volume of micropores increases by 1.3 times in comparison with portland cement (Fig. 1b) and is 0.46 cm^3 / g. This is due to a slower binding of mixing water during hydration of the SHPC, the formation of hydrogelenite and hydrogarnets as well as calcium hydrosilicates of less basicity than with hardening portland cement. The consequence of this is a reduction in the size of capillary micropores and the formation of low-density LD C-S-H with a porosity in the range 3.1 ... 4.2 nm (Fig. 1a).

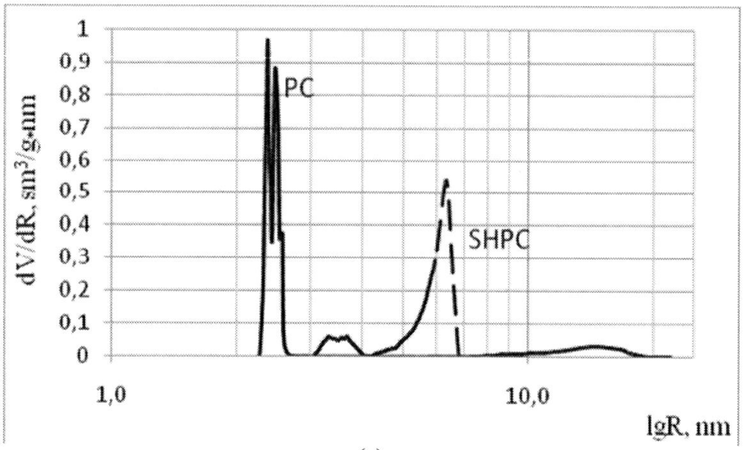

(a)

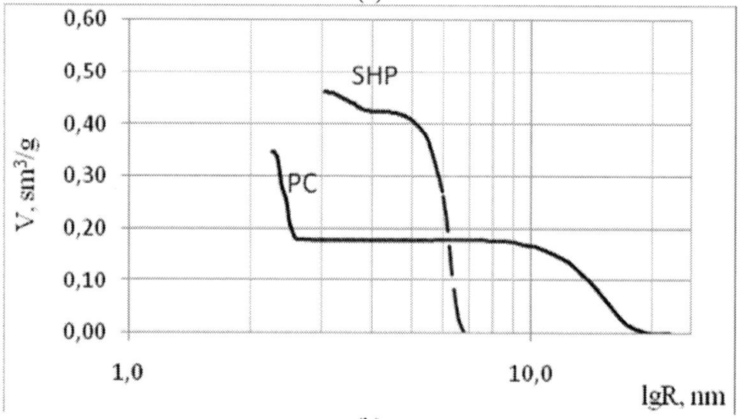

(b)

Fig. 1. Influence of cement type on microporosity of cement stone: (a) pore size distribution; (b) the total pore volume.

The gel microporosity for the SHPC is only 7.6% of the total porosity, the analogous index for the PC is 47%. Consequently, the amount of calcium hydrosilicates formed in the cement stone composition is 40% greater than that of the SHPC for the same hardening time.

The effect of hardening accelerators is based on a change in the solubility of the initial astringent and final products of its hydration due to a change in the ionic strength of the solution [4].

Admixtures that contain ions of the same name with astringents contribute to the formation of embryos of crystalline hydrate neoplasms and at low concentrations they reduce the solubility of astringent and hydrate neoplasms. As the concentration of additives increases to a certain limit, their effect increases somewhat. With an increased concentration of these additives, it is possible to react with calcium hydroxide to form double salts. As a result, the solubility of $Ca(OH)_2$ and the silicate minerals of cement clinker is increased. By varying the solubility of astringent and hydrate neoplasms, it is thus possible to regulate the kinetics of cement hydration.

Admixtures that do not contain the same name with astringent ions, at low concentrations [5] accelerate hardening, and at large, the opposite effect is possible.

The most effective hardening accelerators are alkali metal salts based on Na^+ and K^+ cations with various anions of CO_2^{3-}, Cl^-, SO_2^{4-}, $S_2O_2^{3-}$ and CNS^-, etc.

Effect of sulphates. Sodium and potassium sulphatesare classified as admixtures of the first class [4]. The mechanism of their action is due to the reaction with calcium hydroxide, released during the hydration of cement, with the formation of a two-water gypsum:

$Ca(OH)_2 + Na(K)_2SO_4 + nH_2O = CaSO_4{\cdot}2H_2O + 2Na(K)OH + H_2O.$

The resulting fine-grained gypsum reacts with cement and promotes nucleation and growth of neoplasms. Most effectively as an accelerator, sulfates show themselves on Portland cement and Pozzolan Portland cement. However, in the opinion of Yu.M. Bazhenov [6], sodium sulfate negatively affects the long-term strength of the cement stone.

In addition, Na_2SO_4 hydrolysis retains sodium Na^+ cations in the pore fluid, and the anionic part - $SO_4^{2-}$is bound by alumina-containing phases. Isolation of cations of Na^+ leads to a change in the ionic strength of the solution, i.e. increasing the alkalinity of the medium, causing an increase in the solubility of the silicate constituents and accelerating the hardening [7]. Anions SO_4^{2-} contribute to the formation of sodium-containing hydrosulfoaluminates and calcium hydrosulfoaluminoferrites [8].

The effects of sodium and potassium sulfate on the formation of microporosity during Portland cement hydration are shown in Fig. 2.

When Portland cement is solidified with 2% Na_2SO_4, two pore distribution regions are formed: 2.5...3.3 nm and 5...16.5 nm with maxima at 2.7 nm and 10.5 nm, respectively. The total pore volume increases by 1.2 times as compared with the sample without the admixture and is 0.41 cm^3/g. This character of pore formation of cement stone is associated with a change in the balance between the pores already on the first day of hardening. Sodium sulfate, accelerating the hydration of cement, promotes the rapid formation of capillary micropores in the range 5.5...14.6 nm with a total volume of 0.38 cm^3/g (in the control sample 0.21cm^3/g).

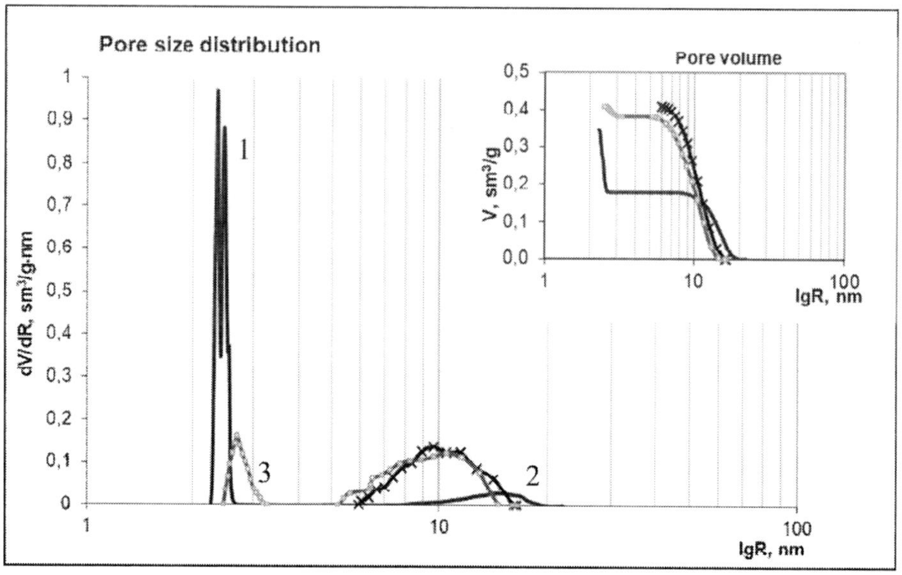

Fig. 2. Effect of sulfate admixtures on the microporosity of the cement stone 1 – PC; 2 – PC+Na_2SO_4; 3 – PC+K_2SO_4.

When Portland cement hydrates in the presence of 2% K_2SO_4, one region of the pore distribution is formed in the interval 6...14.2 nm with a maximum at 10.5 nm, which corresponds to the region of the capillary pores. The total volume of micropores also increases by 1.2 times and corresponds to 0.41 cm^3/g. Potassium sulphate activates the formation of monosulfoaluminate and ettringite, which leads to stresses and deformations leading to a decrease in the strength of the cement stone, an increase in the volume of the solid phase, and the size of micropores.

In the process of hardening slag Portland cement without admixtures, two regions of pore size distribution are formed in the range of 2.9 ... 6.8 nm (Fig. 3). The admixture of Na_2SO_4 contributes to the formation of three regions of the distribution of pores in the range: 1.79...17.8 nm with maxima at 1.9, 2.4 and 8.5 nm. The first two areas refer to the porosity of C-S-H, and the third - to the capillary pores.

This pattern of pore size distribution confirms the formation of high-density calcium hydrosilicates - HD C-S-H (1,75...2,8 nm) in the presence of sodium sulfate. Our data coincide with the results obtained by Ali Noaman Khalid Hussein [9] - two types of calcium hydrosilicates of the composition C-S-H (I) and C-S-H (II) are formed in the hardening slagPortland cement in 1 day. At 7 days, the hydrosilicates have the composition C-S-H (I) and C_2SH (B), the hydrosilicate composition on the 14th day is mainly CSH (II), which also indicates the presence of two types of gel microporosity related to different types of calcium hydrosilicates.

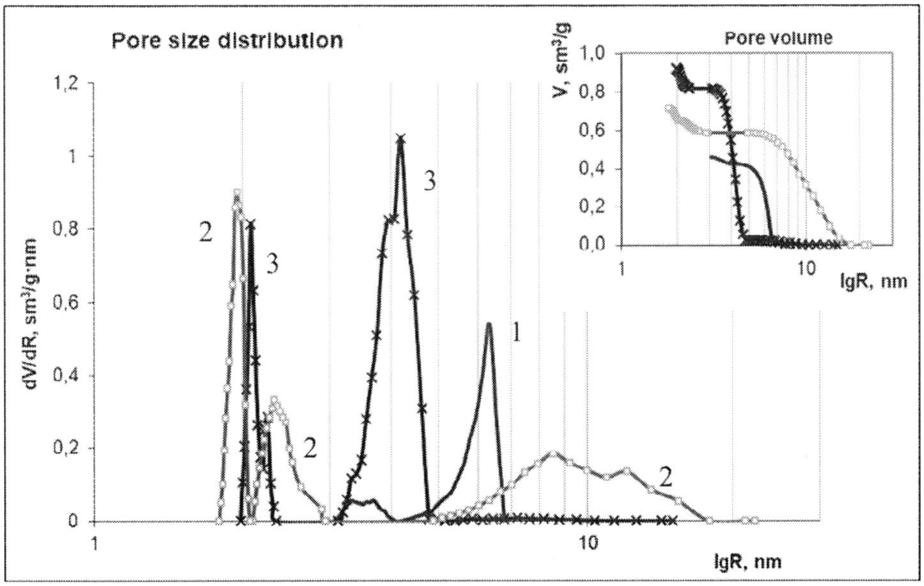

Fig. 3. Microporosity of cement stone with solidification of slag Portland cement with admixtures of sulfates: 1 – SHPC; 2 – SHPC+Na_2SO_4; 3 – SHPC+K_2SO_4.

When hydrating the slagPortland cement with K_2SO_4, two microporous regions are formed in the range of 2...2.5 nm and 3.6...6.6 nm with distribution maxima at 2.3 nm and 5.2 nm. The microstructure of the slag-portland cement stone is represented mainly by the C-S-H phase with the ratio C/S = 1/1.5. The cations Na^+ and K^+can be embedded in such a structure, which increases the strength and resistance to chemical corrosion.

In the opinion of V.B. Ratinov in the presence of additives of sodium and potassium sulfate, new hydrate compounds - sodium-containing hydrosulfoaluminates and calcium hydrosulfoalumferrites - are formed [4]. The reaction products are formed directly in the porous space of the cement stone, filling the pores, which leads to a decrease in their size and volume. O.M. Rosenthal showed that in the presence of Na^+ and K^+ ions alone, the solubility of aluminates increases and the process of their interaction with sulfates develops in the bulk of the liquid phase of concrete with pore filling by crystalline products [10].

Differences in the effect of sulfates can be explained by the fact that for equal valence of compounds, the intensity of their effect on the ionic strength of the solution is higher for Na_2SO_4. This leads to an increase in the solubility of silicate phases of cement, acceleration of its hydration, and an earlier formation of gel microporosity.

Effect of carbonates. In the opinion of V.B. Ratinov and F.M. Ivanov carbonates of potassium and sodium react with calcium hydroxide to form KOH, which will accumulate in the aqueous phase of the solution and contribute to a change in the composition of the pore fluid [11]. This leads to a change in the pH of the medium and accelerates the hydration of the cement.

Hydration of portland cement with admixture of Na_2CO_3 generates two regions of pore distribution in the intervals 2.8...3.22 nm and 4.3...14.5 nm with distribution maxima at 3.02 and 7.5 nm, respectively (Fig. 4). When solidified with K_2CO_3, one region of pore distribution is formed in the range 4...12.4 nm with a maximum at 7.3 nm. The total pore volume when these admixtures are added increases 1.5 times with hydration with Na_2CO_3 and 1.3 times with K_2CO_3.

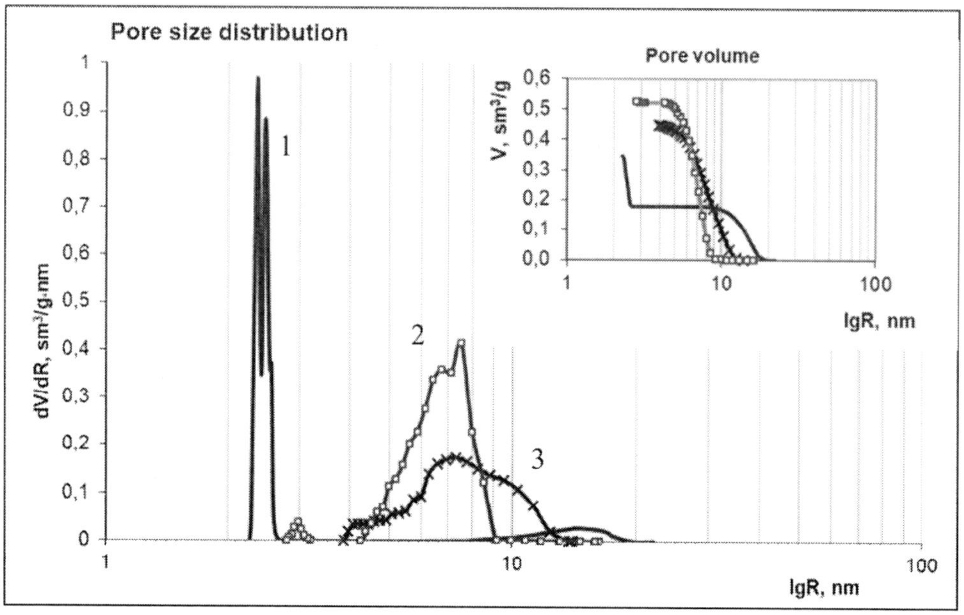

Fig. 4. Effect of carbonate admixtures on the microporosity of cement stone: 1 – PC; 2 – PC+Na_2CO_3; 3 – PC+K_2CO_3.

The accelerating effect of carbonates on the hydration of Portland cement is due to a decrease in the size of capillary pores with simultaneous growth in their quantity. The introduction of Na_2CO_3 contributes to the formation of only a small amount of gel micropores, the total volume of which is 55 times less than that of a sample without an admixture.

Solidification of slag Portland cement in the presence of K_2CO_3 and $Na_2CO_3$is accompanied by the formation of a discrete distribution of micropores in the intervals 3.4...14.5 nm and 2.24...12.4 nm, respectively (Fig.5). In slag Portland cement stone with the addition of potassium carbonate, there is practically no gel micropores. The introduction of sodium carbonate leads to the growth of gel micropores, whose volume is 10% greater than that of the control sample. But at the same time high-density calcium hydrosilicates HD C-S-H (2.2...3 nm) are formed, which has a positive effect on improving the properties of cement stone and concrete.

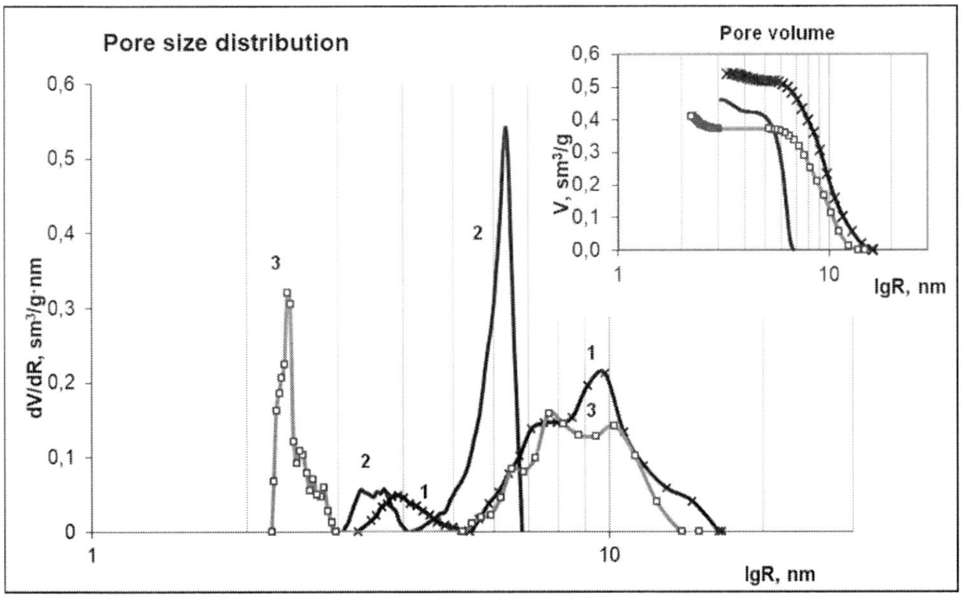

Fig. 5. Effect of carbonate admixtures on the microporosity of cement stone: 1 – SHPC; 2 – SHPC+Na_2CO_3; 3 – SHPC+K_2CO_3.

From these data it follows that the addition of Na_2CO_3 contributes to the early formation of gel porosity. Its formation in the opinion of V.B. Ratinov and T.I. Rosenberg [4] can be associated with the formation of gel-like substances such as sodium hydrosilicates, which provide a more dense structure, which positively affects the physical properties of concrete.

4 Discussion of results

Based on the obtained data on the effect of hardening accelerators in the early stages of hydration of cements and the formation of a microstructure of a cement stone, the following lyotropic series were obtained:

- for Portland cement - K_2CO_3>K_2SO_4>Na_2SO_4>Na_2CO_3;
- for slag Portland cement - K_2SO_4>Na_2SO_4> Na_2CO_3> K_2CO_3.

The ratio of gel and capillary pores allowed to determine the coefficient of efficiency of K_{ef} in the form of the following ratio of dynamic microstructure indices [12]:

$$K_{ef} = -\lg\left(\frac{|V_{cap1} - V_{gel1}|}{V_{tot1}} \cdot \frac{V_{tot0}}{|V_{cap0} - V_{gel0}|}\right), \quad (3)$$

where V_{gel} is the volume of gel pores; V_{cap} is the volume of capillary pores, V_{tot} is the total volume of pores, index 1 corresponds to cement with admixtures, 0 – without admixture.

The results of the efficiency evaluation are shown in Fig. 6.

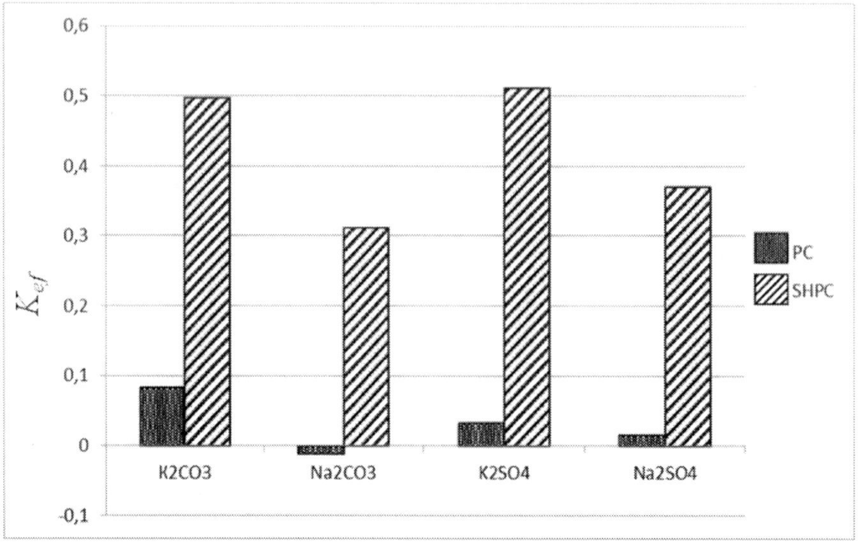

Fig. 6. Efficiency of hardening accelerators.

As can be seen from Fig. 6, for Portland cement, potassium carbonate is the most effective accelerator, and potassium sulfate for Portland slag. It should be borne in mind that the coefficient values are less than zero indicate a negative effect of the additive, at $K_{ef} = 0$ - the additive has practically no effect on the structure formation of the cement stone.

5 Conclusions

Studies of the influence of sulfates and carbonates of potassium and sodium showed that the character of the formation of the microstructure of the cement stone depends to a large extent on the type of cation. Evaluation of the relationship between capillary and gel pores showed that the potassium cation promotes the increase in gel porosity with a decrease in capillary porosity. It has been shown that potassium carbonate for Portland cement is the most effective additive for hardening from the point of view of microstructure modification, and potassium sulfate for slag Portland cement.

References

1 M. Brun, A. Lallimand, J.F. Quinson, and C. Eyraud, Thermochimica Acta., **21**, 59-88 (1977)
2 A.V. Usherov-Marshak, V.P. Sopov, Colloid Journal, **56 (4)**, 600-603 (1994)
3 K.G. Krasilnikov, A.F. Tarasov, Physical and chemical research of concrete and its components, **17**, 100 (1975)
4 G. Dobrolyubov, V.B. Ratinov, T.I. Rozenberg, *Forecasting the durability of concrete with admixtures* (Stroiizdat, Moscow, 1983)
5 V.S. Ramachandran (ed.) *Concrete admixtures handbook* (New Jersey, Noyes Publications, 1984)
6 Yu.M. Bazhenov, V.C. Demyanova, V.I. Kalashnikov, *Modified high-quality concrete* (Publishing House of Association of Construction Universities, Moscow, 2006)
7 D. Manmohan, P.K. Mehta, Cement, Concrete and Aggregates, **3**, 63-67 (1981)
8 W. Kurdowski, *Chemia cementy i betonu* (Wyd. Naukowe PWN, Warszawa, 2010)

9 Ali Noaman Khalid Hussein, *Research hydration and hardening cement stone first condensation moisture*. Thesis for the degree of candidate of technical sciences (Dnepropetrovsk, 2001)
10 O.M. Rozental, F.E. Chetin, *Multilayer structural ordering in heterogeneous processes of ice formation* (Publishing House of the Pedagogical Institute, Sverdlovsk, 1974)
11 V.B. Ratinov, F.M. Ivanov, *Chemistry in construction* (Stroyizdat, Moscow, 1969)

Influence of additives on flexural strength of concrete

Sergii Tolmachov[1,*], *Olena* Belichenko[1], and *Denis* Zakharov[1]

[1]Kharkov National Automobile und Highway University, Department of the Technology of Road Building Material and Chemistry, Yaroslava Mudrogo st. 25, 61002 Kharkov, Ukraine

Abstract. In earlier studies, the effect of chemical and mineral additives on the compressive strength of concretes was considered, but little attention was paid to the flexural strength of concrete. However, monolithic concretes of transport purpose, including road concretes, operate under conditions of tensile loads, which lead to their destruction. Therefore, it is actual to analyze the effect of superplasticizers, especially carboxylate type, as well as mineral additives and their complexes on the flexural strength of concrete. The article shows the results of studies for determining the influence of a superplasticizer of a carboxylate type, microfiller and fiber on the flexural strength of concrete. The influence of the time for maintaining a concrete mixture on the physical and mechanical properties of concrete is shown. The article shown that the addition of the mineral additive into the concrete mixture leads to an increase of flexural strength to 13 %. The use of an organomineral complex leads to an increase the early flexural strength of concrete to 37 %, and at the age of 28 days - to 20 %. The use of the complex of additives and polypropylene fibers results in an insignificant increase of the flexural strength in comparison with concretes containing only a complex of additives.

1 State of the issue

1.1 Analysis of studies and publications

Numerous studies carried around the world have shown the high efficiency of using superplasticizing additives in the technology of heavy cement concretes [1 – 4]. Their use makes it possible to change the nature of the colloid-chemical processes occurring in hardening cement binders. This fundamentally changes the properties of concrete mixtures and the safety of these properties in time. Essentially the properties of concrete are changing, for example, their physical and mechanical characteristics are improved, durability is increased. This is typical for all types of superplasticizers that are currently used. For example, purified lignosulfonates and their complexes with superplasticizers of other types. And also naphthalene- formaldehyde or melamine-formaldehyde superplasticizers, polyacrylate and polycarboxylate superplasticizers. The appearance of polycarboxylates and their wide use made it possible to return to the technology of heavy

* Corresponding author: Tolmachov.serg@gmail.com

concrete with dispersed mineral additives, the use of which was widely studied in the 70-80 s of the last century [5 – 7].

These include additives such as metakaolin and microsilica. The complex of superplasticizer + mineral additive got the name organomineral complex. The use of these complexes alters the rheological properties of concrete mixtures, reduces their delimitation, and increases the density. This also leads to an increase in the strength properties of concrete, however, does not always improve their performance over time. This particularly applies to road concrete and transport concrete, which operate under difficult operating conditions.

In earlier studies, the effect of chemical and mineral additives on the compressive strength of concretes was considered, but little attention was paid to the flexural strength of concrete. However, monolithic concretes of transport purpose, including road concretes, operate under conditions of tensile loads, which lead to their destruction. Therefore, it is actual to analyze the influence of superplasticizers, especially carboxylate type, as well as mineral additives and their complexes on the flexural strength of concrete.

It is known that to increase the tensile strength of concrete, including the flexural strength, you can add fiber to their composition. It is also known that the use of polymeric or basalt fiber leads to a decrease of the compression strength of concrete [8 – 10]. Data on the effect of polypropylene fiber on the flexural strength of concrete are contradictory. In some articles it is shown that fiber increases flexural strength by 10...15 %, in others – by tens of percent. There is no information about the joint effect of organomineral complexes and fiber on the flexural strength of concrete.

1.2 Relevance

This state of the question makes research actual about the effect of superplasticizers of the carboxylate type, mineral additives, polypropylene fibers and their complexes on the flexural strength of monolithic concretes.

2 Materials and methods of research

Were used: the cement of the PC I-500-N (CEM I 52.5N) of the Ivano-Frankivsk plant, crushed stone fr. 5-10 mm, river sand was used.

In concrete mixtures were added:

- chemical additive - superplasticizer of carboxylate type Sika Plast 2508 (SP) produced by the Swiss company Sika (1 % of the mass of cement);
- mineral additive "Universal VM" (MA) produced by the Ukrainian Zaporozhye firm "Budindustria" (7 % of the mass of cement);
- polypropylene fiber (F) produced by the Ukrainian Dnepr company "Spetssnab" (1 kg per 1 m^3 of concrete mix).

The mixture of concrete: cement - 350 kg/m^3, sand - 600 kg/m^3, rubble - 1300 kg/m^3.

The flexural strength of concrete was studied on the samples-beams measuring 4x4x16 and 7x7x28 cm. The compressive strength of concrete was studied on samples-cubes with a size of 7x7x7 cm.

3 Experimental research

Investigations of the effect of various additives on the strength of concrete were carried out on equally mobile concrete mix in the Table 1.Studies have shown that when the mineral additive is added into the concrete, strength increases, but for compressive strength and flexural strength this increase is different. The compressive strength of concrete with a mineral additive increases by only 3...6 % at different times of hardening. At the same time, the flexural strength of concrete with a mineral additive increases by 18, 10 and 8 % on 3, 7 and 28 days, respectively.

Table 1. Strength of concretes with additives

Type of additive	W/C	Flexural strength at the age, days, R_{fs}, MPa			Compressive strength at the age, days., R_{cs}, MPa		
		3	7	28	3	7	28
Without additive	0.5	5.1	6.0	7.0	32.3	39.1	51.2
MA	0.5	6.0	6.6	7.9	34.4	40.0	53.0
CP + MA	0.42	7.0	7.6	8.4	44.4	53.9	68.3
CP + MA+ F	0.43	6.9	7.9	8.8	40.1	50.4	61.7

* the mobility of concrete mixes corresponded to grade P1

This increase of flexural strength, relative to the increase in compressive strength, may indicate that the filler microparticles compact and harden, in the first place, the contact zone "hardening cement paste - aggregate" due to the formation of low-basic hydrosilicates. This zone is responsible for the tension strength of concrete, its hardening leads to an increase of the flexural strength. It should also be noted the tendency to reduce the growth of the flexural strength with increasing hardening time. This can be caused by gradual hardening, both the contact zone and the hardening cement paste both in concretes without additives and with the addition of a microfiller.

The introduction of the organomineral complex leads to an increase of the compressive strength of concrete by 38...34 %, compared to the composition without additives at all times of hardening. The flexural strength increases by 39, 25 and 11 % at the age of 3, 7 and 28 days, respectively. This confirms the information obtained earlier. The increase of compressive strength is due to a reduction in the water content of concrete. There is also a gradual decrease of growth of flexural strength caused by the gradual strengthening of the contact zone in time, regardless of the presence or absence of additives.

The results of our experiment show that in concretes containing an organomineral complex + polypropylene fiber there is a decrease of the compressive strength of concrete, in comparison with the strength of concretes that do not contain fiber. This decrease is 11, 7 and 11 % at the ages of 3, 7 and 28 days of hardening.

At the same time, despite a slight (not more than 3 %) reduction of flexural strength at the age of 3 days, concretes containing an organomineral complex and a fiber, have a higher flexural strength at the age of 7 and 28 days of hardening compared to concrete that does not contain fiber. The difference in strength is 5 and 9 %. This can be explained by the fact that polypropylene fiber reinforces the intergranular space of concrete, which leads to an additional increase of the flexural strength of concrete, in comparison with the effect of only the organomineral complex.

Also, studies were carried out for the effect of maintaining time of the concrete mixture after preparing on the strength of the obtained samples in Table 2.

Concrete mixtures were made and the slump of the concrete mix was measured. After that, samples of concrete were made. Then mixture residue was cured for 45 minutes, then mixed and the slump was measured again, and the samples were also molded.

The same operation was repeated after 90 minutes of maintaining the mixture.

The experiment results showed that the compressive strength of the concrete samples without additives that were made immediately after the making of the mixture was lower than the samples that were made after 45 minutes of maintaining the mixture. At the age of 3 and 7 days, this difference was up to 5 %. It can be noted that by the 28 days of hardening the compressive strength of both samples is the same. The strength of the samples made of concrete mixtures aged 90 minutes is less by 2...4 % compared to the samples made immediately after the preparation of the mixture. This can be explained by the effect of "re-mixing" or "re-vibrating" [11].

Table 2. Influence of the maintaining time of the concrete mixture on the strength of concrete

Measurement time, min	Additive	Slump of concrete mix, sm	Flexural strength at the age, days, R_{fs}, MPa			Compressive strength at the age, days, R_{cs}, MPa		
			3	7	28	3	7	28
5 min	-	8.5	5.2	6.4	7.3	28.3	34.8	44.2
	CP	7	6.6	7.7	8.0	39.7	47.8	61.8
45 min	-	4	5.5	6.6	7.4	29.2	36.7	44.6
	CP	5	6.3	7.5	8.4	43.5	50.4	63.0
90 min	-	1	5.4	6.1	7.4	27.4	34.3	42.5
	CP	4	6.1	7.2	8.2	42.2	48.3	61.7

*Initial W/C in a concrete mix without the additive was 0.55, with the additive – 0.46

It is known that in the process of cement hydration, an unstable, mainly hydroaluminous structure of cement stone is formed. After a certain time of maintaining the mixture and then applying vibration or mixing, this structure breaks down. As a result, a large number of additional crystallization centers appear in the volume of the hardening cement paste. This helps to accelerate the growth of strength, especially in the early stages of hardening. But applying such external influences is necessary at a strictly certain moment, otherwise, this may lead to the opposite result. The decrease in strength can be observed in samples from a concrete mix that has been aged for 90 minutes. The small difference in the strength of the samples from a mixture aged 45 minutes, as compared to the samples that were made immediately after the mixture was made, can also be explained by too late applying of a re-mixing of the mixture than was necessary.

The compressive strength of samples made from mixtures with a superplasticizer and kept for 45 minutes is higher than for samples prepared immediately after the preparation of the mixture. It is 10 and 5 % at the age of 3 and 7 days. But by 28 days of hardening, this difference is reduced to 2 %. The growth of strength between the samples molded from the mixture, which was kept for 90 minutes and molded immediately is much less: 6 and 1 % on days 3 and 7. A small increase in strength can be explained by the retarding effect of the action of carboxylates at the initial stage of hardening.

The flexural strength of samples of concrete without additives that were made at different times is practically the same in all observed hardening periods. In concrete containing the additive of the superplasticizer, it is possible to observe a decrease in flexural strength in samples that were made from pre-aged concrete mixtures. The difference in strength was 7...8 % compared to the strength of the specimens molded immediately. This can be explained by the fact that the zone of contact "hardened cement paste - aggregate" is also destructed during the re-mixing. Since in the initial period of hydration of cement because of the presence of a superplasticizer, the rate speed of formation of the structure in the contact zone slows down, then this deceleration has a greater effect on flexural strength.

4 Conclusion

1. It is shown that adding a mineral additive into the concrete mixture leads to an increase of flexural strength to 13 %, while the compressive strength increases by no more than 6 %.

2. The use of an organomineral complex leads to an increase the early flexural strength of concrete to 37 %, and at the age of 28 days - to 20 %. At the same time, the compressive strength of concrete increases to 37 % during all test periods.

3. The use of the complex of additives and polypropylene fibers results in an insignificant increase of the flexural strength in comparison with concretes containing only a complex of additives – up to 4 %. The compressive strength of such concretes is reduced to 11 %.

4. It has been established that keeping the concrete mixture for 45 to 90 minutes before starting the molding of the samples slightly changes the flexural strength and compressive strength of concrete without additives. But such maintaining primarily leads to a reduction of the flexural strength of concretes containing a superplasticizer to 8 %.

References

1. P.C. Aitchin, *High Performance Concrete* (London: T&F e-Library, 2004)
2. H. Nakanishi, S. Tamaki, ACI SP, **217**, 327-342 (2003)
3. V.S. Izotov, R.A. Ibragimov, Technologies of concrete, **3-4**, 34-34 (2012)
4. S.N.Tolmachev, Highways, **12 (1009)**, 54-57 (2015)
5. V.I. Kalashnikov, E.V. Gulyaeva, D.M. Valiev, V.M. Volodin, A.V. Khvastunov, Construction Materials, **11**, 44-47 (2011)
6. V.R. Falikman, Concrete and Reinforced Concrete, **1**, 48-54 (2011)
7. S.N. Tolmachev, E.A. Belichenko, Bulletin of ODABA, **61**, 392-397 (2016)
8. Milind V. Mohod, IOSR-JMCE, **12**, 28-36 (2015). DOI: 10.9790/1684-12112836/
9. S.N. Tolmachev, E.A. Belichenko, Building materials and products, **1 (91)**, 76-79 (2016)
10. F.N. Rabinovich, *Composites based on dispersed-reinforced concrete. Questions of theory and design, technology, design [Monograph]* (Moscow, ASV, 2004)
11. Kh.V. Abazov, Road concretes on sand-gravel aggregates from re-blended mixtures. Cand. Diss. (Engineering). Kharkov. 143 (2003)

Ecological indices of manufacture of Portland cement clinker and production of the dolomite clinker

Varvara Vinnichenko[1,*], and *Alexander* Ryazanov[2]

[1]Kharkiv National University of Civil Engineering and Architecture, Department of Mechanization of Construction Processes, Sumska st. 40, 61002 Kharkiv, Ukraine
[2]Ufa State Petroleum Technological University, Department of Building Constructions, Pervomayskaya st. 14, 450062 Ufa, Russia

Abstract. It is shown that the production of dolomite clinker in comparison with that of Portland cement is environmentally appropriate. When calcining dolomite for cementitious binder, the pollution of the atmosphere by carbon dioxide is reduced due to its isolation during decarbonization reactions of calcium carbonates. Reducing fuel consumption for clinker burning provides less carbon dioxide emissions from combustion products. Reducing the firing temperature creates obstacles to the formation of nitrogen oxides. The production of binders from dolomite in comparison with the production of Portland cement helps to protect the environment from contamination

1 Introduction

The main direction of industrial development is the reduction of energy costs for the production of products with a simultaneous decrease in the load on the environment. Dolomite rocks are one of the most widespread and insufficiently mastered by the construction industry varieties of mineral raw materials [1]. They can be used for the production of various types of binders and building materials based on them [2 -4]. One of the advantages of such binders is the significantly lower energy costs for firing, compared to the production of Portland cement [5].

2 Results and discussion

2.1 Emissions of CO_2 into the atmosphere from the raw mix

Ecologists warn that if it is not possible to reduce the emission to the atmosphere of carbon dioxide, then our planet is waiting for a catastrophe [6], associated with an increase in temperature due to the so-called greenhouse effect. This phenomenon is explained by the

* Corresponding author: vvinnichenko@ukr.net

fact that carbon dioxide gently slides the sun rays to the Earth, and the heat radiation from the Earth into the space complicates what constitutes the "greenhouse effect".

Carbon dioxide CO_2 is released as a result of decomposition of chemical compounds of calcium and magnesium carbonates, which are part of mixtures for the production of cement clinker. When calcining dolomite for the production of a cement dolomitic binder, the main firing products are magnesium oxide MgO, carbon dioxide CO_2 and calcium carbonate $CaCO_3$. Decarbonization of calcium carbonate is not carried out, since the firing temperature is kept below the start of the decomposition of $CaCO_3$. The process proceeds according to the reaction [5]:

$$CaMg(CO_3)_2 = CaCO_3 + MgO + CO_2 \tag{1}$$

When roasting Portland cement raw meal, calcium and magnesium carbonates dissociate completely to form calcium oxide, magnesium oxide and carbon dioxide by reactions

$$CaCO_3 = CaO + CO_2 \tag{2}$$

$$MgCO_3 = MgO + CO_2 \tag{3}$$

When calcining dolomite by reaction (1), carbon dioxide is released only as a result of the reaction (3) of the decarbonization of magnesium carbonate. The burning of Portland cement clinker is accompanied by the release of carbon dioxide from decomposition of calcium carbonate and magnesium carbonate according to reactions (2) and (3). The yield of chemically bound carbon dioxide from magnesium carbonate and calcium of raw materials per 1 ton of clinker is determined by the formula [7]:

$$G_{CO_2} = \frac{G_c^T \cdot MgO \cdot M_{CO_2}}{M_{MgO} \cdot 100} + \frac{G_c^T \cdot CaO \cdot M_{CO_2}}{M_{CaO} \cdot 100}, \tag{4}$$

where G_c^T – the theoretical consumption of dry raw materials, t;

M_{CO2} M_{MgO} M_{CaO} – the molecular weight of carbon dioxide, magnesium oxide and calcium, respectively, t;

MgO, CaO – the content of magnesium oxide and calcium in the feed mixture, %.

Table 1 presents the chemical composition of the dolomite dropping of Dokuchaevsk Flux and Dolomite plant and the raw mix of cement clinker according to [8]. As follows from Table 1, the content of calcium and magnesium oxides in the cement raw meal is higher than in the dolomite. If we compare the release of carbon dioxide from a Portland cement raw meal – 0.565 t / t of clinker [7] and dolomite – 0.340 t / t of clinker, then the reduction in CO_2 output from the material during calcination of dolomite will be 60%.
Captions should be typed in 9-point Times. They should be centred above the tables and flush left beneath the figures.

Table 1. Chemical composition of raw mix

Material type	**LOI**	**SiO_2**	**Fe_2O_3**	**Al_2O_3**	**CaO**	**MgO**	**Other**	**∑**
Dolomite	45.92	2.15	0.60	0.52	29.83	20.53	0.45	100
The raw mix of cement clinker	38.00	12.92	1.13	3.09	66.19	1.41	0.09	100

2.2 Emissions of CO_2 into the atmosphere as a result of combustion of fuel

Theoretical costs of thermal energy for the formation of dolomite clinker represent the cost of heat for the dissociation of $MgCO_3$, a certain amount of $CaCO_3$, and also the dehydration of impurities [5]:

$$q_{kl} = G^{C}_{CaCO_3} \cdot (\Delta H_t)_1 + G^{C}_{MgCO_3} \cdot (\Delta H_t)_2 + G^{C}_{AS_2H_2} \cdot (\Delta H_t)_3, \tag{5}$$

where $G^{C}_{CaCO_3}$, $G^{C}_{MgCO_3}$, $G^{C}_{AS_2H_2}$ – the content of calcium carbonate, magnesium carbonate and aluminosilicates in the dolomite per 1 kg of clinker, respectively;
$(\Delta H_t)_1$, $(\Delta H_t)_2$, $(\Delta H_t)_3$ – the enthalpy of decarbonization reactions of calcium carbonate, magnesium carbonate, aluminosilicate impurity, respectively.

Calculated by this formula, the value of the theoretical cost of heat energy for the firing of dolomite, which is presented in Table 1, is 225 kcal/kg of clinker. For comparison, the theoretical heat consumption for sintering the Portland clinker of Kramatorsk Cement Plant is 425 kcal / kg of clinker. The actual fuel consumption for firing depends on the moisture content of the raw material and the coefficient of efficiency of the thermal unit. In other words, if the initial moisture content of the raw mix and the efficiency of the furnace are equal, the fuel consumption for calcining the dolomite clinker is 47% lower compared to the Portland cement clinker. During the fuel combustion most of the carbon is converted to carbon dioxide. The formation of CO_2 is directly proportional to the amount of fuel burnt, so the emission of carbon dioxide into the atmosphere during the combustion of fuel will be reduced. Therefore, the production of dolomite clinker in comparison with the production of Portland cement clinker represents a lower environmental risk for the environment in terms of carbon dioxide emissions.

We conducted experimental studies of dolomite clinker and cement [8, 9, 10]. The studies identified rational firing temperature and characteristics of the resulting material. The firing temperature is 700 – 780 ^{0}C. The samples were formed in two ways: with sand (cement in a ratio of 1: 3) and without sand. The closure was carried out with a solution of magnesium chloride six-water diluted with water to a density of 1.2 g/cm^3. When using a mixture of sand the compressive strength of the samples was 22.5 MPa in 28 days. The compressive strength of the samples after 28 days of hardening obtained by molding without sand was 59.5 MPa. The results of physico-mechanical tests are presented in Table 2.

Table 2. The results of physical and mechanical tests

The content of active calcium oxide in the clinker	**Sieve residue**	**Forming method**	**W/C**	**Setting time the hour/min**		**The limit of compressive strength (MPa) in terms**				
	N008			**the beginning**	**the end**	**1 day**	**3 days**	**7 days**	**14 days**	**28 days**
CaO_a-0.61	9.3	1:3 with sand	0.42			6.5	11.0	13.5	18.0	22.5
CaO_a-0.61	9.3	the dough	0.30	0.45	2.35	18.8	25.0	43.8	50.0	59.5

2.3 Formation of nitrogen oxides

Combustion of organic fuels leads to the formation of mainly combustion products in the form of carbon dioxide, water vapor, sulfur oxides (when using solid and liquid fuels) and nitrogen oxides. The most toxic emissions of nitrogen oxides cause the primary need to reduce their emissions into the atmosphere. The amount of nitrogen oxides formed determines the combustion conditions of fuels. Factors affecting the formation of *NOx*: temperature of the flare in the combustion chamber, residence time of the gas in the high-temperature zone, concentration of oxygen and nitrogen in the combustion zone, temperature of heating of the air intended for fuel burning. The greatest influence on the amount of oxides *NOx* produced is exerted by the temperature at which combustion occurs. According to the American Petroleum Institute [11] formation of *NOx* at a flame temperature of 1900 °C is twice as high compared to formation at a temperature of 1300 °C. The temperature of the material during calcination of dolomite is within an interval of about 750-780 °C, and in the production of Portland cement clinker it is 1330-1360 °C. The combustion temperature of the fuel in rotary kilns exceeds the material temperature by 300-400°C. Analyzing these figures, we can conclude that when calcining dolomite, the reduction in the firing temperature will help to reduce the presence of toxic nitrogen oxides in combustion products.

3 Conclusions

It is shown that the production of dolomite clinker in comparison with that of Portland cement is ecologically reasonable:
- the pollution of the atmosphere with greenhouse generated carbon oxides decreases by more than 60%,
- the calcination temperature of dolomite clinker is lower by 420°C - 580°C compared to Portland cement, which worsens the conditions for the formation of toxic nitrogen oxides.

References

1. V. Zyrjanova, G. Berdov, N. Tjuleneva. Current problems in construction and architecture, 189 (2007)
2. G. Shabanova, Proc. scientific. works UkrSART, **138,** 148-154 (2013)
3. C. Mazuranic, J. Amer. Ceram. Soc. **65 (10)**, 523-526 (1982)
4. S. Sorel, US Patent Office, Patent 53/092 (1866)
5. A. Volzhenskij, *Mineral'nye vjazhushhie veshhestva* (Moscow, Stroyizdat, 1986)
6. Climate Change. The Physical Science Basis (2013)
7. H. Vorobyov, D. Mazurov, *Thermotechnical calculations of cement kilns and apparatus*, 134, (Moscow, 1962)
8. V. Vinnichenko, A. Ryazanov, Ecology and industry, **2,** 111 (2013)
9. A. Plugin, V.Vinnichenko, O. Borzjak, Proc. scientific. works UkrSART, **143**, 87-97 (2014)
10. A. Ryazanov, V. Vinnichenko, A. Plugin, Proc. scientific. works UkrSART, **138,** 77-85 (2013)
11. Burners for Fired Heaters in General Refinery Services, **12** (1995)

Technological events in structural evolution of building composites

Valery Vyrovoy[1], *Oksana* Korobko[2,*], and *Oleksandr* Ielkin[3]

[1]Odessa State Academy of Civil Engineering and Architecture, Department of Production of Building Products and Constructions, Didrihsona st. 4, 65029 Odessa, Ukraine
[2]Odessa State Academy of Civil Engineering and Architecture, Department of Architectural Structures, Restoration and Reconstruction of Buildings, Constructions and their Complexes, Didrihsona st. 4, 65029 Odessa, Ukraine
[3]Construction company "BUDOVA", Department of Production of Building Products and Constructions, Osipova st. 22, 65000 Odessa, Ukraine

Abstract. The problems of forming the structure of hardening mineral binders as a result of the implementation of basic events and emerging related processes are considered. It is shown that the self-consistency of events and the phenomena caused by them create a dynamic structural portrait of the developing system. Special attention is paid to the fact that changes in temperature and volume deformations are a "shadow" of events and do not allow us to identify the technological event itself. It is pointed out that the development of background processes predetermines the kinetics of self-development of the structures of developing systems

1 Introduction

Submission of building structures in the form of systems a priori involves their specific structural design. Organization structure depends on the qualitative and quantitative ratios of the starting components, the geometric characteristics of the product and the process conditions and modes of preparation of concrete and its processing into the design system. Thus, the technology is one of the major factors which controlling the structure and thus the properties of composite materials and structures.

In general, the concept of technology (from the Greek techne - art, skill, ability and logos - word, teaching) includes both technology-practice - a set of methods and processes in a particular sector of production and technology, science - scientific description of methods of production and the processes and phenomena that occur in raw material processing to a consumer product. Sakhal D. [1] proposes to use the ideas and methods of system approach for a more complete disclosure of technology concepts. A common feature of the technology, practice, technology, science and technology, the system should be considered as the obligatory presence of a certain set of events, the implementation of which allows obtaining a final product from raw material - in this case, the design system. The events include the chemical, physical, physico-chemical, mechanical, and many other

* Corresponding author: okskorobko71@gmail.com

processes and phenomena associated with the interaction of matter and the organization structure of materials. The implementation of each event in the system, as a rule, is accompanied by a change in the energy condition of the interacting substances, which causes variation of temperature and volume of systems.

2 The change in temperature and volume is a kind of shadow that discards base events

It has historically developed and methodologically stuck that the phenomena which occur as a result of the passing inevitable basic events are treated as information-rich processes, studying which you can retrospectively describe the event itself. In fact, the change in temperature and volume is a kind of shadow that base events cast. These shadows are organized complexly enough, because there may be possible situations in which the shadow includes a "reflection" of various events of different nature, the intensity of symptoms and time of action. The contours of the shadows, combining and merging with each other, are blurred beyond recognition, which does not allow identifying the events that create them. Plato (428 - 348 BC) in the "Sophist" dialogue noted: "... the shadows, when the darkness is mixed with fire, and then the double reflection, when own light of object and strange ones combine and produce a map that makes sense of contrary to the previous causation visibility." "Mappings" of individual events occurring in a distinctive shadow, are not localized either in time or in space. The causal connection between the events and their accompanying processes are hidden beyond averaged last ones. This leads to a reduction, or even to an involuntary falsification of information about the events taking place, making it difficult to identify effective factors for the technological event control. Especially the informational inadequacy between events and processes initiated by them is typical for highly organized and self-assembled objects-systems. At the same time displayed shadows that occurred as a result of basic events become an active background of themselves which initiates the flow of future events. Not excluded situations in which offset in time of manifestation and deepen the flow of events and the emersion and development of related processes gets reduced, which is typical for systems with multifocal sources of realization of certain events. In these cases it is difficult to separate the events from the processes caused by them. In this connection there arises the problem of the analysis of the possible mechanisms of interaction and mutual influence of the basic events and processes caused by them to initiate the formation of structures with the required parameters.

In the initial stages of research in the study as an object cement-aqueous compositions are taken, which are presented in the form of highly lyophobic coarse dispersion systems with freeze interphase boundary [2, 3]. Isolation of cement-water compositions as independent objects of analysis is due to the indisputable fact of their direct influence on the creation of a multi-purpose concrete material. A distinctive feature of the systems under consideration is a polyimineralıty and polydispersity of the particles of dispersed phase, which determines the flow of surface phenomena and related structural transformations of the individual volumes and the entire dispersion system.

The preparation of such compositions is accompanied by a spontaneous re-formation of the interphase interface from S - G into S - L, which leads to a change in the energy state of the system [4]. The average temperature is a faded shade, according to the accepted terminology in the article, of those events that occur in local areas of interphase interactions. Using particles of the same nature, but of different shapes and sizes leads to a temperature change in the local areas of the system. This occurs against occurrence of temperature gradients between the surface and the volume of the particles and between the surface of the particles and the surrounding dispersion medium. Temperature gradients contribute to the development of the internal mass transfer, diffusion processes, activation of surface phenomena, and so on. Thus, in a system comprising monomineral particles of the dispersed phase, local disturbances or non-equilibrium states arise spontaneously, which determines further structural reforms.

Polymineralіty of the dispersed particles assumes a priori the impact of changes in the temperature of each individual particle on changes in the average temperature of the system. For polymineral and polydisperse systems the following situations are possible: - the individual particles are monomineral; - each particle is polymineral formation; - the system is composed of mono- and multimineral particles with different quantitative ratios between them. These cases are typical, with different variations, for real systems and contribute to occurrence of local unbalanced states, the uniqueness of which is hidden in the average temperature changes. It can be concluded that the first peak of heat during the formation of disperse systems is determined by the local exothermic effect which causing local unbalanced conditions. Thereby, the course of further physical and chemical processes of hydration and physical and mechanical processes of structure formation is predetermined. A high concentration of dispersed phase particles suggests that they are located in zones of near or distant coagulation [2]. Under these conditions, all the mono- and polymineral particles based considering their polydispersity, are under the influence of unbalanced forces of interparticle interactions [3]. In the lyophobic systems, the reduction of excess surface energy (Gibbs free energy), which converts a system to the non-equilibrium system, happens by reducing the interfacial area of the interface [4, 5]. The internal reorganization of the system occurs, in which uneven distribution of particles is observed at conservation of their total volume and the total quantity in the system. As a result, the dispersion having fundamentally new structural components - the structural blocks (clusters) and interconnects (inter-cluster) interface [3, 6]. These processes determine the onset and development of subsequent technological events.

The internal distribution of the particles into structural units occurs simultaneously with the lyophilization of their surface. Important and adequately and in detail described in the literature events begin to realize in the system - the physical-chemical processes of hydration of mineral binders [7, 8]. Implementation of the new developments is carried out on the background of the processes resulting from previous events occurred. Schematic diagram of the development of events and their accompanying processes is shown in Fig. 1.

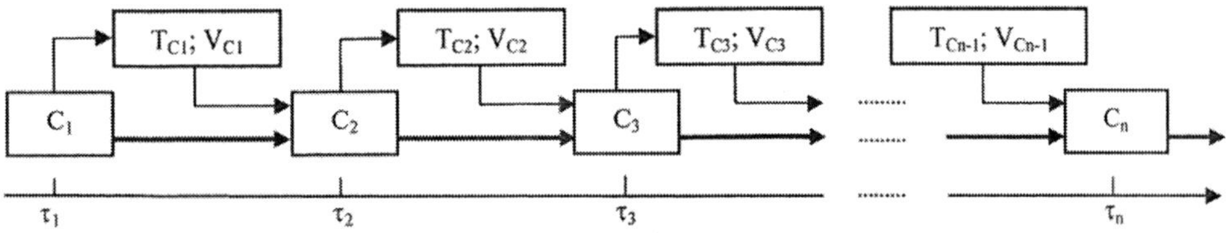

Fig. 1. The scheme of events and their accompanying processes.
$C_3...C_n$ – development of basic events; $Tc_1...Tc_{n-1}$ – temperature change; $Vc_1...Vc_{n-1}$ – volume change.

The inevitability of C_2 event - in our case the cement mineral grains with water - is obvious. Thus the total flow of kinetics hydration processes is determined by the initial temperature of the interacting phases and temperature difference in local areas of the disperse system. In turn, the resulting related phenomena have an impact on the effectiveness of the implementation of new developments $C_3...C_n$, which exacerbates the effects of local temperature changes in the overall process of hydration of mineral binders.

Distribution of poly- and mono-mineral grains of cement by structural units involves multifocal nature of the development of hydration of events with a certain time shift, depending on the chemical activity of each individual mineral of cement. In this regard, the accompanying processes caused by events C_1 and C_2 are a catalyst for the initiation of S_1

events. In turn, the accompanying processes of events S_2 and C_3, integrating, can influence on the kinetics of the developments of events C_3 and M_1, Fig. 2.

It becomes problematic to identify the source of the emergence and development of the processes after a certain period of time. Thus there are a self-support temperature gradients (due to deepening hydration process) between the surface of the particles and their volume, between the surface and the surrounding dispersion medium, and also between the conjugated minerals belonging to the same cement grain [9]. The increased heat release rate of tricalcium aluminate compared to the rate of heat release of tricalcium silicate [7, 10, 11] suggests an uneven distribution of temperature at polymineral cement grain hydration.
New phase ions directionally migrate as a result of thermal diffusion effects manifestations (Soret effect [12]). This leads to local fluctuations in concentration neoplasms products, forming a complex mosaic of interactions of ions of different species that interact to form nuclei of a new phase and create a kind of amorphous component of the cement matrix in which it is difficult to identify the qualitative and quantitative composition. Using the results of a number of authors [7, 8, 13] can be offered to discuss thermal and volume kinetic dependence, taking into account the nature of heat and volume changes of individual minerals and of the whole system, Fig. 3.

Experimental data show that the nature of heat polymineral cement grains (Curve 3, Fig 3) does not allow obtaining fully reliable information about events occurring in certain minerals. Even less information contains data on the kinetics of the heat of hydration of cement compositions as a multimineral and polydisperse systems. Multifocal hydration mechanism initiates energy chaos in the system. It is important that the phenomena that occur as a consequence of the events may be initiated for further chemical processes. Systems with a rather complex initial qualitative and quantitative composition of the dispersed phase spontaneously build path of self-initiation of physical and chemical, physical and physical-mechanical processes of hydration and multiscale structural organization. One of the main places in these processes is the thermal effects that arise as a consequence and are the cause of deepening self-organization phenomena. In these time intervals, the structure completes itself with products of a new phase that are produced by the system. Therefore, such systems should be attributed to autopoiesis systems [14].

To identify the factors of self-motion process control system it is important to consider not total energy processes, but its spontaneous striving for minimization of internal energy by forming organized on the principle of long-range or short-range order of the building blocks of different scales, taking into account the local gradients of temperature and concentration. Using the averaged data on temperature change hardening of mineral binders gives little information and does not allow objectively determining effective prescription and technological factors, which allow controlling complex processes of evolution, taking into account the phenomena of self-organization structure of the system.

Exothermic effects, as inevitable processes, arise when certain events are realized and inevitably lead to a change in the volume of the individual particles of the dispersed phase and the entire system. Besides the initial volume change occurs at the moment of the system forming by forming an interfacial boundary surface due to the adsorption of water molecules on the surface of the particles. Volume changes that occur as a result of certain events initiate development of endodeformations. Endodeformations are deformations that occurs '"inside" the system and initiate the emergence and development of physical, chemical and physics-mechanical phenomena and processes of hydration and structure formation of binders and materials based on them. Endodeformations are attendant processes in the implementation of certain events (for example, adsorption processes, physical and chemical processes of hydration, crystallization processes, mass transfer, etc.). To the events that provoke the development of volume changes, it is necessary to attribute the arising thermal effects. It can be concluded, that some of the endodeformations as "shadow of shadows" capable themselves to initiate the development of thermal effects.

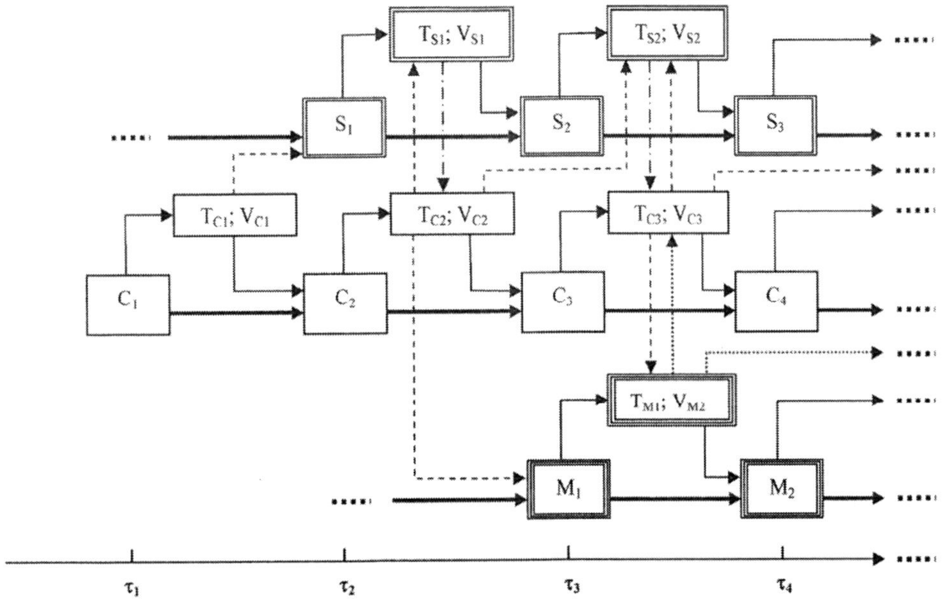

Fig. 2. Development of multifocal interconnectivity of event.
$C_3...C_4$; $S_1...S_3$; $M_1...M_2$ – basic events; T_c; T_s; T_M; V_c; V_s; V_M – the change in temperature and volume at the developments.

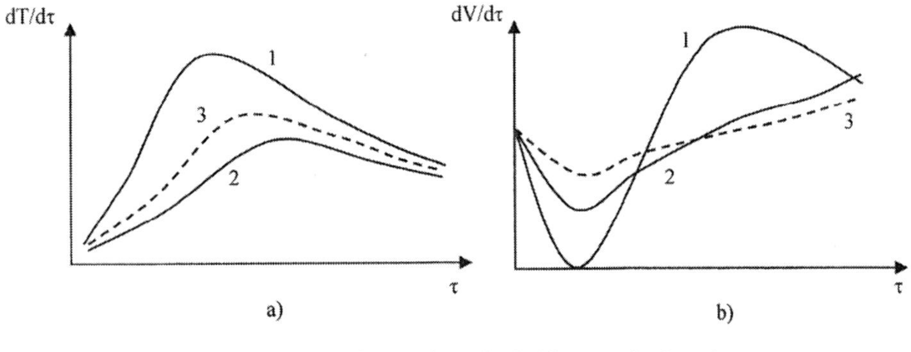

Fig. 3. Kinetics of heat (a) and volume change (b) a single binary and minerals.
1 – tricalcium aluminate (C_3A); 2 – tricalcium silicate (C_3S); 3 – average heat polymineral system (C_3A+C_3S).

Originating as a result of a complex of diverse processes endodeformation predetermine the path of structural adjustment by creating heat transfer currents that in turn stimulates the further development of physical and chemical processes, causing a new deformation flows. The mutual dependence of thermal and deformation processes can not be considered in isolation from each other. Emerging as a result of events that have already happened, thermal effects and endodeformation define future processes by participating in the organization of multiscale structure of the system. However, their participation is associated with the general process of certain vector of development of the system through the development of local events of structural design of the forming and already formed system.

3 Conclusions

Endodeformation, along with thermal effects, are components of the structural "portrait" of the system and represent it in a move that is its natural state. Dynamics of the structure is based on mutual self-activation and the effect of events caused by them on the processes of each other, a kind of behavior in coordination of material component by forming localization directed brooks and streams. The phenomena of self-organization are realized by mutual stimulation. This can lead to the required macro condition of the system. Technological and methodological efforts are needed, which are able to determine the appropriate factors of behavior management for a class of systems with multifocal sources of structural diversity in order to obtain a final product with a normalized set of properties.

References

1. D. Sakhal, *Technological progress: Concepts, models, assessment* (Finance and Statistics, Moscow, 1985)
2. N.B. Uriev, *Highly concentrated dispersion systems* (Chemistry, Moscow, 1980)
3. V.I. Solomatov, V.N. Vyrovoy, *Technological mechanics of concrete* (Riga, Latvia, 1985)
4. Y.G. Frolov, *Course of Colloid Chemistry* (Chemistry, Moscow, 1982)
5. P.A. Rehbinder. *Surface phenomena in disperse* (Nauka, Moscow, 1979)
6. V. Solomatov, V. Vyrovoy, V. Dorofeev, A. Sirenko, *Composite building materials and constructions of the reduced material capacity* (Budivelnik Press, Kiev, 1991)
7. A.A. Pascenco, E.A. Myasnikov, V.S. Vimen, *Cement Theory* (Budivelnik, Kiev, 1991)
8. J. Stark, B. Wicht, *Cement and lime* (Oranta, Kiev, 2008)
9. H. Taylor, *Cement Chemistry* (Mir, Moscow, 1996)
10. A.V. Ursherov-Marshak, *Heat release of cement* (Research Institute for Scientific and Technical Information and Construction Materials Industry, Moscow, 1980)
11. P.V. Krivenko, R.F. Runova, M.A. Sanitsky, N. I. Rudenko, *Alkaline cements* (LLC "Osnova", Kiev, 2015)
12. P.P. Kuzmenko, *Electro transfer, thermal transfer and diffusion in metals,(*Vishcha Shkola, Kiev, 1983*)*
13. O.A. Korobko, V.N. Vyrovoy, collection of scientific works of OSACEA. **3**, 113-117 (2001)
14. U. Maturana, F. Varela, *The tree of knowledge: the biological roots of human understanding,*("Progress-Tradition" Publishing House, 2001)
15. E. A. Petito, Semiofizika and metaphysical realism. Synergetics. Anthology. 1: 329-342. (2013)
16. G., Nicolis, I. Prigogine, *Exploring Complexity: An Introduction* (LCI, Moscow, 2008)
17. V. Vyrovoy, V. Dorofeev, V. Sukhanov, *Composite building materials and constructions. Structure, self-organization, properties* (TES Press, Odessa, 2010)
18. S. B. Fic, V. Vyrovoy, V. Dorofeev, *The processes of self-organization structure of composite building materials* (Politechnika Lublenska Press, Lublin, 2013)
19. V. Vyrovoy, V. Dorofeev, V. Sukhanov, J. Econ. Res. mater., struct., build., constr. **41**, 133-139 (2008)

Stability of material in dynamics of structural organization

Valery Vyrovoy[1], *Vladimir* Sukhanov[1], and *Oksana* Korobko[2,*]

[1]Odessa State Academy of Civil Engineering and Architecture, Department of Production of Building Products and Constructions, Didrihsona st. 4, 65029 Odessa, Ukraine
[2]Odessa State Academy of Civil Engineering and Architecture, Department of Architectural Structures, Restoration and Reconstruction of Buildings, Constructions and their Complexes, Didrihsona st. 4, 65029 Odessa, Ukraine

Abstract. Formation of the scientific ideology is based on the basic model of the structured medium and demands the formalization of the term «structure». The objects-systems are offered to be present in the form of models. It is shown that the mechanism of structure formation on the solid level where gravity prevails differs from mechanism of structure formation at the level where forces of interparticle interactions dominate. It is asserted that geometrical characteristics of products define conditions of the organization of its integral structure. Rather independent coexistence of various solid levels assumes spontaneous realization of various interconnected events which provides the wholeness and safe functioning of a construction-system.

1 Introduction

The content of every paradigm consists of the system of scientific views based on the general model representations defined the entire set of conceptual, methodological, theoretical and experimental regulations [1, 2]. According to I. Prigozhin [3] such a system of natural scientific views forms scientific ideology. He noticed "...the science is an ideology, it is also rooted in the culture" [p. 49, 3]. According to the scientific ideology, a supporter of which I. Prigozhin was, one can argue that with the growing of internal conflicts inside the dominant scientific views the movement trajectory of this paradigm ideas gets in the bifurcation zone. It leads to the shift of paradigm and thereby it causes the displacement of scientific ideologies. A change of ideologies including scientific ones is a lengthy process often painful and tragic. It must be taken into account that it is talked not about replacement "bad by good", "backward by progressive" etc. Drifting or change of paradigm implies the formation of a new system of views based on fundamentally different basic models.

Transition through the point of bifurcation changes the paradigm structure defining its content meaningfulness, the level of its methodological provision, theoretical saturation. The other principles are being formed and scientific schools are being created to promote

* Corresponding author: okskorobko71@gmail.com

the ideas of the new paradigm. The example is the change in the general scientific basic concepts of material (including building) science.

2 Portraits of structure of building composites in dynamics of evolution

Simple, clear, accepted on faith the base model of building materials as the solid medium allows inventing the methodology of their assessment on the level of the average. The explanation of building materials mechanism formation is based on the ideology of reductionism (lat. reduction – return, bringing back) that comes from the assumption that the whole can be understood just if its parts are understood. The methods of designing building products and constructions are based on the model of allowable loads and deformations, which should be below the experimentally determined limit values. On these basic principles many generations of ideologists of unstructured medium, its analysis and calculation and design methods both separate structural elements and structures, buildings and constructions with various kinds and purposes have been taught.

Ideology of unstructured medium implies an ambiguity of perception and evaluation of reality. There is the certain dualism in description of surroundings. There are originality and uniqueness of the processes and objects based on their structural dissimilarity on the one hand. There is unstructured medium on the other hand that is assessed average parameters of temperature, humidity and pressure. They differ in density and strength values, ratios of thermal conductivity, thermal expansion etc. Reality dissolves in the medium quality indicators accepted in the ideology of unstructured medium.

Emerged dualism can be considered as the beginning of the inner conflict within current paradigm. The system approach and synergy in its successful development arise from particular structure existing in objects of different nature and use. The holism ideology is forming (gr. holos – the whole, entire), the main point of it was declared by Aristotle (384 – 322 BC). He stated that the whole includes more than is contained in the simple sum of its constituent parts. The emergence and expression of something new is connected with the structural organization of objects-systems. The key parameter for creating the unexpected one is a structure. Accepting a structure as existing phenomena one can conclude that the ideological scientific positions start to change. Nicolay Gumelyov's words are appropriate in this case "the whole world is discovered again for me". The belief system ideologically oriented to the structure accountability for expressing and maintaining the properties of any object is forming.

In the evolution of different approaches and views a problem of terminology arises. Generally at the beginning existing terminology is used and some additional sense is given. Such a type of term that is widely used in different fields of knowledge is a term structure.

Widespread use of any term spreads its meaning to the blurriness. It is logically to say "language structure", "society structure", "metal structure", "concrete structure" and other ones. The term is used appropriately in biology, medicine, sociology, geology, mechanics, physics, chemistry, materials science and other sciences. The combined principal for objects not similar to their nature is certain hierarchy of their inner structure, inner architecture existed in the synthetic term structure.

Generally the structure (lat. structura – arrangement, order) is understood as a set of stable relations of object satisfied its integrity and identity to itself. In another word it is preserving the basic properties at various external and internal changes [4]. This definition assumes that the structure reflects some fixed condition of the object, which allows qualitatively and quantitatively evaluating and describing its structure taking into account the interaction and the relative position of the totality of the structural elements and groups of them. According to [5] structure can be considered as "stopped motion" or as moment fixed condition of the system in the process of its development, as certain configuration of its components. It is worth to say that term "structure" is used successfully in the description of dissipative systems [6], cooperative phenomena [7], processes of self-organization and adaptation various systems [8] without accurate definition of the essential content. This was the basis to bring into the essential content certain dynamics and self-support in autopoietic system [9], systems whose development goes with an exacerbation [10]. Quite a lot of individual definitions of the term "structure" in separate crystal, metal, polymer and other material descriptions are suggested. Multiplicity of definitions of the same concept assumes its inconsistency to the generalized content as the standard term. Therefore, it seems logical to limit directly the semantic meaning of the term by providing structure in the form of a specific model (lat. modulus – measure, sample) of the object [11]. This method allows focusing on the structural features of the object which are responsible for the manifestation of certain properties according to a specific subject. Choice of model structure depends on the subjective position of a particular investigator – his ideological positions, research objectives, methodology, instrumental and material resources, etc. The subjective nature of the "appointment" of the structure dominant elements is contained in the method of expert evaluations that basically allows estimating prevailing ideology of community of professionals in a time interval. Therefore making a particular model the purpose of the structural description of a particular object (the mechanism, the process) must be defined. One of the purposes of the structural approach is identifying the structural controllable factors which allow the structure to reproduce a required set of structural elements. To do this fixing a large-scale description of the boundaries and establishing a number of priority structure elements are needed to ensure the purpose of the object of analysis.

A many studies are devoted to allocation the scale levels on the size principle in the description of the structure of different nature materials (e.g. [12, 13]). There are supposed to allocate the levels of structural inhomogeneities on the basis of similarity of structural organization mechanisms in the publications [11, 14, 15] that allows to descript a product or constructional integral structure presented in the form of a system of a certain kind [16]. In the process of analyzing the structure organizational mechanism the borders of scales were found taking into account the principal difference in the processes implemented under the influence of force of gravity from the processes that almost do not react to the forces of gravity. V.I. Vernadskiy [17] noticed that macroscopic niche exists where a worldwide attraction reigns. He emphasized that a spatial multiscalement associates with timing multiscalement. This position is similar to H. Bergson's statement [18] that multiple of levels of time and a multiplicity of life worlds exist at different levels. It is pertinently from our point of view to mention E. Petitot's [19] conclusions about emergencity (eng. emergence – appearance of a new one) existence and display two (at least) different levels of organization are required – one is underlying "micro" and another phenomenal "macro" observed at various spatial and temporal scales. The compulsory presence of a time series is noteworthy when the complex behavior of complex systems is considered [20, 21, 22]. A priori we will proceed from the time invariance (lat. invariance – unchanging) at different spatial scale levels without analyzing the nature of time. Within certain time slot (time quantum) the number of taking place and implemented events are fundamentally different, up to impalpability and therefore insensitivity [20].

The principal scheme of different scale structure coexistence is shown in Fig. 1.

The certain scale level, involved in the level of metastable state (MS) where individual particles and their families can reside in a metastable state, is allocated in a separate level. In the case of disintegration of the unit or particles cleavage from it (caused by physic and chemical, physical and physical and mechanical processes) the last transfer at the level where gravity is not dominant. When particles are combined into structural aggregates which weight is higher than force interactions of particles aggregate they fall under the effect of gravity automatically. It can be supposed that MS-level is a link (horizon of events) between two spatial levels. These levels differ in the causes of the occurrence and in the quantity of certain events occurring during a certain time quantum.

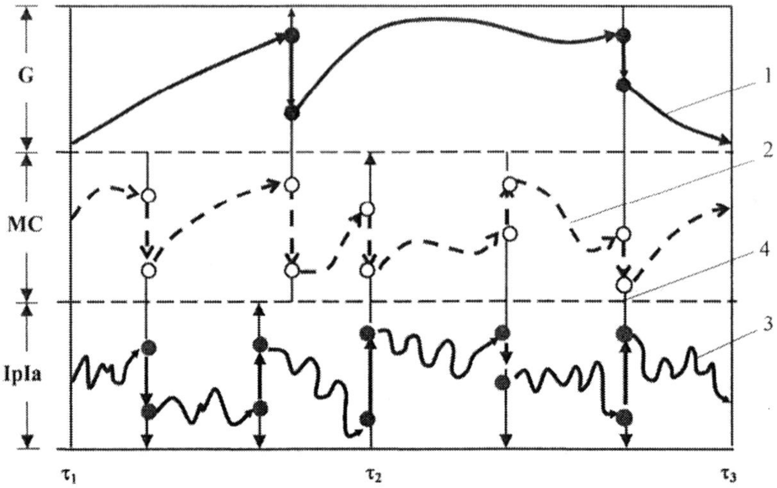

Fig. 1. Diagram of multiscaled interactions. G – scale level where gravity proceeds; MC – scale level where gravity and interparticle interactions can proceed; Ipla – scale level where interparticle interactions proceed; 1, 2, 3 – events at multi- scale levels of structure; 4 – interactions between the structural components at multi-scale levels.

The difference between the mechanisms of structure formation at various spatial levels and the implementation of these acts of structure organization basically not matching in time suggest a certain level of independent existence. Processes occurring at every level subject to the laws manifested at this level, and cannot be carried out at another scale level in most cases. Impact of an event from one level on occurrence of events at another one relates to the rank of external events and occurs usually in mismatched time slots. It means certain autonomy of developments at every level and a certain coexistence of diverse events in the whole system. It can be argued that this coexistence, realized through interaction of diverse events, provides the appearance of new (the other) qualitative characteristics of each scale level and the system as a whole.

Here the events are understood as a set of processes and phenomena of different nature implemented at different spatial scales for different time quantum. The problem arises to associate flows of events related but fundamentally different in essence occurring at different spatial scales and during different time lengths. It needs to identify elements that can be present at all scale levels simultaneously or belong to a specific spatial level and which are capable to provoke the birth of a certain event by their presence, Fig. 1.

In order to determine these elements the task is to trace the formation history of the structure of a particular object. In publications [11, 14, 15] it is underlined that geometric characteristics of a particular product have a significant impact on the structure which components are local and integral field of residual (primary, technological, hereditary) deformations. It allows concluding that directed organization structure with a set of structural elements arises from the impact of the individual characteristics of a particular product. Geometric image of a specific object creates unique qualitative and quantitative ratios of the structural components at different spatial levels with temporal difference of initiation of events displaying. The geometrical parameters of objects should be regarded as significant management factors of the initial structure organization becoming the object-product. Thus the developed model of the structure should be related to a particular object (products, structures, etc.).

3. Functional purpose of structural elements

Structure models of a particular object should contain elements aimed at ensuring sustainable relationships to ensure the integrity and identity of the object under the action of all operational loads. An earlier analysis [14] allowed us to offer a classification of structural elements based on the difference in the speed of response of various elements when object is subjected to external and internal factors. In general case conservative metastable and active elements are marked.

Conservative (lat. conservation – saving) structural elements include elements changed their parameters slow enough during period when object functions. Usually conservative elements belong to the large-scale level where dominant force is gravity and phenomenological approaches of properties evaluation are implemented.

Elements of the structure which are relatively stable under these conditions and able to spontaneously in the stable or unstable state by the action of external or internal factors are referred to the metastable (gr. meta – between, after, through) elements of the structure.

The hallmark of active (lat. activus – activities) structural elements is their ability to respond adequately to the proven impact in a proportion time interval (one tempo rhythm).

Under the structure elements (lat. elementum – element, the original substance) various scale components of the system with specific features ensuring certain functions that contribute to the objectives of each structure level and the system as a whole are suggested for understanding. Structure elements may be specific material formation (e.g., components of the system, the pores and capillaries, cracks, etc.) of the system defining its internal state (e.g., local and integral technological and operational strain, osmotic phenomena, etc.).

System gets the entire set of structural elements in the form of construction products and structures [14] in the process of concrete receiving and its processing into products. At this stage, all the original members are actively involved in multiscale structural design with the formation of new structural elements of the becoming system. The situations are not excluded where creating intermediate structural formations initiate the formation of the intermediate structural birth and development of fundamentally new elements included in the structure of the final product. Equifinality effects are characteristic (lat. aequus – equal proportionate; finalis – final). In our case it means the implementation of the structural organization of the various mechanisms that lead to the emergence of qualitatively identical structure elements. As examples, the origin of technological cracks is considered: - in the formation of discrete structures at the product level and at the level of tumors interaction of initial grain mineral binders; - in the development of gradients of own strains at the interfacial hardening matrix material and fillers; - under the influence of emerging gradient deformation at the level of product structures [11, 14].

A kind of "albums portraits" multiscale structures coexisting on different time scales are created as a result of mutual influence and interaction of all elements at all levels of irregularities. "Albums of portraits" means the permanent dynamics of structural components the totality of which defines a "family portrait" of the structure within a fixed time interval. In this context, a comparison with a "family portrait" suggests that at the same time it has all the elements of the family system. Thus the structure-system comes into the active phase of operation with a set of conservative, metastable and active elements of the structure. Further safe operation of the structure-system depends on the ability of the structural elements by self-consistent interactions and by transforming their own parameters to ensure the preservation of the system properties within continuous operational loads exposure.

Ensuring the stability of the properties for the required time of operation raises the problem of exception and analysis of models and structures in which adaptation processes (lat. adaptation – adjustment) can be implemented by percolation phenomena of self-organization at all various scales and multi time levels of structure-system. Adaptation of concretes [23, 24] should be attributed to the underlying determinants of the viability of production systems. Therefore a number of priority elements that can alter the periods of adverse events should be defined in the model of the structure. This will identify the main governing factors aimed at creating a "family portrait" of the construction-system structure operated under adverse external influences.

4 Conclusions

A structured approach at this stage of its development should be considered as "an introduction to the explanation" of the mechanisms and processes of creation and functioning of systems of various types and purposes. It is based on the becoming scientific ideology based on the ideas and methods of a systematic approach and synergy for which the term "structure" is the dominant concept. There is a paradigms shift based on the basic model of continuous medium in the direction of paradigms based on models of a structured self-developing medium. The fulfilled analysis has led to the conclusion that geometrical characteristics are one of the factors influencing the mechanisms of formation of building products and construction. Therefore, studying the structure of the material without its registration in some constructive forms is futile. In addition, due to blurring and information uncertainty of the generalized definition of the term "structure" it is proposed to develop and analyze models of the structure of concrete products and constructions. In this case, it is necessary to take into account the influence of spatial different-scale levels on the mechanisms of interdependent structural organization of each level of the integrated structure of the construction-system at non-simultaneous realization certain events during the fixed quantum of time. This allows us to consider the structure in a certain dynamics of development both in the establishment of the system and the action of its duty operation. It is proposed to classify the elements of the structure by the time of their response to external and internal influences on the conservative metastable and active elements of individual subsystems and the system itself.

References

1. T. S. Kuhn, *The Structure of Scientific Revolutions* (AST MOSCOW, Moscow, 2009)
2. Br. Latour, *Science in Action: How to follow scientists and engineers through society* (European University in St. Petersburg Press, St. Petersburg, 2013)
3. I. Prigogine, The philosophy of instability, J. Problems of Philosophy, **6**, 45-57 (1991)
4. *Soviet Encyclopedic Dictionary* (Soviet Encyclopedia, Moscow, 1980)
5. E. Soroko, *Golden ratios, processes of self-organization and evolution of systems: Introduction to the general theory of harmony systems* (KomKniga, Moscow, 2006)
6. G., Nicolis, I. Prigogine, *Exploring Complexity: An Introduction* (LCI, Moscow, 2008)
7. H. Haken, *Information and Self-Organization. A Macroscopic Approach to Complex Systems* (Mir, Moscow, 1991)
8. *Synergetics: Research and technologies* (Book House "LIBROKOM", Moscow, 2009)

9. H. R. Maturana, F. J. Varela, *The Tree of Knowledge: The Biological Roots of Human Understanding* (Progress-Traditsija, Moscow, 2001)
10. E. Knyazeva, S. Kurdyumov, *Grounds of Synergetics. Synergetic Vision of the World* (KomKniga, Moscow, 2005)
11. V. Solomatov, V. Vyrovoy, V. Dorofeev, A. Sirenko, *Composite building materials and constructions of the reduced material capacity* (Budivelnik Press, Kiev, 1991)
12. I. Rybev, *Building materials on the basis of binders* (Vysshaja shkola Press, Moscow, 1978)
13. A. Sheikin, *The structure, strength and crack resistance of a cement stone* (Strojizdat Press, Moscow, 1974)
14. V. Vyrovoy, V. Dorofeev, V. Sukhanov, *Composite building materials and constructions. Structure, self-organization, properties* (TES Press, Odessa, 2010)
15. S. B. Fic, V. Vyrovoy, V. Dorofeev, *The processes of self-organization structure of composite building materials* (Politechnika Lublenska Press, Lublin, 2013)
16. V. Vyrovoy, V. Dorofeev, V. Sukhanov, J. Econ. Res. mater., struct., build., constr. **41**, 133-139 (2008)
17. V. Vernadsky, *Philosophical thoughts of the naturalist* (Nauka, Moscow, 1998)
18. J. Bergson, *Creative evolution. Matter and memory* (Harvest, Minsk, 1999)
19. E. A. Petito, Semiofizika and metaphysical realism. Synergetics. Anthology. **1**, 329-342 (2013)
20. A. Alyushin, E. Knyazeva, *Tempo-worlds: Speed of perception and time scales* (Publishing House "LCI" Press, Moscow, 2008)
21. I. Prigogine, I. Stengers, *Time, chaos, quantum* (Progress Press, Moscow, 1994)
22. S. W. Hawkins, *Three books about space and time* (Amphora, St. Petersburg, 2014)
23. L. Sheynich, K. Pushkareva, *The processes of self-organization of structure of build composites* (Gamma-Print Press, Kiev, 2009)
24. V. Cherniavsky, *Adaptation abiotic systems: concrete and reinforced concrete* (DNURT, Dnipropetrovsk, 2008)

The influence of repeated loading on work of the steel fiber concrete drainage trays and pipes on the roads

Oleksandr Andriichuk[1*], *Volodymyr* Babich[2], *Ivan* Yasyuk[1], and *Serhii* Uzhehov[1]

[1] Lutsk National Technical University, Civil Engineering Department, 43018 Lutsk, Lvivska St. 75, Ukraine
[2] National university of water management and nature resources use, Civil Engineering and Engineering Structures Department, 33000 Rivne, Soborna St. 1, Ukraine

Abstract. The drainage system is one of the components of the road design. The condition of the subgrade and pavement depends on its effectiveness. The main structural elements of the drainage system on the roads are gutters and pipes. They are made of concrete or reinforced concrete. Under the influence of climatic factors and fluctuations of the vibration caused by the vehicles movement on the surface, it occurs destruction: formation of cracks, potholes, husking of concrete, destruction of protective layer of concrete, etc. It should be noted that these structures perceive the dynamic and thermal effects. The low fracture materials toughness poses the issue of searching ways of its increase. One solution of this problem is the use of dispersion-reinforced concrete gutters and pipes. The article presents the results of research strength, crack resistance and deformability of gutters and pipes using steel fiber reinforced concrete under the action of repeated loads

Non-pressure pipes and street gutters on the roads used in the construction of drainage systems for drainage of rain, snowmelt and groundwater from road sections with a high degree of load, highways, walkways, airports and terminals. It should be noted that future traffic on the roads I-A and I-B categories, according to calculations takes more than 10000 vehicles per day. The throughput of an individual vehicles of Ukraine even exceed these standards, in particular, the highway Kyiv-Boryspil traffic is about 40 thousand vehicles per day. A huge stream of traffic through the cross section of a road of this class creates significant dynamic vibration oscillations, which act on road transport facilities. While the low fracture toughness of the materials of gutters and pipes puts the issue of finding ways to improve it.

One solution in this direction is using of concrete in these construction of gutters and pipes with the addition of reinforcing elements looks like a short steel cuts – steel fiber reinforced concrete. The combination of rigid (and because of this with considerable reserves of strength) fibers with the matrix (concrete) allows to localize the danger associated with brittle destruction of the matrix and to realize the basic properties of fibers: greater potential tensile strength and high modulus of elasticity.

[*] Corresponding author: aleklutsk@gmail.com

Application efficiency of steel fiber concrete in building structures is achieved by reducing labor costs for reinforcement works, combining technological operations of preparation, reinforcing, laying and compaction steel fiber reinforce concrete mixture, extending the life of structures and reducing costs for various types of maintenance. Use it instead of the concrete significantly reduces the consumption of construction while simultaneously improving operating properties, durability, temperature resistance, water resistance [1-3].

Despite a number of qualitative advantages, steel fiber reinforced concrete (SFRC) is still not fully investigated material. It should be noted that the relevance of using SFRC for the manufacture of the roadside gutters and pipes for wastewater are described in the article [4-5].

In research is established that steel fiber reinforced concrete as an effective material should be used for the manufacture of gutters and pipes of the drainage systems; the use of gutters and pipes from SFRC for water and wastewater instead of standard reinforced concrete is allows to: completely abandon the use of rebar; reduce the consumption of materials of construction, and as a consequence, their weight; to reduce labor costs for production of structures and significantly increase the durability of constructions. In other papers [6-8] presented the results of the researching of SFRC pipes and gutters to the action of a single load.

The article presents the results of empirical researches of strength and deformation characteristics, stress-strain state of non-pressure pipes and roadside gutters made of steel fiber reinforced concrete to the action of repeated short-term loads. These studies were conducted in the building laboratory of Lutsk NTU.

To perform the research was made several gutters and pipes. In the process of testing samples is compared concrete, reinforced concrete and SFRC samples.

For concreting pipes and gutters of type "Half-pipe" was designed and manufactured a special easily-dismountable metal formwork (Fig. 1). It provided the evenness of the surface of experimental samples with a wall thickness t=40 mm, an inner diameter d_{in}=300 mm (d_{out}=380 mm) and the length of the elements l=300 mm.

Fig. 1. General view of the formwork to producing test samples of gutters.

Wave fiber with a diameter of 0.8 mm and a length of 50 mm is applied to obtain steel fiber concrete as a particulate reinforcement of research gutters (Fig. 2).

Testing of prototypes is performed by using a metal traverse with concentrated load. The lower part of the element leaned on a rigid base. For this test will be used the hydraulic press PSU-125. To improve the measurement accuracy of the current effort is used an exemplary dynamometer, that allows to measure the loadings with an accuracy of 50 N. In this case, the loading is created using hydraulic jack. Gutters researches are carried out according to the scheme presented on Fig. 4, and the researches of pipes on Fig. 5.

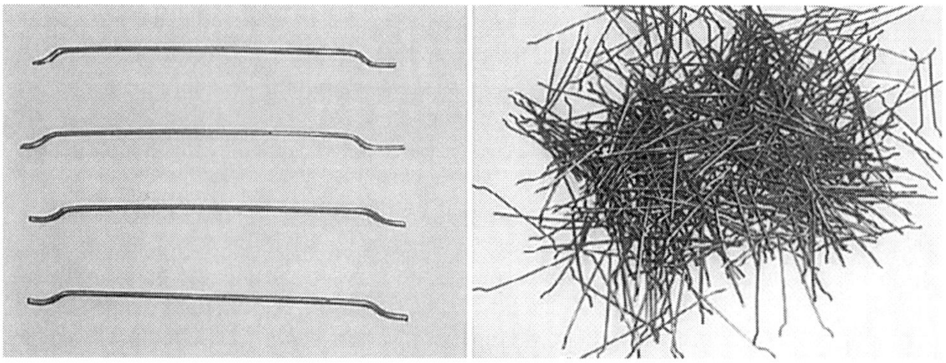

Fig. 2. General view of the steel anchor fibers for the manufacture of SFRC.

On Fig. 3 presented SFRC gutters and pipes after demolding.

Fig. 3. General view of produced SFRC gutters and pipes.

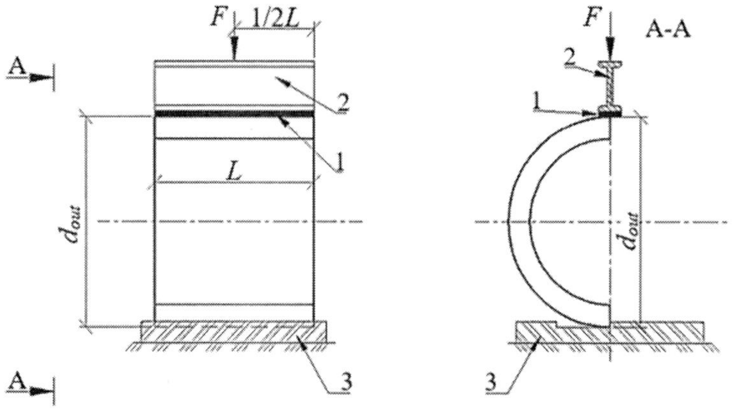

Fig. 4. Gutter's test scheme: 1 - rubber carpet; 2 - metal traverse; 3 - motionless basis.

During the research of gutters and pipes the loading is applied through the steps of 8–12% of the breaking stress, that defines by the theoretical method during the calculation. After each stage of loading the pause is withstood for 5–7 minutes, in which the necessary data was taken (indicators measurements, strain indicators of the complex and measured crack widths (a_{crc})).

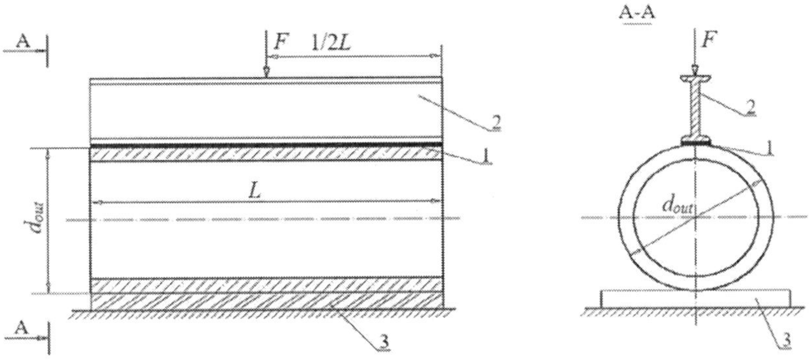

Fig. 5. Pipe's test scheme: 1 - rubber carpet or cement mortar; 2 - metal traverse; 3 - motionless basis.

For measuring displacement of the prototype walls that resulting from the appropriate loading of gutter, used the dial indicator MIG-1 with a scale division of 0.01 mm, which is attached by using of metal holders, which in turn stick to the concrete surface with an epoxy glue. This ensures their reliable position relative to the sample during the measurement of the displacements of the gutter walls in the diametrical vertical direction.

Crack widths will be determined with a microscope MPB-3 with a scale division of 0.02 mm. In order to improve the visual monitoring of the occurrence and growth of cracks, lime mortar is applied to inner and outer surfaces of elements before researches.

For measuring the strain of concrete and SFRC strain gauges with a working base of 50 mm are glued on the inner and outer surfaces of the pipes and gutters, whose data are recorded by tensometric measurement complex. General view of the test pipes and gutters is shown in Fig. 6 and Fig. 7.

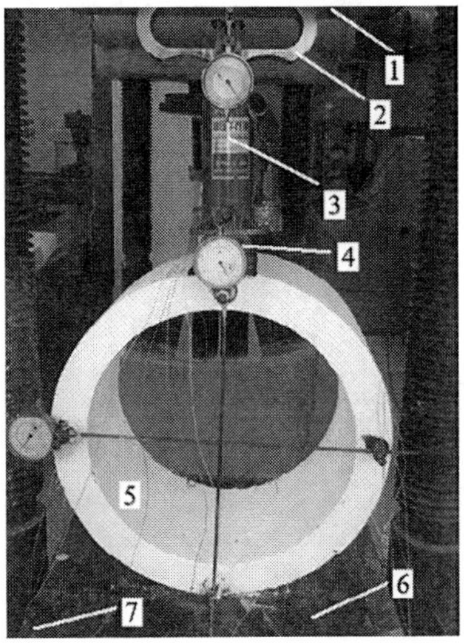

Fig. 6. General view of the test elements of annular cross-section (pipe): 1 - upper support press plate of PSU-125; 2 - testing dynamometer; 3 - hydraulic jack; 4 - metal traverse; 5 - test element of annular cross section; 6 - rubber carpet; 7 - lower support press plate of PSU-125.

A detailed description of the methodology of experimental research SFRC road drainage gutters presented in paper [9], and the methodology of research SFRC pipes presented in paper [10].

Researches of reinforced concrete gutters were carried out (labelling: RCg–1, RCg–2 and RCg–3) with the percentage of reinforcement steel frames μ=2. During the research, gutters RCg–1-3 for 10 cycles were loading to level η_{cyc}=0.6 of the breaking stress, and at 11th cycle were brought to destruction.

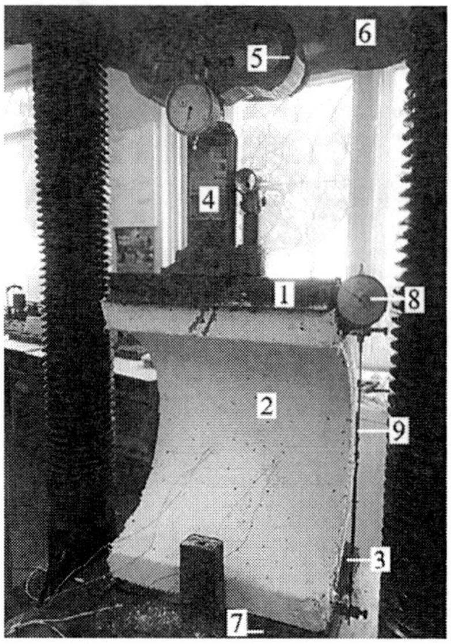

Fig. 7. General view of the drainage gutters research: 1 - metal traverse; 2 - research drainage gutter; 3 - motionless basis; 4 - hydraulic jack; 5 - testing dynamometer; 6 - upper press plate of PSU-125; 7 - lower press plate of PSU-125; 8 - displacement sensor; 9 - upright guides.

Deformation of the cross section gutters RCg–1-3 (Δl) within F=0-6.33 kN occurred linearly and reached the maximum load on this section (F=6.33 kN) value Δl=0.94 mm. Within F=6.33-9.33 kN deformation of reinforced-concrete gutters increased nonlinear to values Δl=0.94-2.78 mm (the process of cracking). Further deformation of the samples got a specific linear pattern, but with a large increasing.

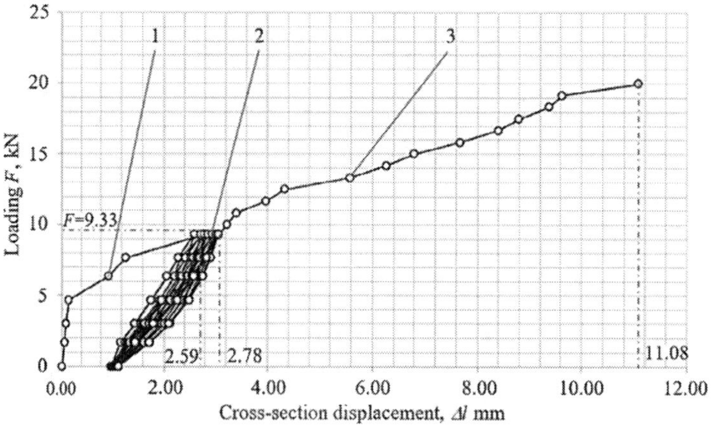

Fig. 8. Deformation of RCg gutters under the action of repeated loads: 1 - during the 1st cycle; 2 - during the 2nd-10th cycles; 3 - during the 11th cycle.

The first visible cracks in the samples RCg–1-3 occurred in the first cycle at the load F=7.67 kN (a_{crc}=0.23 mm). At the maximum load in the first cycle, which was F=9.33 kN, the cracks opened to the a_{crc}=0.38 mm. At the unloading the samples to F=0 kN residual crack widths were within a_{crc}=0.28-0.43 mm. Form of cracks in the RCg gutters is shown on Fig. 9.

Researches of SFRC gutters were carried out (labelling: SFRCg–1, SFRCg–2 and SFRCg–3) with the percentage of fibers μ=2. During the research gutters SFRCg-1-3 for 10 cycles were loading to level η_{cyc}=0.6 of the breaking stress, and at 11th cycle were brought to destruction.

Fig. 9. The form of cracks in the RCg gutters: area of tension and compression.

At the loading on the cycles of maximum effort (F=9.33 kN) with each cycle of SFRC–1…3 samples deformation is constantly increasing in the interval Δl=0.47-1.07 mm (Fig. 10).

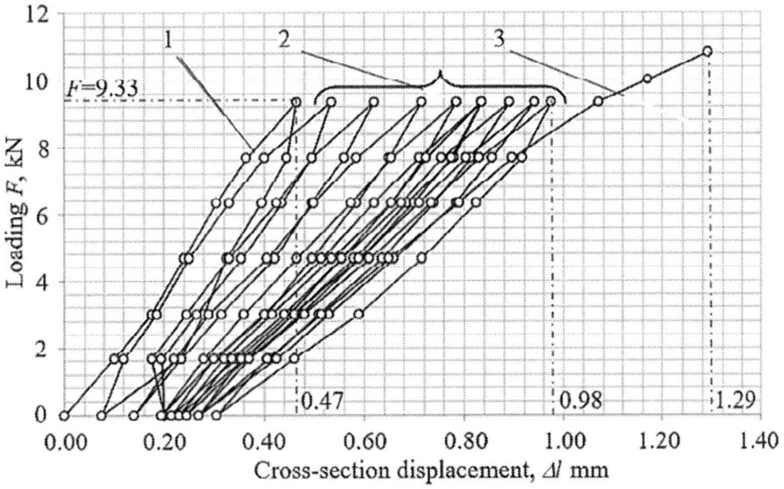

Fig. 10. Deformation of SFRCg gutters under the action of repeated loads: 1 - during the 1st cycle; 2 - during the 2nd-10th cycles; 3 - during the 11th cycle.

In the elements SFRC–1-3 the first visible cracks appeared at 10 cycle at F=9.33 kN and width a_{crc}=0.05 mm. At the 11th cycle at F=9.33 kN, crack widths already were at a_{crc}=0.1 mm. View of cracking of the SFRCg gutter (Fig. 11).

Studies of non-pressure pipes under the action of repeated loads was conducted on concrete samples (RCg-1, RCg-2 and RCg-3) and SFRC samples (SFRCg–1, SFRCg–2 and SFRCg–3). Elements were reinforced with steel frame (RC) and anchoring fibers (SFRC) with the percentage of reinforcement μ=2.5.

The samples during ten cycles were loaded by degrees to the level η_{cyc}=0.6 from the breaking stress, and on the 11th cycle was brought to destruction. Unloading of samples on cycles were carried out the same steps as in the load. Cycles 4th, 6th and 8th were intermediate, another words, they were loaded once to its maximum value, to F=7 kN, and then unloaded to F=0 kN.

During the study, the pipes move sections of samples in vertical and horizontal directions were fixed, and the determined deformation of the concrete in a compressed and stretched areas using strain gauges with a base of 50 mm, which were connected to strain-gauge station "Izmeritelnaya tenzometricheskaja sistema VNP-8".

Fig. 11. The form of cracks in the SFRCg gutters: area of tension and compression.

Determined by using dial indicators, averaged displacement of the cross sections under the action of repeated loads in concrete samples (RCg–1-3) are presented in Fig. 12, and in the SFRC samples (SFRCg–1-3) are presented in Fig. 13.

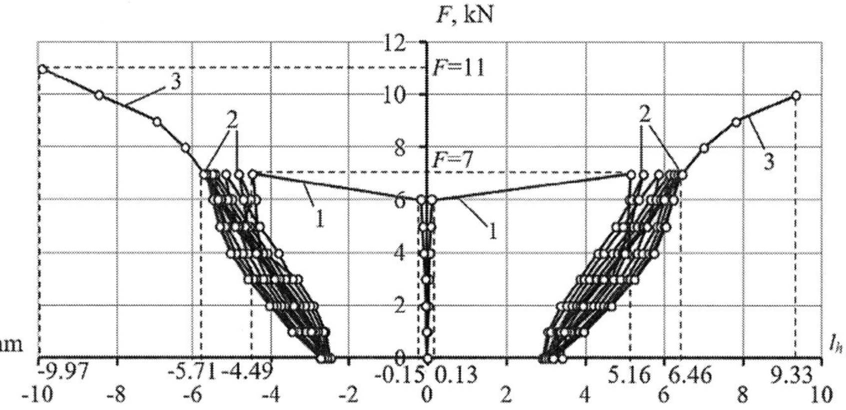

Fig. 12. Averaged displacement cross-sections of samples RCg–1-3 in horizontal (Δl_h) and vertical (Δl_v) directions: 1 - during the 1st cycle; 2 - during the 2nd-10th cycles; 3 - during the 11th cycle.

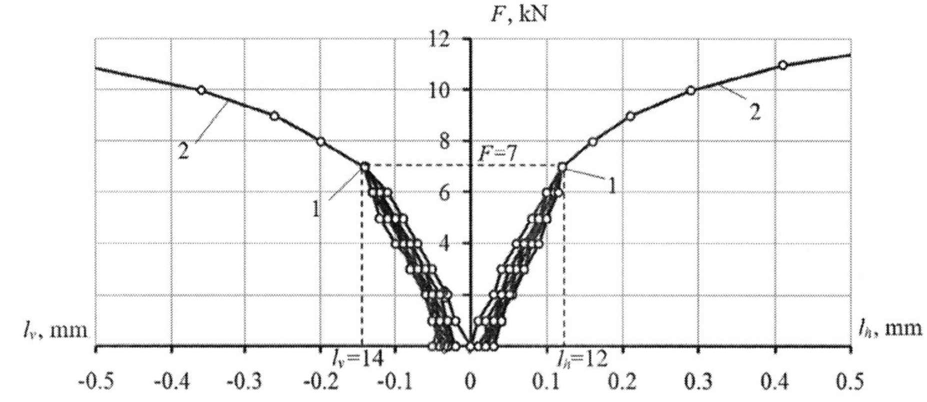

Fig. 13. Averaged displacement cross-sections of samples SFRCg–1-3 in horizontal (Δl_h) and vertical (Δl_v) directions: 1 - during the 1st cycle; 2 - during the 2nd-10th cycles; 3 - during the 11th cycle.

Deformation of concrete in samples of RCg–1-3 within F=0÷6kN has developed almost linearly and reached a maximum value of F=6kN, $\mathcal{E}_{max}$=13×10^{-5} (Fig. 14). During the initiation of the effort F=7kN in the stretched areas of the elements RCg–1-3 began to appear cracks. The cracking led to failure of the strain gauges that were placed in these areas, both outside and inside. Therefore, curve F–$\mathcal{E}$ in the area stretching ends with F=6kN. In the zone of compression when loading effort F=7kN strain increased from $\mathcal{E}_c$=13×10^{-5}(F=6kN) to values $\mathcal{E}_c$=106×10^{-5} (more than eight times).

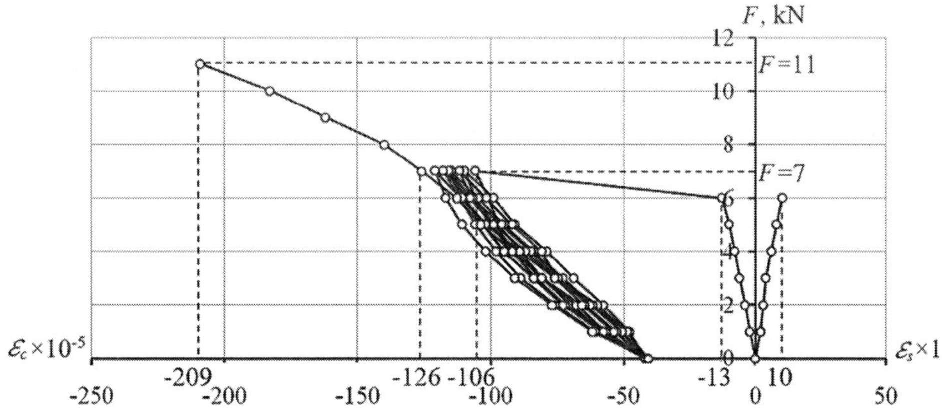

Fig. 14. Deformation diagram of compressed ($\mathcal{E}_c$) and stretched ($\mathcal{E}_s$) reinforced concrete under the action of repeated loads at RCg–1-3.

At the 1st cycle of loading, the values of the full and residual strain of concrete in compression zones were: $\mathcal{E}_{c,max}$=6×10^{-5}and $\mathcal{E}_{c,pl}$=42×10^{-5}. Over the next 9 cycles before the 10th inclusively, in the areas of compression was moderate growth of total deformation to $\mathcal{E}_{c,max}$=126×10^{-5}. Residual deformation is almost not increased and at the 10th cycle was $\mathcal{E}_{c,pl}$=42×10^{-5}. On the next 11th destructive cycle, due to increasing the load above the level η=0.6 occurred an increase of total deformation to $\mathcal{E}_{c,max}$=209×10^{-5}(at F=11kN).

Deformation of SFRC samples SFRCg–1-3 (unlike strains in the same samples of RCg–1-3) was fully tracked during all cycles (Fig. 15), as in the zones of compression and tensile zones – the process of fracturing has not led to failure of the strain gauges, which were placed in the tensile zones.

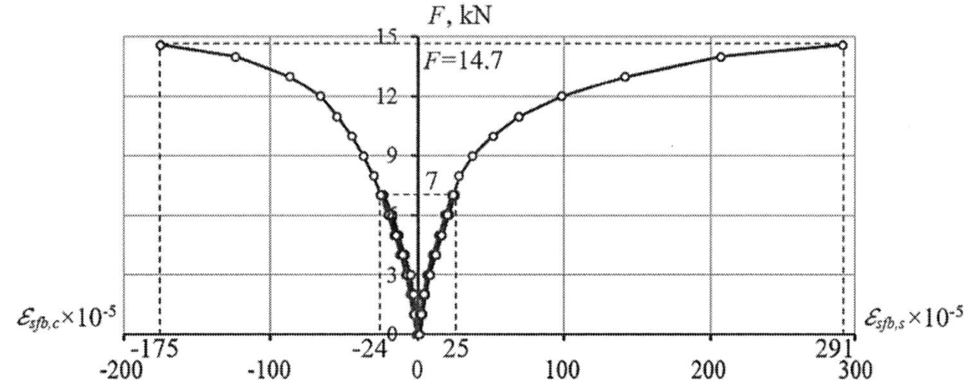

Fig. 15. Deformation diagram of compressed ($\mathcal{E}_{sfb,c}$) and stretched ($\mathcal{E}_{sfb,s}$) SFRC under the action of repeated loads at SFRCg–1-3.

Conclusions

On the basis of conducted research, it can be argued that drainage gutters and non-pressure pipes made of SFRC under repeated loads have a high hardness and crack resistance in comparison with similar items which is made of ordinary reinforced concrete.

At the repeated loads in SFRC gutters for 10 times loading deformation of the cross-sections were (at maximum load cycle) Δl=0.47÷0.98mm, while at typical RC elements these values were (from the 1st to the 10th cycle) is several times greater Δl=2.59÷2.78mm.

At the repeated loads in SFRC pipes for 10 times of horizontal load (Δl_h) and vertical (Δl_v) movement diametrical of the cross sections did not change and was at a maximum load cycle of 0.12÷0.14mm, while in typical RC elements, these values constantly (from the 1st to the 10th cycle) was increased Δl_h=5.16÷6.41mm and Δl_v=4.49÷4.64mm, respectively. In addition to the increments of load (ΔF=1kN) average moving sections in SFRC samples was 0.01÷0.02mm, and RC - 0.1÷0.3mm. At repeated loads, which ranged η_{cyc}=0.6 of the destructive, during 10 times the load in the pipes from the SFRC full and permanent deformation virtually did not change, while in RC samples they increased by 10÷25 %.

At the repeated loads, level of which does not exceed 70 % of the destructive, SFRC pipes works elastic – moving in diametrically opposite directions do not exceed 0.05% of the diameter, and the cracks occurs only when uses additional loading.

Cracks in the experimental gutters and pipes of reinforced concrete occurred at the 1st cycle but in the samples from SFRC only on the 11th. The first visible cracks in the samples from SFRC in most cases were grown at one or two steps of the load from destructive forces, and in the samples of reinforced concrete at loads equal to 0.3-0.4 of maximum carrying capacity of the section. They were recorded in all samples only in the zone with maximum bending moment.

In a typical reinforced concrete elements cracks were almost straight, with a clear gaps, and in SFRC samples, the cracks were with fuzzy contours and expressed netting.

References

1. L. Dvorkin, O. Dvorkin, Y. Ribakov. *A method for optimal design of steel fiber reinforced concrete composition*. Materials and Design, Elsevier (2011)
2. R. Kinash, V. Bilozir. *Deformational calculation method of bearing capability of fiber-concrete steel bending elements*. Czasopismo Techniczne (2015)
3. K.V. Talantova. *Stroytelnye konstruktsyy s zadannymy svoistvamy na osnove stalefybrobetona*. Barnaul, Russia (2009)
4. Y.M. Babych, O.V. Andriichuk. Resursoekonomni materialy, konstruktsii, budivli ta sporudy, Rivne, **18**, 119–126 (2009)
5. O.V. Andriichuk, I.M. Yasyuk. Naukovi notatky, Lutsk, **45**, 7–14 (2014)
6. O.V. Andriichuk, Y.M. Babych. *Stalefibrobetonni beznapirni truby*. Lutsk, Ukraine (2012)
7. Y.M. Babych, O.V. Andriichuk. *Proektuvannia ta vyhotovlennia beznapirnykh trub iz stalefibrobetonu*. Lutsk, Ukraine (2012)
8. O.V. Andriichuk, I.M. Yasyuk. Resursoekonomni materialy, konstruktsii, budivli ta sporudy, Rivne, **18**, 371-378, (2015)
9. Y.M. Babych, O.V. Andriichuk. Fizyko-khimichna mekhanika materialiv, **4**, 59-62 (2016)
10. O.V. Andriichuk, I.M. Yasyuk. Resursoekonomni materialy, konstruktsii, budivli ta sporudy, **30**, 86-95 (2015)

Calculation of reinforced concrete ceilings with normal cracks accounting the Chebyshev approximation

Taljat Azizov[1*], *Olexiy* Melnyk[1], *Olga* Orlova[1], *Angelina* Kalenchuk-Porkhanova[2], and *Larisa* Vakal[2]

[1]Pavlo Tychyna Uman State Pedagogical University, Department of Professional and Technological Education, 20300 Uman, Ukraine
[2]VM Glushkov Institute of Cybernetics of NAS of Ukraine, 03187 Kiev, Ukraine

Abstract. The article shows the influence of torsional rigidity of reinforced concrete elements on the spatial work of bridges, overlappings, building frames and other complex statically indeterminate systems. It is shown that the determination of torsional stiffnesses by the existing methods assumes the obligatory presence of spatial spiral cracks, and torsional stiffness in the presence of normal cracks is not investigated. A method for determining the torsional rigidity of reinforced concrete elements is described in the presence of normal cracks in them. It is shown that this approach allows calculating the torsion of reinforced concrete elements of any cross-section and taking into account the nonlinear properties of concrete. The article also describes the use of approximation methods, in particular, the apparatus of the best Chebyshev approximation. In Table 2, the displacements obtained as a result of the approximation with the displacements obtained directly from the calculations using the Lira software using volumetric finite elements are compared. In column 6 of the table, the displacement values obtained by software package (Lira software) and in column 7, the displacements obtained on the basis of approximation in the Matlab environment are given.

It is known that the torsional rigidity of their elements exerts a significant influence on the spatial work of the plate-ribbed systems. In reinforced concrete slab-ribbed systems (bridges, ribbed monolithic and prefabricated ceilings), flexural and torsional rigidity is affected by various cracks [1,2,3,4].

When local loads are applied to prefabricated or monolithic reinforced concrete floors, cracks may appear in individual beams or slabs, in others they may be absent. In this case, the torsional and flexural rigidity in the beams without cracks and with cracks will differ. It was shown in [2,3,4] that the redistribution of the local load depends in practically the same manner on both the flexural and torsional stiffnesses of the individual elements. This dependence is significant. Consequently, the definition of bending and torsion stiffness is an important and urgent task. Despite this, most calculations for the design of various structures, including well-known powerful software complexes, such as ANSYS, NASTRAN, LIRA,

* Corresponding author: oleksiy.melnyk@udpu.edu.ua

are carried out without taking into account the change in torsional stiffness resulting from the formation of normal cracks.

The torsional stiffness of individual elements with normal cracks can be taken into account in the program complexes mentioned above, when using volumetric finite elements. However, when calculating, for example, the skeleton of a multi-store building, a bridge structure or an overlap consisting of many elements, modeling [4,5,6].

Each element (column, beam, plate, etc.) from volumetric finite elements with the inclusion of reinforcement elements is very, very laborious and practically impracticable.

To date, there are a lot of studies on the flexural rigidity of reinforced concrete elements with cracks and a very small number of studies of the torsional stiffness of such elements. Most of the studies relating to torsion in reinforced concrete are devoted to the study of the strength of such elements. The existing methods for determining torsional stiffness relate mainly to reinforced concrete elements with spiral cracks under the action of torsional bending, although experimental studies have established a significant effect of normal cracks on the torsional stiffness of reinforced concrete elements. In addition, in these works we consider simple types of sections: a rectangle with a symmetric reinforcement, a ring, and cylindrical elements. Considering the fact that in practice there are a variety of types of sections: T-waves, I-beams, hollow triangles, box-shaped, etc., some studies have been carried out on this topic at the Odessa State Academy of Construction [7,8]. However, these works are only at the initial stage of research. The works devoted to the investigation of the torsional stiffness of reinforced concrete elements with normal cracks [7,8] have an approximate and particular character.

The objective of this article is developing methods for determining the torsional stiffnesses of reinforced concrete elements with normal cracks using algorithms and software complexes for processing and compressing arrays of experimental data by replacing them with a certain analytical expression (approximant) with few coefficient parameters using the known approximation methods.

Qualitatively new approach when performing such a replacement is the use of intellectualized methods of approximation of the function by the best Chebyshev (uniform) approximation, which is much more effective and universal than the interpolation and rms approximation methods.

The main advantage of the Chebyshev method of approximation in comparison with other methods of approximation is to ensure the accuracy of the approximation obtained at a certain set of points of the approximation interval at all points of this interval.

The advantages of the Chebyshev approximation allow us to solve with high accuracy not only the obtaining of an approximant and, as a result, the compression with large (several orders of magnitude) data compression coefficients of a discretely given functional dependence (the direct approximation problem), but also the task of restoring the values of the dependence on "unlit" (Inverse approximation problem) (for more details, see [9,10]).

In the works of the authors of [8], a series of studies of the torsional stiffness of reinforced concrete elements of rectangular and T-sections with normal cracks was carried out. In these works it was shown that the main part of the problem of determining the torsional stiffness of a reinforced concrete element with normal cracks is the determination of mutual displacement of the crack edges. This problem can be solved using both the approximate method and the finite element method. One of the disadvantages of solving this problem, as mentioned above, is the condition of using a large number of volumetric finite elements, which complicates both the creation of the calculation scheme and analysis of the calculation results, especially since this is only part of the solution of the general problem of the stress-strain state of a reinforced concrete element with normal cracks in torsion [4]. At the same time, the use of the methods of elasticity theory is probably not possible for all cases of solving the problem.

Following [7], it can be assumed that the mutual displacement of crack edges will be a clear function of the height of the compressed zone, the height and width of the beam section and the distance between the cracks, which allows us to apply the methods of approximation to determine this dependence. Let us first consider the dependence of the displacements in a beam of a rectangular cross-section on two parameters-the width b and the section height h-for a fixed value of the distance between the cracks and the height of the zone compressed from the bend. In this case, perhaps not for all cases, it is possible to construct a graph of such a dependence in the form of a certain surface.

Below are examples to explain the derivation of the approximation dependencies.

Example 1. Let there be a beam of a rectangular cross-section made of a material with a modulus of elasticity E=32500 MPa, a shear modulus of 10000 MPa. The remaining parameters have such values: distance between cracks l_{crc}=400 mm; height of the compressed zone x=30 mm, torque T=1000 N·cm.

Approximation of the required surface in the Matlab environment is quite simple. The authors have previously carried out a series of calculations on the program complexes using volumetric finite elements. Variable parameters and movements are shown in Table 1.

Table 1. Variable parameters and movements between cracks.

#	Dist. between cracks l_{crc}, mm	Height of the compressed zone x, mm	b, mm	h, mm	Displacement of the crack edges (according to program complexes) $\Delta\cdot 10^3$ (mm)
1	400	30	60	100	3.104
2			80	100	1.668
3			100	100	1.043

The approximate surface obtained by processing the initial data in the Matlab environment is shown in Fig. 1.

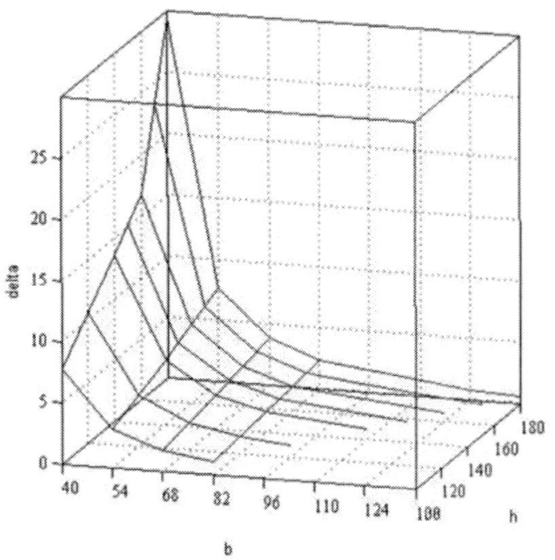

Fig. 1. The dependence of the displacement of the crack banks (delta) on Width (b) and height (h) of the beam section.

In Table 2, the displacements obtained as a result of the approximation with the displacements obtained directly from the calculations using the Lira software using volumetric finite elements are compared. In column 6 of the table, the displacements values obtained by software package (the Lira software) are given, and in column 7 the displacements obtained on the basis of approximation in the Matlab environment. As can be seen from the table, the values in columns 6 and 7 coincide with a sufficiently high accuracy, which confirms the correctness of the adopted approach to solving the problem.

Table 2. Comparison of displacements of crack edges obtained by the Lira software using volumetric finite elements and the approximation method.

#	Dist. between cracks l_{crc} (mm)	Height of the compressed zone x (mm)	b (mm)	h, (mm)	Displacement $\Delta\cdot10^3$ (mm)		Error (%)
					Finite element method	Approximation	
1	2	3	4	5	6	7	8
1	400	30	60	140	5.2469	5.3575	2.11
2			80	140	2.72018	2.7735	1.96
3			100	140	1.641188	1.6743	2.02

A similar approach is fairly simple to apply for elements with any other shape of sections (T-shaped, I-beams, box-like, etc.), and also with inclined cracks. The number of variables for the approximation can be different.

Creating a library of function approximants (similar to Table 2, Fig. 1) will greatly simplify the solution of many problems of determining the stiffness parameters of reinforced concrete elements with cracks that can enter as a separate block into existing software systems.

When determining the stiffness parameters of a reinforced concrete element with normal (or inclined) torsion cracks, to determine the movements of any points, it is necessary to compile approximate expressions (based on a certain number of calculations using finite element method), in which the cross-sectional dimensions b and h, the height of the compressed zone (The height of the zone through which the torque is transmitted), as well as the length of the block separated by normal cracks, the angle of the fracture (in the case of inclined cracks), etc.

For more complex sections, the number of variable parameters will be larger. For example, when the height of the compressed zone is within the edge of the reinforced concrete element of the I-section (Fig. 2), the displacement of the crack sides (the angle of mutual rotation of the two blocks separated by a normal crack) will be a function of seven variables:

$$\Delta crc = f(b_1 h_1, b_2, h_2, t, h, x) \quad (1)$$

To approximate such functions of several variables, one can use a software implementation of one of the algorithms of the apparatus of the best Chebyshev approximation, described in [9,10].

It should be noted that the database of values of these parameters of approximants can be obtained for specific values of the elastic modulus E and the shear modulus G of concrete. Taking into account that this problem is linear, to obtain displacements in an element with other values of elasticity and shear module, it is easy to multiply them by the ratio of the corresponding values of the parameters of the considered construction (or the corresponding iteration stage) and the values of the parameters given in the database.

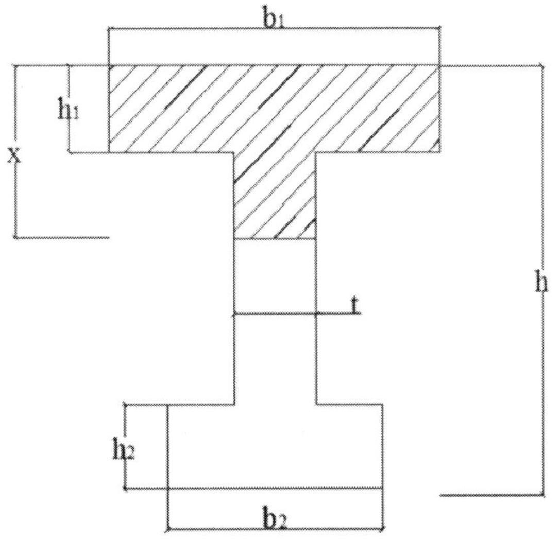

Fig. 2. Scheme to determine the parameters of the approximant for a reinforced concrete I-section with a normal crack.

Example 2. Consider the concrete case of calculating the torsional stiffness of hollow-core slabs according to the proposed technique. The section of a hollow-core plate can be represented with sufficient accuracy in the form shown in Fig. 3.

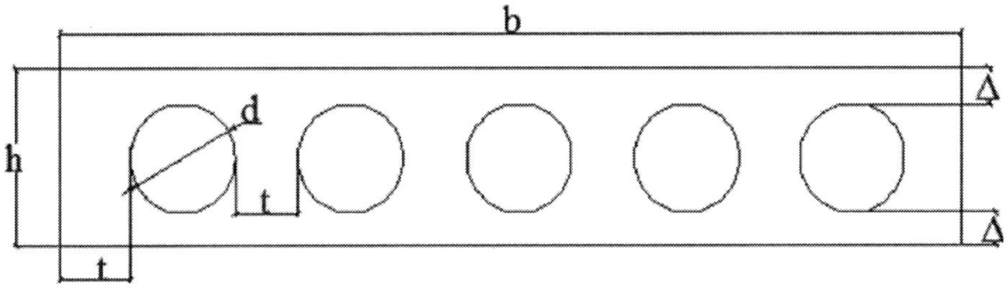

Fig. 3. Scheme to determine the approximation parameters for a hollow-core plate with a normal crack.

The function of mutual displacement of the crack banks in this case will have the form:

$$\Delta_{crc} = f(b,h,d,t,\Delta,n) \tag{2}$$

where n is the number of voids; The remaining notations are shown in Fig. 3.

In order to approximate the numerical experiment data in this case, it is effective to use the apparatus of the best uniform (Chebyshev) approximation using generalized polynomials [9,10]. Thus, for example, using this apparatus for a T-element with a width of the upper flange $b1$, thickness h_1, width of the edge b_2, height of the edge h_2, based on a series of Lira calculations using volumetric finite elements, the result of a numerical experiment for the angle of mutual rotation φ of two blocks separated by a normal crack in the form:

$$\varphi = \frac{0.00011792 b_1 h_1 + 0.1886524}{b_2 h_2 + 16.95905} \tag{3}$$

Formula (3) makes it possible to calculate the angle of mutual rotation of the crack edges for any values of the geometric parameters of the T-element. It should be noted that this formula was obtained on the basis of data from a numerical experiment with varying the geometric data of the T-section at a specific height of the compressed zone (in this case it is the thickness of the shelf of the brand). If the height of the compressed zone also needs to change, then the rotation angle function φ will contain not four variables (as in formula (3)), but five, including the height of the compressed zone (the zone through which torque is transmitted from block to block). But in any case, once obtained (although based on a fairly complex set of data from a numerical experiment), this function can be used in the design practice as many times as you like.

It should be noted that the advantages of the above method of determining the torsional stiffness by creating a database of approximant parameter values is the possibility to take into account the nonlinear properties of concrete. Since the calculation taking into account the nonlinear properties of materials is carried out using iterations, then new deformation characteristics should be adopted at each step of iterations. In the database, the mutual displacement of the shores of the normal crack is a function of the geometric parameters and the height of the zone compressed from the bending (normal crack height) of the expression (1) obtained for specific fixed values of the elastic modulus E and the shear modulus G of the concrete. Therefore, having experimental data of these parameters for a particular case under consideration, at each step of the iteration we use an expression of the type (1) with correction for the value of the elasticity and shear module at the current iteration to these parameters accepted in the database.

Creating a library of approximants will greatly simplify the solution of many complex problems, where the number of such elements would be much smaller than the number of finite elements using the traditional finite element method.

A new approach to the determination of torsional stiffnesses of reinforced concrete elements with cracks is proposed, which allows solving torsion problems of reinforced concrete elements of any cross-section using methods for approximating the displacement functions at the crack location obtained from solving a number of problems using the finite element method. It should be noted that the use of the best Chebyshev approximation in this apparatus makes it possible to improve significantly the efficiency of solving problems. Also the significant effect of torsional stiffnesses on the spatial work of bridge elements and overlapping under the action of local band loads is shown.

In the long term, it is proposed to define functions of type (2) or (3) for solving the problems of reinforced concrete elements torsion of different cross sections and their various sizes, as well as extending the proposed approach for the calculation taking into account the nonlinear properties of reinforced concrete.

References

1. T. Azizov, Rigidity of Reinforced Concrete Elements, **10**, 576-590 (2009)

2. T. Azizov, *Spatial work of reinforced concrete floors. Theory and methods of calculation* (Poltava, 2006)

3. O. Melnyk, *Stiffness and strength of reinforced concrete hollow triangular section elements with normal cracks torsional* (Odessa, 2013)

4. T. Azizov, International science Ukrainian edition, **12,** 55-59 (2010)

5. N. Sribniak, *Torsional stiffness of concrete elements overlap with normal cracks* (Odessa, 2009)

6. Bernardo L.F.A., Sérgio M. R. Lopes L.F.A., Luís, Materials and Structures **41**, 1143-1167 (2008)

7. T. Azizov, Bulletin of the Odessa State Academy of Construction and Architecture, **39**, 4-9 (2010)

8. T. Azizov, Bulletin of the Odessa State Academy of Construction and Architecture, **33,** 4-11 (2009)

9. A. Kalenchuk-Porkhanova, L. Vakal, Computer means, networks and systems, **7,** 32-38 (2008)

10. A. Kalenchuk-Porkhanova, L. Vakal, Cybernetics and systems analysis, **6,** 155-164 (2009)

Efficient construction of the motorway and highway bridge superstructure (experimental studies)

Vladimir Babayev[1], *Mark* Bekker[1], *Valeriy* Shmukler[1*], *Sergey* Bugayevskiy[2], *Roman* Kaplin[1], and *Yuriy* Krul[2]

[1]O.M.Beketov National University of Urban Economy in Kharkiv, 61002, Kharkiv, Marshal Bazhanov str., 17, Ukraine
[2]Kharkiv National Automobile and Highway University, 61002, Kharkiv, Yaroslava Mudrogo str., 25, Ukraine

Abstract. The design of the span structure of the bridge was presented. Its structure was formed by the creation of a composite steel and concrete element consisting spatial steel perforated boxlike-section blocks, a system of shear links of a special type and a monolithic reinforced concrete slab containing inside emptiness-formers The novelty of the design and methods of its creation predetermined the need for experimental testing. It was implemented in full-scale conditions and had, in addition to research, acceptance nature. In this connection, the article presents the results of tests of the composite steel and concrete structure of the superstructure of the bridge over Sukhoy Torets in Barvenkovo. The nature of its deformation under the static and dynamic loads is shown. During the tests, the components of the stress-strain state were determined at the distinguished points of the structure, fixed under static load. The amplitude and frequency response of the bridge unit were established for various dynamic impact options. Based on the obtained data, the structural resistance features under load impacts was analyzed, the design model of the bridge was verified, and the compliance of the accepted design parameters with the limitations of the applicable reference documents was confirmed.

1 Introduction

Most bridges of Ukraine were built in the 1950s and 1960s. Due to the considerable service life and abnormal operation, a large part of these structures have defects and damage to the superstructure elements, which significantly reduce the durability and load-carrying capacity of the entire structure. Many bridges have a non-standard span under the current classification. When such span structures need to be replaced, the issue arises of choosing an alternative design, which generally should be superior at least to the known standard solutions of the past years.

* Corresponding author: konstruktsii@gmail.com

1.1 Initial data

The bridge over the river Sukhoy Torets was built in 1955. The structure was designed as a 4-span reinforced concrete plate-girder structure. The main girders of the span structure are made of reinforced concrete, with the nonstandard span 14.06 m. The overall width of the roadway is 7.0 m. The bridge piers are solid, concrete, with a steel and concrete pier cap. The distance between the pier axes is 14.1 m. The total length of the bridge was 56.4 m. The design load at the time of construction of the structure was N-13, NG-60. The overall view of the bridge is shown in the figure 1.

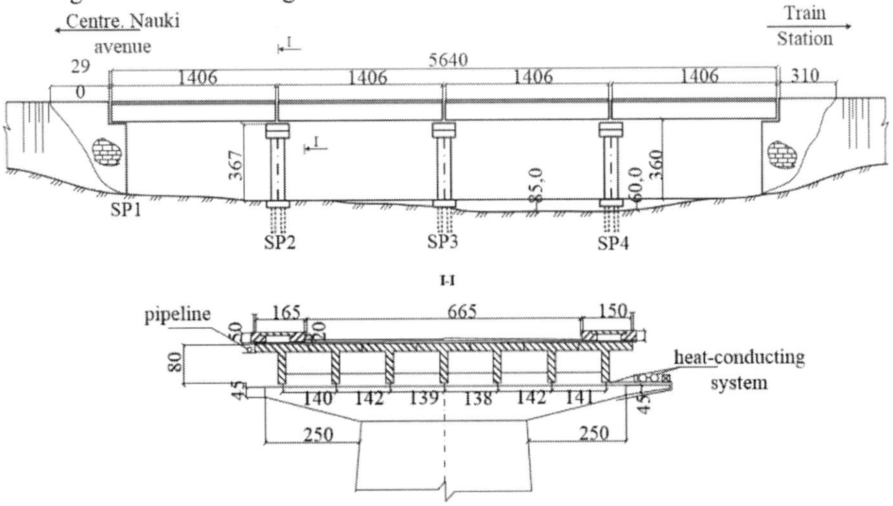

Fig. 1. The diagram of the overall view of the bridge over Sukhoy Torets in Barvenkovo.

Heavy military equipment that moved on the bridge resulted in a deflection up to 15 cm in the span structure, although it did not collapse. Cracks with a width of more than 10 mm appeared on the main girders of the superstructure. The construction was classified as the 5th operational condition, i.e. the bridge was estimates as non-operable. Major overhaul with the replacement of all superstructures became necessary for its further operation. Since the bridge piers were in satisfactory condition, the girders were replaced on the existing piers, with their partial restoration and reinforcement.

1.2 Construction solution

Due to the non-standard length of the span, a two-component system was suggested for use as a new superstructure, which could be produced according to individual dimensions without significant extra labour intensity during the manufacture and installation of the structure. The superstructure is a steel-reinforced concrete structure consisting of perforated metal units of the boxlike-section and a steel-reinforced concrete carriageway slab laid on the non-removable formwork made of corrugated sheet [1]. This construction is a modification of the shell-type bridge superstructure system [2]. Unlike the prototype, in the caisson shell cross-section of this superstructure, four blocks with transverse spreading were installed at a distance of 2.4 m (between the axes of adjacent units). The units were combined into a single superstructure by arrangement of a 200 mm thick reinforced concrete slab. The slab is made of C20/25 grade concrete. The supports made of the rectangular pipe and inclined reinforcing bars are welded to the upper unit flanges to ensure the joint performance of metal units with a reinforced concrete carriageway slab. These elements, together with the connected reinforcing mesh of the slab, are a spatial shear link system that ensures efficient performance of the steel-reinforced concrete structure.

The cross-section and the overall view of the new superstructure are shown in Figure 2.

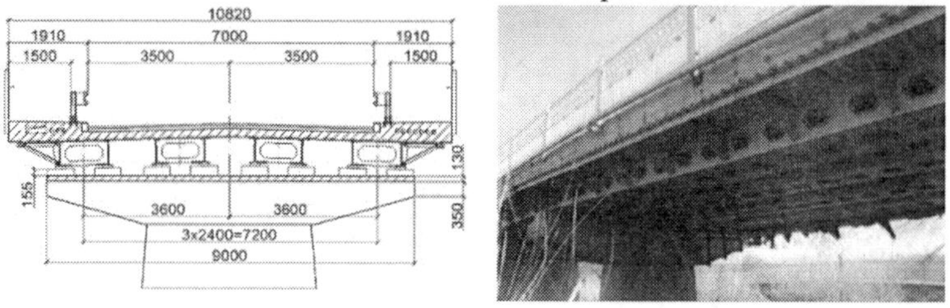

Fig. 2. Steel-reinforced concrete superstructure of the bridge over the river Sukhoy Torets in Barvenkovo.

As indicated above, the superstructure units have a boxlike section (Fig. 3). Elements of the units were made of perforated sheet steel of grade 15ХСНД. Non-waste technology was applied for manufacture of all perforated elements of the units. The unit structure includes two walls (main girders), a lower plate and transverse diaphragms. The diaphragm spacing is 1m. The unit wall thickness is 10 mm, the bottom plate wall thickness is 30 mm, and the diaphragm wall thickness is 12 mm.

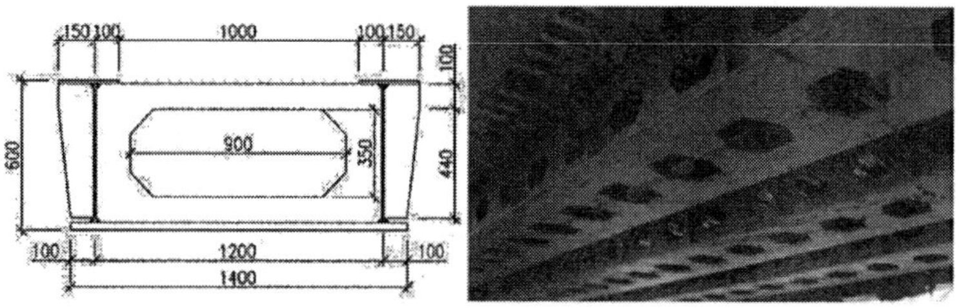

Fig. 3. Construction of metal units of the steel-reinforced concrete bridge superstructure.

The working design of the steel-reinforced concrete bridge superstructure was elaborated at the Department of Building Constructions of Beketov National University of Urban Economy (Kharkiv) for A-15, NK-100 load. All construction work was performed by Stalkonstruktsia LLC.

In Table 1, material consumption required for production of one span of the steel-reinforced concrete superstructure is summarized. Specifications of the steel and concrete (standard design series 3.503-14) and metal (standard design series 3.503.9-43/89) superstructures are also given here for comparison.

Table 1. Materials consumption for the bridge superstructures.

Superstructure design option		Materials consumption for m^2 of the superstructure	
		Steel spread, t/m^2	**Concrete spread, m^3/m^2**
1	Steel and concrete superstructure	0.14	0.55
2	Metal superstructure	0.19	0.17
3	Suggested superstructure	0.17	0.2

2 Tests

Researchers of the Department of Building Constructions of Beketov National University of Urban Economy in Kharkiv performed static and dynamic tests for study and acceptance of the construction.

Test object – steel-reinforced concrete superstructure of the bridge over the river Sukhoy Torets in Barvenkovo.

Test objective – evaluation of stress-strain state components in the distinguished points of the superstructure unit under static load, as well as determination of amplitudes and frequencies of self-induced and forced vibrations of the unit under dynamic impact.

The following research tasks were fulfilled to achieve the above test objectives:
- measuring deflections in the distinguished points of the construction;
- measuring deformations and determination of actual stresses occurring in the distinguished unit cross-sections;
- comparing experimental and theoretical data for the stress-strain state components (verification of the design finite element model); and
- measuring frequencies and the corresponding amplitudes of their self-induced and forced vibrations, and consequent establishment of the possibility of resonance occurrence.

2.1 Loading system

Two loaded KAMAZ–6520 dump trucks were accepted as a test load. The vehicles were weighted before testing. The reference weight of each vehicle was Q=19.2 t. During testing, the vehicles were placed on the superstructure in accordance with the previously proved and accepted loading patterns (within the testing program).

During the static part of testing, four load arrangement patterns were provided:
- placing a dump truck with its rearmost axle in the middle of the span near the right safety fence (Fig. 4);
- placing two dump trucks with their tailgates adjacent to each other, near the right safety fence (Fig. 5);

Fig. 4. First static load position pattern.

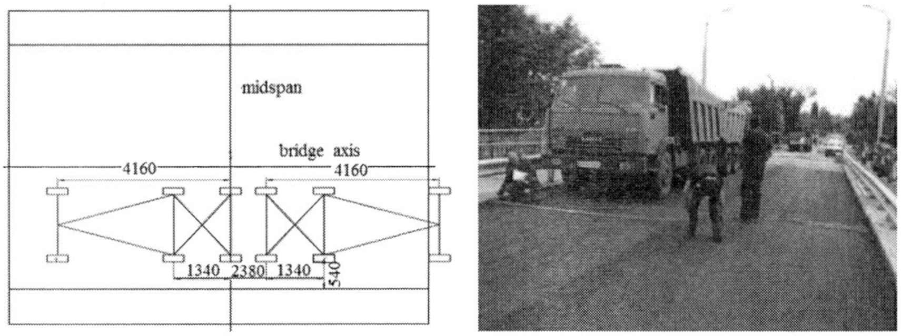

Fig. 5. Second static load position pattern.

- placing two dump trucks with their tailgates adjacent to each other, along the bridge axis (Fig. 6);

- placing two dump trucks with their tailgates adjacent to each other, near the left safety fence (Fig. 7).

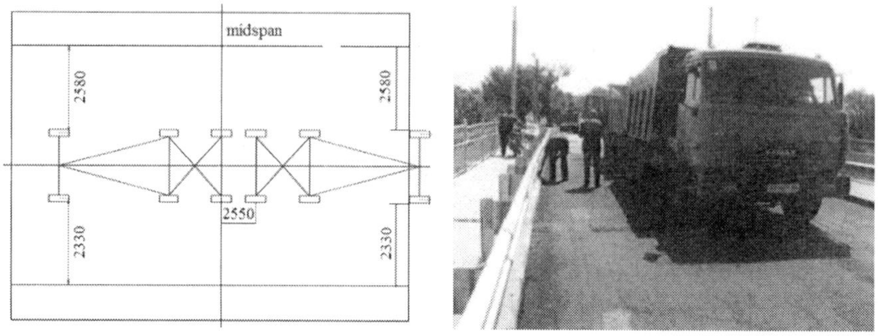

Fig. 6. Third static load position pattern.

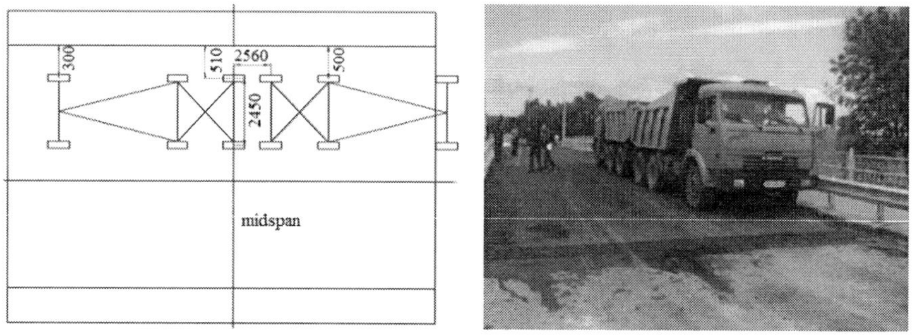

Fig. 7. Forth static load position pattern.

Once the vehicles were placed, the tests performing researchers were monitoring closely the readings of the meters in order to stop loading, if necessary (case of emergency). Changes in the state of the bridge were recorded without load, during loading and during subsequent unloading. The final measurements were made to determine the permanent deformations in the structure.

During the dynamic part of testing, the characteristics of self-induced and forced vibrations of the superstructure were determined. Self-induced vibrations per se occur when the span structure is operating in the free vibration mode after the shock impact. The dynamic effect of the shock load was studied by dropping a weight of 1 t (85x48x55 cm foundation block) from the height of 1.2 m onto the carriageway.

A temporary load in the form of vehicles which corresponds to the actual mode of operation of bridges was used to determine the characteristics of forced vibrations. The test program provides the following options of dynamic impact:
- movement of a column of two KamAZ-6520 dump trucks at a speed of 40 km/h and a 2-second traffic interval;
- movement of a single dump truck at a speed of 40 km/h over the road surface with an obstacle (a wooden bar 20 cm wide, 4 cm thick and 3.5 m long, palced in the middle of the span); and
- movement of a single dump truck at a speed of 40 km/h and sharp braking in the middle of the span.

2.2 Measurement system

During the static part of testing, values of vertical displacements (deflections), as well as axial and transverse deformations of the elements of the metal unit of the steel-reinforced concrete superstructure were recorded.

Deflections of the structure were recorded using 6-PAO-0,1 Maximov mechanical deflection meters. The principle of operation is that the measured movement of the monitored object is transmitted via the transmission chain (string) to the driving unit and further via the transmission mechanism to the arrow (Fig. 8). One end of the string is fixed directly on the monitored structure or on a rigid frame, hitched with at least one loop over the drive unit and pulled by the load. Deviation of the arrow relative to the scale and the arrow rotation indicator enable to determine the value of the measured movement. The transmission chain assembled from the weights and threads was maintained in constant tension to prevent the influence of external force factors and was checked before each reading.

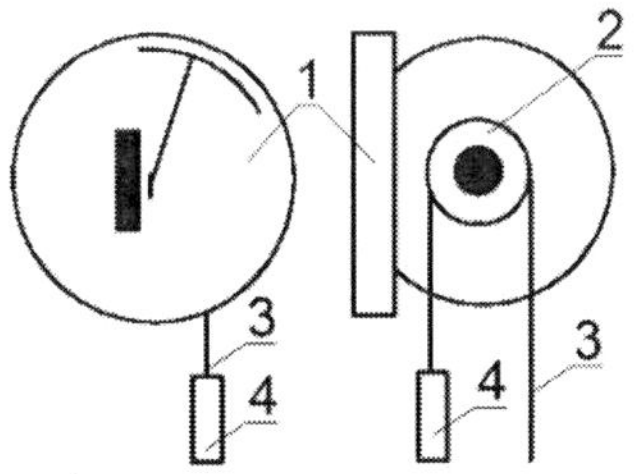

Fig. 8. Instrument arrangement diagram: 1 - Maximov deflection meter, 2 - roller, 3 - thread, 4 – load.

The deflection meters were fixed to the unit in the bearing parts, in the quarters and in the middle of the span. The instruments were arranged on the preliminarily prepared horizontal grounds using mount pillars and clamps. Two instruments were arranged in the quarters and in the middle of the span to record possible twisting on the unit. Deflection meter arrangement diagram and overall view are shown in Figure 9.

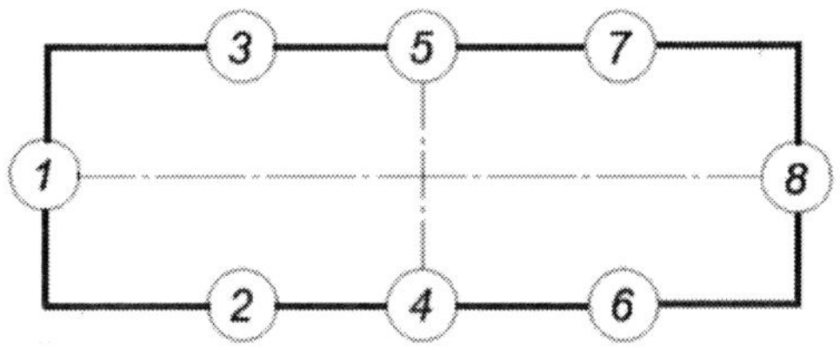

Fig. 9. Deflection meter arrangement diagram and overall view.

Gugenberger mechanical strain gauges were used to record deformations occurring in the walls of the metal unit. The gauges were arranged around the orifices, in the bearing part and in the middle of the span. The paint coat was removed on the prearranged grounds for installation of the instruments to detect deformations of the metal walls of the unit. Gauge arrangement diagram and overall view are shown in Figures 10-11.

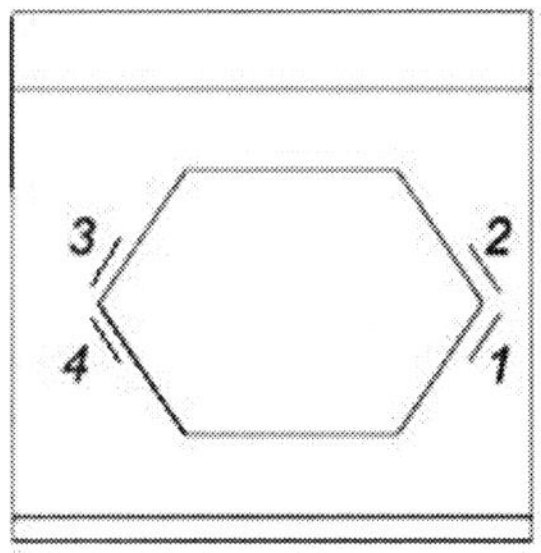

Fig. 10. Gauge arrangement diagram and overall view in the bearing part of the unit.

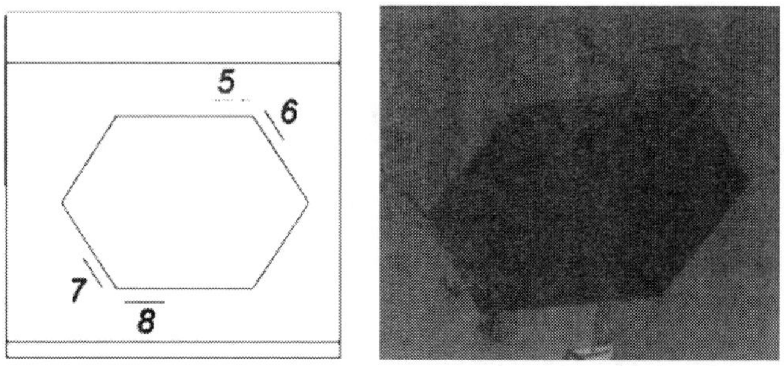

Fig. 11. Gauge arrangement diagram and overall view in the middle of the span.

Self-induced and forced vibration characteristics of the superstructure were determined using AGAT-M portable vibration analyzer. Using this instrument as a data collector together with DIAMANT-2 software enabled to obtain the amplitude frequency response (AFR) of the superstructure under different dynamic impact conditions by means of the built-in function of complex vibration expansion using discrete Fourier transformation.

2.3 Preparatory stage of the experimental testing

The following preparatory works were performed before testing:
– study of the physical and mechanical characteristics (properties) of the constituent materials of the constructions;
– theoretical analysis of the construction behaviour under action of the static test load.

The non-destructive method with OMSh-1 scratch hardness tester was used to determine concrete characteristics. The determined concrete grade is C20/25. Concrete deformation diagram is allowed to be presented as seventh power polinomial [3]:

$$\sigma = \alpha_1\varepsilon + \alpha_3\varepsilon^3 + \alpha_5\varepsilon^5 + \alpha_7\varepsilon^7 \tag{1}$$

Unknown coefficients in (1) can be expressed in terms of the the material deformation characteristics [3]:

$$\alpha_3 = \frac{\varepsilon_u^4}{2\left(\varepsilon_u^2-\varepsilon_R^2\right)^2}\left[\frac{1}{\varepsilon_R^2}\left(\frac{R}{E_0\varepsilon_R}-1\right)\left(5-7\frac{\varepsilon_R^2}{\varepsilon_u^2}\right)-\frac{1}{\varepsilon_R^2}\left(\frac{\varepsilon_R^2}{\varepsilon_u^2}-1\right)+\frac{2\varepsilon_R^4}{\varepsilon_u^6}\left(\frac{\beta R}{E_0\varepsilon_u}-1\right)\right]$$

$$\alpha_5 = \frac{\varepsilon_u^4}{2\left(\varepsilon_u^2-\varepsilon_R^2\right)^2}\left[\frac{1}{\varepsilon_R^4}\left(7\frac{\varepsilon_R^2}{\varepsilon_u^2}-3\right)\left(\frac{R}{E_0\varepsilon_R}-1\right)-\frac{1}{\varepsilon_R^4}\left(1-\frac{\varepsilon_R^4}{\varepsilon_u^4}\right)+\frac{4\varepsilon_R^2}{\varepsilon_u^6}\left(\frac{\beta R}{E_0\varepsilon_u}-1\right)\right]$$

$$\alpha_7 = \frac{\varepsilon_u^4}{2\left(\varepsilon_u^2-\varepsilon_R^2\right)^2}\left[\frac{1}{\varepsilon_R^4\varepsilon_u^2}\left(\frac{R}{E_0\varepsilon_R}-1\right)\left(3-5\frac{\varepsilon_R^2}{\varepsilon_u^2}\right)-\frac{1}{\varepsilon_R^4\varepsilon_u^2}\left(\frac{\varepsilon_R^2}{\varepsilon_u^2}-1\right)+\frac{2\varepsilon_R^4}{\varepsilon_u^6}\left(\frac{\beta R}{E_0\varepsilon_u}-1\right)\right]$$

where E_0 – initial stress-strain modulus; ε_R – compressive strain in the concrete at the peak stress; ε_u – ultimate strain; R – strength limit; βR – stress corresponding to the ultimate strain. Laboratory tests of the standard metal test plates were performed to evaluate mechanical characteristics of the steel. Steel grade was determined to be 15ХСНД, which proved the design characteristics.

Concrete and steel deformation diagrams were plotted using the obtained data (Fig. 12).

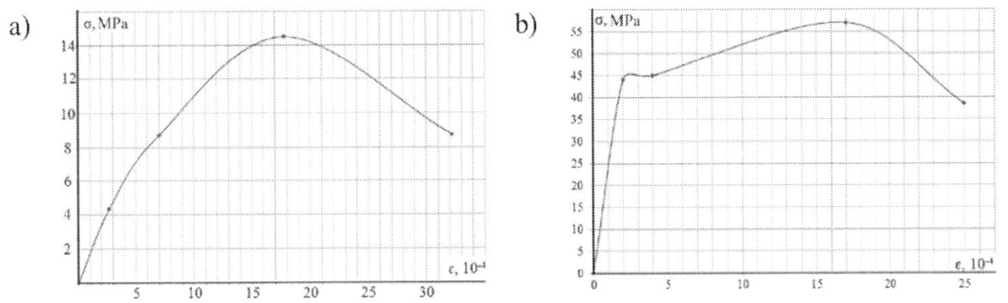

Fig. 12. Material deformation diagrams: a) for concrete; b) for steel.

Theoretical calculations were performed in the SCAD Office environment, with preliminary building of a finite-element model of the superstructure. The superstructure construction was simulated with shell elements of zero Gaussian curvature, bumpers – with special rod elements. Reference mobile loads (A-15 and NK-100) and static test load were chosen as loads in accordance with the accepted loading patterns. The structural design of the suggested superstructure is shown in Figure 13.

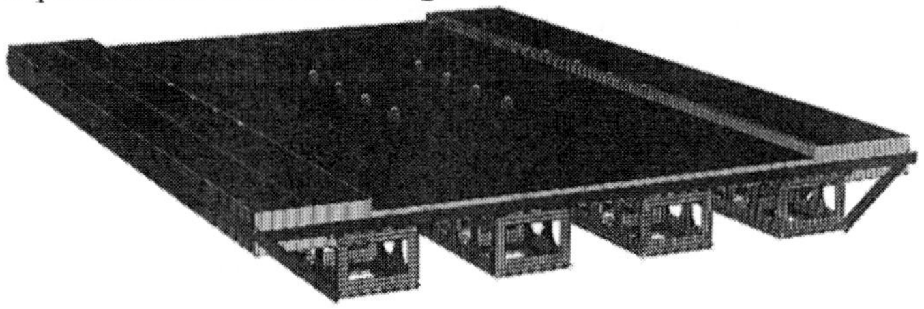

Fig. 13. Structural design of the superstructure.

With the span length 14.06 m, the greatest stress in the superstructure will evidently be caused by load NK-100 located in the middle of the span along the bridge axis. The maximum deflection of the construction, with an allowance for the dead weight, is 28.1 mm, which is 1/498 of the span length (Fig. 14). The deflection caused by the dead weight of the construction was 17.1 mm. Maximum stress (in the lower flange) was 2,384 kg/cm^2, minimum stress (in the upper flange) was 550 kg/cm^2 (Fig. 15).

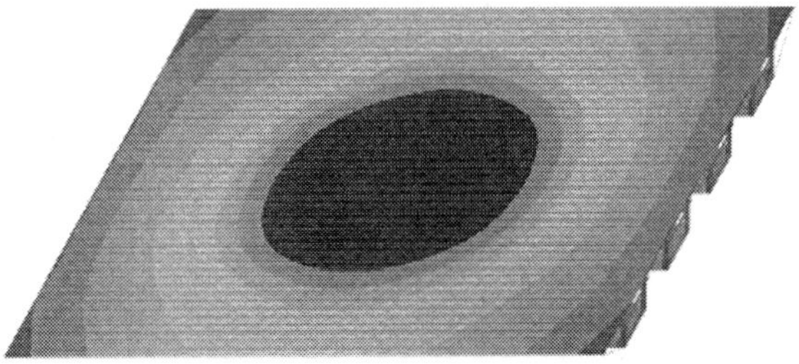

Fig. 14. Isofield of vertical displacement of the construction.

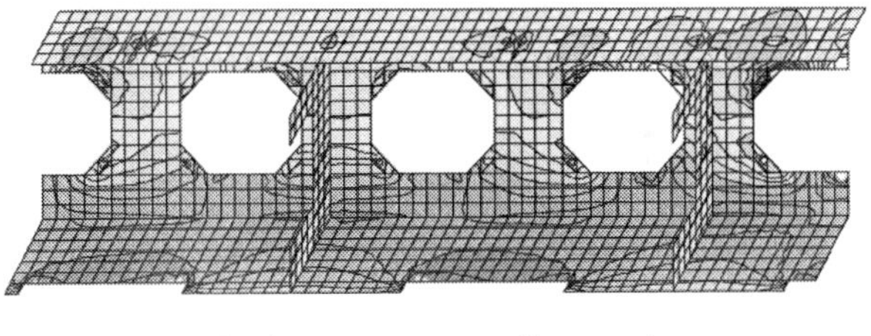

Fig. 15. Isofield of normal stresses in the superstructure (fragment).

Values of vertical displacements of the constructions and principal stresses in the cross-sections of the unit were obtained during calculation of the superstructure for the test load. The obtained theoretical computation data are summarized in Table 2 as isofields of vertical displacement and isofields of main stresses.

Maximum deflection of the construction in the middle of the span of 3.04 mm was obtained with the second static loading pattern. The stress values at the corners of orifices and in the concentration points did not excess 250 kg/cm^2.

3 Test data

3.1 Static test data

Construction deflection lines were plotted for all 4 static loading patterns based on processing of the values measured with deflection meters and theoretical data obtained during computation (Fig. 16 to 19).

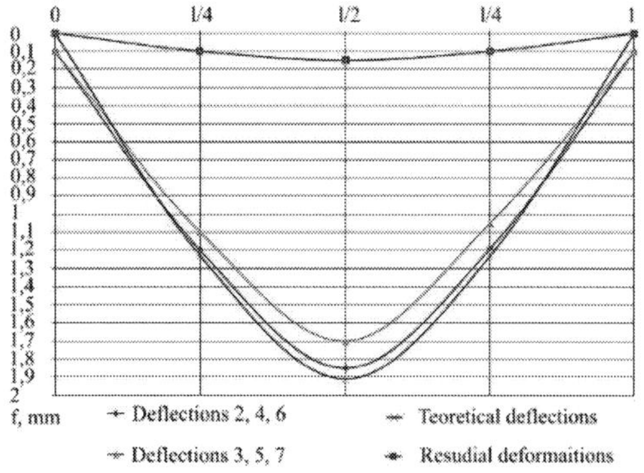

Fig. 16. Structural deflections (first pattern).

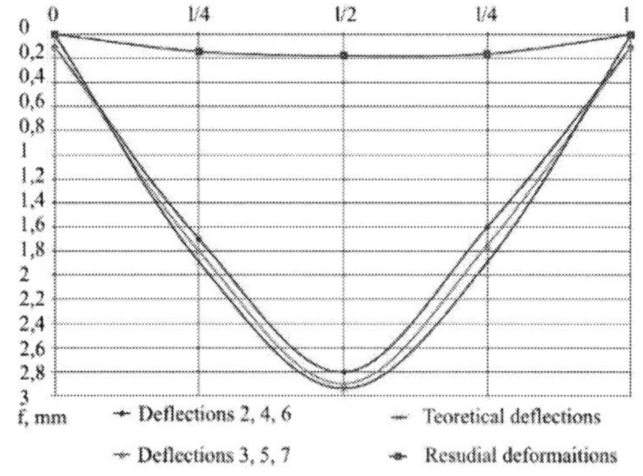

Fig. 17. Structural deflections (second pattern).

Maximum test deflection was recorded with the 2nd load pattern *f*=2.9 mm. No unit twisting was recorded. Permanent deformations recorded after unloading (for each load) did not exceed 10%. Discrepancy of the obtained test data with the theoretical deflections did not exceed 5%.

The deformations measured with mechanical strain gauges are summarized in Table 2.

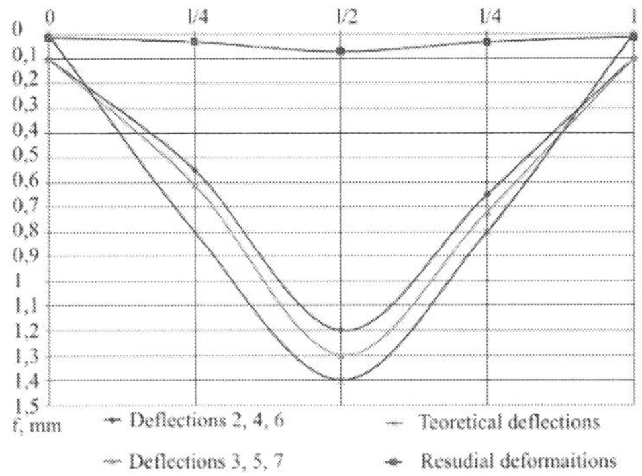

Fig. 18. Structural deflections (third pattern).

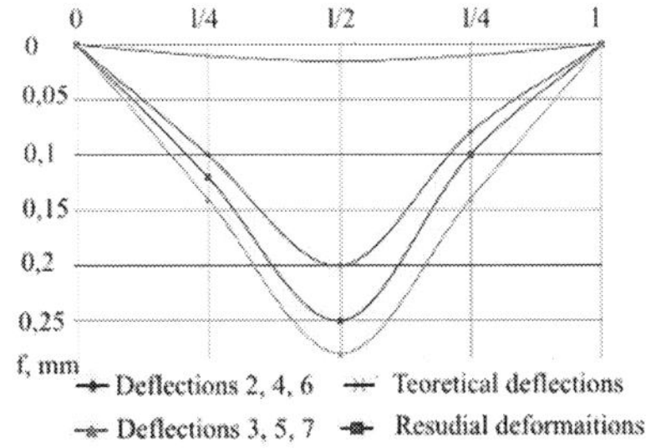

Fig. 19. Structural deflections (forth pattern).

Table 2. Deformation values measured with mechanical strain gauges.

Load pattern	ε_1, mm·10^{-4}	ε_2, mm·10^{-4}	ε_3, mm·10^{-4}	ε_4, mm·10^{-4}	ε_5, mm·10^{-4}	ε_6, mm·10^{-4}	ε_7, mm·10^{-4}	ε_8, mm·10^{-4}
Pattern 1	0.6	0.6	0.6	0.6	0.78	0.78	0.78	0.78
Pattern 2	0.82	0.79	0.78	0.79	1	0.8	1	0.8
Pattern 3	0.67	0.67	0.67	0.67	0.79	0.79	0.79	0.79
Pattern 4	0	0	0	0	0.25	0.25	0.25	0.25

Actual stress values occurring in the tested cross-sections of the unit were calculated using the "σ-ε" diagram (Table 3). Point numbering was accepted as shown in Figures 10 and 11.

Table 3. Stress values obtained using the diagram "σ-ε".

Load pattern	σ_1, kg/cm^2	σ_2, kg/cm^2	σ_3, kg/cm^2	σ_4, kg/cm^2	σ_5, kg/cm^2	σ_6, kg/cm^2	σ_7, kg/cm^2	σ_8, kg/cm^2
Pattern 1	150	150	150	150	180	180	180	180
Pattern 2	200	185	180	185	250	195	250	195
Pattern 3	160	160	160	160	185	185	185	185
Pattern 4	0	0	0	0	50	50	50	50

Maximum relative deformation value $1\cdot10^{-4}$ mm corresponds to the stress value of 250 kg/cm^2. The error in comparison of the experimental and theoretical stress values was 5-10%.

3.2 Dynamic test data

During the dynamic part of testing, amplitude frequency response (AFR) of the superstructure unit under different dynamic impacts were obtained. The diagram describing the dependence of amplitudes from the self-induced vibration frequencies is shown in Figure 20.

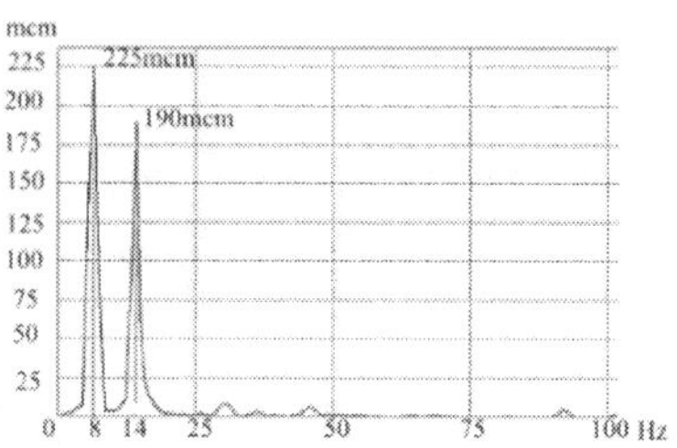

Fig. 20. Self-induced vibration frequency range.

Two peaks at vibration frequency 8 Hz (amplitude 225 μm) and 14 Hz (amplitude 190 μm) can be seen on the self-induced vibration frequency range. These frequencies correspond to period 0.019 sec and 0.011 sec, respectively, which do not fall into the prohibited range $0.45 \leq T \leq 0.6$ sec.

Any gradually damped oscillatory motions of the system after the impulse excitation can be described with the formula:

$$y(t) = A_0 e^{-\delta t} \sin(2\pi f t) \quad (2)$$

where $y(t)$ – value of displacement of the oscillating body; A_0 – initial vibration amplitude; f – vibration frequency; δ – attenuation coefficient.

However, attenuation coefficient cannot describe completely the attenuation intensity. Therefore, as it is known, logarithmic decrement of vibrations is introduced to characterize the attenuation intensity [5]. In view of the connection of the attenuation coefficient with the frequency $\delta = d \cdot f_0$ and attenuation coefficient with the amplitude frequency response of the vibratory system $\delta = \pi \cdot \Delta f$, logarithmic decrement can be expressed as follows:

$$d = \pi \frac{\Delta f}{f_0} \quad (3)$$

where f_0 – vibration frequency at the maximum amplitude; Δf – bandpass of the vibratory system at 0.707 of the maximum.

Characteristics of the self-induced vibrations are given in Table 4.

The obtained values of the self-induced vibration decrement are consistent with those for the steel-reinforced concrete bridges.

The diagram describing the dependence of amplitudes from the forced vibration frequencies under different dynamic impact options with the mobile load are shown in Figures 21-23.

Table 4. Characteristics of the self-induced vibrations of the superstructure.

Self-induced vibration frequency, Hz	Self-induced vibration period, sec	Self-induced vibration decrement
8	0.019	0.043
14	0.011	0.049

Only one strongly pronounced peak stands out on the forced vibration frequency ranges. Also, forced vibration frequencies vary within the range of 5 to 12.5 Hz, while the recorded amplitudes do not excess 80 mm.

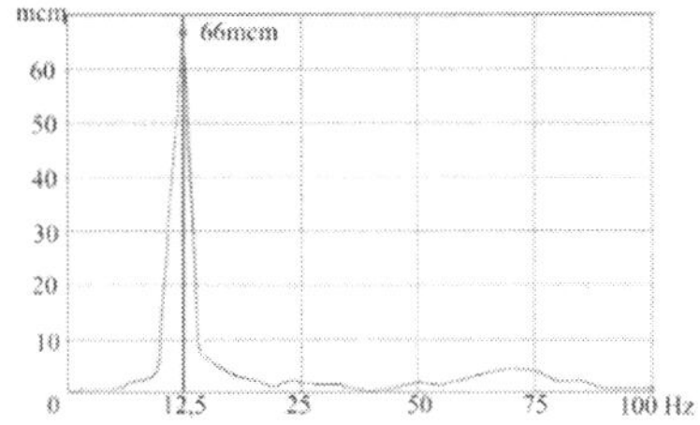

Fig. 21. Forced vibration frequency range (pattern 1).

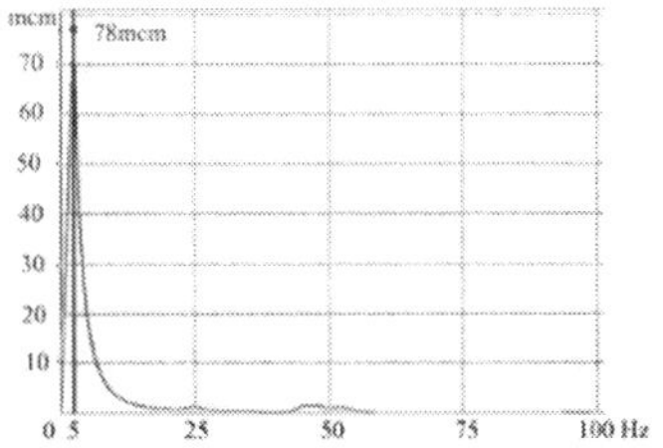

Fig. 22. Forced vibration frequency range (pattern 2).

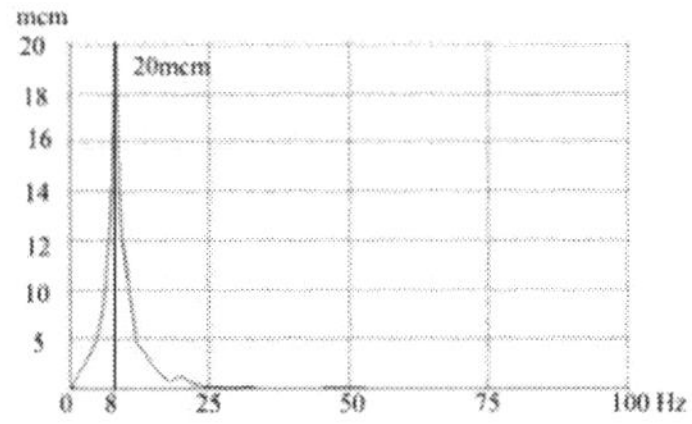

Fig. 23. Forced vibration frequency range (pattern 3).

The frequency range defining virtual resonance vibration modes, in its turn, can be assigned as ±20% change of self-induced vibration frequencies (at least, for the fundamental period). Thus, the prohibited range for the forced vibration frequencies is:

$$0{,}8\omega \le f \le 1{,}2\omega \tag{4}$$

where f – forced vibration frequency; ω – self-induced vibration frequency.

Substituting successively the measured self-induced vibration frequency values into the in inequation (4) we can notice that:

- the prohibited forced vibration range lies within the range of 6.4 to 9.6 Hz and 11.2 to 16.8 Hz;
- the forced vibration frequencies of the bridge superstructure unit are within the prohibited range when loaded by the first and third mobile dynamic load pattern.

Nevertheless, occurring resonance can be considered non-dangerous, if the deflection from the static load and dynamic extra load does not exceed the estimated deflection from the reference load multiplied by the dynamic coefficient [6]. Based on the quoted approach we can write the safety condition for the resonance:

$$y_{L/2} + A << y(1 + \mu) \tag{5}$$

where A – frequency amplitude; y – maximum deflection of the construction from the reference load; $y_{L/2}$ – static deflection in the middle of the span from the considered load; $1 + \mu$ – amplification factor (rated value for the steel-reinforced concrete bridges is accepted as $1 + \mu = 1.4$).

Is it was mentioned above, maximum estimated deflection of the construction caused by reference load NK-100 is (without accounting for the dead weight of the construction) y=11 mm. Static deflection when a vehicle is placed in the middle of the span is $y_{L/2} = 2\text{mm}$.

Substituting the data obtained for the first and third dynamic loading patterns to the left and right parts of the inequation (5), we can obtain:

– for the first pattern $y_{L/2} + A = 2 + 0.066 = 2.066\,\text{mm}$ and $y(1+\mu) = 11 \cdot 1.4 = 15.4\text{mm}$, i.e. $2.066\text{mm} \ll 15.4\text{mm}$;

– for the third pattern $y_{L/2} + A = 2 + 0.02 = 2.02\text{mm}$ and $y(1+\mu) = 11 \cdot 1.4 = 15.4\text{mm}$ and, as consequence, $2.02\text{mm} \ll 15.4\text{mm}$.

The inequations are evidently fulfilled very safely. It shows that none of the mentioned dynamic impacts cannot sway the bridge up to dangerous amplitude values.

Based on the conducted studies, we can conclude that:

- the values of actual deflections of the structure of the steel-reinforced concrete superstructure in the quarters and middle of the tested span were determined. The maximum experimental deflection was recorded for the 2nd loading pattern f = 2.9 mm. No unit twisting was recorded. The permanent deformations recorded after unloading at each loading stage did not exceed 10%;

- the values of the actual stresses occurring at the corners of the orifices on the walls of the metal unit were calculated based on the processed deformation values obtained using mechanical strain gauges and laboratory tests for determination of mechanical characteristics of steel. The stress value of 250 kg/cm^2 corresponds to the maximum value of relative deformations of $0.15 \cdot 10^{-4}$ mm. The permanent deformations recorded after unloading at each loading stage did not exceed 10%;

- the error in comparison of the experimental and theoretical stress values was 5-10%. Thus, the calculated finite element model of the superstructure was verified;

- during the dynamic part of testing, amplitude-frequency response (AFR) of the superstructure unit were obtained for different dynamic impact options and diagrams of the spectral characteristics of self-induced and forced vibrations were built. The vibrations are damped. The logarithmic decrement of self-induced vibrations of the superstructure is within the range 0.043-0.049. The fulfilment of the safety condition of the resonance occurring in the first and third dynamic impact patterns was estimated. The safety condition is fulfilled with large tolerance.

The obtained results enabled to commission the bridge in May 2016.

References

1. Y.N. Krul: Kommunalnoe khozyaystvo gorodov, **101**, 31-40 (2011).
2. V. Shmukler, M. Shutkin, B. Shutkin, Y. Krul Patent **74599** MPK (2012.01) E01D 1/00 Ukraine (2012).
3. V. Shmukler Zbirnik naukovyh pratc Ukrainian state university of railway transport, **167**, 54-69 (2017)
4. N. Bezukhov, O. Luzhin, N. Kolkunov A manual for specialized construction universities. 3-rd edition, Moscow, High scool, 264 (1987).
5. G. Atabekov Textbook for universities. 3-rd edition, ispr. i dop., Moscow: Energy, 319, (1966)
6. L. Khazanov Issledovanie moctovykh i tonnelnnykh soorygeniy, Moscow, 101–111 (2006).

New constructive solutions for building of transport construction facilities

Vladimir Babayev[1], and *Valeriy* Shmukler[1*]

[1]O.M.Beketov National University of Urban Economy in Kharkiv, 61002, Kharkiv, Marshal Bazhanov str., 17, Ukraine

Abstract. New structural systems for civil and transport engineering were examined. The basis for the formation of the proposed reinforced concrete structures is the ideology of reducing its dead weight, with a given bearing capacity, the realization of which is accomplished by burial during concreting large-sized liners of a given shape and manufactured from lightweight, inexpensive composite materials. The process of erecting these systems is presented in two forms: for flat structures - using self-tightening concrete, and for curvilinear ones - by using shotcrete technologies. The second direction is presented by steel-reinforced concrete structures. These structural systems were created on the basis of innovative component and methods of rationalization of parameters. The basis of the above methods is a compiler which includes the finite element method, adaptive evolution method and special iterative procedures. Experimental verification of structural solutions and formation procedures for suggested systems was performed. Comparison between theoretical and experimental data is given. Suggested systems were implemented in a number of building companies.

1 Introduction

Search for new advanced constructive solutions and efficient technologies for their implementation is predetermined with the complex of requirements for state-of-the-art construction projects combined with limited resources. At the same time, the dual "design-technology" task means constructing systems that provide a wide variability of special planning solutions, contain necessary properties of environmental positivity, resistance to environmental variability, and economic expediency.

Nevertheless, the processes occurring in the world around us (climate transformations, seismic activity, abnormities of artificial origin, etc.) and the population increase bring to the foreground the issues which have hitherto had rather local, than global solution.

They include, first of all, the need to allocate better lands for agricultural and forestry needs, reconstruction of existing and introduction of new sources of energy, saving of drinking resources and, consequently, creation of "intelligent" buildings and structures [1]. This is the basis for the integrated design of individual elements and structural systems as a whole in a logical combination with engineering support.

*Corresponding author: konstruktsii@gmail.com

The above said is regarded as a reason for creation and implementation of new constructive systems, the main features of which are discussed below.

1.1 General

The need to improve the covering and flooring design, as well as pedestrian bridges, predetermines the search for fast-erecting systems built from large factory-assembled elements. The above is consistent with R. Descartes' concept involving the thesis that any saving is ultimately a time saving.

Steel-reinforced concrete constructions are known to have certain positive prospects here [2]. However, heavy weight of the reinforced concrete part worsens the amplitude frequency response of the structure, thus contributing to increase in dynamic forces (including seismic) and/or occurrence of resonance modes. In addition, the criterion worsens characterizing the ratio of the share of the resource spent for perception of its own weight and those spent for perception of the payload.

The above was the reason for creation of a system that combines efficient design and technological solutions.

1.2 Design and technological process features

The suggested flooring construction is a spatial flat parallel slab-and-rod system (structure) made of two materials: reinforced concrete and steel [3, 4, 5, 6]. The specific feature of the reinforced concrete slab is a profiled steel sheet used as external reinforcement, which is also a non-removable formwork. The sheet is fixed in the plane of the upper boom of the structure by special shear connections and either self-tapping screws or electric plugs. Another fixing element here is cold-drawn wire mesh reinforcement. Lightweight concrete is usually used. Using prefabricated reinforced concrete slabs of the "per cell" size is possible.

The specific feature of the metal structure is its manufacture of flat elements of the "span" size, which simplifies significantly the design of the unit and the on-site assembly, minimizing the time and labour intensity of the assembly. The feasibility of this structure is limited to the span of 60.0 m. The essence of the design solution of this dual aggregate is as follows: formation of a structure consisting of metal modular elements, upper and lower booms, and reinforced concrete monolithic or prefabricated slabs (Fig. 1, 2, 3).

Each modular element is a truss with parallel chord. Truss booms are made from single angles rotated at 45° to the vertical axis. Lattice rods generally have closed cross-section and are attached to metal gusset plates. During the on-site assembly, modular elements are installed rotated at an angle of 45° to the horizon and bolted together. If necessary, longitudinal girders made of metal profiles located perpendicularly to the modular elements can be arranged above and below the resulting fold system. In this case, the upper girders together with the booms of the modular elements form closed sockets, into which the reinforced concrete prefabricated slabs or profiled sheet and reinforcing mesh are laid. In the latter case, concrete is poured on the mesh reinforcement. The above girders in combination with the angle elements and mesh reinforcement are shear connections that ensure the joint performance of the composite (monolithic) reinforced concrete disc with the rods of the spatial metal structure. The lower booms unite the modular elements along the lower plane of the structure. Additional positive feature of the suggested solution is the fact that the angle element (shear coupling) is manufactured using the non-wastes technology for triangle plates (Fig. 3). The principal unit of the modular element is shown in Figure 1. In particular, a non-figured truss can be formed placing a gusset plate between the feathers of the flange angle. This significantly improves the architectural expressiveness of the system being created. In this case, the nodal element is a "table" consisting of a rectangular slab and three triangular bearing plates welded into the boom angle from inside. Furthermore, a positive fact here is a complete two-sided closure of the tubular cross-section of the brace, which minimizes significantly the corrosion processes in these elements.

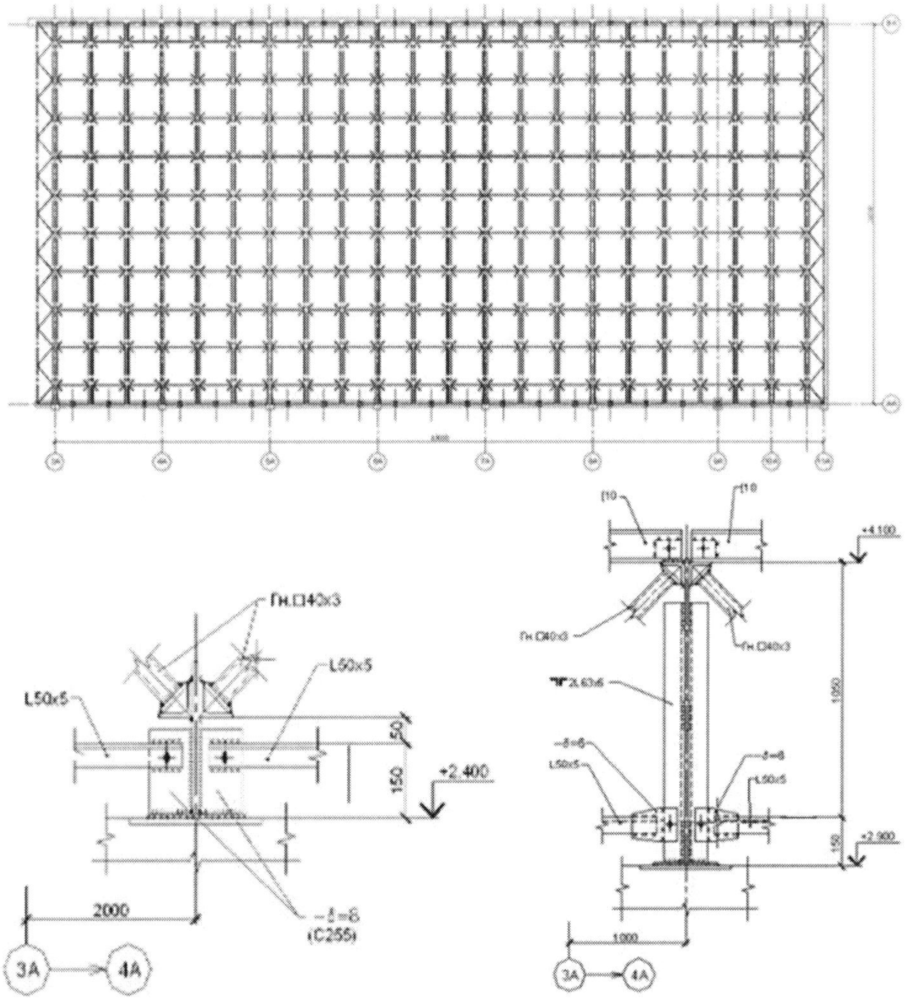

Fig. 1. Flooring arrangement. Flooring diagram. Cross-sections.

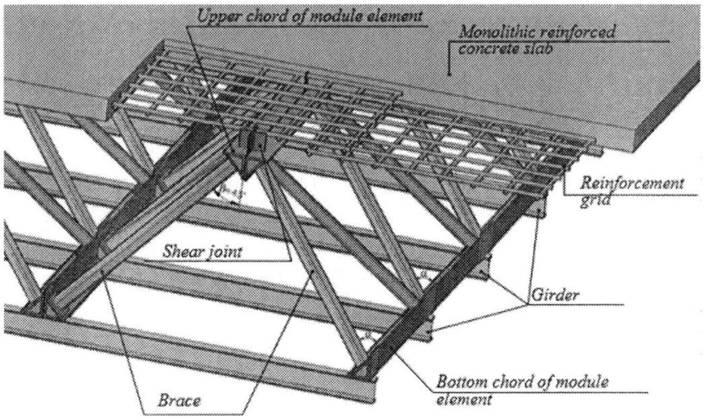

Fig. 2. Suggested flooring design.

The reliability of the suggested design is provided by the same (standard) solution of the connection joints of both upper and lower chords, reduction of welded joints performed on the site, absence of welded joints between the slabs located in the plane of the upper boom and the lattice rods. Simplification, accuracy and speed of assembly are achieved due to rotation of the booms of modular elements by 45° to their vertical axis, which, consequently, ensures a tight fit of the upper and lower booms of modular elements to each other during assembly. Thus, the considered design is simultaneously a conductor for its own assembly. Rational use of the applied materials is ensured by the appropriate location of the reinforced concrete (slab) and the metal (lattice) parts of the structure.

This is due to the fact that the monolithic reinforced concrete slab or prefabricated slabs are located in the upper chord plane and are completely in the compressed zone of the structure (Fig. 2, 3).

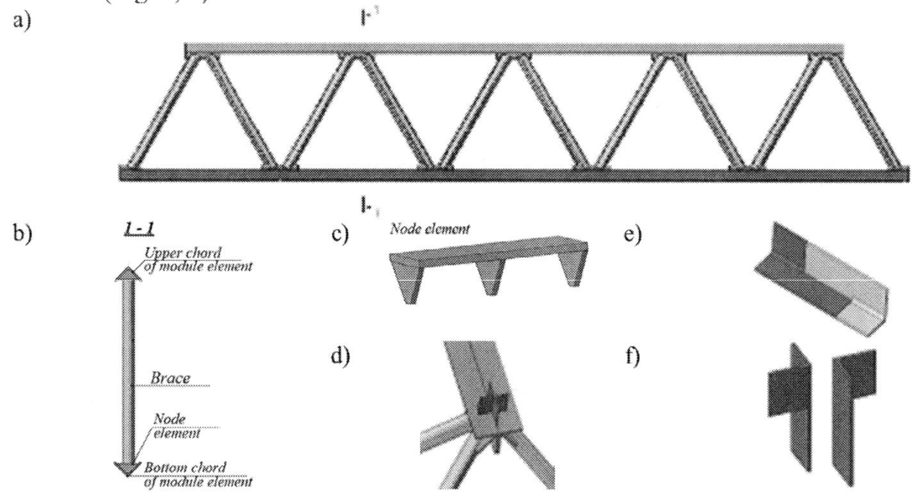

Fig. 3. Properties of the flooring construction: a) modular element; b) cross-section 1-1; c) beating "table"; d) assembly of shear connection; e,f) construction of shear connection.

The assembly of the metal-concrete spatial flooring is reduced to the direct connection of modular elements (trusses). They are connected first using bolts, and then welding, in the limited number of points.

Liners from a light, inexpensive material (expanded polystyrene, mineral wool, etc.) are laid inside the reinforced concrete slab when concreting to reduce its dead weight. Such a solution allows reducing the weight of this element to 50%, without significant complication of the technological process [7].

1.3 Simulation of a stress-strain state and rationalization of parameters

The theoretical basic of algorithms and procedures for determining rational parameters of the considered construction is the suggested methodology for constructing iterative computational sequences [8].

It is based on a compilation that synthesizes the finite element method and the adaptive evolution method [9, 10, 11]. Combination of the finite element method and the method of regulation by the system parameters, especially in the form of the adaptive evolution method, enables to construct a direct determination procedure for physical and geometric characteristics of the structures. The developed algorithm assumes the determination of internal system parameters as a result of the sequential construction of a system with equal energy strengths. The optimized, equable system in known to minimize the construction material consumption [10].

The above said defines the concept of a rational design as a system for which in all elements the condition $e = const$ (e – deformation energy density) is fulfilled. Also, for the least material intensive systems $e = e_u$ (e_u - ultimate deformation energy density) [9].

The above compiler predefines iterative clarification of geometric and/or physical and mechanical system parameters in accordance with the conditions $e \to e_n$ (e_n - rated value of deformation energy density). In this case, the finite element method is used at each step (of the iteration process), while the evolution adaptation means the possibility of accounting the values of calculated parameters of any previous steps at the i th step (hereditarity).

With the dependence «σ - ε» (σ – stress, ε - strain) and ignoring the load alleviation and elastic recovery, we can determine the strain energy density using the formula:

$$e = \int_0^{\varepsilon^*} \sigma(\varepsilon) d\varepsilon \le e_n, \tag{1}$$

where ε^* - current strain value.

In turn, the unknown internal parameters affect the configuration and volume, and are determined based on the hypothetical expression:

$$h^{i+1} = h^i \left(e^i e_n^{-1}\right)^\gamma, \tag{2}$$

where h – corresponding internal parameter, for example, the cross-sectional height of the slab or the cross-sectional area of the rods;
i – number of the internal (current) iteration;
γ – adaptive evolution parameter, while, γ€[0,1] and is determined based on the simulation experiment [10, 13].

The general procedure for solving the formulated problem now is written as:

$$A_i \in \left[A_i^{\min}; A_i^{\max}\right]; \delta_r^{m,n} \in \left[\delta_{\min}; \delta_{\max}\right] \Rightarrow \left\{q^{m,n}\right\} =$$

$$\left[R^{m,n}\right]^{-1} \left\{P^{m,n}\right\} \Rightarrow e_r^{m,n} \Rightarrow\Rightarrow \delta_r^{m,n+1} = \left(e_r^{m,n} e_n^{-1}\right)^\gamma \cdot \delta_r^{m,n};$$

$$A_r^{m,n+1} = (e_r^{m,n} \cdot e_n^{-1}) \cdot A_r^{m,n}, \tag{3}$$

$$U^L = \inf_\alpha U(\alpha)$$

$$m=1,2,\dots L; n=1,2,\dots S$$

where: $R^{m,n}$ - system stiffness matrix; $\{q^{m,n}\}$ - displacement vector; $\{P^{m,n}\}$ - outside force vector; A_i – cross-sectional area of the rod (internal parameter); δ_r - characteristic dimension of the r-th finite element (internal parameter); $U_r^{m,n}$ - potential strain energy of r-th finite element; V_r – volume of r–th finite element; $A_i^{\min}; A_i^{\max}$; h_{min}; h_{max} - known invariables; $e_r^{m,n}$ - potential deformation energy density in the r-th finite element; S – number of internal cycle iterations; determined in the counting order until the inequation is fulfilled:

$$|V^{n+1} - V^n| / \min\left\{V^{n+1}; V^n\right\} \le \mu, \tag{4}$$

L – number of external cycle iterations; μ – known invariable; V – total adduced element volume.

To determine external parameters of the construction α_i (i=1, 2, 3…m), let us write the system of balance equations as the finite element method:

$$R(\alpha_i)\cdot q = P. \tag{5}$$

The system (5) contains n unknown nodal displacements and m unknown external system parameters, which are included in (5) in the non-linear manner. In [9], it is suggested to choose (at each iteration step) from the whole set of possible values of unknown parameters those at which potential energy of strain (PED) reaches the lower limit, i.e.

$$U = \inf U^k \text{, k=1, 2….,} \tag{6}$$

where U - potential energy of strain of the system, k – variant number.
Let us accept inclination angles of the braces of the structure as external parameters and calculate the value of potential energy of strain of the construction $U = U(\alpha,\beta)$ (α,β - inclination angles of the braces in the mutually perpendicular planes, external parameters). Let us write the system of nonlinear equations

$$\begin{cases} \dfrac{\partial U(\alpha,\beta)}{\partial \alpha} = 0 \\ \dfrac{\partial U(\alpha,\beta)}{\partial \beta} = 0 \end{cases} \tag{7}$$

Its solutions determine α and β values, corresponding to $\inf U$ and define minimum consumption of materials. The system of (7) equations is simplified for the considered construction, since the angle between the modular elements is known before and is β=45°. This fact is predetermined with the flooring design. Thus, for the parameters included in (1.7), we have:

$$U = \sum_{i=1}^{k} \frac{N_i^2 L_i}{2E_i A_i}, \quad A_i = |N_i|[\sigma]_i^{-1},$$

$$U = \sum_{i=1}^{k} \frac{1}{2E_i}[\sigma]_i |N_i| L_i, \tag{8}$$

and (7) acquires the form:

$$\frac{\partial U}{\partial \alpha} = U_o \frac{\partial \varphi(\alpha)}{\partial \alpha} = 0, \quad U = U_0 \varphi(\alpha), \tag{9}$$

where N_i – stress in the i–th rod; L_i – rod length; $[\sigma]_i$ – permissible stress; E_i – stress-strain modulus of the 1st genre of the rod material; A_i – cross-sectional area of the rod; k – number of rods in the system; U_0 – known invariable; $\varphi(\alpha)$ – dimensionless function of the inclination angle of the brace α.
Calculations for modular elements of different height demonstrate that the function $\varphi(\alpha)$ around the extreme point hardly deviated form the right line parallel to the axis of abscissas

(Fig. 4). This fact provides for using discrete assortment without significant increase of consumption of materials. However, expressions (8), (9) defining minimum weight of the construction are valid only if its constant for all rods of the deformation modulus system of the 1st genre and permissible stresses, i.e. $E = const$, $[\sigma] = const$. This fact is a consequence of connection linearity for statically determinable trusses between the minimum volume of the rod system and the minimum potential energy value:

$$U = fV, \tag{10}$$

V – total volume of the rod material; $f = \frac{[\sigma]^2}{2E}$ - known invariable.

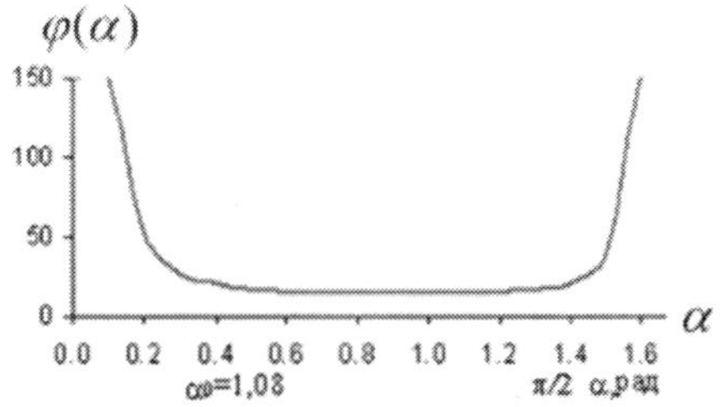

Fig. 4. Connection between the potential deformation energy of the system and inclination angles of the brace.

Thus, the given algorithm implements definition of the indicated inclination angle of the rods at the given height of the construction. The external parameter can be estimated using discrete search of the height.

The solution obtained with (3), in general, does not satisfy limitation for displacements, crack width, self-induced vibration frequencies, technology etc. This fact results in the necessity to increase cross-sections of the element with equal energy strengths.

In this connection, an additional task is formulated as follows:

$$\Delta V(\{\alpha\}) \to \inf, \tag{11}$$

$$\omega^{\min} \geq \omega_{adm};\ a_{crc}^{\max} \leq a_{adm};\ f^{\max} \leq f_{adm}, \tag{12}$$

where ΔV – extra volume of the material; {α}– internal parameter vector; $\omega^{\max}, \omega_{crc}^{\max}, f^{\max}$ - minimum (maximum) value of self-induced vibrations or crack width or deflection; $\omega_{adm}, a_{adm}, f_{adm}$ - known invariables, normally accepted from the national standards.

The problem (11) - (12) is solved by search for a new, in relation to the found one, equal-energy state of the system, at which $e < e_n \wedge e = const$. Topology of the system with equal energy strengths ($e = e_n$) is accepted as first approximation.

Further, fulfillment of limitations (12) is checked for this solution. If they are fulfilled, this solution can be consider final. In general, first approximation will not be a solution (11)-(12). Then a notion of conditional ultimate strain energy density e_{nm} is introduced (m – number of inequations in (12)).

Then the procedure is constructed as the Kiffer-Johnson's algorithm to reach the boundary of the region of feasibility [13]. As a result of the operation of the considered iterative algorithm, the geometry and/or physical and mechanical characteristics of the structure are determined.

1.4 Experimental testing

Experimental verification of the elaborated system and methods for its calculation was carried out. The non-destructive tests were conducted at an ambient temperature of 30°C and humidity of about 60%. A research object was a fragment of the superstructure of a pedestrian bridge with a length of 7.0 m, a width of 2.0 m, and a construction height of 0.5 m was adopted (Fig. 5). The hydraulic testing (water loading) method was adopted as a loading method [13]. Therefore, a tank was arranged on the upper surface of the slab, allowing short-term and long-term loading on the entire structure, its half and quarter. Loading was performed in steps from 1 kPa to 8 kPa in increments of 1 kPa, followed by unloading (low-cycle loading). The instruments included 78 resistance sensors, 13 deflection meters, 4 dial indicators, 10 displacement sensors sensors controlled with SIIT-3 information and measuring system and a PC with the appropriate software installed (Fig. 7).

Fig. 5. Loading of the entire construction.

Deformations and displacements were measured in the horizontal and vertical direction of the construction, as well as on individual elements (such as slab, booms, braces). In addition, a possible shift of the reinforced concrete slab relative to the metal structure was to be determined. The results of the measurements are illustrated by the diagrams in Figures 5, 6 and 8.

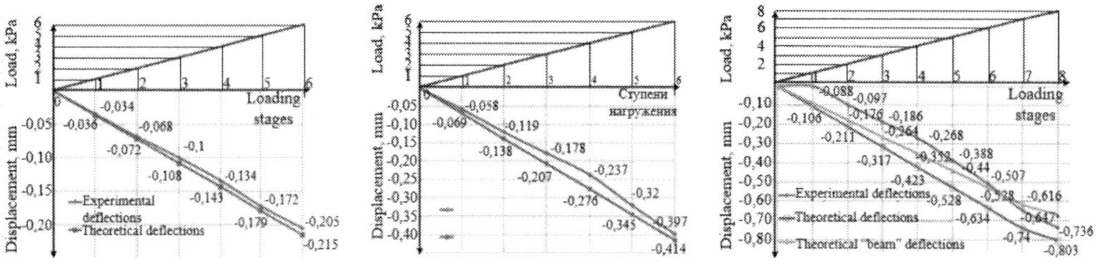

Fig. 6. Maximum displacements in the middle of the span at short-term and long-term loading.

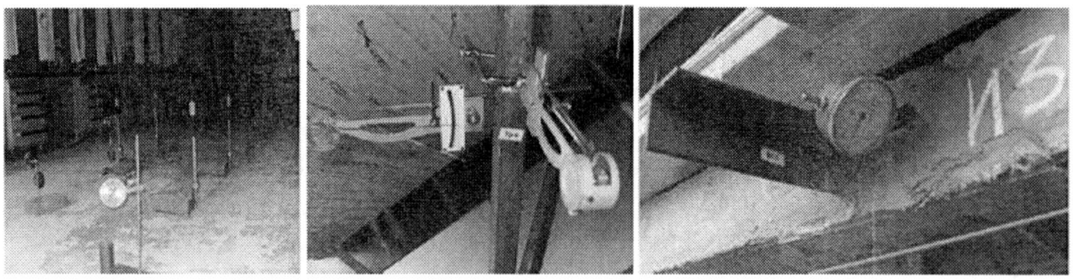

Fig. 7. Available instruments.

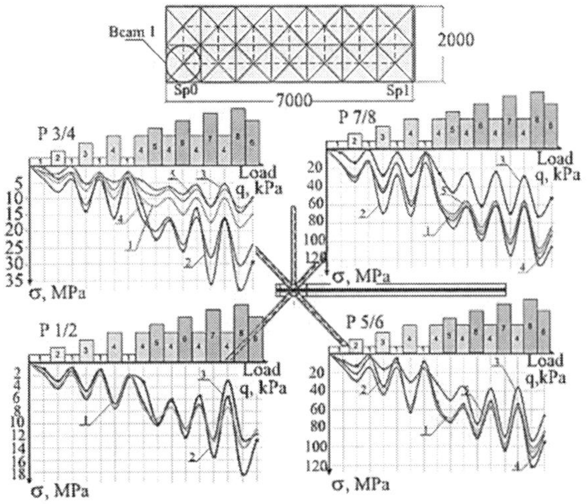

Fig. 8. Stresses in the braces of the superstructure when the entire span is loaded.

As a result of the conducted full-scale tests of the superstructure of the pedestrian bridge, it was established as follows:
- the suggested discrete-continuum system of new type shear connections ensured the joint perfromance of metal structural elements (booms and braces) and a reinforced concrete slab at all loading stages;
- the nature of the system deformation loaded with hydraulic low-cycle short-term and long-term loads corresponds to its resistance data, and the quantitative and qualitative parameters coincide with a sufficient degree of accuracy with the theoretical data of measured values obtained using the constructed finite element model of the design. The discrepancies between the theoretical and experimental data of stresses are within the range of 10-15%;
- vertical displacements under the action of the operational load did not exceed: $f / L = 0.0012 < 0.0025 \cdot L$, L - span;
- the permanent deformation values after unloading at each loading stage did not exceed 5%, which indicates the elastic performance of the structure;
- no cracks were detected in the reinforced concrete slab, which would characterize the loss of the carrying capacity and the continuity of the structure;
- the period of self-induced vibrations of the structure does not fall within the range prohibited for pedestrian bridges (0.45 to 0.6 sec);
- reinforcement of the structure by including the reinforced concrete slab in joint performance with the upper boom resulted in an increase of rigidity of the system, on the one hand, and significant increase of its weight, on the other hand. In turn, the increase of the total weight of the structure causes a drop in self-induced vibration frequency. At the same time, it should be noted that an increase in frequency can be achieved by the arrangement of the suggested efficient reinforced concrete slab with internal liners;
- comparison of experimentally obtained dynamic characteristics of the superstructure of the pedestrian bridge with the corresponding theoretical data, as well as in the case of static loading, characterizes the design model as sufficiently correct (the spread of theoretical and experimental data of the studied parameters did not exceed 13-16%).

The considered designs have been implemented in a number of Ukrainian building companies. The reasons for wide use of this system are as follows:
- a variety of applications (floorings, coverings, bridge superstructures, walls, etc.) (Fig. 9);
- limited height of the structure $(0.04 \div 0.05)L$;

- the possibility of using it for horizontally, vertically and obliquely installed structures;
- limited consumption of materials;
- low labor intensity of the assembly, which enables to minimize the assembly time.

Fig. 9. Facilities where the technology was implemented.

2 Efficient system of reduce-weight reinforced concrete elements

2.1 General

Reducing the weight of the building is one of the priority issues of the theory of constructions. Minimization of the weight of the structure according to B. Fuller determines the degree of its perfection, comprising such notions as environmental positivity, seismic stability, cost-effectiveness, etc. At the same time, flooring is known to make the principal contribution to the weight of buildings and, in particular, high-rise ones (up to 60 to 70%). In this connection, search for rational solutions for these constructions is relevant.

2.2 Design features

In [7, 15], solutions of monolithic and precast reinforced concrete coverings and floorings, as well as vertical elements, such as walls, pylons, and nuclei, which are structurally anisotropic systems, are suggested. In general, these constructions have the structure of simple external and complex internal geometry (Fig. 10). The main idea here is sinking of liners of agreed shape and sizes made of light, cheap materials such as polystyrene foam, mineral wool, etc. during concreting. Thus the monocoque topology of the constructive element is formed which is similar to bionic systems.

The efficiency of such a design is also determined by the symmetry of its cross-sections which enables its arbitrary support, thus generating a sign-variable field of moments of deflection (Fig. 11).

The above stated is a consequence of the difference in concrete resistance to stretching and compression, on the one hand, and the maximum efficiency (in terms of material consumption) of one-sided ribbed systems, on the other hand. Thus, the construction of the considered slabs consists of two (upper and lower) casings, internal ribs and block-out liners (Fig. 10). Furthermore, if necessary, internal ("hidden") capitals and crossbars are organized inside the element. Reinforcement is a system of flat carcasses located in the ribs (crossbars) and flat grids reinforcing the shells (capitals) (Fig. 11). The outer enclosure of elements, holes and openings are framed with a cut sheet. The installed spacer and internal connection system (usually made of composite materials) provides necessary protective layers of concrete and prevents liners from floating. Both light-weight and heavy concrete can be used for constructive elements.

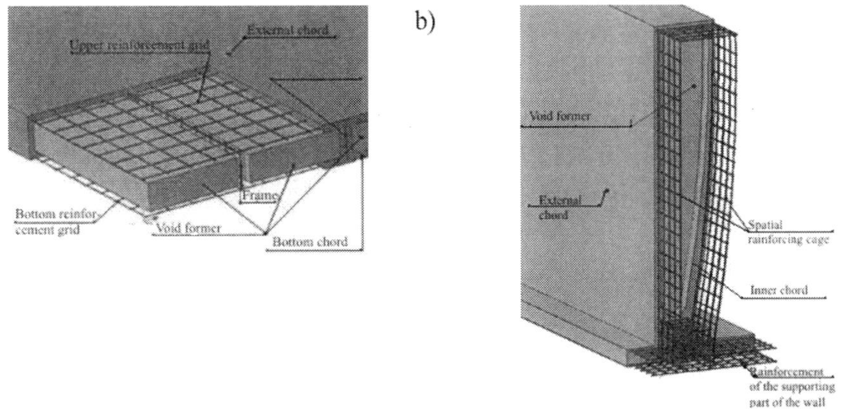

Fig. 10. Topology of the suggested horizontal (a) and vertical (b) efficient elements

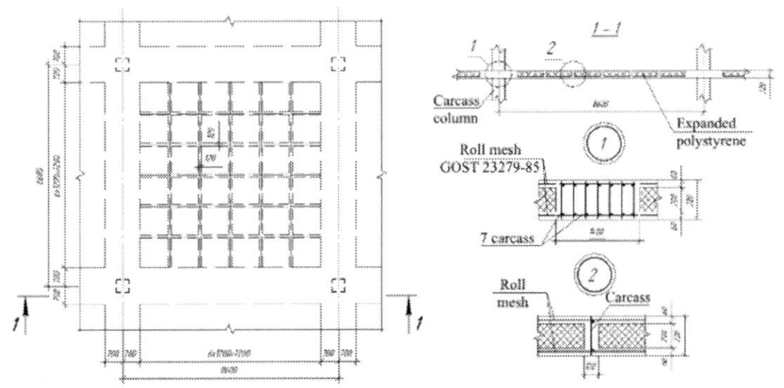

Fig. 11. Fragment of the continuous flooring slab

2.3 Rationalization of parameters

The obtained structure with equal structural thicknesses of the solid and the suggested elements enables to save concrete (up to 40-45%) with the same consumption of reinforcement. Also, the shape and size of the formed internal pockets have significant impact on the nature of the stress-strain state of the system, which explains the expediency of searching for its internal geometry by methods of the control and optimization theory [9, 13, 16].

In connection with this, we introduce the criterion of rationalization consistent with [9, 13] in the form:

$$U \to inf\, U(\vec{\alpha}^{k}), \quad k = 1, 2, \ldots \infty, \tag{13}$$

where U – potential strain energy (PDE); k – number of comparison variant; α – geometric parameter vector; $\alpha \in M$; M - set of permissible values of external geometric parameters.

Let us accept requirements of the equal-energy state of the system (construction) as an additional condition to (13), i.e.

$$e(\vec{\alpha}) = const, \tag{14}$$

where e – potential strain energy density (PDED).
Solving of practical problems with the criterion (13), to some extent, proves the efficiency of its operation, as well as possibility to regulate not only geometric, but also physical and mechanical characteristics of the system.

Consistently with [9], let us assume that we understand external extensive parameters as parameters which determine resistance of the system without change of the total volume of the material (dimensions of the system elements, bearing connection coordinates and points of load application, etc.), and the internal intensive parameters determine topology, volume, as well as property of the materials.

Equal-energy state caused by (14), as a rational form of existence of artificial elements, ensures the most appropriate distribution of the material, as well as its efficient performance in the construction. If the iterative approach is applied to (14) based on the adaptive evolution method (AEM), for any geometric or physical and mechanical parameter of the construction a formula can be written as follows

$$h_{i+1} = h_i [e_i e_{iu}^{-1}]^{\gamma}, \tag{15}$$

where h – geometric (physical and mechanical) system parameter; i – iteration number; $\gamma \in (0,1)$ - characteristic of adaptive evolution reflecting convergence rate of AEM procedure.

The introduced provisions allow not only to designate a reference solution, but also construct an energy profile of the structure, which is critical. The energy profile of the structure is understood as 3D construction and mapping of the energy density field of deformation with simultaneous determination of the PDE. This fact, along with the definition of the rational system topology (not only form, but also content) predetermines the establishment of the limiting state both for the structural elements of the construction (for example, finite elements) and for the entire system. Moreover, the unloaded (parasitizing) elements of the system are filtered out subject to convergence of procedure (15).

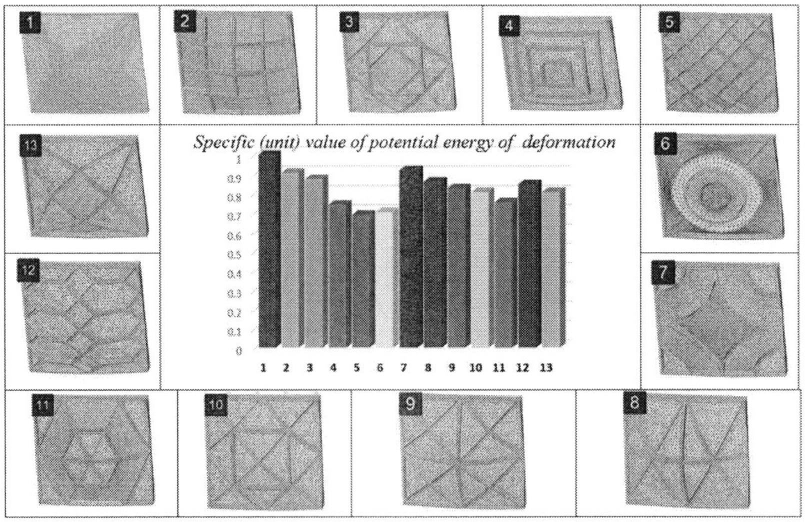

Fig. 12. Illustrated example.

It should be noted that, in addition to minimizing the consumption of materials, the criteria (13) and (14) can be used for approximate estimation of some critical mechanical characteristics. In this connection, let us consider an illustrative example. Let us consider 13 square bending constructively anisotropic slabs of the same volume (material consumption). All slabs are supported along the perimeter with hinges and carry uniformly distributed single load. Exclusive geometry of the reinforcing ribs is defined for each slab.

In consistency with the introduced criteria, the task to determine the topology of the edges is set, for which, firstly, the criterion (13) is fulfilled, and secondly, the fulfillment of the limitation (control) is estimated (14). The considered problems were solved in LIRA software environment [10]. The solution results are shown in Figure 12.

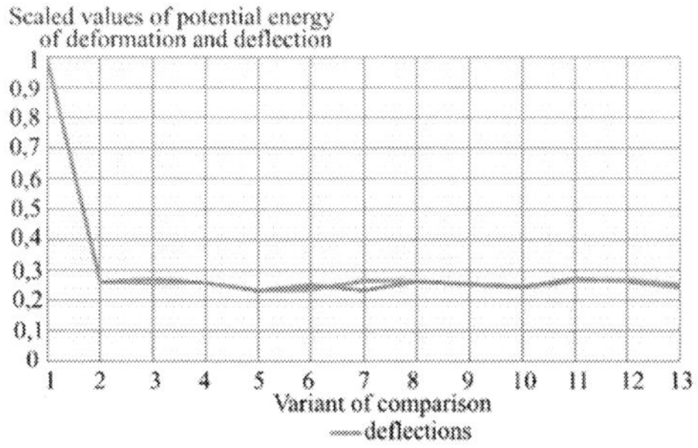

Fig. 13. Maximum deflection and PDE for comparison options.

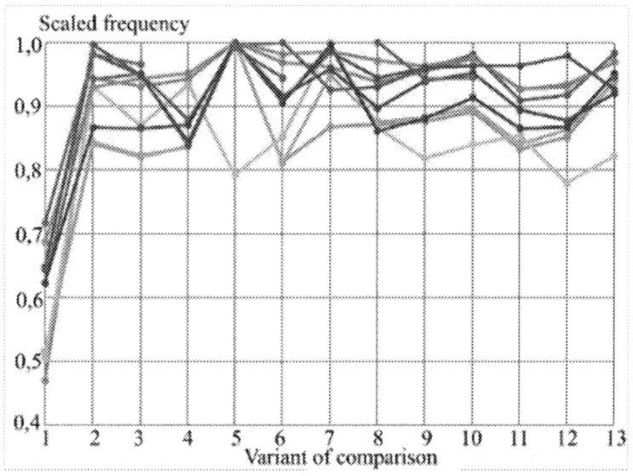

Fig. 14. Self-induced vibration frequencies of the considered slabs.

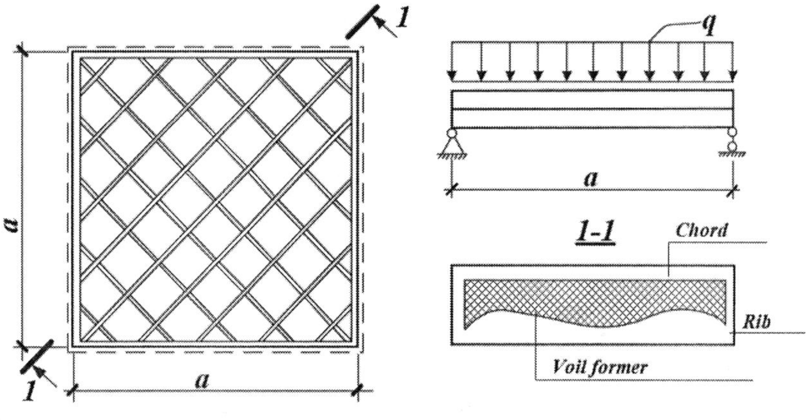

Fig. 15. Ribbed slab with ribs of variable stiffness.

The diagrams in Figures 13, 14 demonstrate, that the fifth option of the slab topology corresponds not only to the minimum value of the PDE, but also to the maximum value of the self-induced vibration frequency of the natural pitch and practically of overtones, as well as the minimum deflection value, which indicates the maximum bending stiffness of this slab. Thus (although with a certain degree of caution) we may state that the criterion (13) can be used as a universal one. The condition (14) is fulfilled, inter alia, by variable rigidity of reinforcing ribs [16, 17, 18]. Also, in-depth block-out liners are used to ensure the manufacturability of the solutions (for example, for reinforced concrete or steel-reinforced concrete structures) (Fig. 15) [6]. For metal elements, in this case, installation of structural elements with variable cross-section and/or use of metals of different grade can be appropriate [6].

2.4 Features of the technological process

Building of the considered systems in monolithic design assumes the use of traditional formwork of the "table" type. The implementation procedure involves concreting in lifts, performed in accordance with the following procedure: installation of the lower mesh and the rib frames (Fig.16a) ÷ concreting of the bottom shell with concrete compression using the concrete vibrator (Fig.16b) ÷ laying of the liners and the upper mesh (Fig.16b) ÷ concreting the ribs and the upper frame (Fig. 16d) with consequent compression with a deep vibrator and beam vibrator [17, 18]. The work is carried out with pours moving "towards oneself", and the selected mixture composition enables movement on fresh concrete (Fig. 16).

Fig. 16. The building sequence of efficient flooring.

Moulds with opening sides are used in the manufacture of prefabricated elements, while the technological manufacturing process is practically similar to the indicated one (Fig. 17).

Fig. 17. Efficient prefabricated flooring slab.

There is an advanced alternative technological process using the concrete spraying technology (Fig. 18). It is appropriate to note here that minimization of waste of concrete is ensured with the selected mixture composition and the concreting procedure itself [19]. Nevertheless, off-form concreting is the most important fact. In this case, inserts fixed to the reinforcing special frame play the role of the screen.

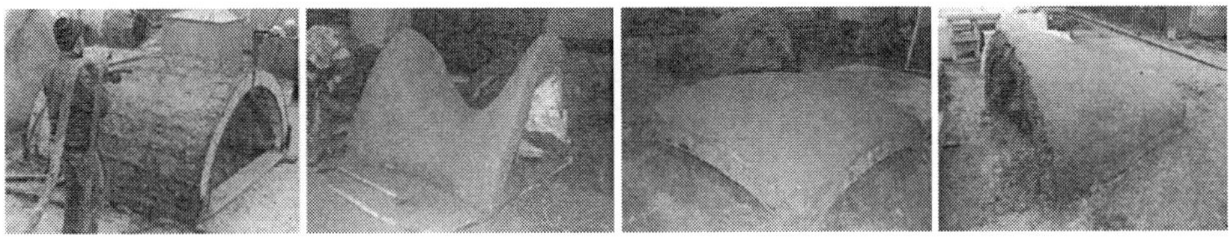

Fig. 18. Off-form concrete spraying (completion).

2.5 Experimental verification

Constructive solution, as well as its formation methods were verified experimentally [20]. An object of study was a new type reinforced concrete flooring of monotropic design built above the first floor of a sports and fitness complex under construction. The flooring was a continuous cantilever slab 9.4 m wide and 43.0 m long. One side of the slab rested on the caisson supporting wall, and the other rests on the 6 m span girders that conditionally divide the slab across into the 7.5 m span and the 1.9 m console. The structural thickness of the slab is 0.3 m, including the upper and lower plating 0.06 m and 0.05 m thick, respectively, and the rest of its height is filled with polystyrene foam liners (Fig. 19). Longitudinal ribs with a width of 0.1 m are formed inside the plate along the span, located at intervals of 0.75 m. The height of the ribs is equal to the total height of the slab.

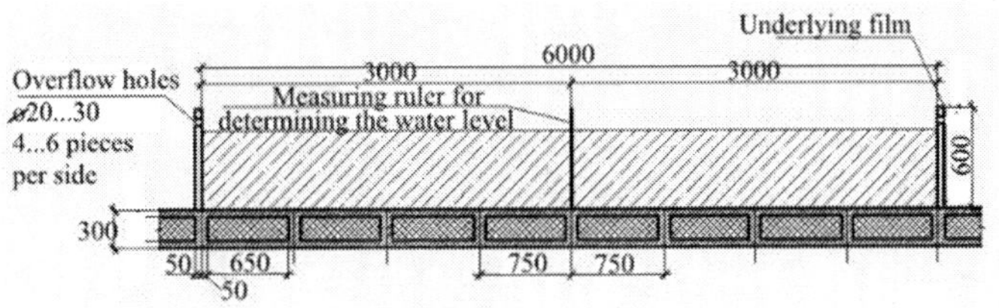

Fig. 19. Flooring cross-section and loading systems.

A new method of full-scale testing of shallow slabs and shells was used to test the efficient flooring [13, 14]. The loading tanks were assembled from the PERI formwork sections, while waterproofing was ensured with laying three layers of the polyethylene film. The assembled containers were filled with water through the supply pipe to the design level. The overall view of the tank on the flooring is shown in Fig. 20. The water level was regularly measured along the perimeter of the reference enclosure, in the quarters and in the centre of the pool with water to determine the exact nature of loading.

Fig. 20. Overall view of the loading system.

During the experiment, deflections of the characteristics points of the slab were measured using SIIT-3 system. The measurement point arrangement diagram is shown in Fig. 21.

 DOI: 10.1051/matecconf/201711602004

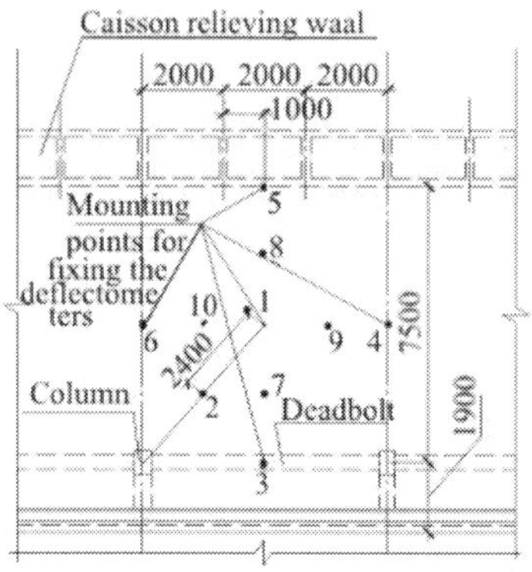

Fig. 21. Sensor arrangement plan.

Loading was performed in incrementing steps of 100 kg/m^2 each (10 cm water). Instrument readings were recorded after each supply of the water volume. Then measurements were performed in 10 min (until deformations was stabilized).
The deflection measurement findings at each loading step of the flooring fragment along and across the axes of the tank are given in Figure 22.

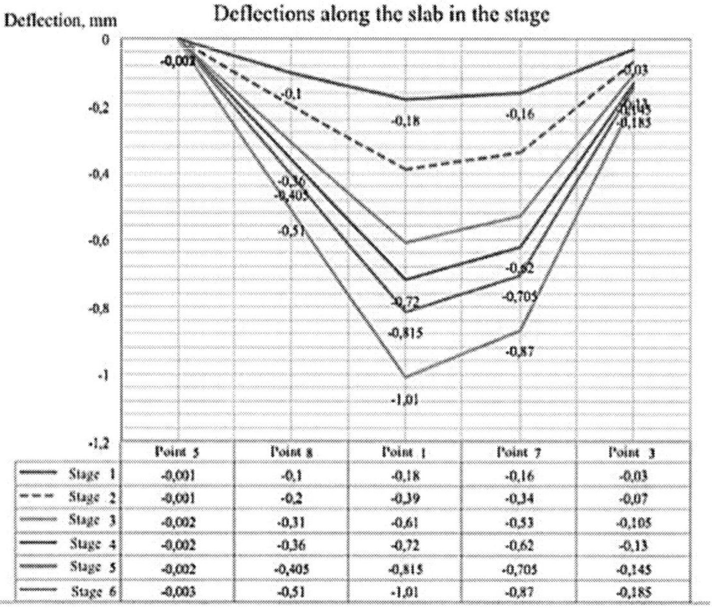

	Point 5	Point 8	Point 1	Point 7	Point 3
Stage 1	-0,001	-0,1	-0,18	-0,16	-0,03
Stage 2	-0,001	-0,2	-0,39	-0,34	-0,07
Stage 3	-0,002	-0,31	-0,61	-0,53	-0,105
Stage 4	-0,002	-0,36	-0,72	-0,62	-0,13
Stage 5	-0,002	-0,405	-0,815	-0,705	-0,145
Stage 6	-0,003	-0,51	-1,01	-0,87	-0,185

Fig. 22. Deflections of the flooring fragment at loading steps 1-6 along the main instrument arrangement lines: a) deflection lines along the slab span.

Mathematical simulation of the stress-strain state of the research object was performed before testing. Simulation was performed in 3D-finite element form, while the actual calculation was performed in Lira software environment [10]. The analysis of stress-strain state and comparison of theoretical and experimental data confirmed the qualitative and quantitative adequacy of the model (discrepancy within the range of 15%) (Fig. 23). Moreover, the established nature of the system deformation demonstrated expediency of using the direct design ideology in the suggested form and the above mentioned economic efficiency of the solutions obtained.

DOI: 10.1051/matecconf/201711602004

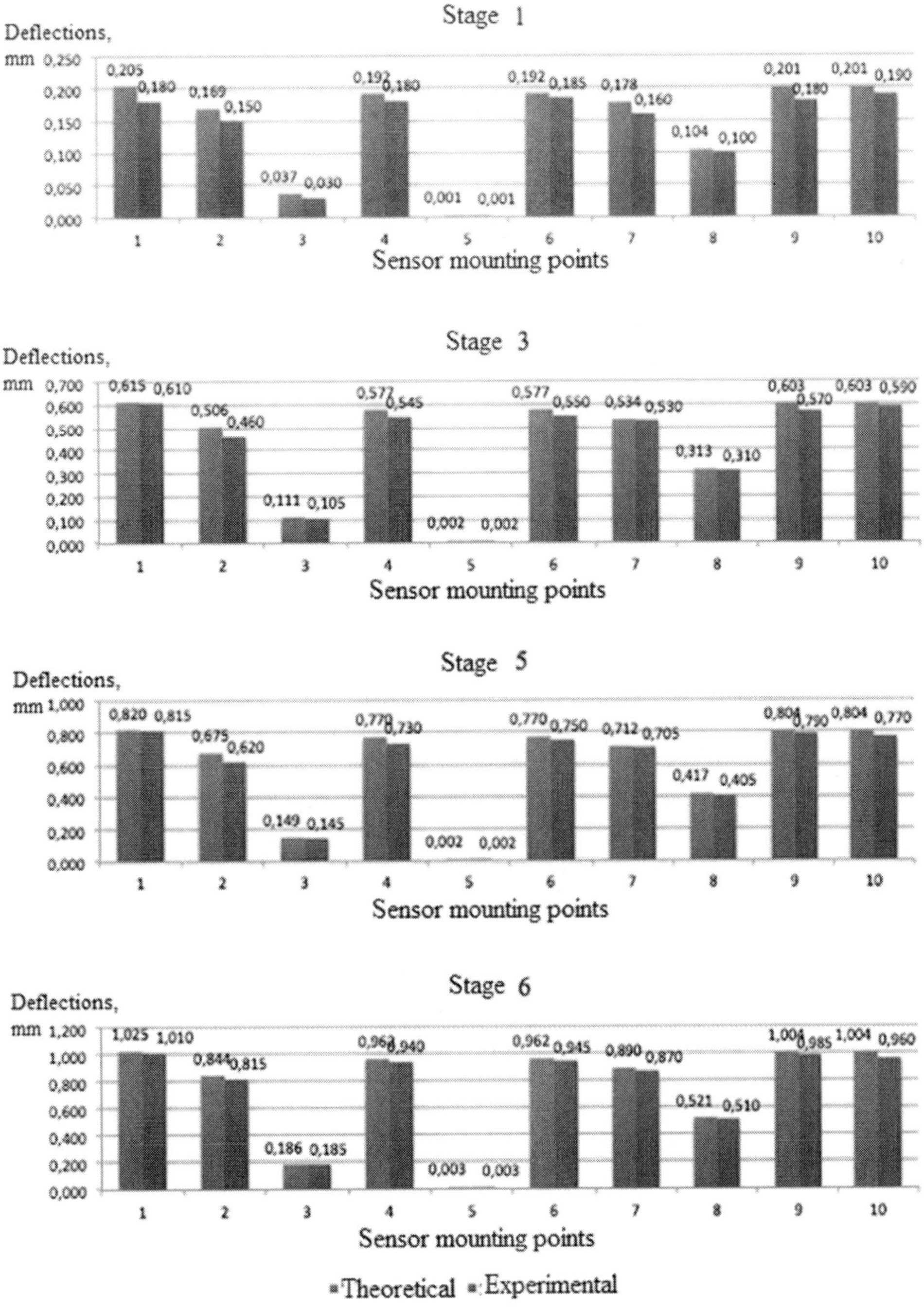

Fig. 23. Comparison graphs for theoretical and experimental deflections in the characteristic points of the slab.

Today, the elaborated designs and technologies have also been introduced and successfully operated in a number of Ukrainian building companies (Fig. 25).

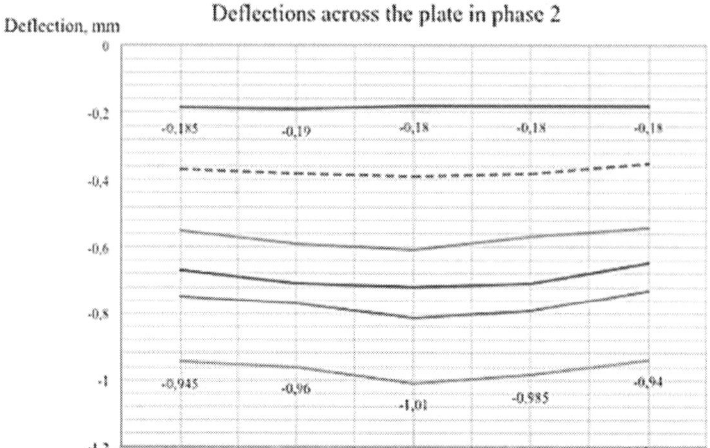

	Point 6	Point 10	Point 1	Point 9	Point 4
Stage 1	-0,185	-0,19	-0,18	-0,18	-0,18
Stage 2	-0,37	-0,38	-0,39	-0,38	-0,35
Stage 3	-0,55	-0,59	-0,61	-0,57	-0,545
Stage 4	-0,67	-0,71	-0,72	-0,71	-0,65
Stage 5	-0,75	-0,77	-0,815	-0,79	-0,73
Stage 6	-0,945	-0,96	-1,01	-0,985	-0,94

Fig. 24. Deflections of the flooring fragment at loading steps 1-6 along the main instrument arrangement lines: b) deflection lines across the slab span.

Fig. 25. Facilities containing efficient floor slabs of the suggested design.

In conclusion, it should be noted that the need to save all kinds of resources has always been a dominant factor in the aspiration to improve artefacts. At the same time, the highest integral positivity can be achieved in reasonable transfer to them of solutions borrowed from the wildlife. In turn, the features of the suggested systems and the natural dualism reflected in them are examples of application of concepts given by Galileo:
- if you do not know how to do it - ask Nature;
- no structure can be operated reliably if the form of its ruin (destruction) is unknown.

References

1. Intelligent buildings: integrated systems and control. Dr Gerard Healey [Electronic resource] // issinstitute.org.au/.../ISS-FEL-REP...-low-res.pdf.
2. *EN 1994-1-1-2009 Eurocode 4. Design of building structures*, Brussels (2004)
3. V. Shmukler, N. Burak, A. Duker, Vassim Ismail. Patent № 84560, MPK (2006), E04D 3/24, Ukraine (2008).
4. V. Shmukler, E. Beregnaya, V. Gerasimenko, Vassim Ismail, O. Kalmikov. Patent № 51336, MPK (2009), E04D 3/24, Ukraine (2010).

5. V. Shmukler, A. Fuksman, V. Gerasimenko, Vassim Ismail, O. Kalmikov. Patent № 63846, MPK (2011.01), E04B 1/00, E04D 3/24 (2006/01), Ukraine (2010).
6. V. Shmukler, A. Kislov, E. Krasnova, E.Petrova *Proceedings of the 15th Conference for Lithuania Junior Researchers SCIENCE-FUTURE OF LITHUANIA*, Vilniuse (2012)
7. V. Shmukler, V. Bedim, I. Shmukler, D. Bedim. Avtorskoe svidetelstvo № 1738962, E04C 2/26, E04B 5/02, USSR (1992).
8. V. Reinboldt, J. Ortega. *Iterative methods for solving nonlinear systems of equations* Moscow: Mir, 558, (1975).
9. G. Vasilkov. *Evolutionary problems of structural mechanics. Synergetic paradigm* Rostov-na-Donu: Infoservis (2003).
10. A. Gorodetskiy, V. Shmukler, A. Bondarev. *Information technology of calculation and design of building structures.* Kharkiv 889 (2003).
11. V. Shmukler. ISEC-03 Third International structural Engineering and construction Conference, Shunan, Japan, 539-545 (2005).
12. V. Shmukler, K. Mohamad. Proceedings of the 2th International Congress FIB, Naples, Italy, 110-127 (2006).
13. V. Shmukler, U. Klimov, N. Burak. *Light-weight frame systems*, Kharkov, 336, (2008).
14. V. Shmukler, A. Chuprynin, R. Abbasi Patent № 44125. MPK (2009), G01M 19.00, G01N 3.00, G01M 5.00, Ukraine (2009).
15. V. Shmukler, O. Kalmikov, E. Lugchenko, A. Goishin, V. Zinchenko. Patent № 79485, MPK, E02D 29/02 (2006.01), Ukraine (2013).
16. V. Shmukler. Global Construction, Dundee, Scotland, 183-190 (2005).
17. V. Shmukler, M. Pomazan. Patent №69346, MPK (2012.01), E04C 2/00, E04B 5/00, E04G 21/00, Ukraine (2012).
18. V. Shmukler, M. Pomazan. Patent №75556, MPK (2012.01), E04C 2/00, E04B 5/00, E04G 21/00, Ukraine (2012).
19. S. Bugaevskiy. «Naukovo-technichni problems suchasnogo zalizobetonu» **78 (2)**, 478-586 (2013).
20. I. Steblovskii, E. Beregnaya Zbirnyk naukovych pratc «Budivelni konstruktsii», **79**, 347-355 (2013).

Estimation of flexures of the reinforced concrete elements according to the National Ukrainian & European standards

Volodymyr Babich[1], *Volodymyr* Dovbenko[1], *Lyudmyla* Kuzmych[1,*], and *Tetyana* Dovbenko[1]

[1]National University of Water and Environmental Engineering, Department of Water engineering and Reclaimed Land Operation, 33028 Rivne, Soborna St. 11, Ukraine

Abstract. The objective of this work is analysing of the calculating results of theoretical values of the beam flexure by the Ukrainian National Standard, European Standard and by the experimental data of concrete beams testing.The results of the experimental investigations of flexures of the reinforced concrete beams are shown. The theoretical comparison of the flexures of the reinforced concrete beams according to the Ukrainian National Standard and European Standard is done. It was proved that the flexures by these methods have an identical convergence with experimental data. The algorithm for determining deformations of materials at the known bending moments by using the method of simple iteration or method of successive approximations is proposed. The solution of this problem allows practical use of the National Standard and European Standard for calculating flexures. Analyzing of the experimental flexure and the calculated value is showed; the significant difference can be explained by the empirical coefficient.

1 Introduction

The design of concrete structures in Ukraine is carried out according to National Standards [1, 2], which is harmonized with European Standard Eurocode 2 [3], but their use causes some difficulties in estimation of structures.

The basis of the National and European Standards contains calculations by the limit states, but the algorithms have some differences. Use of the algorithm of these standards should be compared with experimental results. However, this is a major problem in the design of concrete structures, namely the difference between the calculation results obtained by the National Standards [1, 2] and the European Standard Eurocode 2 [3].

Practical calculations to determine the limit states for the carrying capacity (the first group uls) of concrete bending elements by the Standards [1, 2] are partly covered in [4, 5, 6, 7, 8], but the results of calculations by the limit states of the availability for operation (the second group sls) are very scarce, except for a few works [9, 10, 11, 12, 13].

[*] Corresponding author: l.v.kuzmych@nuwm.edu.ua

The objective of this work is to continue analyzing of the calculating results of theoretical values of beam flexure by the Standards [1, 2, 3] and by the experimental data of concrete beams testing.

2 Methods and techniques

Experimental study of the reinforced concrete elements were conducted at the beams, twin – samples, the cross-section size is 100×200 mm, the total length 2000 mm. Heavy concrete of class C25/30 was used to manufacture samples. Reinforcing was made of the welded frame with a longitudinal reinforcement 2Ø14 mm of class A500 and transverse Ø6 mm of class A240 fitting increments 100 mm (fig. 1).

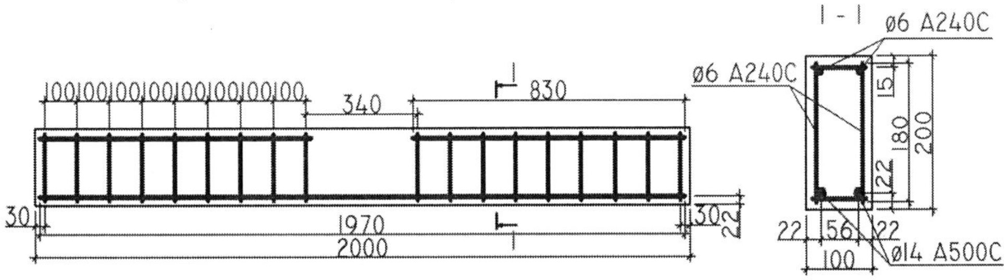

Fig. 1. The reinforcement scheme of the experimental beams.

Sample 150×150 mm cubes, 150×150×600 mm prisms and crescent reinforcing samples of profile Ø14 mm, class A500, length 400 mm were tested to determine the strength and deformation characteristics of materials, along with the study of concrete beams.

Experimental studies of beams were conducted using the bench transferring the load from a hydraulic jack to a traverse. So the beams were tested by the free lying beam scheme, which was established on two supports. The test bench for beams is shown on fig. 2.

A load on the beams was increasing in increments of 0.05 to 0.1 of the maximum to the destruction. During testing at each stage of loading there was a prolonged exposure for 10 to 15 minutes, during which the parameters of devices were recorded and the visual inspection of the samples was performed. We also determined the nature of crack width and crack opening.

Fig. 2. The install for researching of experimental beams

3 Results and Discussion

Before testing the beams, we determined the mechanical properties of materials experimentally. The strength of concrete cubes was $f_{ck,cube}$ =30.2MPa, of prisms - $f_{ck,prism}$=27.8MPa, the initial modulus of elasticity of concrete E_{ck} =20.9GPa, the relative deformation of concrete ε_c = 1.78‰.

The test of reinforcement class A500 Ø14 mm has showed the following characteristics:

- strength of the fluidity limit f_{yk} =617.3MPa;
- tensile strength f_{tk} =727.7MPa; E_s =196.509MPa;
- reinforcement deformations, which fit σ_y: ε_s =3.12‰.

The destruction of the beams was a result of puncturing compressed zone of concrete. By the results of three twin - beams the average destruction load is $2F_{ULS}$ =79.5kN (destruction moment M_{ULS} = 23.9kN·m). The results of experimental investigations are noted in Table. 1, and the fracture of beams is shown in fig. 3.

Table 1. Results of the test of experimental beams.

beam mark	destruction load		deformations		moment of the first cracks M_w, kN·m	the maximum width of the crack opening w_{max}, mm	the maximum flexure f_{max}, mm
	M_{ULS}, kN·m	$2F_{ULS}$, kN	concrete ε_c, ‰	armature ε_s, ‰			
b-1	23.7	79.1	2.11	2.93	2.4	0.59	17.2
b-2	24.5	81.7	2.47	2.93	2.4	0.31	16.9
b-3	23.3	77.8	2.13	2.58	2.4	0.28	18.8
aver.	23.9	79.5	2.24	2.81	2.4	0.39	17.6

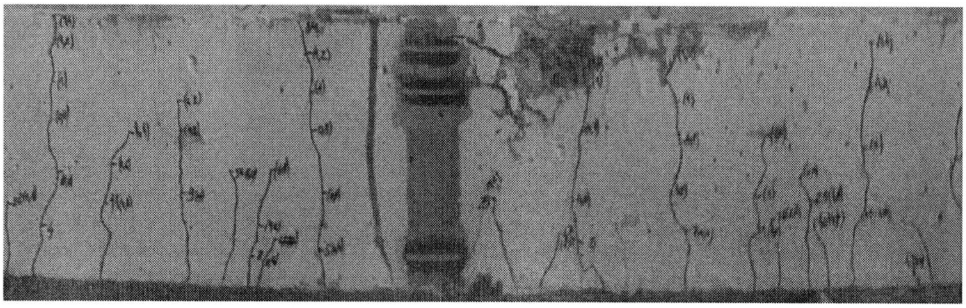

Fig. 3. The fracture of beams.

Flexures of reinforced concrete beams are determined by the general rules of structural mechanics based on the bending deformation characteristics of the element sections by its length. Value of flexures of reinforced concrete beams is determined by the curvature of the element, which depends on the deformation of materials and determines by formula [2]:

$$\aleph = \frac{1}{r} = \frac{(\varepsilon_{cm}+\varepsilon_{sm})}{d}, \quad (1)$$

ε_{cm} - deformation of compressed concrete extreme fiber;
ε_{sm} - averaged strain stretched reinforcement.
d - height of the section of the beam.

Curvature of an element is determined by the formula (1) according to National Standard [2], but the algorithm for determining the values of deformation of materials at known bending moments are not created.

For solving this problem we use a simple iteration method or the method of successive approximations to the condition of equality efforts at the compressed and stretched zones of the cross section:

$$S_{cc} = S_{ct} + S_s, \tag{2}$$

S_{cc} - internal efforts into concrete of the compressed zone;
S_{ct} - internal efforts into concrete of the stretched zone;
S_s - efforts into stretched fixture.

Using the deformation model as a fifth-degree polynomial [1], the equilibrium condition can be written as (fig. 4):

$$f_{ck}bx\sum_{k=1}^{5}\frac{a_k}{k+1}\left(\frac{\varepsilon_{c_I}}{\varepsilon_{c1}}\right)^k =$$

$$= f_{ctm}b(h-x)\sum_{k=1}^{5}\frac{a_k}{k+1}\left(\frac{\varepsilon_{sr}}{\varepsilon_{ctu}}\right)^k + E_sA_s\frac{\varepsilon_{ctu}}{h-x}(h-x-c), \tag{3}$$

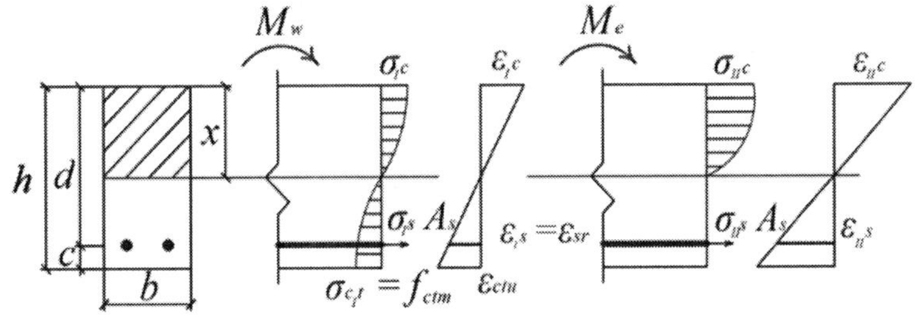

Fig. 4. Stress-strained state of reinforced concrete elements in the limit state: "no cracks" (a); "with cracks" (b).

Given the stress-strained state of reinforced concrete elements of rectangular section and using the hypothesis of flat sections, the ratio between the strains can be written as:

$$\varepsilon_{sr} = \varepsilon_{s_I} = \frac{\varepsilon_{ctu}}{h-x}(h-x-c), \tag{4}$$

$$\varepsilon_{c_I} = \frac{\varepsilon_{ctu}}{h-x}x, \tag{5}$$

To determine ε_{c_I} and ε_{s_I} it needs to find the height of the compressed zone of concrete, which is determined based on the equilibrium conditions of internal efforts into a cross-section.

With the help of dependencies (4) and (5) pre having area's compressed height, by the method of successive approximations, we reach the condition of equilibrium of internal forces in section (2) to within 2%. As a first approximation we have $x = 0{,}5d$ taken in the future to change the condition of equilibrium within ± 2%.

Flexure is determined by a known formula from structural mechanics:

$$f = k_m\chi l_d^2, \tag{6}$$

k_m - coefficient that depends on the design of a beams scheme (in the case of experimental conditions $k_m = 0.10648$) (Table. 5.5 [2]);
χ - curvature of the element
l_d - estimating step of beams (experiment $l_d = 1800$ mm).

Determination of flexures refers to the calculation of limit states for availability to operation, therefore flexures comparing better to perform in the stage of operation. Conditionally assumed operational load M_{SLS} =0.6·M_{ULS}. =0.6,·23.9=14.4 kN·m.

Estimated flexure for beams at the operating moment M_{SLS}=14.4 kN is f_{DSTU}=6.3 mm. The discrepancy between the experimental value of the flexure of reinforced concrete beams (f_{exp}=9.3 mm) with M_{SLS}=14.4 kN·m and calculated (f_{DSTU}=5.62 mm) is 39.6%. Such a significant discrepancy between experiment and calculation can be explained by the fact that into the formula of the definition of the flexure there is an empirical coefficient, which may give some error.

Algorithm for estimation of the flexure shown in Chapter 7 (par. 7.4.3) of the European Standard [3] is to determine the curvature of the elements of the formula:

$$\aleph = \xi \aleph_{II} + (1-\xi)\aleph_I, \tag{7}$$

$\aleph$ - total curvature of the beam;
ξ - coefficient of distribution is determined by the formula:

$$\xi = 1 - \beta\left(\frac{\sigma_{sr}}{\sigma_s}\right)^2, \tag{8}$$

β – coefficient of the impact of the term of load or stress repetition of the load on the mean deformation:
1.0 (for a short single load);
0.5 (for frequent or many cycles of repeated loads);
σ_{sr} - stress in the tension reinforcement, which is calculated for the section "no cracks" if the load at the moment of formation of the first crack;
σ_s - stress in the tension reinforcement, which is calculated for the section "with crack";
$\aleph_I$ – "no cracks" beam curvature;
$\aleph_{II}$ – "cracked" beam curvature.

There is the same situation as into the National Standard [2], a clear sequence of the determination of "no crack" and "cracked" curvatures isn't created. Using the already known method of simple iteration (successive approximation), we find the values $\aleph_I$, $\aleph_{II}$.

According to the formula (7) we find the common curvature. By the formula (8) a theoretical flexure of control beams at M_{SLS} =14.4kN·m is f_{EN} =5.57mm, the disagreement with experiment is 40.1%.

The findings of experimental and calculated flexures during the operating bending moment (M_{SLS} =14.4kN·m) are shown in Table 2.

Table 2. Comparison of experimental and calculation flexures.

M_{SLS}, kN·m	$f_{exp.}$, mm	f_{DSTU}, mm	Δ, %	f_{EN}, mm	Δ, %
14.4	9.3	5.62	39.6	5.57	40.1

Thus, the Table 2 demonstrates convergence of estimated flexure, which was determined by the National [2] and European [3] Standards is almost the same. Particular attention is attracted the significant difference between the experimental and estimated flexure, which is about 40%.

Comparative diagram of flexures is shown in Fig. 5. Such significant difference between experiment and estimation can be explained by the fact that the formula (6) contains the empirical coefficient k_m for determining flexure, which can give an error.

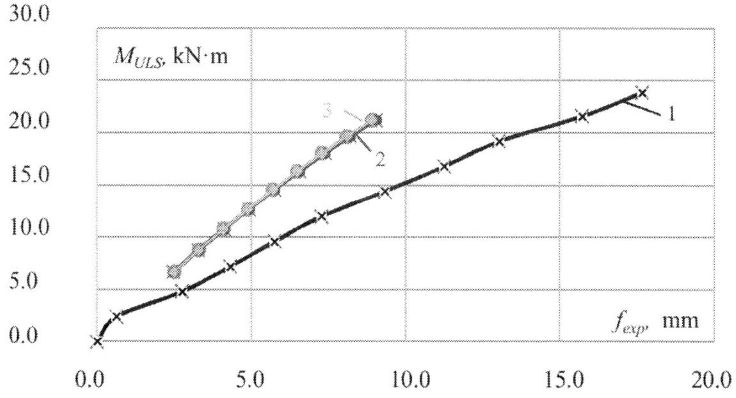

Fig. 5. Diagram of flexures: 1 - experimentally; 2 - estimated by DSTU; 3 - estimated by Eurocode 2.

A similar problem is considered in articles [9-12], where there is almost the same as the convergence estimation of flexures by the National [2] and European [3] Standards.

4 Conclusions

The algorithm for determining deformations of materials at the known bending moments by using the method of simple iteration or method of successive approximations is proposed. The solution of this problem allows practical use of the National [2] and European [3] Standards for calculating flexures.

Analyzing of the experimental flexure and the calculated value showed, that they have a significant difference 39.6 to 40.1%. This difference can be explained by the empirical coefficient.

The flexures, which are theoretically determined by the National [2] and European [3] Standards are the same.

In the following publications, the authors plan to continue comparing experimental and estimated data on other test items to determine the limit states.

References

1. *Concrete and reinforced concrete structures. Key provisions: DBN V.2.6-98: 2009.* Minregionstroy of Ukraine (2011)
2. *Concrete and reinforced concrete structures of heavy concrete. Design Rules: B V.2.6-156 ISO: 2010.* Minrehionbud of Ukraine (2011)
3. *CEB-FIV Eurocode 2: Design of Concrete Structures. Part 1: General Rules and Rules for Buildings. ENV 1992-1-1,* Brussels, CEN (1991)
4. Murashko L.A., Kolyakova V.M., Smorkalov D.V. *The calculation of the strength of normal and oblique sections to the longitudinal axis of the bending concrete elements for GSN V.2.6-98: 2009: Training. Manual,* Kiev, KNUBA (2012)
5. A.M Pavlikov, A. Boyko *Calculating the strength of the normal sections of beam elements with nonlinear deformation model (based DBN V.2.6-98: 2009).* Poltava, PoltNTU (2012)
6. Byby E.V., Narayanan R,S. *Recomendation for Evrocode 2: Designing of concrete structures: guidance for proektyrovschykov for EN 1992-1- 1 and EN 1992-1-2. Evrocode 2: Designing of concrete structures. Common rules and rules for buildings.* MGRS (2013)
7. Bliharskyy S.J., I. Karhut, Struk R.F. *Calculation and design of normal and sloping sections of concrete elements: Training. Manual.* Lviv, "Lviv Polytechnic" (2014)

8. Babaev V.M., Bambura A.M., Pustovoytova O.M., Reznik P.A., Stoyanov E.G., Shmukler V.S. *Practical calculation of elements for concrete structures DBN V.2.6-98, 2009 compared with estimates for the SNIP 2.03.01-84 * and EN 1992-1-1(Eurocode 2)*, Kharkov, Golden Pages (2015)
9. Kuzmych L., Kvasnikov V. Advances in Intelligent Systems and Computing (AISC), **543**, 659-663 (2016)
10. Babich V.E. Scient. Coll. Series: Tech. Science and architect, **101**, 532-540 (2011)
11. Babich V.E., A.E. Polyanovska, Boreychuk L.M. Economical materials, structures, buildings, Coll. Science. Works, **32,** 121-128 (2016)
12. Babich V.E., Boreychuk L.M. Modern technologies and calculation methods in construction: Coll. Science. Works, **5**, 192-198 (2016)
13. Babich V.E., Dovbenko V.S., Rebesh G.I., Tkach N.V. Economical materials, structures, buildings. Coll. Science. works, **33**, 121-128 (2016)
14. Yehia N.A.B. Engineering Structures, **31**, 404–416 (2009)

Shear strength of reinforced concrete beams strengthened by P.B.O. fiber mesh under loading

Zinoviy Blikharskyy[1], *Roman* Khmil [1*], and *Pavlo* Vegera[1]

[1]Lviv Polytechnic National University, Department of Building Constructions and Bridges, Bandera str. 13, Lviv, Ukraine, 79013

Abstract. This article presents experimental study of sheer strength of reinforced concrete beams without transverse steel reinforcement, which strengthened by composite materials. The feature of tests is that the beams' strengthening is made under simultaneous action of loading. The research program involves a series of test beams with size 2100 × 200 × 100 mm and which contains control sample and three reinforced samples by reinforcing FRCM system. FRCM system consisting of two components: mineral mortar based on modified cement Ruredil X Mesh M750 and reinforcing P.B.O. fiber mesh Ruredil X Mesh Gold (Italy). The strength research of test samples was carried out with the shear distance to effective depth ratio $a/d = 2$. The strengthening loading levels were selected at 0.0, 0.3, 0.5 from shear strength of non strengthened control sample. As a result of experimental studies we found that during strengthening design the inclined cross section of beams we should take into account the existing level of loading. Using the strengthening system Ruredil X Mesh Gold the strengthening effect is reduced at 2.8 to 2.9 times while the existing level of loading increase from 0 to 50%.

Introduction

Nowadays, many industrial and civil buildings have already exceeded their estimated lifetime. Corrosive influence leads to significant corrosion damage of reinforcement of the constructions, sometimes even to its full destruction. Demolition costs for these buildings or new construction costs are extremely high. Therefore we consider strengthening of old structures to continue their lifetime and bring them in accordance with the requirements of actual codes and standards.

A lot of studies of shear strength of concrete structures have been conducted, concerning the study of stress-strain state design principles, methods of strengthening using steel or concrete elements. Using a significant amount of consumables leads to finding the strengthening material, which would have provided high shear strength increasing of RC beams with lower cost of labor [1-2]. The composite materials made from carbon composite and others fibers in a cloth or ribbons belong to these materials. The simplicity and quickness of installation of the reinforcement allows us to gain good results without major delays.

* Corresponding author: roman.y.khmil@lpnu.ua

Analysing researches of shear strength with different types of composite materials reinforcing, it can be concluded that the study of stress-strain state of shear reinforced is a key issue, with ongoing development goals and objectives, which are set by the researchers [3-5]. However, few studies are conducted on reinforced beams under simultaneous action loading and reinforcing. Therefore, the establishment of real sheer strength of RC beams with damaged sheer reinforcing and strengthened by composite materials under action of loading is a topical scientific and practical issue.

1 The testing program

In this article we research the beams without transverse reinforcement, which serve as a beams model with full destructive stirrups after corrosion influence. The research program involves a one series of test beams with size 2100×200×100 mm which contains control sample and three reinforced samples by reinforcing FRCM system. FRCM system consisting of two components: mineral mortar based on modified cement Ruredil X Mesh M750 and reinforcing P.B.O. fiber mesh Ruredil X Mesh Gold (Italy) [6].

The strength research of test samples was carried out with the shear distance to effective depth ratio a/d = 2. The strengthening loading levels were selected at 0.0, 0.3, 0.5 from shear strength (V_{ed}) of non-strengthened control sample.

RC beams designed that the beams destruction after strengthening took place by shear. To accomplish it these beams were designed with a significant margin of tensile rebar. A method of experimental research involves alternate testing of two inclined sections of reinforced concrete beams, which reduces the cost of materials and necessary instrumentation and detailed presented at [7].

2 Experimental study of shear strength

2.1 Research results of control beams

At the first stage of research, control sample was tested. Beams mark follows BO – beam ordinary, the first digit – serial number, the second digit – prototypes number and the third digit – section number. For example, BO 1.2-2 means that tested example from the first series of the second beam of the second section.

The limit value of concrete compressed deformation at cross section above the inclined crack was taken as the criterion of shear strength [8, 9]. The shear strength of the RC beams equated with physical destruction of compressed concrete above the inclined cracks. Both limit states (ULS and SLS) of control beam occur in the following sequence:

- crushing compressed concrete and plastic deformation of reinforcement cage (ULS).
- opening limit width of the inclined crack (w_{max}=0.4 mm) in concrete (SLS);

The first inclined crack disclosed in mid-height cross section, at the load – 50 kN at an angle equal to 45°. At the next stage the inclined crack spread to the bottom surface of the beam. With increasing loading the width of crack increased and the crack propagation occurred to the top surface of the control sample (Fig. 1).

 DOI: 10.1051/matecconf/201711602006

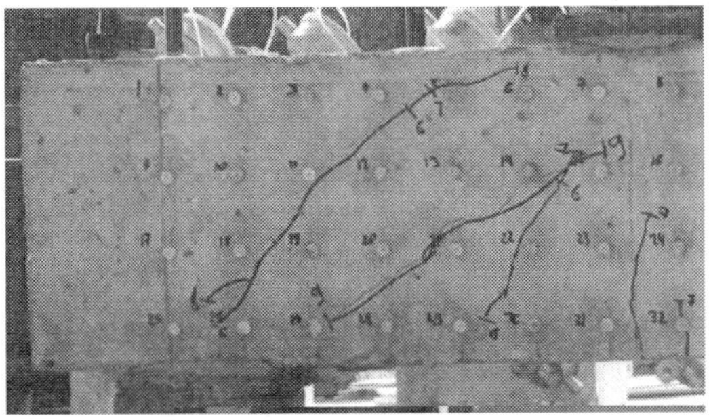

Fig. 1. Distribution of inclined cracks in the control sample.

On occurrence of limit width of crack w_{max}=0.4mm, another crack was opened in the direction from the loading point to the edge of beam support. This crack was destructive one. Inclined crack spread throughout the height of the beam, then occurred the crushing of compressed concrete. Longitudinal tensile reinforcement received typical plastic bending deformations. Shear strength of RC beams BO-1.1 was V_{ed}=95 кN. General view of the tested control beam is shown in Fig. 2.

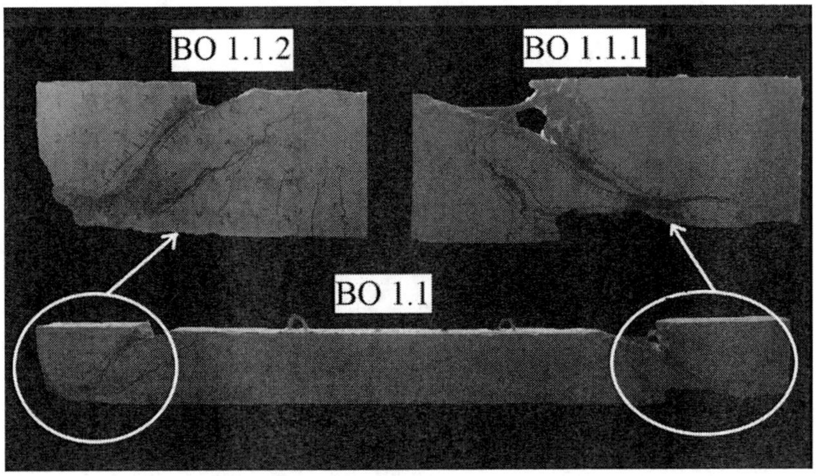

Fig. 2. The tested control beam BO-1.1

For each cross section of test specimen at shear area there was obtained area of distribution concrete deformation by using rappers of the comparator measurement (Fig 3).

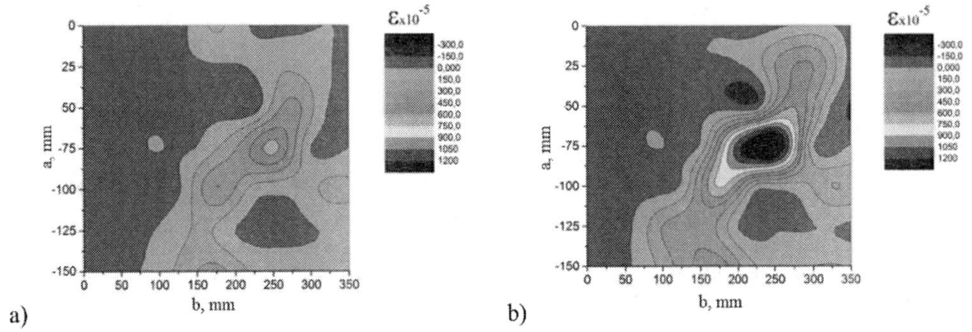

Fig.3. The distribution of concrete deformation in the shear area for beam BO-1.1: a) the area before the opening inclined crack; b) the area before the exhaustion of ULS

Measurement of concrete deformation was performed using rappers of comparator with grid spacing 50×50 mm. Grid frame of comparator located on the front edge of specimen (Fig. 2). The distance between the rappers were measured with an accuracy of 0.1 mm.

Tensile deformations fixed with cracks width in concrete. This method of measurements provides a distribution of strains from the support point to the loading point. As we can see (Fig. 3b) the tensile deformations are in the direction of distribution inclined cracks, and compression deformations are around them.

2.2 Results of the study of strengthened beams

According to the testing program, three specimens were strengthened by reinforcing FRCM system with 70 mm wide strips (Fig. 4-6). These beams mark follows BSC – beam strengthened by composite material; the first digit – serial number, the second digit – prototypes number and the third digit – section number. Index 0...0.5 means the loading level under which the strengthening was performed and defined from shear strength (V_{ed}) of non-strengthened control sample.

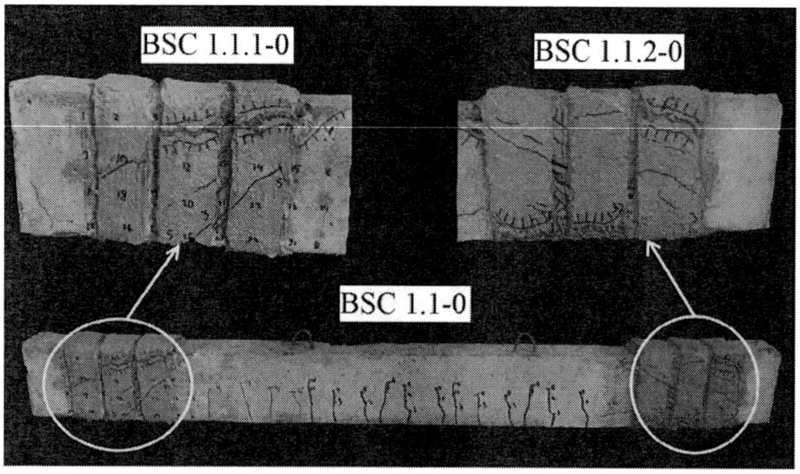

Fig. 4. The tested strengthened beam BSC 1.1-0

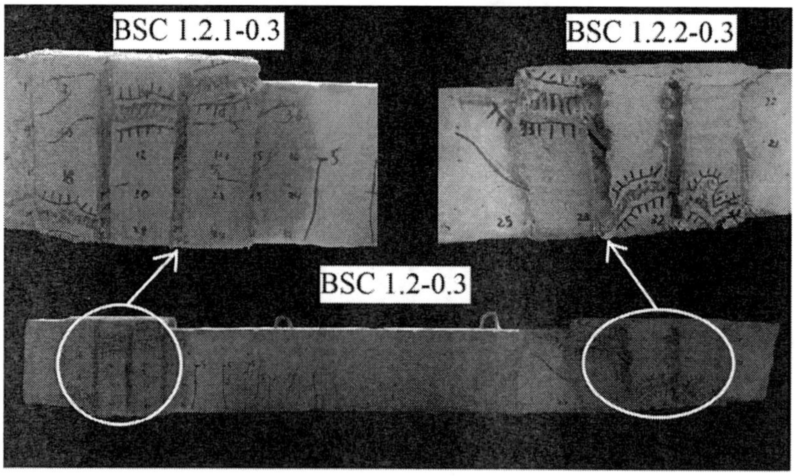

Fig. 5. The tested strengthened beam BSC 1.2-0.3

The criterion of achievement shear strength was the same as for the control sample. The shear destruction of RC beams strengthened by composite materials took place in the following order:

- opening limit width w_{max}=0.4 mm of inclined crack on the concrete surface;
- distribution inclined crack to the compressed concrete and appearance mesh of cracks with width w_{max}=0.2 mm on the surface of strengthening system;
- concrete damage in the area of the main tensile stress and exfoliation strengthening system in this area;
- plastic deformation of rebar and crushing of compressed concrete, large deformation reinforcing materials, which can be seen due to the damage of the cover layer.

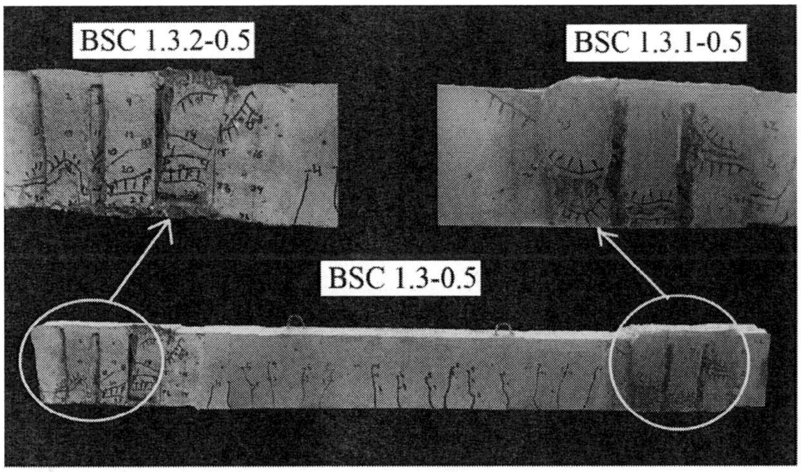

Fig. 6. The tested strengthened beam BSC 1.3-0.5

With further increasing loading the materials detachment from beam continues and there occurs a violation of anchoring. The shear strength of samples strengthened without any loading was amounted to $V_{Ed} = 137.5$ kN. Average shear strength for samples strengthened with some loading level were: for beam BSC 1.2-0.3 – $V_{Ed} = 120$ kN and for BSC 1.3-0.5 – $V_{Ed} = 110$ kN.

A type of fracture was changed in strengthened beams, it was more plastic, without any concrete particles and plastic deformation of steel reinforcement cage. The distribution of concrete deformations is shown in Fig. 7. As we can see, the tensile deformations occupy a larger area.

For the sample strengthened without initial loading, stress concentration was held at half of the beam height, similar to the control sample. Samples, strengthened with some initial loading, have another type of stress distribution that is caused by the inclusion strengthening materials at working similar to the transverse steel reinforcement.

Concrete deformation with a width of opening inclined cracks, the nature of the growth are the same as for the control samples but reach much higher values. With the exhaustion of shear strength, strengthening materials received significant deformation, which led to the loss of its original appearance, but the gap was not observed.

According to recommendation [10] the deformation of composite materials should be 40% of its limit tensile deformation. What we can see according to experimental data, this type of FRCM system strengthening is effective one for shear strengthening.

DOI: 10.1051/matecconf/201711602006

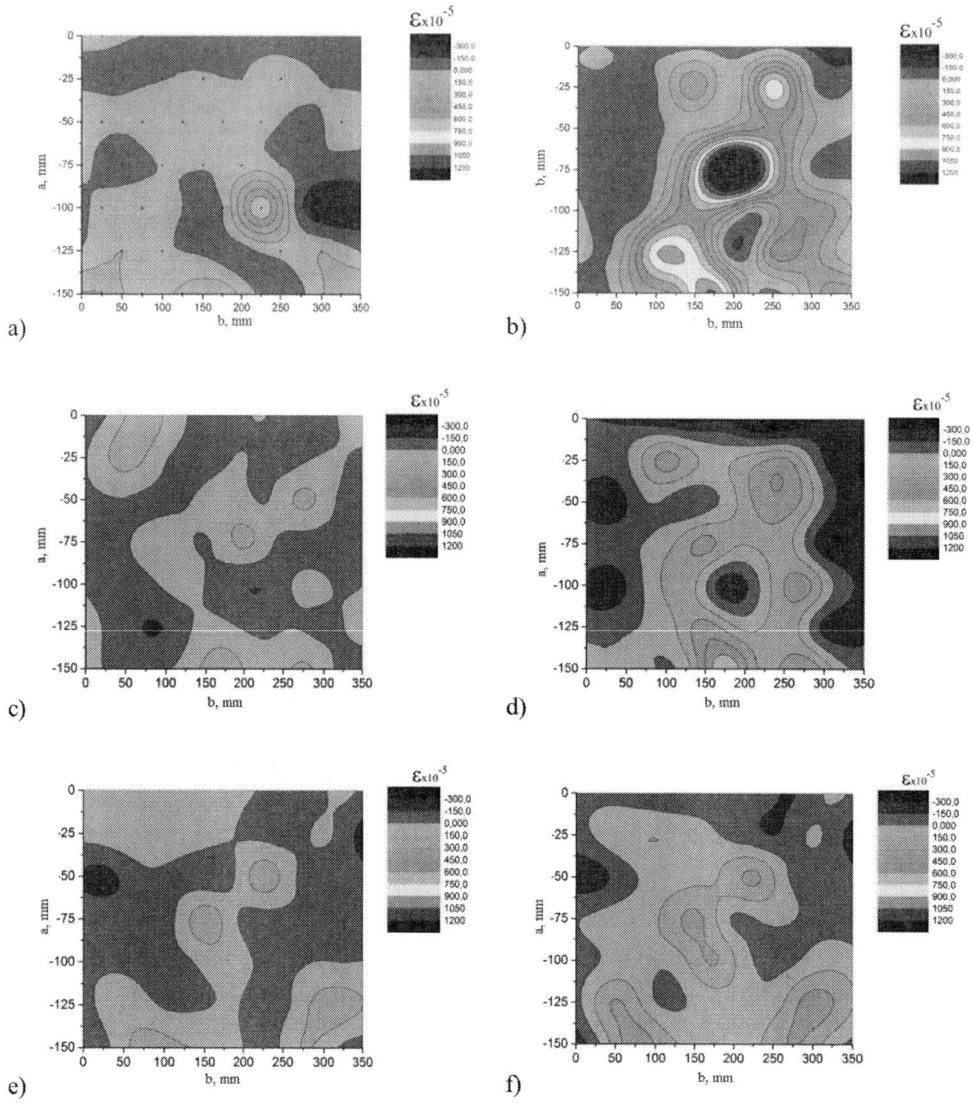

Fig.7. The distribution of strain in the beam: area of deformation before opening inclined cracks in the beam: a) BSC 1.1-0; c) BSC 1.2-0.3; e) BSC 1.3-0.5; area of deformation before exhaustion of ULS for beams: b) BSC 1.1-0; d) BSC 1.2-0.3; f) BSC 1.3-0.5

2.3 Analysis of experimental results

The maximum effect of increasing the shear strength was 45% for samples strengthened without initial loading. It should be noted that with increasing loading level up to 0.5 of shear strength of control beam ($0.5V_{ed}$) the effect of strengthening decreased to 16%. For beams reinforced at the level of 0.3·effect decreases to 21%.

Namely for samples, according to the level loading, the effect of strengthening decreases by 2.8 times, from 45% to 16%, so to perform strengthening at the level of more than 0.5 is impractical.

The results of the comparative analysis of the shear strength of strengthening RC beam on the shear are shown in Table 1.

Table 1. The ultimate limit state (ULS) inclined cross sections of testing beams.

Types of beams	Types of the beams' cross section	Cross section *bxh* mm.	Span of the beam l_0 mm	Experimental shear strength V_{Ed}^{exp}, кN	Average experimental shear load, V_{Ed}^{exp}, кN	Increasing shear strength $\frac{V_{ed}}{V_{ed}^{BO1.1}}$
BO 1.1	BO 1.1.1	201x101	1900	97	95	-
	BO1.1.2		1550	93		
BSC 1.1-0	BSC 1.1.1-0	199x100	1900	130	137.5	1.45
	BSC 1.1.2-0		1650	145		
BSC 1.2-0.3	BSC 1.2.1-0.3	200x100	1900	126	120	1.26
	BSC 1.2.2-0.3		1650	117		
BSC 1.3-0.5	BSC 1.3.1-0.5	201x98	1900	116	110	1.16
	BSC 1.3.2-0.5		1650	114		

The inclined crack width w_{max}=0.4mm was the criterion for serviceability limit state. The results of the experimental data of the shear strength for serviceability are in Table. 2.

Table 2. The serviceability limit state (SLS) inclined cross sections of testing beams.

Types of beams	Types of the beam cross section	Maximum inclined crack width, mm	Experimental shear strength V_{Ed}^{exp}, кN	Average experimental shear load, V_{Ed}^{exp}, кN	Increasing shear strength $\frac{V_{ed}}{V_{ed}^{BO1.1}}$
BO 1.1	BO 1.1.1	0.4	80	80	-
	BO 1.1.2		80		
BSC 1.1-0	BSC 1.1.1-0		120	110	1.38
	BSC 1.1.2-0		100		
BSC 1.2-0.3	BSC 1.2.1-0.3		100	100	1.25
	BSC 1.2.2-0.3		100		
BSC 1.3-0.5	BSC 1.3.1-0.5		90	90	1.13
	BSC 1.3.2-0.5		90		

Character and value of increasing shear strength by SLS for all samples reinforced composite system is close in meaning to ULS. But the fixed disclosure of limit crack width w_{max}=0.4 mm inclined at 20 kN was lower than the loss of the shear strength of inclined section. Also, the maximum values of crack opening constitute the 0.5...0.6 mm at sample BSC 1.2-0.3 and BSC 1.1-0 and 0.7 mm at BSC 1.3-0.5. On the surface of the material limit the width crack disclosure constitutes the w=0.2 mm, which is considerably less than the limit one.

Fig. 8. Distribution of cracks on the surface of the strengthening

The sharp increase in the width of the crack opening exposing the composite materials is observed with physical destruction in specimens. It should be noted that the inclined crack width was measured on the concrete surface between strengthening elements. In the cover layer of strengthening material opening width of cracks was less than 0.25 mm, but on the surface of a there were a mesh of cracks (Fig. 8).

3 Conclusions

1. The maximum effect of increasing the shear strength (ULS) is 45% for samples strengthened without initial loading. In accordance with the increase of the initial loading the strengthening effect decreases. For beams reinforced at the level of $0.3V_{ed}$ the effect decreases to 21% and for BSC 1.3-0.5 to 16%.

2. Character and value of increasing shear strength by SLS for all samples reinforced composite system is close in meaning to ULS. But the fixed disclosure of limit crack width w_{max}=0.4 mm inclined at 20 kN was lower than the loss of its shear strength. The maximum effect of ULS is 38% for samples strengthened without initial loading. For beams reinforced at the level of $0.3V_{ed}$ the effect decreases to 25% and for BSC 1.3-0.5 to 13%.

3. For samples, according to the level loading, the effect of strengthening falls to 2.8-2.9 times, from 45% to 16% and from 38% to 13%, so to perform strengthening beams like were tested at the level of more than 0.5 is unpractical.

4. This peculiarity that strengthening effect depends on loadings level should be considered when we design the strengthening systems.

References

1. E. David, C. Djelal, F. Buyle-Bodin, *2nd Int. PhD Symposium in Civil Engineering,* 1 (1998)
2. L. Ombers, Compos. Struct. **№94,** 143 (2011)
3. R.S. Chavan, P.M. Pawar, IJERA, Vol.**3**, Issue **6**, 554 (2013).
4. E. A.-M. Ahmed, *Shear behavior of concrete beams reinforced with fibre-reinforced polimer (PFR) stirrups* : dis. doctor sciences: 05.23.01, 291 (2009).
5. A. Alzate, A. Arteaga, A. de Diego, D. Cisneros, R. Perera, Mater. Construcc., **63(310)** 251 (2013).
6. http://english.ruredil.it/Ruredil_Prodotti_chimici_per_edilizia/cataloghi.asp
7. P. Vegera, R. Khmil, Z. Blikharskyy, JCEEA, T. XXXII, z. **62** (nr 4/2015), 447 (2015).
8. National Standard of Ukraine, *Concrete and reinforced concrete structures: DBN B.2.6 - 98: 2009*, 84 (2011).
9. Europian Norm, *EN 1992-1-1:2004 Eurocode 2: Design of concrete structures - Part 1-1: General rules for buildings*, 225 (2004).
10. *Externally bonded FRP reinforcement for RC structures. Technical report*, FIB, 130 (2001).

Experimental study of properties of heavy concrete with bottom ash from power stations

Victor Bondar[1], *Volodymyr* Shulgin[1], *Oksana* Demchenko[1], and *Ludmila* Bondar[1]*

[1]Poltava National Technical Yuri Kondratyuk University, Department of technology building designs, products and materials, Pershotravnevyi Avenue, 24, Poltava, 36011, Ukraine

Abstract. This article deals with the influence of cement quantity, plasticizing additives and compaction time on the strength and water consumption of concrete during its manufacturing using bottom ash from a thermal power station. The study was carried out using three factorial experiments. Variables varied on three levels. The obtained pattern functions characterize a relationship between strength, water consumption and variable factors. These factors include cement quantity, plasticize additives and compaction time. Compilation of Pareto effect charts allowed estimation of the significance of function indexes. Analysis of surface pattern function has revealed the optimal correlation between additive quantity and compaction time, cement quantity and additive quantity, cement quantity and compaction time. Compression strength of concrete was taken as the pattern in the pattern function. When analyzing the pattern function with water consumption as a pattern, optimal correlations between additive quantity and compaction time, cement quantity and additive quantity, cement quantity and compaction time were revealed. Application of STATISTICA 12 software has specified values of factors when the maximum strength is achieved. Correlations of components which have an impact on water consumption have been determined. The conclusions contain the quantitative findings of the study.

1 Introduction

Significant amount of bottom ash is produced during coal burning for generation of thermal and electrical energy at thermal power stations. One of the ways of safe disposal of this waste is its use in the building industry for heavy concrete manufacturing.

An optimal quantity of bottom ash in a heavy concrete mixture is calculated by compiling the concrete composition provided that all necessary indicators of its quality are reached [1]. Concrete mixtures with bottom ash possess a higher viscosity, better mobility, lower water separation, and segregation. It makes bottom ash based mixtures perfect for use in monolith concreting. Slag use decreases a necessary quantity of cement, and therefore, decreases the exothermy of concrete. That allows prolonging the time before the solidification start and, therefore, prolongs the period of time when the concrete is available for use [2-6]. The

* Corresponding author: lvbondar06@gmail.com

features of this kind of concrete are not investigated enough. As well as the influence of bottom ash on mechanical properties of concrete [7].

2 Research results

2.1 Materials for concretes with use of ash and slag

This research aim is to determine the effect of concrete composition, which includes bottom ash, on its physical and mechanical properties. Another task is to compile an optimal composition of concrete with strength class C25/20, using ash and slag as additives.

For the research, we used a concrete mixture of the following composition. Cement CEM I 42.5 N is characterized by high early strength, high content of C_3S, and low content of C_3A [8].

Ash and slag (bottom slag) from Starobeshivska thermal power station the true density of which is 2.6 g/cm^3; bulk density is 1.206 g/cm^3; coarseness factor is 1; water requirement is 24%. Granite gravel was used as coarse-grained filler, its fraction was 5 to 10 mm, void ratio is 47%, bulk density is ρ_b = 1350 kg/m^3, true density is ρ_t = 2.71 g/cm^3. Polycarboxylate superplasticizer GLENIUM 51 served as the additive.

2.2 Mathematical experiment planning

The research planning based on the 3-factorial experiment on 3 levels. Cement quantity, concrete mixture compaction time, and quantity of plasticizer additive were the variable factors. In order to ensure a proper porosity at each level, water-cement (V/C) ratio was 0.45. It also accounted for peculiarities of initial materials structure. If V/C is lower than 0.45, it will be impossible to provide a proper compaction index foe the mixture. There were control specimens manufactured for the research. Quartz sand used for manufacturing the specimens, and its strength was 31 N/mm^2.

2.3 Regression analysis of concrete strength research

According to concrete strength research results, pattern function was obtained (1) using STATISTICA 12 software

$$Y_R = 30.07 + 0.525\,x_1 + 0.78\,x_2 - 2.3\,x_3 - 0.34\,x_1^2 + 0.44\,x_2^2 +$$

$$+ 0.24\,x_3^2 - 1.18\,x_1\,x_2 - 0.75\,b_{13}\,x_1\,x_3 - 0.57\,x_2\,x_3 \qquad (1)$$

The obtained model is adequate according to lack-of-fit test. Pattern function surfaces is plotted in accordance with equation (1), it showed in figures 2-4. Pareto effect chart shows (fig. 1) that the equation indexes at x^3 are the most significant in the model. This means that pattern function even in a stationary area has got some curvature and can be described by a polynomial function of the second degree. The results allow estimation the adequateness of the model since there were three parallel experiments conducted at level 0.

Analysis of surface shape shown in Figure 2 reveals that the maximum strength was obtained when compaction time was 3 minutes and additive quantity was 8 liters. Further increasing the additive quantity could not lead to any increase in concrete strength.

Analysis of surface shape in figure 3 shows that the maximum strength is reached at 390 kg of cement and 8 l of the additive.

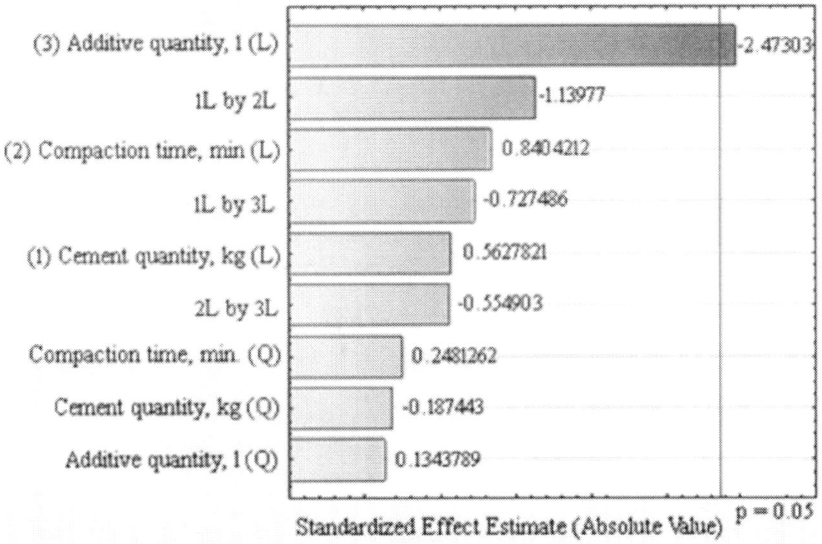

Fig. 1. Pareto effect chart for estimation the statistical significance of the regression equation indexes of concrete strength.

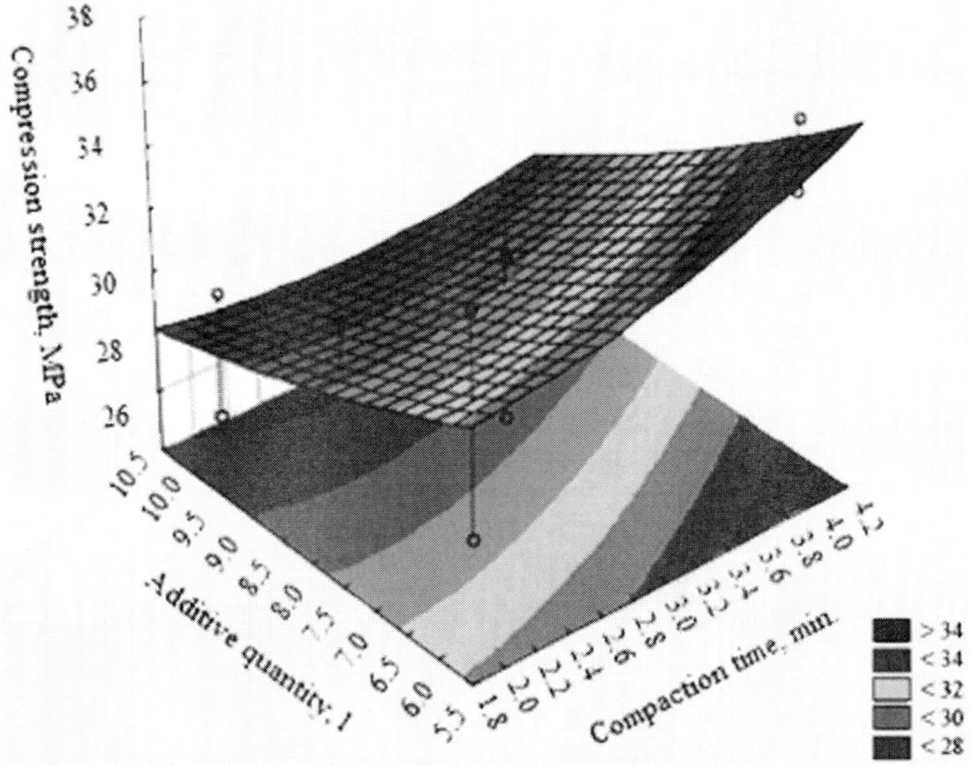

Fig. 2. Overall view of pattern function surface for limit compression strength of concrete dependent on additive quantity and compaction time of concrete mixture.

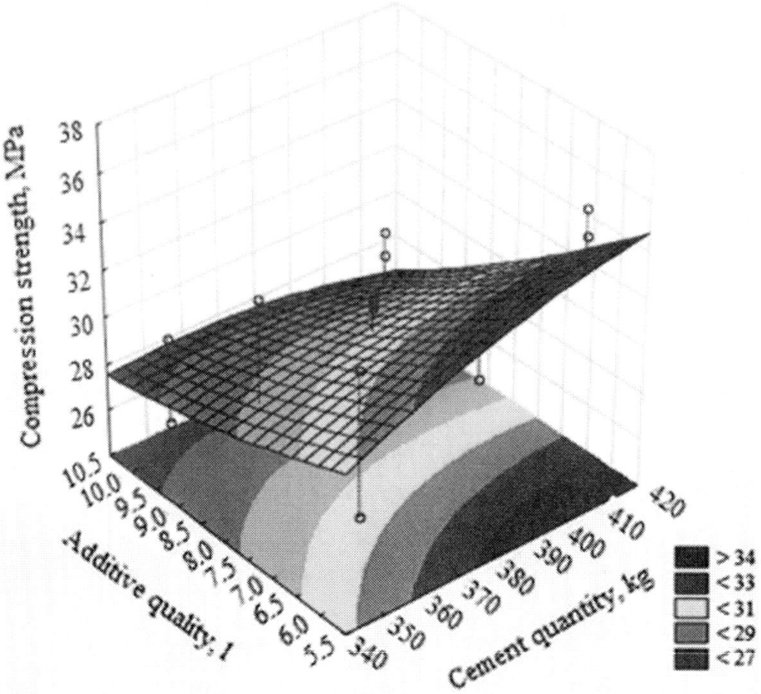

Fig. 3. Overall view of pattern function surface for limit compression strength of concrete dependent on additive quantity and cement quantity.

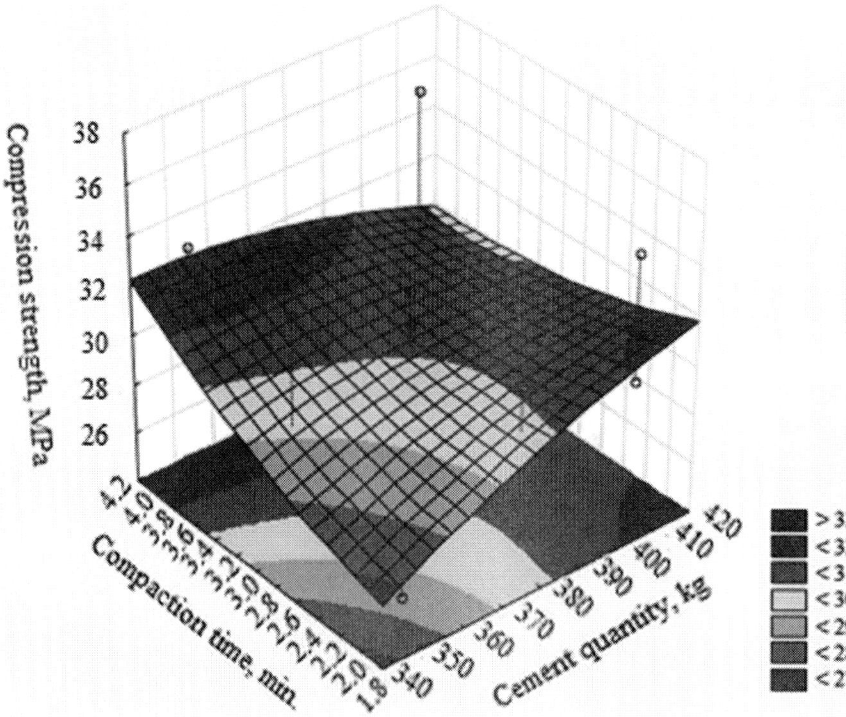

Fig. 4. Overall view of pattern function surface for limit compression strength of concrete dependent on cement quantity and compaction time of concrete mixture.

Analysis of surface shape in Figure 4 shows that the maximum strength is reached at 3 min. of compaction time and 400 kg of cement. Further increasing of the quantity of additive doesn't lead to any increasing in concrete strength.

Visual analysis of the surfaces reveals only approximate values of optimal relations of components. However, STATISTICA software allows determination of the factors at which the investigated value reaches its maximum magnitude (Table 1).

Table 1. Critical values of variables.

Factor	Observed minimum	Critical values	Observed maximum
Cement quantity, kg	350.000	350.056	410.000
Additive quantity, l	6.000	12.990	10.000
Compaction time, min.	2.000	2.410	4.000

2.4 Regression analysis of concrete water consumption

According to concrete water consumption research results, pattern function has been obtained (2)

$$Yw = 2.23 + 0.06\,x_1 + 0.4\,x_2 - 1.4\,x_3 - 1.54\,x_1^2 + 0.23\,x_2^2 +$$
$$+ 0.95\,x_3^2 - 0.18\,x_1 x_2 - 1.16\,b_{13}\,x_1 x_3 - 0.63\,x_2 x_3 \quad (2)$$

Pareto effect chart shows (fig. 5) that the equation indexes of combined cement and additive quantities are the most significant in the model.

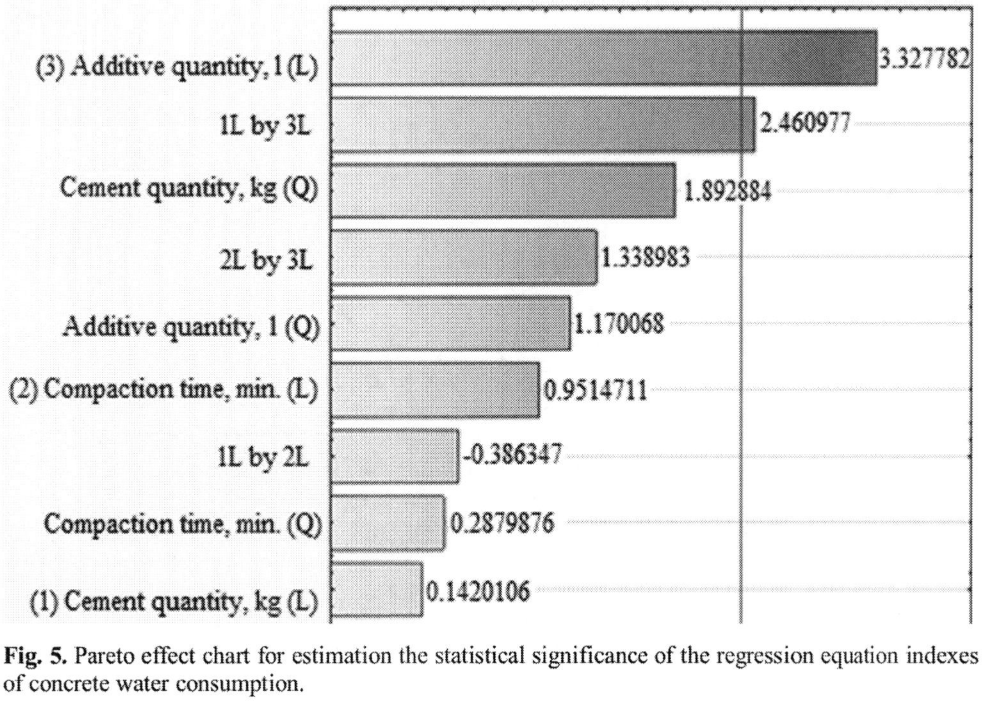

Fig. 5. Pareto effect chart for estimation the statistical significance of the regression equation indexes of concrete water consumption.

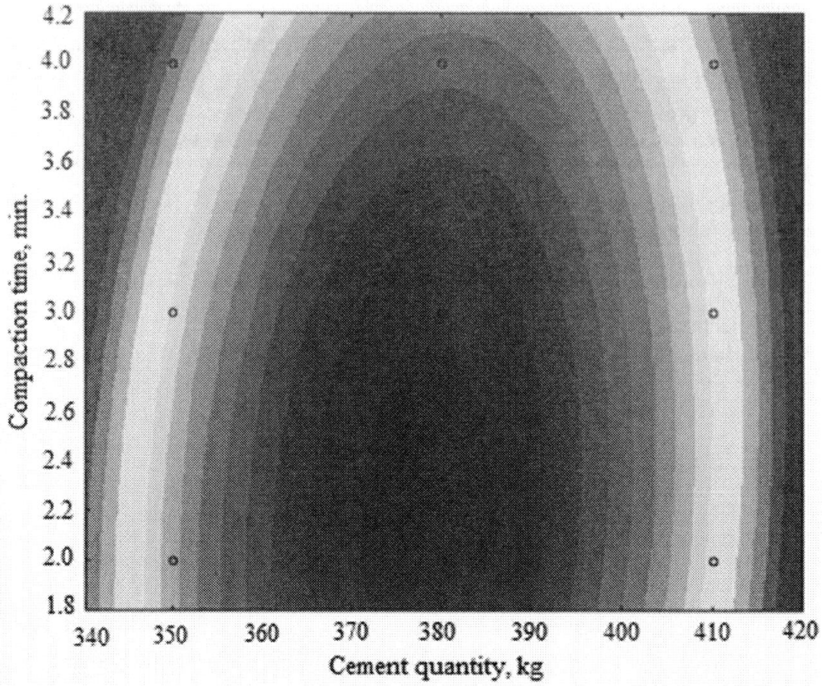

Fig. 6. Isolines of pattern function surface of water consumption against quantities of cement and additives.

Analysis of isolines of pattern function surface shows that the minimum water consumption could be reached at cement quantity about 390-400 kg, and 8 liters of additive.

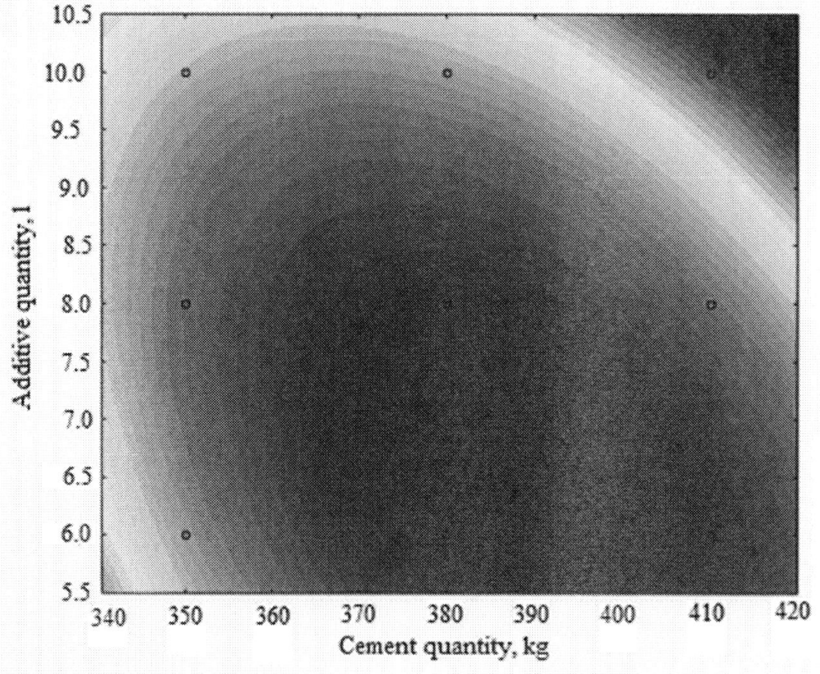

Fig. 7. Isolines of pattern function surface of water consumption against quantity of cement and compaction time.

Analysis of isolines of pattern function surface shows that the minimum water consumption could be reached at cement quantity about 390 kg, and 3 minutes of compaction time.

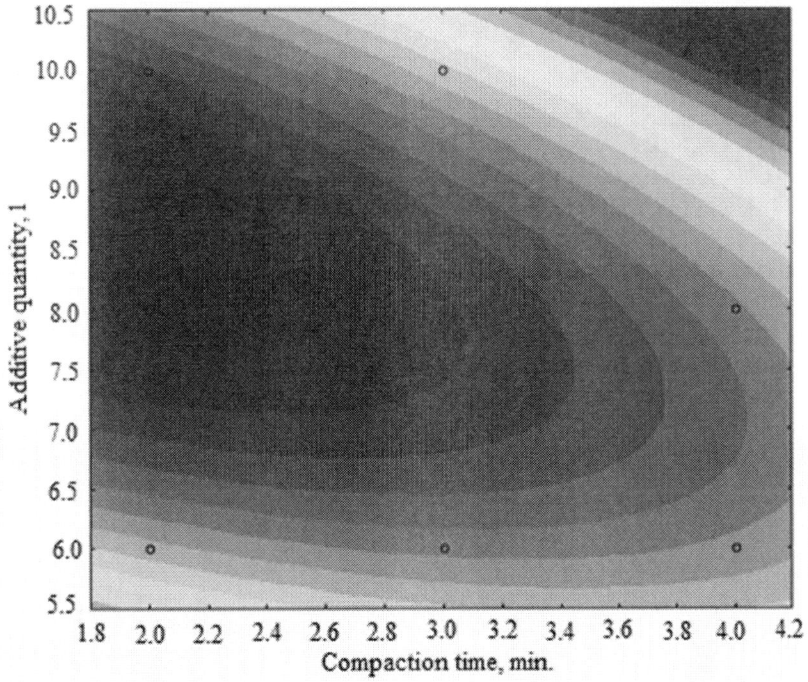

Fig. 8. Isolines of pattern function surface of water consumption against quantity of additive and compaction time.

Analysis of isolines of pattern function surface shows that the minimum water consumption could be reached at additive quantity about 7.5 to 8.1 and 3 to 3.6 minutes of compaction time.

The values of obtained relations of components during the study of its influence on water consumption is shown in table 2.

Table 2. Critical values of variables.

Factor	Observed minimum	Critical values	Observed maximum
Cement quantity, kg	350.000	379.100	410.000
Additive quantity, l	6.000	7.750	10.000
Compaction time, min.	2.000	1.800	4.000

3 Conclusion

Bottom ash usage as a fine filler allowed us to obtain concrete of 36 N/mm^2 strength. The strength of concrete with quartz sand was 31 N/mm^2. Therefore, bottom ash with additive provides 14% higher strength of concrete and could be used instead of sand in concrete mixtures.

Improvement of cement quantity causes increase of compaction index. It was determined that optimal compaction time for concrete is 3 minutes. The minimum value of water consumption is 2.0 %. This value was reached when the quantities of cement and plasticizer additive are 390 kg and 8.1 respectfully.

Water consumption of control specimens with sand filler is 0.6 % higher than that of concrete with bottom ash.

References

1. P.V. Krivenko, E.K. Pushkareva, V.I. Gotz, G. Y. Kovalchuk, *Tsementy i betony na osnovi palyvnykh zol ta shlakiv*, Kiev (2012)
2. Nurul Izzati Raihan Ramzi Hannan, Shahiron Shahidan, Noorwirdawati Ali, Mohamad Zulkhairi Maarof, MATEC WEB CONF, 97 (2017)
3. V.M. Malhotra, P.K. Mehta, High-Performance, High-Volume Fly Ash Concrete, Inc., Canada (2002)
4. Seyoon Yoon, P.J.M. Monteiro, D.E. Macphee, F.P. Glasser, Mohammed Salah-Eldin Imbabi, CONSTR BUILD MATER, 54, 432-442 (2014)
5. V.M. Malhotra, High-Performance, Concrete International 24(7), 30-34 (2002)
6. P.K. Mehta, International Workshop on Sustainable Development and Concrete Technology, 3-14 (2004)
7. N. Zaichenko, A. Serduk, Visnyk DNABiA, 1(99), 137-144 (2013)
8. M. Thomas Concrete Thinking, Portland Cement Associations, 1-24 (2007)

Study of the stress-strain state of compressed concrete elements with composite reinforcement

Yurii Bondarenko[1,*], *Karina* Spirande[1], *Maryna* Iakymenko[1], *Mikhail* Mol'skyj[1], and *Dimitry* Oreshkin[2]

[1]Kharkov National University of Civil Engineering and Architecture (Kharkov), Department of reinforced concrete and stone structures, 61002, Sumskaya Str., 40, Kharkov, Ukraine
[2]CEO Technological Group EKIPAGE (Kharkov), ACE-1 (Assembly and Construction Enterprise), Enakievskaya Str., 4, Kharkov, Ukraine

Abstract. The efficiency analysis of the application of glass composite reinforcement in compressed concrete elements as a load-carrying component has been performed. The results of experimental studies of the deformation-strength characteristics of this reinforcement on compression and compressed concrete cylinders reinforced by this reinforcement are presented. The results of tests and mechanisms of sample destruction have been analyzed. The numerical analysis of the stress-strain state has been performed for axial compression of concrete elements with glass-composite reinforcement. The influence of the reinforcement percentage on the stressed state of a concrete compressed element with the noted reinforcement is estimated. On the basis of the obtained results, it is established that the glass-composite reinforcement has positive effect on the strength of the compressed concrete elements. That is, when calculating the load-bearing capacity of such structures, the function of composite reinforcement on compression should not be neglected.

1 Introduction

At present, the interest to applying of composite reinforcement in bearing structural elements of buildings and structures has noticeably increased due to its increased corrosion resistance, high tensile strength, low specific gravity, chemical and magnetic inertness, dielectric properties, radio transparency, and low heat transfer coefficient.

Despite of the above advantages, the composite reinforcement also has some disadvantages and special features: a relatively low modulus of elasticity compared to that of steel reinforcement, brittleness at failure (lack of fluidity), anisotropic properties of the material (low shear strength and axial compression), and low fire resistance.

In general, composite reinforcement is used in bent elements, but low modulus of elasticity leads to large transitions, so it is especially effective with prestressing. At the same time, compressed reinforced concrete elements are one of the main bearing vertical

* Corresponding author: bonya8947@gmail.com

structures of buildings and constructions. The insufficiency of data on the mechanical and deformation characteristics of composite reinforcement during compression restrains its use in real structures as a working armature and requires additional experimental and theoretical studies. At the same time, the operation efficiency of composite reinforcement as load-carrying one in compressed elements is unambiguously assessed by various studies.

2 Analysis of the recent studies and publications

One of the problems with the use of composite reinforcement is the uncertainty in obtaining its reliable mechanical and deformation characteristics under compression due to a variety of different failure mechanisms. The nature of the destruction of samples can be expressed both in the loss of stability of the entire sample and in its individual fibers. Researchers of strength composite rods indicate the different compressive strengths they have obtained, on average 30 to 50% lower than their tensile strengths [1-2].

There are either no single standard methods for determining the modulus of elasticity and the Poisson's ratio for compression, which are necessary for the calculation and design, of compressed concrete elements reinforced with glass-composite reinforcement. Meanwhile, the composite reinforcement is an anisotropic material and, consequently, a different modular one (the module of elasticity under compression and stretching may differ), which causes different deformation and destruction in its nature and direction. The data on the deformation-strength characteristics of compressed concrete elements reinforced with composite reinforcement are insufficient.

Some foreign design standards [3-6] do not recommend the use of composite reinforcement as a load-bearing one compressed in neither compressed elements nor even compressed bent elements; or it is allowed to use it, but with zero calculated compressive strength [7]. Nevertheless, as it was shown in [8, 9], the compressive strength of composite reinforcement should not be neglected.

In Ukraine, the design standards for structures using composite reinforcement are regulated by state standard [10]. Calculations are recommended to be performed according to the current norms for calculating reinforced concrete structures, replacing the steel reinforcement with non-metallic (composite), taking into account the linear diagram of the work and its design strength. These documents do not stipulate any special estimated dependencies. Standards associated with the testing of such reinforcement have not been developed yet at the state level.

3 Main part of the study

3.1 Experimental studies of composite reinforcement under compression

To determine the deformation-strength characteristics of composite reinforcement during compression in the testing laboratory of the Department of Reinforced Concrete and Stone Structures of Kharkov National University of Civil Engineering and Architecture (KNUCEA), a series of relevant studies of glass fiber reinforcing rods was carried out [11]. The glass-composite reinforcement (ACGR) manufactured by CEO Technological Group EKIPAGE (Kharkov) was admitted as the studied. At the same time, an attempt was made to evaluate (by numerical methods) the possible influence of the surrounding concrete medium on the magnitude of the indicated characteristics (modulus of elasticity and temporary resistance of the reinforcement) under compression experimentally and theoretically.

The samples with a nominal diameter of 10 (GCA-10) and 25 (GCA-25) mm, embedded in test couplings, with a working section length of 60 mm and 125 mm respectively were tested (Fig. 1).

Fig. 1. Schemes of testing composite samples for axial compression

Measurements of deformations were carried out by strain-gages with a base of 10 mm: longitudinal - S1, S2; transverse (annular) - S3, S4 (Fig. 2).

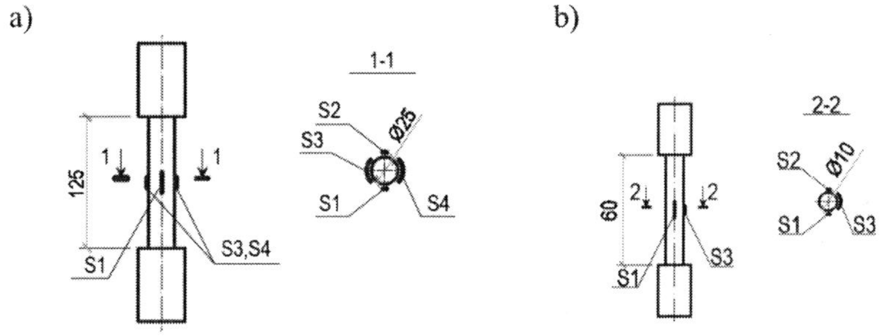

Fig. 2. Scheme of sensors arrangement on the samples of the series: a - GCA-25; b - GCA-10

Experimental samples were tested for compression by a single static load up to physical destruction. The load was applied in steps of 5% ... 10% of the destructive forces with an exposure of up to 1 min., increased smoothly, without jerks at a constant speed. It has been experimentally established that the strength of the EKIBAR glass-composite reinforcement for compression (483 MPa - Ø10, 294 MPa - Ø25) is less than in the tensile (950 MPa - Ø10, 600 MPa - Ø25) by approximately 2 times. The mechanism of destruction of the reinforcement is caused by the loss of stability of the compressed fibers within the matrix and, correspondingly, its fracture in the transverse direction. On the basis of results of the tests, diagrams of the state of the glass-reinforced plastic reinforcement during compression were constructed (Fig. 3).

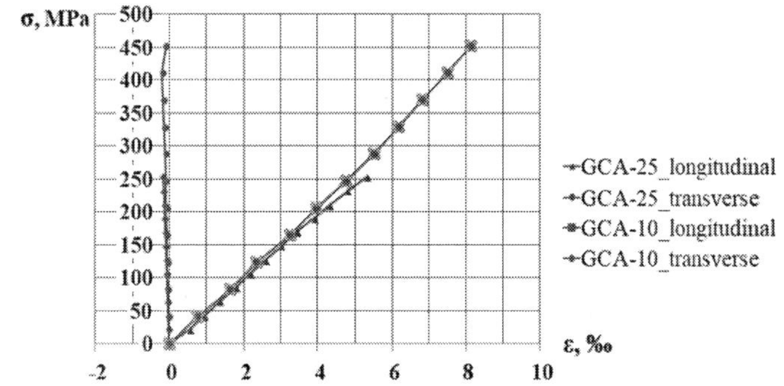

Fig. 3. Deformation of glass-composites reinforcement under compression

DOI: 10.1051/matecconf/201711602008

The values of Poisson's ratios are obtained as follows 0.02 and 0.033 and relative shortening at a destructive load of 1.02% and 0.69% for rods with a diameter of 10 and 25 mm, respectively. A linear relationship between stresses and strains is established up to failure, the modulus of elasticity at compression (47.9 GPa - Ø10, 48.7 GPa - Ø25) is less than in tension (51 GPa - Ø10.52 GPa - Ø25) by 6%.

3.2 Experimental studies of compressed concrete cylinders

Experimental studies of concrete compressed elements were carried out, the program of which included manufacturing and testing series of three compressive samples at static loading up to physical destruction.

The experimental samples were concrete cylinders with a diameter of 100 (CC-100), 150 (CC-150) and 200 (CC-200) mm with a ratio of diameter to height of 1:3. Dial gauges with a division rate of 10^{-3} mm were used to measure the longitudinal and transverse deformations of concrete. Longitudinal ones were measured with the arrangement of the instruments at an angle of 120^0 along the circumference of the samples on the basis of 100 mm in the middle of the part, the transverse ones - along the diameter of the samples with the corresponding base.

As a result of the tests, fracture formation and fracture patterns of the compressed cylindrical concrete samples under static loading were obtained; graphs of longitudinal and transverse deformation of concrete under loading were made (Fig. 4).

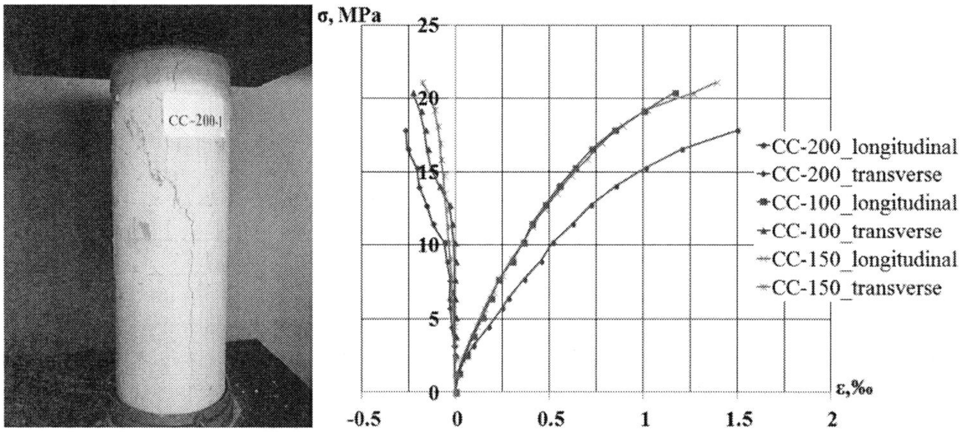

Fig. 4. Nature of fracture and deformation of concrete cylinders under compression

3.3 Experimental studies of compressed concrete elements with composite reinforcement

In order to study the strength and deformation of compressed concrete elements reinforced with longitudinal composite rods, experimental studies of testing series of concrete cylinders 100, 150, 200 mm in diameter reinforced with glass-composite reinforcement with diameters of 10 and 25 mm, installed in the centre of cross section were made.

The experimental samples were tested - concrete cylinders of 100, 150 and 200 mm in diameter, reinforced with EKIBAR glass-composite armature with nominal diameters of 10 and 25 mm (CCA-100-10 series, CCA-100-25, CCA-150-10 and CCA-200-25). Measurement of longitudinal and transverse deformations of concrete was carried out similarly to concrete samples. In order to measure the longitudinal deformations of the ACGR reinforcement, strain-gages with a base of 10 mm were glued at 120^0 along the circumference of the rods in the middle part of the samples in amount of 6 pieces (Fig. 5).

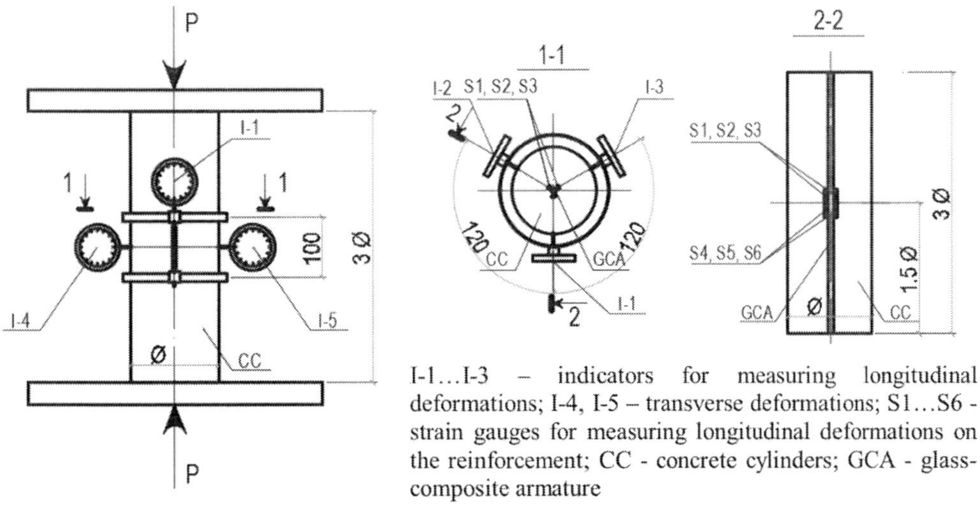

I-1…I-3 – indicators for measuring longitudinal deformations; I-4, I-5 – transverse deformations; S1…S6 - strain gauges for measuring longitudinal deformations on the reinforcement; CC - concrete cylinders; GCA - glass-composite armature

Fig. 5. Scheme of testing concrete samples with composite reinforcement of the CCA series

As a result of the tests, fracture formation and fracture patterns of compressed samples of glass-composite-concrete elements under static loading were obtained (Fig. 6).

Fig. 6. General view of tests and destruction nature of concrete samples (series CCA) with glass composite reinforcement under compression

As a result of comparison of the test data for the samples of CC and CCA series, an increase in the bearing capacity of the reinforced samples was defined: by 6.7% with a reinforcement percentage of 1.51% and by 13.7% with a reinforcement percentage of 6.04%. Analysis of the destruction schemes showed that, regardless of the percentage of elements reinforcement, the fracture is of a similar shape. The criterion for the destruction of glass-composite concrete (as well as reinforced concrete under similar conditions) elements under compression is the limiting compressibility of concrete. Obviously, these data should be taken into account when choosing the value of the design resistance of the glass-composite reinforcement.

3.4 Numerical studies of the stress-strain state of compressed concrete elements with composite reinforcement

Numerical studies of glass-composite cylindrical reinforcing rods Ø 10 and 25 mm with heights equal to free lengths of experimental samples without couplings (Fig. 7) were carried out using the LIRA-SAPR software. There are considered loadings, which completely rebuilt the loading levels being created during the tests, which allowed to compare the results of calculations and experimental data.

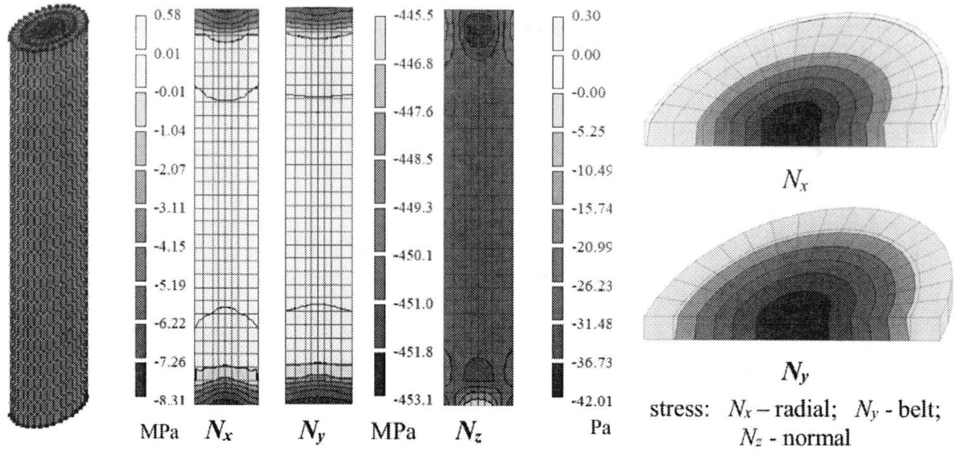

Fig.7. Investigations of the stress-strain state of glass-composite reinforcing rods: finite element calculation model (a); nature of the change in stress along the height (b) and the width (c) of the element

Based on the calculation results, the dependencies of "stresses on longitudinal and transverse deformations" were constructed (Fig. 8).

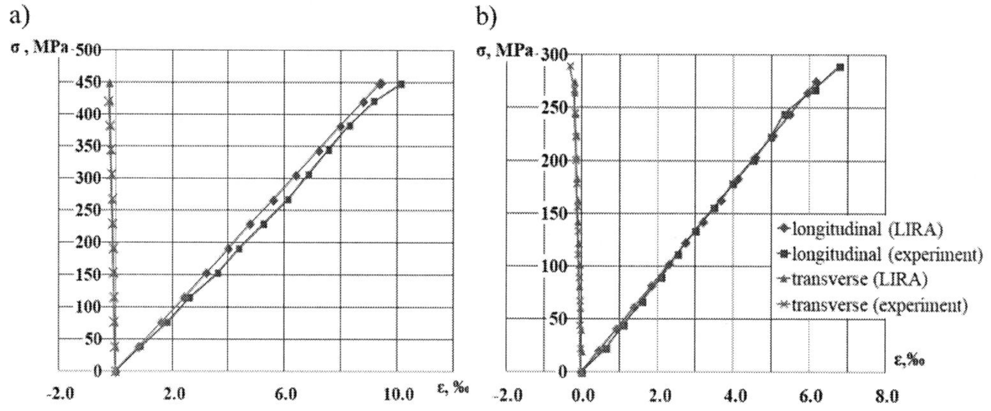

Fig.8. Diagrams of the state of the glass-composite reinforcement Ø 10 mm (a) and Ø 25 mm (b)

The data of this calculation allow us to conclude that the dependencies are close to those obtained during the tests.

Concrete cylinders 100, 150 and 200 mm in diameter, and similar ones with glass composite reinforcement Ø10 and Ø25 mm, were also subjected to numerical comparative analysis, which made it possible to assess the influence of the reinforcement percentage on the stressed state of the elements and their bearing capacity, without additional experimental studies. Five loading levels have been considered - 5, 25, 50, 75, 100% of the destructive values obtained during the tests. Numerical studies make it possible to observe changes in the nature of the stressed state, to identify areas of stress concentration, to estimate the influence of the percentage of reinforcement on the stress state of the elements, for example, a sample with a diameter of 150 mm (Table 1).

Table 1. The nature of the change in stresses in samples of Ø150 mm

	CC-150	CCA-150-10	CCA-150-25
N_x	-5.18 to +0.229 MPa	-4.95 to +2.87 MPa	-4.25 to +2.57 MPa
	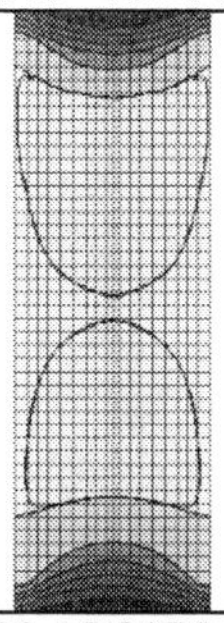	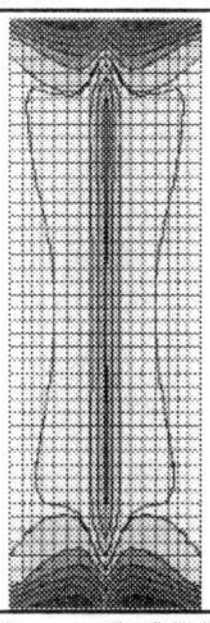	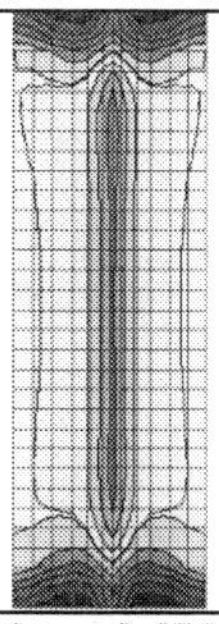
N_y	-5.33 +0.317 MPa	-5.33 to +2.87 MPa	-4.76 to +2.57 MPa
	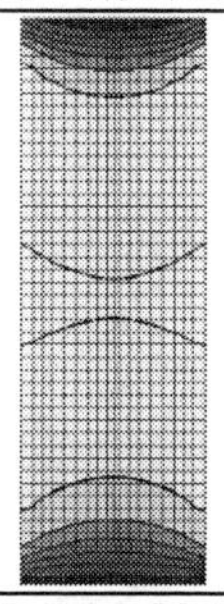	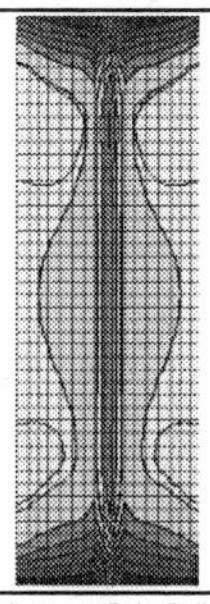	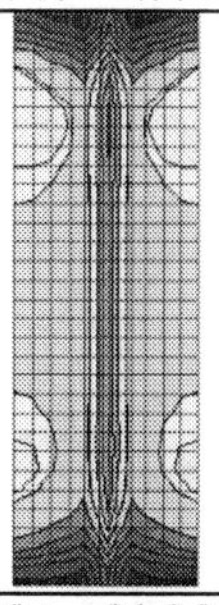
N_z	-27.4 -21.5 MPa	-42.2 to -21.5 MPa	-37.5 to -21.2 MPa
		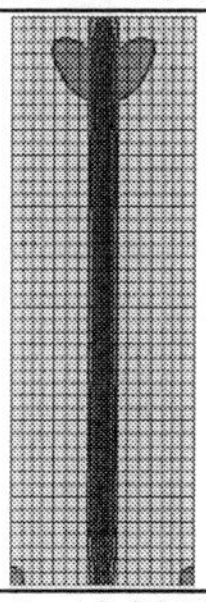	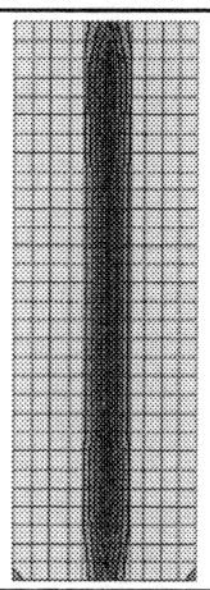
N_x	+2.05 to +0.0235 kPa	+2560 to +0.00131 kPa	+2300 to +7.89 kPa
	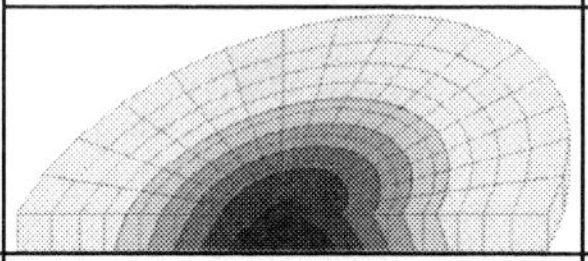	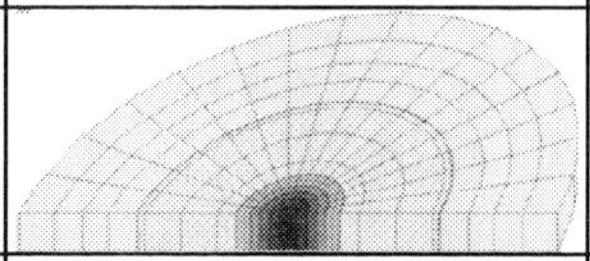	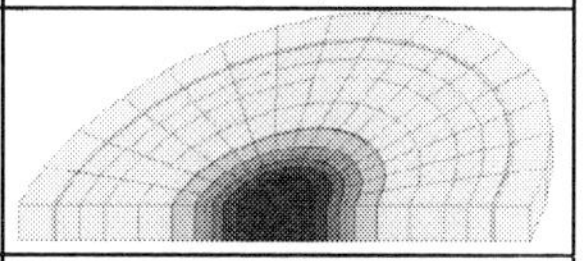
N_y	-0.926 to +2.05 kPa	-583 to +2560 kPa	-1090 to +2300 kPa
	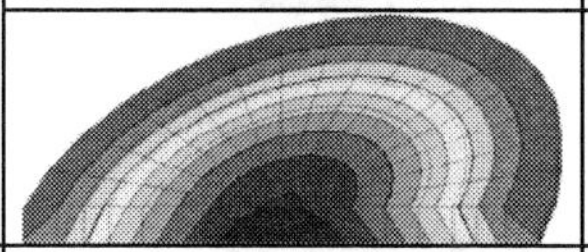	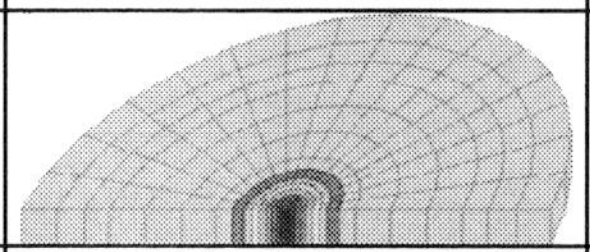	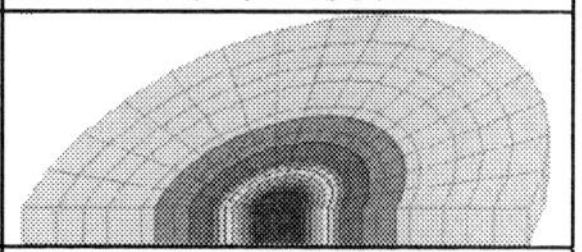
N_z	-23.9 to -23.9 kPa	-40.1 to -23.8 kPa	-39.5 to -23.4 kPa
	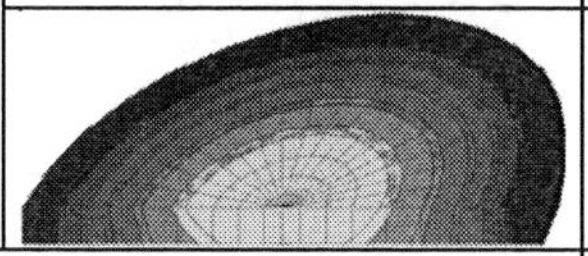	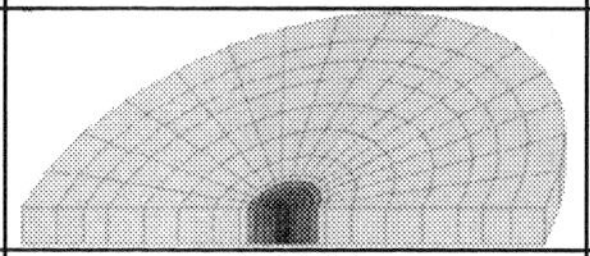	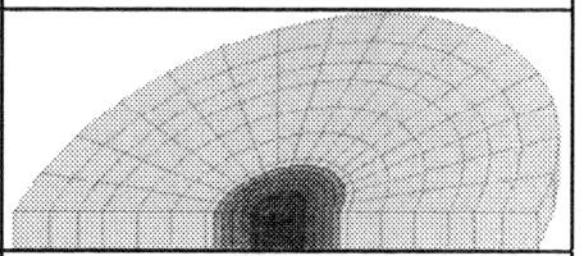
The percentage of reinforcement		0.44 %	2.78 %
N_x - radial; N_y - belt; N_z – normal stresses; positive sign (yellow-brown) - tension			

Thus, it can be stated that the investigated glass-composite reinforcement works in compressed elements similarly to metal. In connection with the fact that the initial modulus of elasticity is higher than that of concrete, it takes on a part of the stresses in proportion to its modulus of elasticity. So an increase in the bearing capacity is observed with a reinforcement percentage of more than 0.5%. The obtained results make it possible to use such modeling in the numerical study of glass-composite concrete cylinders, specifying the characteristics of materials obtained from experimental studies.

4 Conclusions

The experimental and numerical studies of the stress-strain state of concrete cylinders with longitudinal glass-composite reinforcement suggest that the reinforcement in compressed concrete works in a manner similar to that of steel, and can be taken into account as a working one in calculating the strength of elements.

The findings obtained in this study should be regarded as exploratory, opening the possibility of more accurate and purposeful planning experimental studies to establish the efficiency of using composite rods as working reinforcement in compressed concrete elements, refining its physical and mechanical characteristics when compressed in a concrete medium and, as a result , the creation of relevant regulatory documents.

References

1. A. Lapshinov. Proceedings of Moscow State University of Civil Engineering. **10,** 96-105 (2015).
2. V. Plevkov, I. Baldin, K. Kudyakov, A. Nevskiy. Vestnik of Tomsk State University of Architecture and Building. **5(58)**, 91-101 (2016).
3. ACI 440.1R-06. *Guide for the Design and Construction of Structural Concrete Reinforced with FRP Bars*. American Concrete Institute Standard and Commentary, Detroit (2006).
4. CNR-DT 203/2006. *Guide for the Design and Construction of Concrete Structures Reinforced with Fiber-Reinforced Polymer Bars*. Italian Research Council, Rome (2006).
5. CAN/CSA-S806-12. *Design and Construction of Building Components with Fibre Reinforced Polymers*. Canadian Standards Association, Toronto (2002).
6. Fib Bulletin 40. *FRP Reinforcement in RC Structures*. International Federation for Structural Concrete, Stuttgart (2007).
7. JSCE Guidelines for Concrete. *Recommendation for Design and Construction of Concrete Structures Using Continuous Fiber Reinforced Materials*. Japan Society of Civil Engineers, Tokyo (1997).
8. M. Afifi, H. Mohamed, B. Benmokrane. J. Compos. Constr. **18**, 04013017 (2014).
9. A. Nevskii1, I. Baldin, K. Kudyakov IOP Conf. Ser.: Mater. Sci. Eng. **71**, 012037 (2015).
10. Yu. Klimov, A. Andriukevych. Science & Construction. **1(7)**, 18-21 (2016).
11. DSTU-N B V.2.6-185:2012. *Instruction of application and production of concrete construction with non-metallic composite reinforcement based on basalt and glass roving*. State standard of Ukraine, Kyiv (2012).
12. D. Oreshkin, Yu. Bondarenko, K. Spirande, M. Mol'skyj. Scientific Bulletin construction. **2(84),** 250-258 (2016).

Basics of modeling and technology of creating reinforced concrete elements of "Monofant" system construction

Sergey Bugayevskiy[1]*, *Vladymir* Gerasymenko[2] , *Alexander* Konyukhov[3] , and *Valeriy* Nikulin[3]

[1]Kharkiv National Automobile and Highway University, Department of Bridges, constructions and structural mechanics, 61002 Kharkiv, Yaroslav Mudry st., 25, Ukraine
[2]Kharkiv National University of Civil Engineering and Architecture, Department of Graphs, 61002 Sumska str., 40, Kharkiv, Ukraine
[3]Zhilbud-2 Company in Kharkiv, Company management, 61145 Kharkiv, Space Str., 12, Ukraine

Abstract. In architectural and construction system "Monofant" a self-supporting skeleton consisting of a spatial curvilinear framework and unrecoverable hollow-core inserts made of expanded polystyrene was used to erect buildings and structures of curved shapes from reinforced concrete lightweight structures. For this, an internal, non-removable formwork of complex shape makes up given geometry of the structure and becomes a certain screen on which a shotcrete mixture is sprayed. The problem of creating a technology for manufacturing polystyrene inserts for the erection of cylindrical shells has been solved. At the same time we achieved reducing the waste while cutting a flat rectangular sheet of expanded polystyrene and obtaining the minimum number of elements in the assortment for manufacturing triangular inserts used to produce a screen that provides shotcreting curvilinear reinforced concrete structures with the method of sprayed concreting. The obtained results confirm the effectiveness of using lightweight "Monofant" system designs that have a reduced width of not more than 40-50% from the structures with solid width.

1 Introduction

The idea of optimizing the shape of the cavities and increasing the dimensions of voids in reinforced concrete products and structures remains one of the most dominant concepts of reducing the material consumption of buildings and structures. Its implementation is related to both computationally constructive and technological problems. Application of non-extractable inserts-cavities in the technology of producing reinforced concrete lightweight structures allows to reduce the weight of structures by 20-40% without reducing their strength characteristics. In the result of introducing more complicated process of manufacturing reinforced concrete lightweight structures in comparison with solid cross-section, minimizing the cost of manufacturing liners, due to the use of recycled plastics or the use of cheap materials still remains a highly important problem.

* Corresponding author: bugaevskiysa@gmail.com

2 Analysis of publications

In the global practice, unified volumetric modules made of polymer materials of various shapes are widely used as non-extractable insert-fillers in the technologies of production of reinforced concrete lightweight structures [1-4]. Due to the lack of the necessary production facilities they cannot be commonly used in the Ukrainian civil engineering.

The arrangement of liners inside concrete and reinforced concrete structures and products is strictly regulated, and in most cases is connected with the construction of spatial reinforcement skeletons or with one of the dimensions of a structure or product, for example, with the width of the overlap. Accordingly, the size of the liner (diameter) of a spherical shape is determined from the condition providing two widths of the protective overlap layer. In general, the unified arrangement of the same type of liners is used when concreting a building or any other construction, which in most cases does not ensure the rationality of creating the internal geometry of lightweight structures (the shape of internal ribs, their width and pitch).

An alternative approach is the use of expanded polystyrene as the most economically justified material for the manufacture of hollow-forming agents [5, 6]. The production of void-formers can be performed according to given geometric parameters or by cutting a necessary form from expanded polystyrene blocks. However, in our opinion, it doesn't solve the problem of creating liners for reinforced concrete lightweight structures with a surface which differs from the surface of zero Gaussian curvature and manufactured by non-traditional methods of concreting with the use of vibration, but by methods of pneumatic spraying (shotcreting).

The architectural and construction system "Monofant" [7, 8] was developed in Ukraine under the guidance of prof. V.S. Shmukler. In it a self-supporting skeleton consisting of a spatial curvilinear framework and unrecoverable hollow-core inserts made of expanded polystyrene or mineral wool was used to erect buildings and structures of curved shapes from reinforced concrete lightweight structures. For this, an internal, non-removable formwork of complex shape makes up given geometry of the structure and becomes a certain screen on which a shotcrete mixture is sprayed onto both sides with the help of a wet shotcreting method. To create a solid screen, the gaps between the liners that form internal ribs of lightweight structures, are closed on one side with grids, e.g. wire mesh or with a perforated sheet. The uniqueness of the suggested method of erecting such structures is confirmed by the Ukrainian patent for the invention [9].

Geometric shape of the reinforced concrete lightweight structure influences the geometry of the internal non-removable formwork.

3 The purpose and objective of the problem

The purpose of this work is modeling and technology of producing an internal non-removable formwork for concreting curved lightweight structures made of reinforced concrete on the "Monofant" system.

The objective of the study is determining the shape of the inserts and their production from expanded polystyrene for erecting reinforced concrete lightweight structures of cylindrical form.

4 Determination of the shape and creation of a cavity-maker

Various architectural sites can be constructed using reinforced concrete lightweight cylindrical structures without applying an expensive individual formwork (Figure 1).

The manufacturing technology of reinforced concrete structures with curvilinear shape described in the article was tested while erecting a planned fragment of a 2.2x2.2 m cylindrical shell (Figure 2). The width of the reinforced concrete shell was 26 cm, where 16 cm stand only for the width of the non-removable hollow-core inserts made of expanded polystyrene, and 5 cm from the outer and inner sides – for the reinforced concrete bindings.

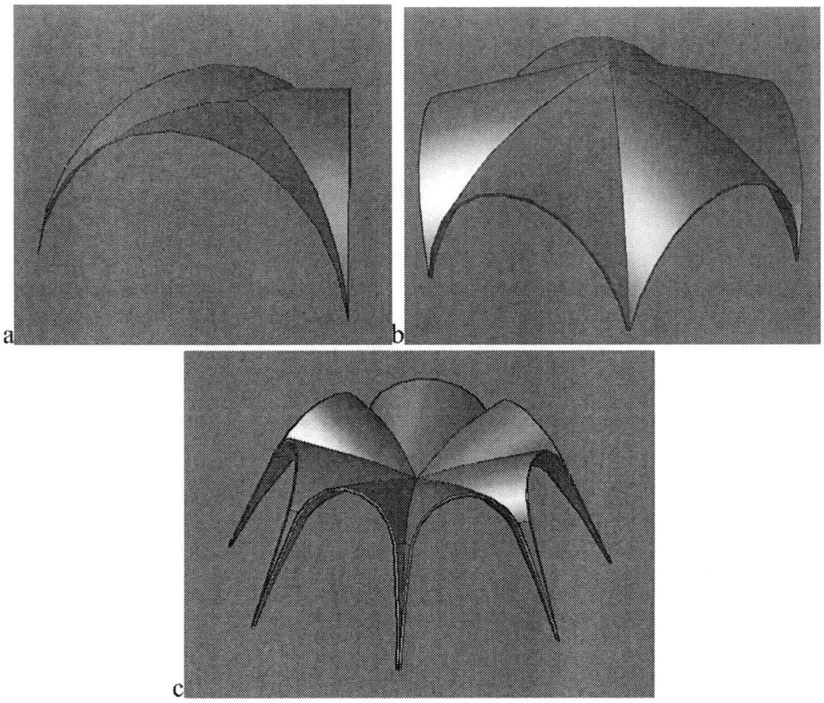

Fig. 1. Bearing structure of interpenetrating cylindrical surfaces: a – three surfaces; b – five surfaces; c – seven surfaces.

The choice of geometric and typological characteristics of non-extractable void generators for producing reinforced concrete structures and products should be made considering the optimization of design and technological solutions. For those certain requirements for ensuring the strength of lateral and internal ribs to shear along the inclined and longitudinal sections must be observed; elimination of stress concentration in the areas of conjugation of the shelf and ribs; requirements to ensure rigidity and crack resistance in normal and inclined sections, and, finally, fulfillment of all structural requirements for maintenance with the protective layer.

To make a curvilinear-shaped insert of the construction, we are offering a method for its production from individual flat elements that can be cut from a rectangular flat sheet of expanded polystyrene with a machine tool (Figure 2, a).

At the same time, the main issue is reducing the waste when cutting a flat rectangular sheet of expanded polystyrene and the maximum reduction in the number of elements in the assortment for the manufacture of the liner used as a non-removable formwork of cylindrical shape. When constructing the diagonally inner edges of the shell, four inserts of two type sizes should be made (Figure 3).

The cabbling of each insert individually into compound elements of a secant parallel to one of the sides of the shell will result in a large number of types of elements with the same shape, but different length ranging from 230 to 2030 mm. When dividing each liner with a secant parallel or perpendicular to the larger side of the liner, we get an even larger increase in the types of elements and their clear separation for two types of liner size (Figures 4, 5).

Fig. 2. Manufacturing a self-supporting skeleton: a – an element of the liner; b and c – the construction of the skeleton.

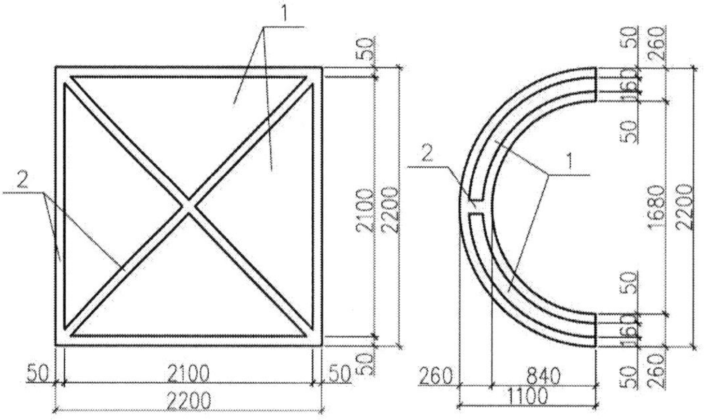

Fig. 3. Shape of liners in a cylindrical shell: 1 – insert; 2 – internal ribs.

To optimize the cutting of the inserts, we are offering an option for the production of four inserts at once as a single whole liner, followed by cutting it into four parts diagonally (Figures 6, 7). This cutting option allows you to reduce the number of sizes up to two elements.

When cutting along the shell, all elements of a rectangular outline with the length of 2100 mm (two types of size) are obtained, but both with a trapezoidal cross section, which will require additional cutting operations on each individual element. The disadvantage of the final prefabrication of a non-removable formwork is the possible significant deviation in the shape of the resulting structure from a cylindrical one and the additional use of templates to correct inaccuracies connected with the peculiarities of making liners in this way.

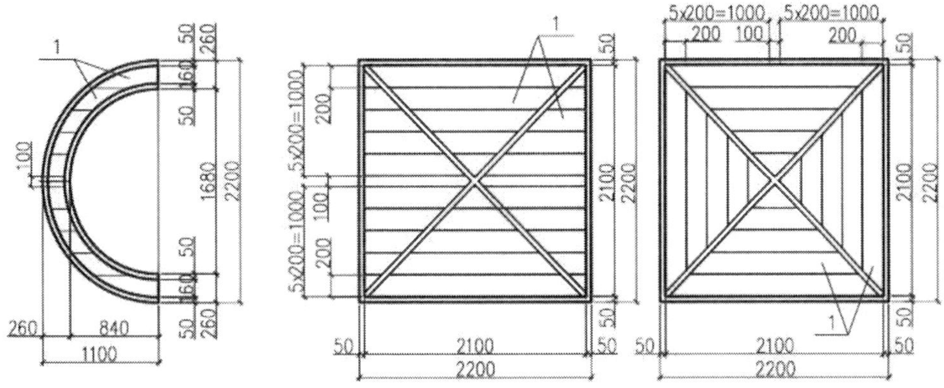

Fig. 4. Cutting each liner separately (option 1): 1 – types of liner elements.

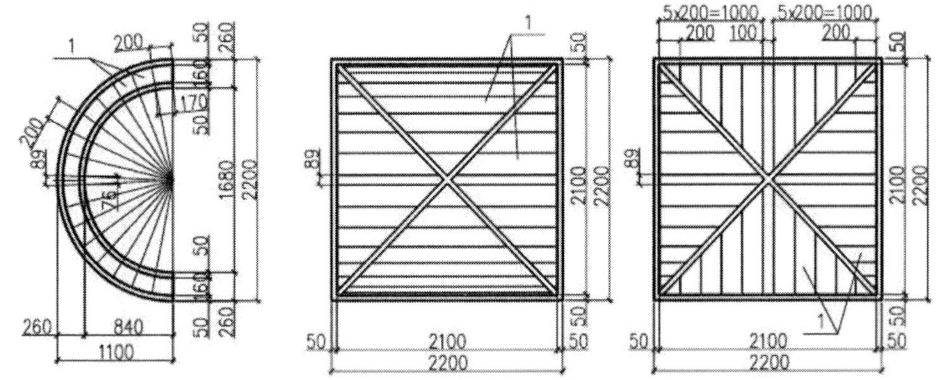

Fig. 5. Cutting each liner separately (option 2): 1 – types of liner elements.

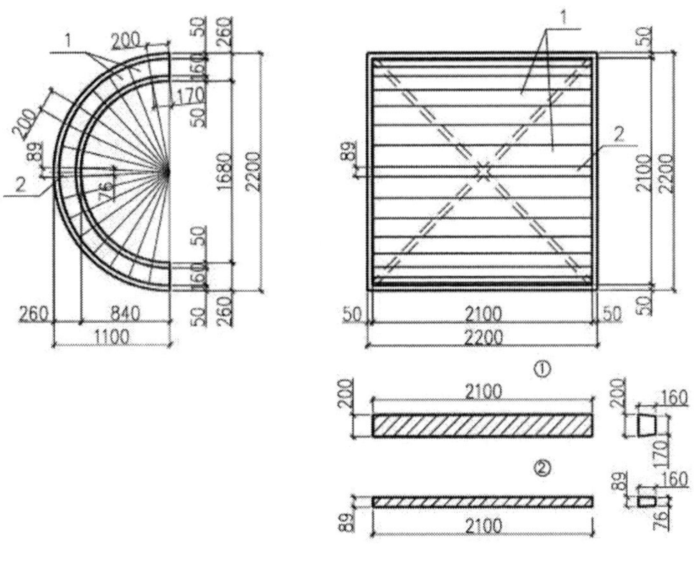

Fig. 6. Cutting four liners as a single element (alongside) 1, 2 – types of liner elements.

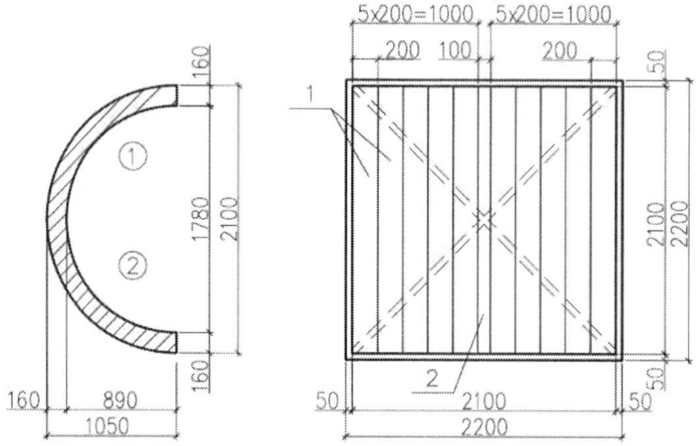

Fig. 7. Cutting four liners as a whole element (crosswise): 1, 2 – types of liner elements.

When cutting across the shell, the elements into which the non-removable formwork is divided have a plan dimension of 1050x2100 mm and a significant curvature, which results in a large waste of materials during their manufacture (Figure 8). At the final prefabrication of the non-removable formwork, the accuracy of manufacturing cylindrical liners is ensured.

To optimize the cutting of elements with the planned dimension of 1050x2100 mm, an additional cabbling of them into parts 1/2 and 1/4 of the arc length is suggested (Figure 9).

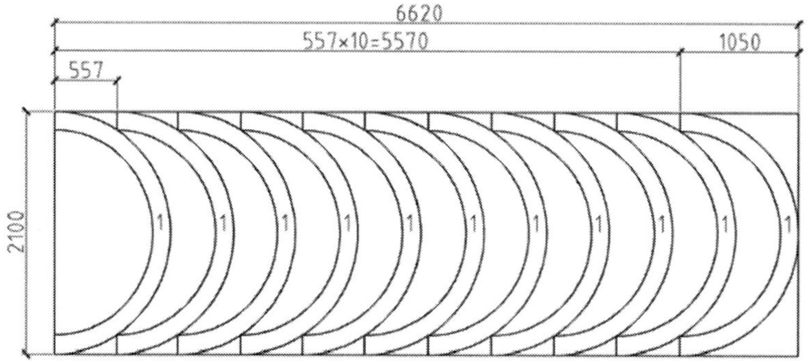

Fig. 8. Layout of solid elements.

Interchange in the prefabrication of fixed formwork elements 2+2 and 3+2+3 will ensure that the seams do not fit as well as the integrity of the structure made of expanded polystyrene.

The next option of cabling the inserts in the radial direction (Figure 10), will further reduce the consumption of materials. Assembling liners of non-removable formwork with alternating elements 4+5+6, 5+6+4, 6+4+5, will ensure the overlap of seams through three rows of segments. The lap length of the seams can be calculated from the length of the chord L:

$$L = 2R(\sin\alpha / 2) \quad (1)$$

where R – radius of a circle; α – central angle.

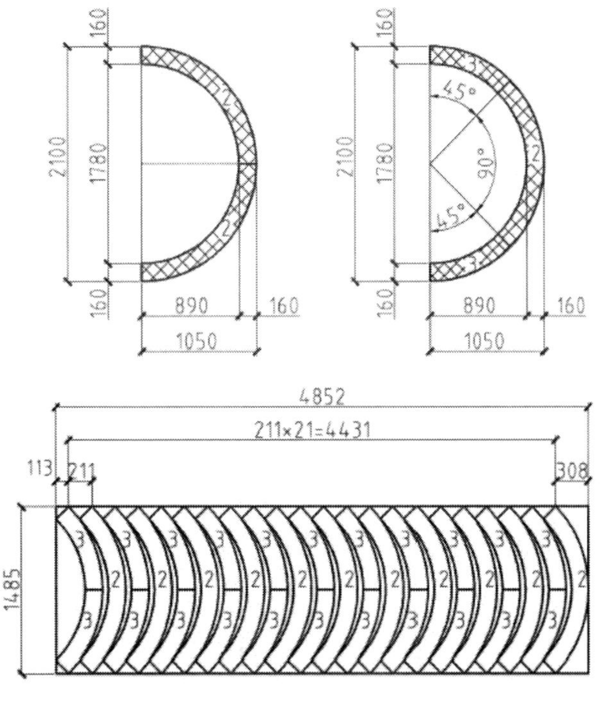

Fig. 9. Layout of compound elements: 2, 3 – types of liner insert elements.

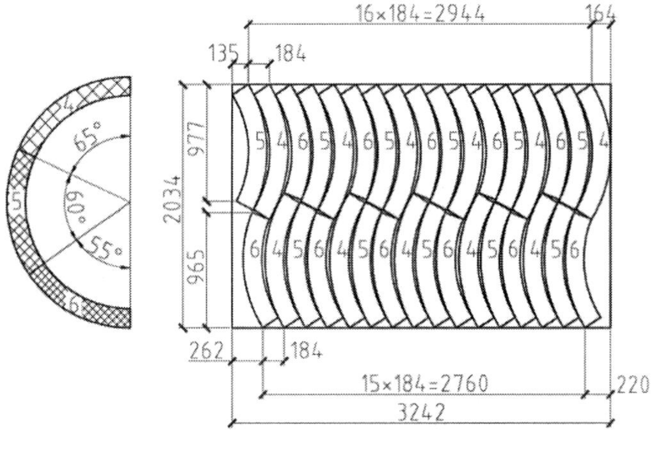

Fig. 10. Layout of compound elements: 4, 5, 6 – types of liner insert elements.

The options of cabling into elements (Figures 9, 10) significantly reduce the cost of material when cutting a sheet of expanded polystyrene of a rectangular shape (Table 1).

As it can be seen on Table 1, the cutting of liners according to Figure 10 is more economical in terms of material consumption and packing density.

In order to unify the production of liners during the erection of constructions on the "Monofant" system, we're offering production of a prismatic body tiling with multifaceted elements in the shape of a truncated pyramid with a base in the shape of regular polygons. There are three correct tilings of the plain: triangular parquet, square parquet and hexagonal parquet (Figure 11).

Table 1. Areas of void generators and waste in the process of cutting liner inserts from a rectangular sheet.

Figure	Dimensions of expanded polystyrene sheet ($b \times h$)= area, $m \times m$=m^2	Insert element / amount	The area of one insert/ the area of inserts of this position, m^2/m^2	Waste area, m^2	Packing density, %
8	6.621x2.1=13.9	1/11	0.487/5.363	8.54	38.58
9	4.855x1.485=7.21	2/11	0.244/2.682	1.85	74.5
		3/22	0.122/2.682		
10	3.248x2.034= 6.605	4/11	0.176/1.937	1.24	81.2
		5/11	0.163/1.788		
		6/11	0.149/1.639		

Fig. 11. Examples of proper tiling of a plain.

Figures 12, 13 presents various options of such a tiling for the cylindrical structure shown in Figure 2.

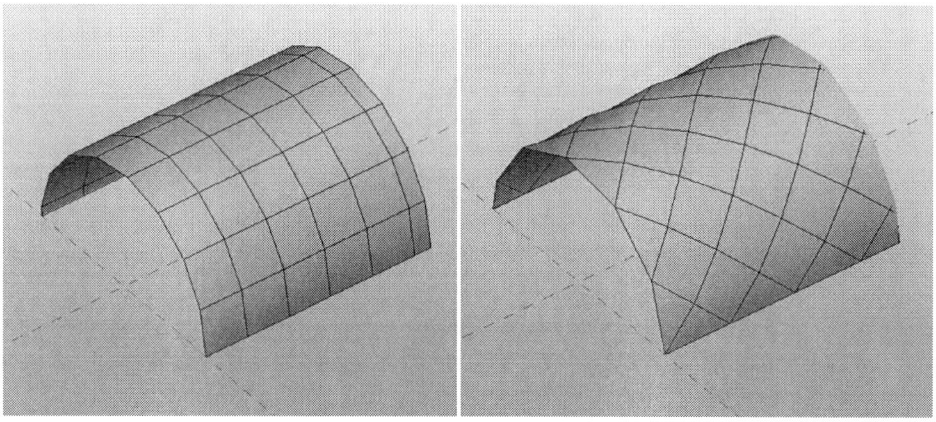

Fig. 12. Examples of tiling a cylindrical surface with regular polygons.

The presented parquet tiling is suggested to be made of sheet material and, with small widths of the structure, it is allowed to produce them not in pyramidal but in prismatic shape, which greatly facilitates the cutting process. The packing density of such cavities will be close to 100%, which will significantly reduce material waste.

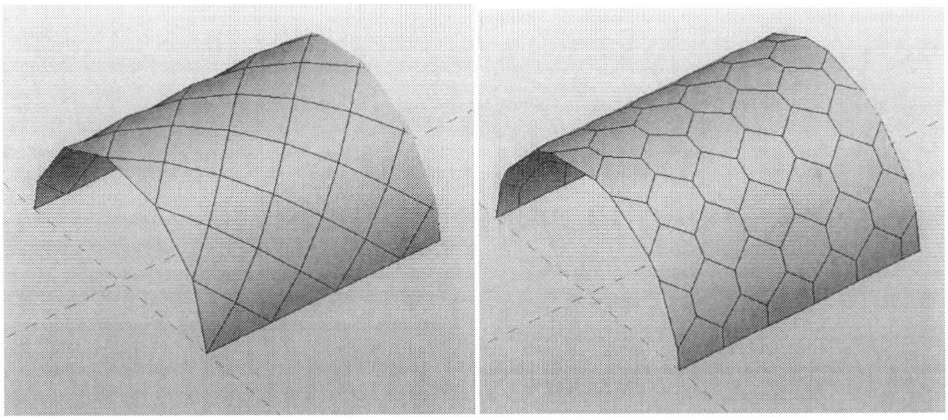

Fig. 13. Examples of tiling a cylindrical surface with regular polygons.

Conclusions

The problem of creating a technology for manufacturing polystyrene inserts for the erection of cylindrical shells has been solved. At the same time we achieved reducing the waste while cutting a flat rectangular sheet of expanded polystyrene and obtaining the minimum number of elements in the assortment for manufacturing triangular inserts used to produce a screen that provides shotcreting curvilinear reinforced concrete structures with the method of sprayed concreting. The obtained results confirm the effectiveness of using lightweight "Monofant" system designs that have a reduced width of not more than 40-50% from the structures with solid width.

References

1. C. Albrecht, *Proceedings of The 9th fib International PhD Symposium in Civil Engineering*, 85-90 (2012)
2. M. Mota, Concrete International, **32**, 41-45 (2010)
3. A. Churakov, Construction of Unique Buildings and Structures, **6**, 70-88 (2014)
4. O. Meneiliuk, K. Shavva, V. Taran, Modern Industrial and Civil Construction, **9**, 221-229 (2013)
5. V. Shmukler, Y. Klymov, M. Burak, *Framework of hollow core types* (Golden Pages, Kharkiv, 2008)
6. S. Bugayevskiy, Building Construction, **78**, 478-586 (2013)
7. V. Babaev, V. Shmukler, S. Bugayevskiy, V. Nikulin, Journal of Civil Engineering and Construction, **5**, 143-156 (2016)
8. V. Shmukler, S. Bugayevskiy, V. Herasymenko, O. Stefan, *Mat. III Int. Sc. Materials III International Scientific Conference "Safety in transport and manufacturing – education, science and practice"*, 282-289 (2016)
9. V. Shmukler, V. Babaev, S. Bugayevskiy, A. Bih, *Method of construction of elements of structures of curvilinear shape*. Patent of Ukraine № 113669 (2017)

The study of carrying capacity of timber slabs with use the finite elements method

Vitalii Demeshok[1*], *Alina* Zalevs'ka[1], *Evgen* Tychenko[2] and *Yana* Zmaga[1]

[1]Cherkassy Institute of Fire Safety named after Chernobyl Heroes National University of Civil Protection of Ukraine Science Department 18000 Cherkassy, Ukraine.
[2]Educational-methodical Center of Civil Protection and Life Safety in Cherkasy Region, 18000 Cherkassy, Ukraine.

Abstract In the article, the results of the study of behavior of timber slab under influence of fire with the standard "time-temperature" curve are presented. The finite element method was used for it. For the calculation we constructed a grid models of timber slabs. As a result of solution of the thermal problem was obtained temperature distribution and the graphs of maximum deflection of timber slabs and its slew rate depending on the time of the test. The obtained graphs allow to obtain data on the occurrence of the limit state of loss of bearing capacity by comparing current values of displacements and velocities with the maximum allowable. Analysis of the graphs shows that the criteria limit state of loss of bearing capacity does not occur. Calculation method of evaluating the fire resistance of timber slabs was developed. For it use database about strain-stress state of this slabs in conditions of influence of the fire.

The construction of buildings, including multi-stored ones, with a large number of people in them often used wooden bearing structures. Such structures are made of timber slabs, but unless elements of wooden frames. The wood has a high strength, low specific weight and good decorative qualities. Wooden constructions provide durability of buildings and structures during exposure to fire. This must confirm with the requirements of the building regulations [1]. They are distributed for fire resistance and flame spread prevention. In the design of buildings these measures should be ensured through appropriate design solutions, incorporating the relevant safe geometric dimensions and measures of fire protection. At the stage of design of fire-resistant building structures calculated methods are used. Currently, theoretical and methodological framework of this approach is contained in the valid Ukrainian standard [2]. These methods are flexible and allow to consider all features related to the material properties, geometric dimensions and boundary conditions. However, they are much less labor-intensive and expensive for experimental methods.

Among the existing calculated methods, we can distinguish two types such us: simplified and clarified [3]. Simplified methods are simplified mathematical models. They are derived from a simple calculation formulas with the introduction of coefficients. Obtained during the synthesis of a large amount of experimental material. For development engineering calculation methods are applied refined methods. They are based on

*Corresponding author: scorpiomz@ukr.net

mathematical modeling of the behavior of wooden structures in fire conditions at the macro level. For this methods the finite element methods in modern calculation of software complexes, such as ANSYS, Abaqus, SAFIR and others were involved [4, 5]. Taking into account it, the study was aimed at creating a reliable accurate mathematical models of the behavior of wooden structures in fire conditions is important.

Analysis of recent achievements and publications

In works [4, 5] an approach to study of the behavior of timber slabs during a fire, consisting in the carrying out mathematical modeling by the finite element method (hereinafter FEM) is offered. This approach refers to refined calculation methods and allows to describe accurately the behavior of timber slabs in fire. The obtained mathematical model in these works with the large volume of collected data have a significant disadvantage in the consideration of the deformation of timber slabs only in the elastic region. This situation is due to the fact that modern software systems should provide the difference of the elastic-plastic properties of wood in tension and compression. There are still open questions of the distributions of the parameters of the tensely-deformed state (hereinafter – TDS) in cross-section, deformation scheme, distribution of defects, as well as information regarding the mechanism and causes of destruction of wooden floor slabs during a fire.

In this regard the purpose of the study is formulated.

Formulation of the problem and its solution

The objective of this work is to develop calculation methods for evaluating the fire resistance of timber slabs of ceiling on the basis of certain key parameters of the tensely-deformed state, charts of deformation, distribution of defects in the application of the finite elements method.

On fig. 1 chart of structural elements of wooden flag that is examined is given.

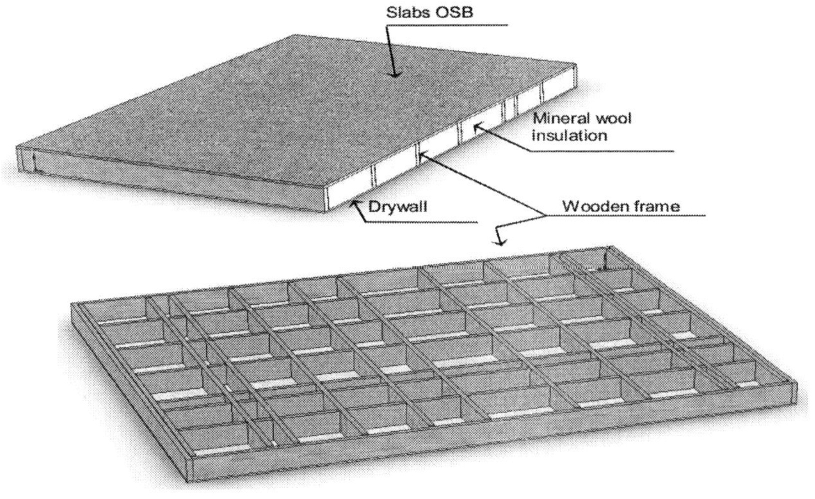

Fig. 1. Structural diagram of the wooden plate.

To solve this problem the formulation of set of input data is required, which include the properties of the components of the material wooden plate, parameters, boundary conditions, that take into account the applied loads and thermal effects. We used the thermal properties of the wood of the frame plates which are described in the European standard [3] concerning the calculation of wooden constructions for fire resistance.

Used features meet the requirements of the European standard [3]. The mechanical characteristics are a set of diagrams "stress-strain" with the descending branches for certain values of the heating temperature of the material. For the calculation of it a mathematical model, whose parameters are given in tab. 1 was used.

Table 1. The basic mathematical models for calculation of the plate on fire.

Physical process	Used mathematical model (method)	J-lo
Heating engineering task		
Heat-conducting	Equalization of non-stationary heat-conducting, approximated by means of FEM	[5]
Maximum terms	III kind	[5]
Physical nonlinearity	Iterative method Newton-Raphson	[5]
Static task		
TDS	FEM	[5]
Plastic deformation	The associative theory of plastic deformation von Mises	[7]
Physical and geometric nonlinearity	Iterative method Newton-Raphson	[7]

For the boundary conditions was used the parameters of the European standard [3].

For the calculation we constructed a grid models of timber slabs are shown in Fig. 2. When constructing grid models for the heat task it was assumed 55,676 elements, and for static task, it was assumed 8 142 elements.

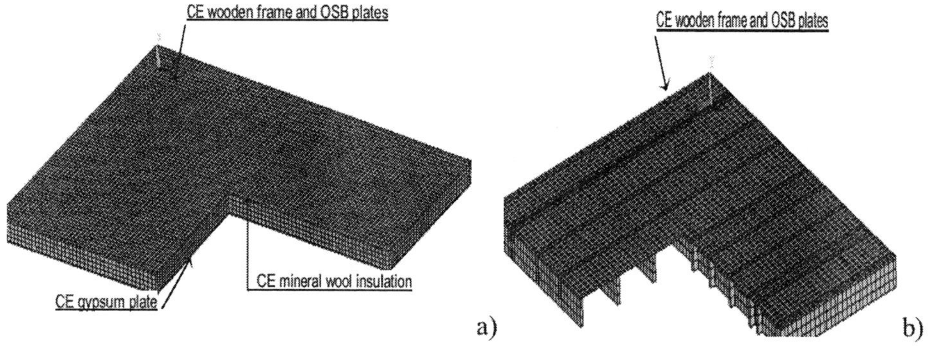

Fig. 2. Grid models: the thermal problem (a); to the static task (b).

When we solved the statistical problem, we have received the graphs of maximum deflection of timber slabs and its slew rate depending on the time of the test. The obtained curves are shown in Fig. 3.

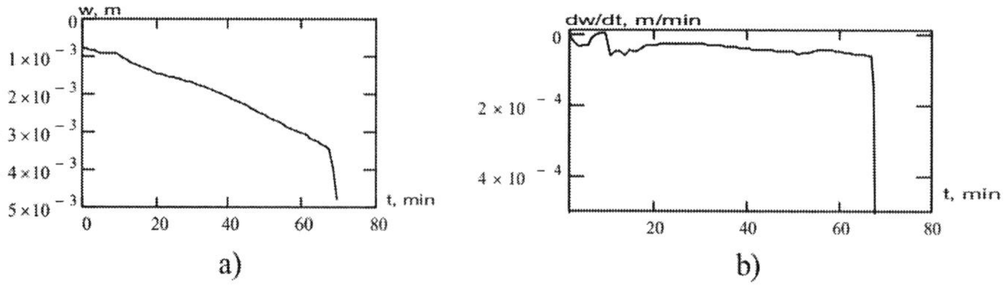

Fig. 3. Graph of maximum deflection of a wooden plate (a) and slew rate (b) depending on the duration of the test.

The obtained graphs allow to obtain data on the occurrence of the limit state of loss of bearing capacity by comparing current values of displacements and velocities with the maximum allowable, which are respectively defined by the formulas [2]:

$$D = \frac{L^2}{400 \cdot b} = 196.76 \text{ mm} \quad (1)$$

$$\frac{dD}{dt} = \frac{L^2}{9000 \cdot b} = 8.79 \text{ mm} \quad (2)$$

where L is the length of the plate between supports, mm, b – height of cross section, mm.

Analysis of the graphs shows that the criteria (1) limit state of loss of bearing capacity does not occur. Nevertheless, the abrupt increase of the deformations 68 min indicates the formation of plastic hinges in timber frame elements, which, in turn, shows the onset at the moment of loss of bearing capacity. Analysis of the graphs in Fig. 3 shows that the loss of bearing capacity occurs on the 68 minutes of fire with a standard temperature regime. This means that the fire resistance class of the investigated of timber slabs is not less than REI 60. The fire resistance class is the largest of the regulated [1].

Considering the conducted study we can draw the following conclusions:

1. Numerical study of the behavior of wooden plates during the operation of the fire with a standard temperature regime.
2. It was shown that wood slab at a certain time interval reduces its maximum deflection, then the value stabilized for nearly 40 minutes.
3. It was shown that the reduction of the maximum deflection of a wooden plate at a certain time interval, and further stabilization due to thermal expansion of the lower layers of the wooden beams of the slab.
4. It was revealed that the presence of gypsum plasterboard from below the ceiling allows to increase significantly the fire resistance of these boards due to loss of carrying capacity for fire resistance class REI 60.
5. The main provisions of the calculation methods of fire resistance assessment forr wooden slabs based on finite element method were developed.

References

1. *DBN V.1.1-7-2002 Fire protection. Building structures fire safety*, Kiev, (2003).
2. *DSTU B V.1.1-4-98, Fire protection. Building structures. Fire resistance test methods. General requirements, ISO 834: 1975,* Ukrarchbudinform, Kiev, (1998).
3. *EN 1995-1-2:2004. Eurocode 5: Design of timber structures: General-Structural fire design* Brussels (2004).
4. J. Konig L. Walleig Timber frame assemblies exposed to standard and parametric fires, Stocholm, (2000).
5. P.R. Collier, A. H. Buchanan, Fire Technology, **38**, 125-145 (2002).
6. U. Franchuk *Tables of thermal properties of building materials*, (NII of building physics, 1972).
7. V. S. Sakharov *Method finite element in solid mechanics* (Higher school, 1982).

The bearing capacity experimental determination of the keyed joints models in the transport construction

Oksana Dovzhenko[1*], *Volodymyr* Pohribnyi[1], *Ilona* Yurko[1], and *Iryna* Shostak[1]

[1]Poltava National Technical Yuri Kondratyuk University, Building Department, Pershotravneviy sq., 24, Poltava, 36011, Ukraine

Abstract. The joints ensure the joint performance of the load carrying structural systems and they are the most responsible and important elements. Keyed joints are widely used in construction. They are characterized by an increased resistance to shear. On these grounds the structural concepts of keyed joints need further improvement. The article presents the research results of experimental test pieces five series in the form of single keys and one-keyed joints. Those samples have been tested in Poltava National Technical Yuriy Kondratyuk University. Follow strength factors have been varied: geometric parameters of joints (depth, height, width) and their ratio; angle of support surface (rectangular, trapezoidal and triangular key); level of compression; reinforcement (quality of reinforcement and the nature of its location); jointing width. The samples were made of heavy-weight, expanded clay and fibre concrete. The experiments program includes influence study both of one of these factors and their combinations. The deformations, nature of failure, the ultimate load have been studied. Structural parameters of keyed joints which ensure the efficient behaviour have been installed.

1 Introduction

One of the most important elements of the roadway of bridges and overpasses are the keyed joints of their bearing elements. Under significant shear force action keyed joints are the most effective because they have high shearing resistance. Keyed joints are used in the formation of composite sections of precast and cast-in-place and precast constructions (here, as a rule, the contact seam of is used), as well as in the reconstruction of reinforced concrete road bridges. There is a known method of reinforced road bridges widening by means of superimposed slab during the reconstruction. In this case slabs are connected by tension-keyed joint. Thus, keyed joints are widely used in transport construction and they require further detailed study.

The effectiveness of constructive solutions of compounds largely depends on the accuracy of accounting for their design of all factors determining the strength. The carrying capacity of keyed joints is affected by the class and type of concrete, the size of their

* Corresponding author: o.o.dovzhenko@gmail.com

reduction and the reinforcement percentage. Researchers have a consensus on the issue of the qualitative nature of the influence of these factors, differing only in quantitative evaluation. What can be said about the dependence of resistance to the destruction of a number of geometric parameters of joints. These include the ratio of the depth of the key to its height, the shape of the profile (the angle of inclination of the loaded face of the key) and the width of the joints. The ambiguity of their accounting leads to unreasonable reserves of carrying capacity and a significant increase in the material intensity of the joints.

In [1-3] is proposed to take into account the shape of the profile of the key way junction in the evaluation of the carrying capacity of connections of concrete and reinforced concrete elements. Increase in the value of the ultimate load of reinforced and crimped trapezoidal and triangular keys in comparison with rectangular ones is established.

However, in [4, 5] this factor is neglected. The researchers also do not have a common vision regarding the dependence of the strength of the joint on the width of the joint. The analysis of the studies [6, 7] does not introduce unambiguity in the solution of this question. Thus, along with the statement about a significant reduction in the carrying capacity with an increase in the joint width, an opinion is expressed that the effect is not significant. This predetermines the need to obtain new results.

2 Experimental program

The code of the first four series samples includes: lettering; the ratio l_k/h_k; the profile shape (P – rectangular, Tr – trapezoidal, T – triangle); concrete type (H – heavy, C – expanded clay, F – fibre concrete); the level of compression σ/f_c; percentage of reinforcement ρ_w; 1 or 2 – the reinforcement location in one or two levels; NT – the presence of friction on the loaded faces (in the remaining samples grip on the abutting surfaces is eliminated). For the fifth series, after the letter designation, the width of the seam is shown.

First series (12 samples were from heavy- concrete and expanded clay concrete) has been investigated to determine the areas of the failure cases implementation of the rectangular concrete keys depending on their sizes ratio l_k/h_k [8].

The influence of the compression value σ on the keys strength at the shear action has been studied during testing the second series (29 samples). In this series height of the keys has been varied within the scope of h_k = 50 – 150 mm, and the depth was constant l_k = 50 mm. This led to the change of parameter l_k/h_k from 0.3 to 1. The compression value was σ/f_c =0 – 0.47 [9].

Within the limits of the third series (21 samples made from heavy-weight and expanded clay concrete) the reinforcing effect on the keys strength under shear action has been studied [9]. Samples lateral surfaces have been extended to provide the reliability of bar anchorage. Loading devices were 50 mm (l_k) lengthwise. That corresponds to the l_k/h_k = 0.25 ratio, while the height was h_k = 200 mm. In the plane of shear keys have been reinforced. Two types of cages have been used: reinforcement placed in the same level with the middle of the key's height, as well as the two-level placement. The reinforcement ratio has been varied between ρ_w = 0 – 1.03 %.

There are single-keyed contact joints (17 samples) with rectangular, trapezoidal and triangular keys with l_k/h_k = 0.5 in the fourth series [9]. Support surfaces angle influence on the carrying capacity has been studied. The keyed joints have been reinforced in the plane of the shear in the middle of their height by 2 bars Ø8, f_{yk} = 240 MPa (ρ_w = 0.67 %). Samples with the trapezoidal fibre concrete keys have been tested in order to define the influence of the reinforcement ratio ρ_w = 0 to 2.89 % and the nature of the cross reinforcement placement along the key's height (one or two levels) on the samples strength [10].

The fifth series (7 samples made from expanded clay and fibre concrete) included single-keyed joints with rectangular keys and variable jointing width t_j = 25, 50, 100, 150 and 200 mm [9]. The influence of the t_j on the ultimate loading and the nature of the joints failure at l_k/h_k = 0.25 – 0.5 was studied.

Both loading schemes moment and no-moment have been implemented during the testing. The load has been transmitted directly to the point of keys or the outer surfaces of the component parts of keyed joints. Manufacturing process of samples and load transmission scheme are described in [9].

3 Analysis of the results and discussion

Primarily the nature of failure and maximum load of the keys depend on the ratio of depth l_k to height h_k. When l_k/h_k = 0.1 – 0.2 shearing has been observed under the loading platforms. It has been complemented with the spalling of a part of element along the inclined plane (in other words, shear has been realized on inclined plane). Samples with the l_k/h_k = 0.3 – 0.5 ratio collapsed along the whole concrete section near the shear plane. When l_k/h_k = 0.6, failure of the key had flexural nature and was brittle along the tension area (fig. 1).

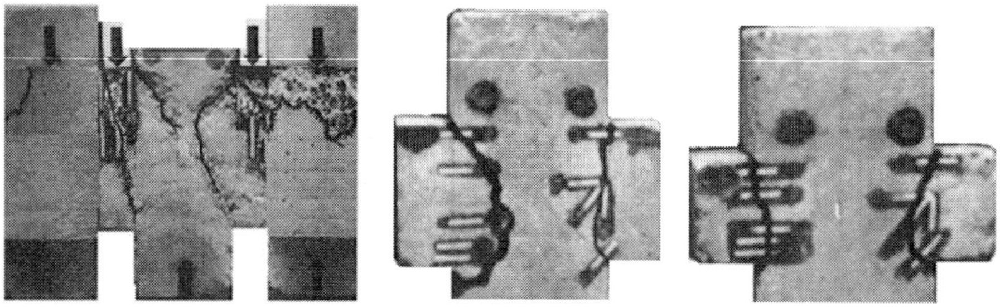

K-0.2-R-C K-0.4-R-C K-0.6-R-C

Fig. 1. The nature of failure of the concrete rectangular keys made from expanded clay concrete depending on l_k/h_k ratio.

The failure resistance of concrete keyed joints at increasing firstly, and then goes down with the increase of the l_k/h_k ratio. The strength increases by 25% when l_k/h_k ratio is changing from 0.5 to 0.3. However, the rules [4, 5] don't take it into account, although the decline in the influence of the bending moment is obvious. Maximum strength has been obtained: for keyed joints made from expanded clay concrete when the ratio is l_k/h_k = 0.2, for keyed joints made from heavy-weight concrete when the ratio is l_k/h_k = 0.3 in agreement with the data [11].

Compression of keyed joints σ to $0.5f_c$ expands the limits of shear failure to l_k/h_k = 1 and prevents the displacement of keys in the horizontal direction (fig. 2). Significant increase of the ultimate load with the growth of compression has been observed [12, 13]. The failure surface of the reinforced concrete samples has been delineated by a curve close to vertical that is due to the accepted ratio l_k/h_k = 0.5.

The strength of the samples increased twice with the increase of ρ_w to 1.03%. The value of limit load of the keyed joints with two-level reinforcement placement was approximately 10% higher compared to the keyed joints with central reinforcement placement (see Table 1).

Tests show that single-keyed contact joints (fig. 3) angle of support surface of keys does not change the nature of their failure, but affects the magnitude of the fracture failure.

Keys with the triangular profile have the highest strength, and keys with the rectangular profile – the lowest one. The difference between them is 15% (see Table 2).

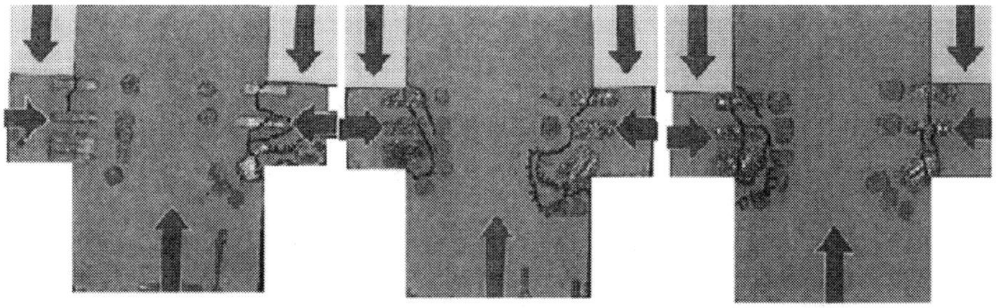

CK-0.8-R-C-0.23 CK-0.8-R-C-0.34 CK-0.8-R-C-0.35

Fig. 2 The nature of failure of the compressed keys made from expanded clay concrete when $l_k/h_k = 0.8$ and $\sigma/f_{c,prism} = 0.23 - 0.35$.

Table 1. Parameters of experimental samples of single keys.

№ series	Samples code	Concrete strength, MPa		σ, MPa	Cross reinforcement		f_{sh}, MPa
		f_c	f_{ct}		amount	f_{yw}, MPa	
	K-0.2-R-H						2.36
	K-0.4-R-H	18.4	1.9				4.13
1	K-0.6-R-H						2.89
	K-0.2-R-C			0			4.51
	K-0.4-R-C	22.5	1.8				3.42
	K-0.6-R-C						2.37
	CK-0.8-R-C-0.23			5.38			5.85
	CK-0.8-R-C-0.34			7.96			7.25
	CK-0.8-R-C-0.35	23.4	2.3	8.19			7.18
	CK-1.0-R-C-0.32			7.49			6.36
	CK-1.0-R-C-0.33			7.72			7.00
	CK-0.4-R-H-0.09			2.03	–	–	7.46
	CK-0.4-R-H-0.14			3.04			8.66
2	CK-0.6-R-H-0.09			2.03			5.95
	CK-0.6-R-H-0.18			4.05			9.27
	CK-0.6-R-H-0.32			7.09			10.32
	CK-0.8-R-H-0.42	22.5	2.25	9.55			7.22
	CK-1.0-R-H-0.29			6.47			5.76
	CK-1.0-R-H-0.30			6.64			5.53
	CK-1.0-R-H-0.35			7.79			6.09
	CK-1.0-R-H-0.41			9.22			6.52
	CK-1.0-R-H-0.47			10.60			7.51
	RK-0.25-R-C-0.34-1				2 Ø8	260	4.09
	RK-0.25-R-C-0.39-2				4 Ø6	280	5.30
3	RK-0.25-R-C-0.76-1	23.5	1.51	0	2 Ø12	250	6.56
	RK-0.25-R-C-0.67-2				4 Ø8	260	6.85
	RK-0.25-R-C-1.03-1				2 Ø14	240	6.86
	RK-0.25-R-C-1.03-2				4 Ø10	255	7.36

According to the authors, this is due to a change in the trajectory of principal normal compressive stresses as a result of inclination of loaded key verge, the presence of the load horizontal component and reinforcement including into working process on the earlier stages of loading, which leads to the cage effect, and the dominant influence of the cut compared to the break-away (lateral force compared to the bending moment).

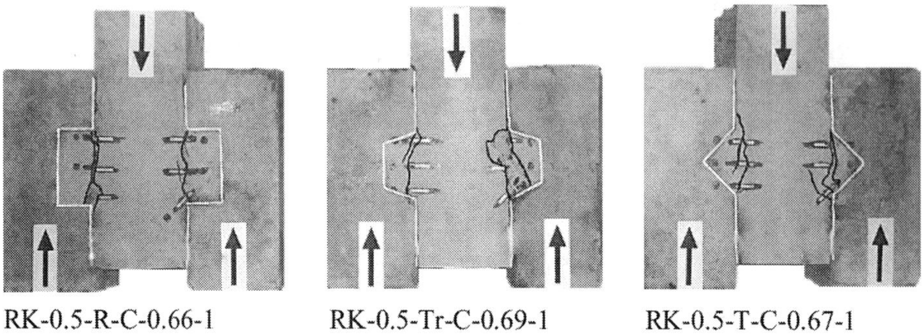

RK-0.5-R-C-0.66-1 RK-0.5-Tr-C-0.69-1 RK-0.5-T-C-0.67-1

Fig 3. The nature of failure of the contact joints made from expanded clay concrete with different key profile.

Table 2. Parameters of experimental samples of one-keyed joints.

№ series	Samples code	ψ,°	Concrete strength, MPa		Cross reinforcement		t_j, mm	f_{sh}, MPa
			f_c	f_{ct}	amount	f_{yw}, MPa		
4	KJ-0.5-R-C-0.68-1	0	16.8	1.45	–	–	0	4.73
	KJ-0.5-R-C-0.66-1'		15.7	1.28				4.38
	KJ-0.5-Tr-C-0.67-1	17	16.8	1.45				5.0
	KJ-0.5-Tr-C-0.69-1'		15.7	1.28				4.59
	KJ-0.5-T-C-0.69-1	45	16.8	1.45				5.82
	KJ-0.5-T-C-0.67-1'		15.7	1.28				4.97
	KJ-0.5-Tr-F-1.99-2	17	12.0	1.48	4 Ø10	230		4.42
	KJ-0.5-Tr-F-1.81-1		12.0	1.48	2 Ø14	210		4.95
	KJ-0.5-Tr-F-2.28-1		14.0	2.0	2 Ø16	220		4.92
	KJ-0.5-Tr-F-2.89-1		14.0	2.0	2 Ø18	236		5.65
5	KJW-50-0.25-R-C-0.67-1	0	6.4	0.78	2 Ø8	240	50	28.0
	KJW-100-0.25-R-C-0.64-1						100	30.0
	KJW-150-0.25-R-C-0.65-1						150	26.0
	KJW-200-0.25-R-C-0.66-1						200	20.0
	KJW-25-0.5-R-F-0.67-1		12	1.48		260	25	53.0
	KJW-100-0.5-R-F-0.67-1						100	47.0
	KJW-200-0.5-R-F-0.67-1						200	38.0

Statement of the equal strength of triangular and rectangular keyed joints has been connected with the inclusion in the sampling for the comparative analysis of the test results of rectangular keys with a different ratio l_k/h_k, that leads to a significant spread of strength and as a result to a cautious estimate. While changing only the profile of keyed joints the higher strength of triangular keys has been set in comparison with the rectangular keys [1-3].

Three cases of failure for single-keyed joints with variable width have been observed (fig. 4): when t_j = 25 mm – keys inclined cracks shear by vertical section has been noticed; when t_j = 50 mm and t_j = 100 mm – keys shear by the inclined plane within the height of the key; when t_j = 150 mm and t_j = 200 mm – inclined cracks that started at the top corner of a key and distributed in space between keys. The key's strength decreases with the increase of joint's width.

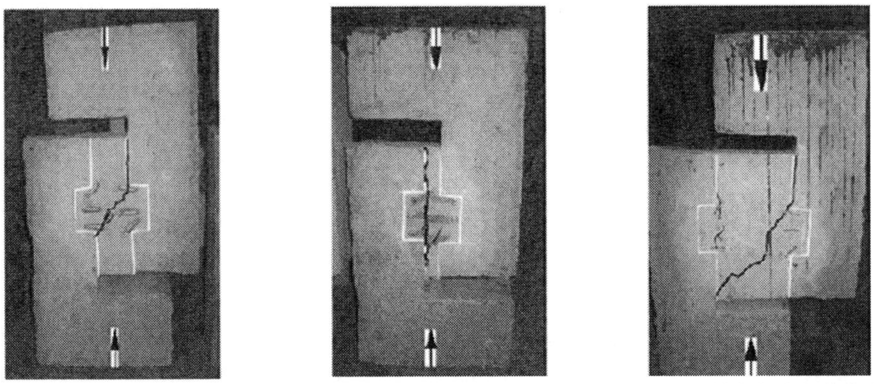

KJR-25-0.5-R-F-0.69-1 KJR-100-0.5-R-F-0.67-1 KJR-200-0.5-R-F-0.67-1

Fig. 4. The nature of failure of the single-keyed joints with variable joint's width made from fiber concrete when l_k/h_k = 0.5 and ρ_w = 0.67 %.

Failure surfaces of the concrete, reinforced and compressed keys at the shear action consist of the tensile and compressed areas. Sizes of the last ones rise with the increase of the compression value and the reinforcement ratio. When loads are close to the ultimate ones, concrete deformations reached value of (180 – 290)×10^{-5} in the most compressed areas (in the bottom incoming corners of the key). Moreover, maximum deformations were noticed in samples with higher compression level, reinforcement ratio and spaced-apart reinforcement.

Analysis of studies results of different types of keyed joints has been made on the basis of the evaluation of influence of separate factors and it allowed to explain a number of existing different interpretations of influence assessment of ratio of keys l_k/h_k, the width of the seam t_j, the profile shape on keyed joints strength.

Experimental assessment of the strength of keyed joints research and data from other studies [1-3, 6, 7, 11-13] found theoretical substantiations and corroboration of [14], which are based on vibrational methods, the concept of a rigid body and the principle of virtual velocities.

4 Conclusions

The obtained data allow to make the following conclusions:

1. Ultimate load value and the nature of failure of the concrete keys (crushing, shear, tear, failure) depend on ratio of their sizes in the key plane. For maximum joint strength it is recommended to admit l_k/h_k = 0.25 ratio.

2. Compression and reinforcing significantly increase keyed joint's strength (up to 2.5 times), spread shear limits of the failure form (till l_k/h_k = 1) and increase concrete plastic properties.

3. Two-level reinforcement placement along the key's height increase jointing strength to 10% and deformations of the concrete compressive zone – to 50%. The dowel effect appears in the bottom reinforcing rods.

4. Keys with the triangular profile have maximum strength at l_k/h_k = 0.5 (15% comparing with the rectangular keys).

5. The sharp increase of the concrete and reinforcement deformations was noticed. Also plastic constituent of the deformation was noticed in the phase before shear failure of the joint. At $\rho_w \leq 1\%$ reinforcement stress reach the yield point at the moment of failure.

6. The use of the fiber reinforced concrete for the joint's grouting increases crack resistance and strength of the joints, and changes the nature of failure from brittle to plastic.

7. Concrete keys are recommended to design with a rectangular profile with a ratio of depth to height $0.2 \leq l_k/h_k \leq 0.5$, reinforced concrete and compressed – trapezoidal with the ratio l_k/h_k in the above range and triangular with $l_k/h_k = 0.5$. This ratio may to be increased to value 1 when compression value is $\sigma/f_c = 0.4$ or percentage transverse reinforcement is $\rho_w = 1$ %). Reinforcement in the range of 1% can be considered effective.

References

1. Y. Naotaka, T. Norimono, K. Katori, S. Hayashi, Struct. Constr. Eng., Japan, **10(1)**, 441–442, (1997)
2. S.I. Izni, H.P. Khairul, H. Mansoor, A. Bady, A.A. Saim, N.N. Sarbini, Malaysian Journal of Civil Engineering, **26(3)**, 414–430, (2014)
3. Ch.H. Lee, W.J. Chin, E.S. Choi, Y.J. Kiml, Journal of the Korea Concrete Institute, **23(2)**, 235–244, (2011)
4. *Design of Concrete Structures – Part 1: General rules and rules for buildings, EN 1992-1-1: Eurocode 2*, Brussels, (2007)
5. *Betonni ta zalizobetonni konstruktsii z vazhkogo betonu. Pravila proyektuvannya, DSTU B V.2.6-156: 2010*, National Standard of Ukraine, (2011)
6. G.N. Ashkinadze, M.E. Sokolov, L.D. Martynova, *Zhelezobetonnyye steny seysmostoykikh zdaniy. Issledovaniye i osnovy proyektirovaniya*, Moscow, (1988)
7. I.A. Gerasimova, Beton i zhelezobeton, Moscow, **4**, 72–76, (1987)
8. V.V. Pogrebnoy, *Prochnost' betonnykh i zhelezobetonnykh elementov pri sreze*, Ph.D thesis in Engineering science, Poltava, (2000)
9. O.O. Dovzhenko, *Mitsnist' shponkovykh z'yednan' betonnykh i zalizobetonnykh elementiv: eksperymental'ni doslidzhennya*, Poltava, (2012)
10. I.A Yurko, *Mitsnist' shponkovykh stykiv iz fibrobetonu na osnovi syntetychnykh volokon*, Ph.D thesis in Engineering science, Poltava, (2012)
11. A.V. Yashin, Beton i zhelezobeton, Moscow, **6**, 17–18, (1978)
12. G. Rombach, Technical University Hamburg, Germany, **2**, 56–60, (2002)
13. S.H. Rizkalla, R.L. Serette, J.S. Heuvel, E.K. Attiogbe, PCI Journal, **3–4**, 104–120, (1989)
14. O.O. Dovzhenko, V.V. Pohribnyi, Y.V. Chursa, Bulletin of the National University "Lviv Polytechnic", Lviv, **755**, 111–117, (2013)

Improvement of the mathematical model of the diagram of deformation of the compressed composite steel and concrete structures

Stanislav Fomin [1*], *Yuriy* Izbash[1], *Iryna* Plakhotnikova[1], *Serhii* Butenko[1], and *Ruslan* Shemet[1]

[1]Department of reinforced concrete and stone structures, Kharkiv National University of Civil Engineering and Architecture, 61002, Kharkiv, 40, Sumskaya Str., Ukraine

Abstract. The mathematical model of the stress-strain ratio of compressed concrete at elevated temperatures for composite structures described in Eurocode EN 1994-1-2:2005 Eurocode 4 divided in two stages: Stage 1 - represents the ascending branch and is defined by two parameters: compressive strength $f_{c,\theta}$ and corresponding deformations $\varepsilon_{cu,\theta}$, stage II - the descending branch which is accepted for numerical methods. The design practice showed deficiencies in the provisions on models of deformation diagrams given in Eurocode 2. Part 1-2 General provisions. Structural Fire Design. (EN 1992-1-2:2004), (ENV 1992-1-2:1995), which were reflected in the national standards in the process of harmonization. A technique for determining the refined diagrams "$\sigma_{c,\theta} - \varepsilon_{c,\theta}$" was developed for the design of composite reinforced concrete structures in case of fire. Diagram parameter values were adjusted for concretes based on silicate filler at high temperatures, the mathematical stress-strain model of the concrete under compression and high temperatures was specified.

1 Introduction

The order of Ministry of Regional Development of Ukraine with the participation of GP GNIISK KNUCEA has developed national standard DBN V.2.6 - 160:2010 for the design of composite steel and concrete structures [1], and standard DSTU-N-P B B.2.6 - 159:2010 [2] on structural fire design by now. The design practice has shown the shortcomings of the provisions on the deformation diagram models given in Eurocode 2. Part 1-2: General rules – Structural fire design (EN 1992-1-2:2004 [3]), (ENV 1992-1-2:1995 [4]), which were reflected in national standards [5] in the course of harmonization.

2 Analysis of recent studies

In this study [6], we developed a method for determination of $\varepsilon_{cu1,\theta}$ criterion of carrying capacity of concrete to calculate the fire resistance of concrete structures of buildings, based on which the data on strength and deformation properties of concrete at higher

*Corresponding author: sfomin@ukr.net

temperatures given in Eurocode EN 1992-1-2:2004 and ENV 1992-1-2:1995 were clarified. The comparison of a diagram according to DBN V.2.6 - 98 [7] at normal temperature with a similar diagram of the standard [2] shows a significant overestimation of the value $\varepsilon_{cu1,\theta=20^\circ} = 0.0200 = 20\ ^0/_{00}$ (based on [7] $\varepsilon_{cu1,ck} = 2.4$ to $4.5\ ^0/_{00}$, depending on the concrete class). These diagrams were obtained on special presses provided with heating device and servocontrol of pressure in the press cylinder, which allows to pass along the descending branch practically to zero voltage. In this case, the descending branch is no longer an area of deformation of whole concrete specimen, but the area of deformation of individual parts of crushed concrete. So, the values of deformations in columns 4 and 7 of Table 3.1 [2] are not maximum concrete deformation $\varepsilon_{cu1,\theta}$.

DSTU-N-P B V.2.6-159:2010 (EN 1994-1-2:2005, MOD) [2] harmonized with EN 1994-1-2:2005 [4] strengthened erroneous concept of Eurocodes specified, which proposed linear approximation of descending branches of the concrete deformation diagram with silica filler under high temperatures.

The paper purpose is to improve the mathematical stress-strain model of concrete under compression and high temperatures. The specification of the criteria of the carrying capacity of concrete in the calculation of the fire resistance of composite structures in EN 1994-1-2:2005.

3 The results of the theoretical studies performed

In section 3, of the standard [2], we indicated that the strength of concrete and the mechanical properties at high temperatures can be determined from the stress-strain dependences given in EN 1992-1-2 [5], and as shown in Fig. 3.2 [2]. The stress-strain dependencies on the ascending branch shall be determined by two parameters: strength for compression $f_{c,\theta}$ and deformation $\varepsilon_{cu,\theta}$, which corresponds to $f_{c,\theta}$.

$$\sigma_{c,\theta} = f_{c,\theta} \left\{ 3\left(\frac{\varepsilon_{c,\theta}}{\varepsilon_{cu,\theta}}\right) \Big/ \left[2 + \left(\frac{\varepsilon_{c,\theta}}{\varepsilon_{cu,\theta}}\right)^3 \right] \right\} \tag{1}$$

Using formula (1), we would obtain data on the elasticity modulus $E_{ck,\theta}$ during heating that are absent in the standard, but it is complemented by Table 3.1 [4] of the values of the diagram parameters. In this table, the deformations $\varepsilon_{c1,\theta}$ are taken the same for all classes of concrete, including for silicate (column 4) and carbonate (column 7) fillers. Therefore, depending on the concrete class in the formula (1), we can vary only the value of the compression strength f_c, so these diagrams, even on the ascending branch, incorrectly reflect the actual physical properties of non-linearity.

In DBN [7] (at 20 °C) it was proposed to use two equations for describing the relationship between σ_c and ε_c for short-term axial load. The equation (2) that is used in EN 1992-1-1:2005 [8]:

$$\frac{\sigma_n}{f_{ck}} = \frac{k\eta - \eta^2}{1 + (k-2)\eta} \tag{2}$$

where:

$$\eta = \varepsilon_c / \varepsilon_{c1,ck} k = 1.05 E_{ck}\, \varepsilon_{c1,ck} / f_{ck,prism} \tag{3}$$

and equation (4) in the form of a polynomial of the fifth degree, which is based on the results of numerous experimental studies of NIISK, the statistical processing of which allowed us to offer a more complete regulatory framework.

$$\sigma_{\tilde{n}} = f_{(ck),(cd)} \sum_{k=1}^{5} a_k \eta^k \tag{4}$$

The limits of use were determined for both formulas:

$$0 < |\varepsilon_c| < |\varepsilon_{cu1,ck}| \tag{5}$$

where $\varepsilon_{cu1,cK}$ – nominal maximum concrete deformations.

Fig. 1 shows complete diagrams "σ_c - ε_c" at 20 °C for concrete class C 16/20 according to the formula (4) of DBN [7], formula (2) of Eurocode 2 [8] and the ascending branch according to the formula (1) DSTU, EN 1992 [4]. As we can see from this figure, the diagrams according to DBN and EN 1992-1-1 on the ascending branch practically coincide, the diagram according to the formula (1) of DSTU [2] has significant deviation - the curve according to DSTU is flatter. This arises due to large values $\varepsilon_{c1,ck,\theta}$ = 2.5 ‰ (0.0025) compared to the baseline.

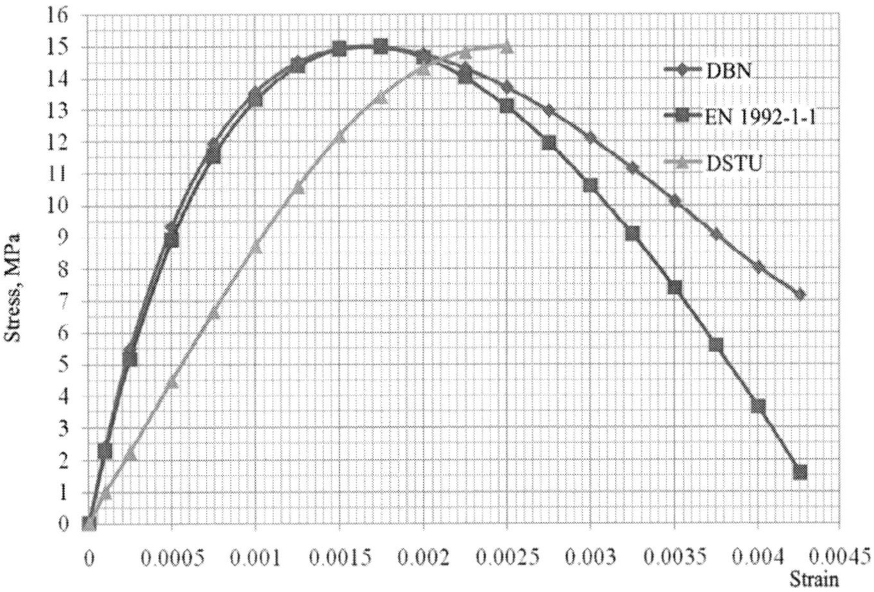

Fig.1. Comparison of diagrams "σ_c - ε_c" according to the formula (2) of DBN, formula (4) and formula (1) of DSTU [4] for concrete class C 16/20 at θ = 20 °C.

If using this curve, we will calculate the value of the elasticity modulus, we will obtain a value E_{ck} = 0.4·15/0.00065 = 9.23 GPa, which is 2.5 times lower than the standard E_{ck} = 23 GPa. Therefore, the formula (1) of Eurocode [2] is unsuitable for determination of modulus of elasticity.

The replacement of deformation $\varepsilon_{c1,\theta}$ with $\varepsilon_{cu,\theta}$, and deformations $\varepsilon_{cu1,\theta}$ – with $\varepsilon_{ce,\theta}$ (maximum deformations during fire on fig. 3.2 [2] is uncomprehended.

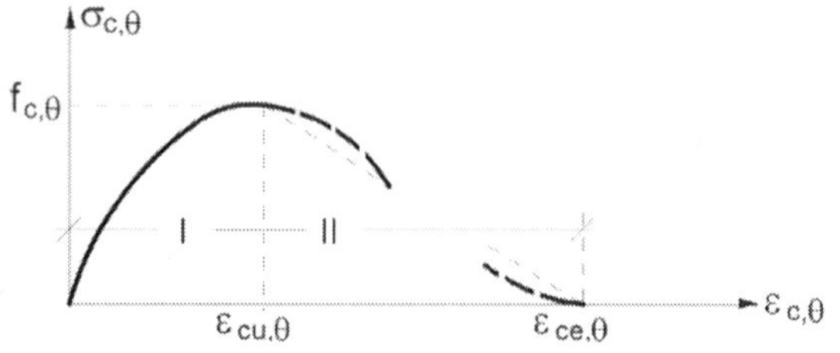

Fig.2. Mathematical model [2].

In addition, stress-strain relations for concrete of silicate filler under high temperature were proposed in Appendix V. At the linearly descending section (shown in Figure V.1 of the appendix), the maximum deformation $\varepsilon_{ce,\theta}$ = 47.5 ‰. They correspond to the mathematical formulation shown in Fig. 3, and the data in Table 1 [2]. The allowable range and the recommended values of the deformation $\varepsilon_{cu,\theta}$ for $f_{c,\theta}$ (Fig. 1) is proposed to determine from the Table 2 (V.1 [2]). Table 2 does not differ from Table 3.1 EN 1992-1-2 [5] for ordinary concrete based on silicate filler except for that deformations $\varepsilon_{c1,\theta}$ (second column) in is were replaced with deformations $\varepsilon_{cu,\theta}$, and deformations $\varepsilon_{cu1,\theta}$ – with deformations $\varepsilon_{cu1,\theta}$ – (third column).

To perform elastic and non-elastic calculations of the fire resistance in program complexes, the values of the elasticity moduli of the compressed $E_{tc,\theta}$ and extended $E_{tc,\theta}$ concrete, deformations $\varepsilon_{c1,\theta}$ and $\varepsilon_{tc1,\theta}$, relevant compression $f_{c,\theta}$ and tensile strengths $f_{tc,\theta}$, limit deformations $\varepsilon_{cu,\theta}$, $\varepsilon_{tcu,\theta}$ and limit stresses relevant to them $\sigma_{cu,\theta}$ and $\sigma_{tcu,\theta}$ are needed.

Actual limit deformations $\varepsilon_{cu1,\theta}$ determined based on the energy approach and are shown in adjusted Table of the paper [6]. In the paper [9], we developed a method for determination the heave concrete elasticity modulus under high temperature heating, criterion of carrying capacity of concrete $\varepsilon_{cu1,\theta}$ to calculate the fire resistance of concrete structures of buildings, based on which the data on strength and deformation properties of concrete at higher temperatures given in EN 1992-1-2:2004 and ENV 1994-1-2:2005.

The significant issues arose in the considered model on the ascending branch. One is that the concrete classes are not fully taken into account on the diagrams. With concrete class, the value of strength f_{cm} only increases, and deformation ε_{c1} remain the same and equal ε_{c1} = 2.5 ‰. This, as evidenced by Fig. 2, leads to large errors in the determination of the elastic modulus of concrete and "stress-strain" dependence is incorrectly described. Another is that such a stress-strain dependence is unused, which not properly represents the nature of work of concrete considered. Diagrams "σ_c - ε_c" of concrete in the basic normative documents take into account the necessary reference points: in the formula (2) of the Eurocode 2 [8] the coefficient $k = 1.05E_{ck}\varepsilon_{c1,ck}/f_{cm}$ is introduced, in the formula (3) of DBN [7] the polynomial coefficient are selected based on the conditions [9]:

- the curve passes through the origin;
- the secant modulus at stresses $\sigma_c = 0.3f_{ck}$ is equal to the elesticity modulus $E_{ck,}$ included into current norms;
- at deformations $\varepsilon_c = \varepsilon_{c1,ck}$ the stresses in the concrete reach maximum values and are equal to the prism strength $\sigma_c = f_{ck,prism}$;
- in the entire range of the curve existence $\varepsilon_c = 0$ to ε_{cu1} it has a single extremum at $\varepsilon_c = \varepsilon_{c1,ck}$.

When deriving the formula (1) [DSTU], the reference points described were not taken into account.

The processing of the stress-strain diagram of the compressed concrete under high temperatures provided in ENV 1992-1-2:1995 and using affinity effect between the diagrams in the cold (at 20 °C) and diagrams at heating [10] allows us to calculate "σ_c - ε_c"

diagrams of the predetermined class concrete based silicate filler at high temperatures by any formula built on the above assumptions.

For example, let us construct full diagrams for compressive concrete class C 16/20 according to the formula (2) of Eurocode 2 [8]. We will take data from Table 3.1 of EN 1992-1-1:2005 [8] for $\theta = 20$ °C: $f_{cm} = 24$ MPa, $E_{cm} = 29.000$ MPa, $\varepsilon_{c1,ck} = 0.0019$, $\varepsilon_{cu1,ck} = 0.0035$.

Then, the stress-strain dependence (2) will look like:

$$\sigma_{c,\theta} = f_{cm,\theta} \frac{k\eta - \eta^2}{1 + (k-2)\eta} \tag{6}$$

where:

$$\eta = \varepsilon_c / k_\theta = 1.05 E_{cm}\, \varepsilon_{c1,\theta} / f_{cm,\theta} \tag{7}$$

This expression is fair at $0 < |\varepsilon_c| < |\varepsilon_{cu1,\theta}|$, where $\varepsilon_{cu1,\theta}$ - nominal limit deformations at temperature.

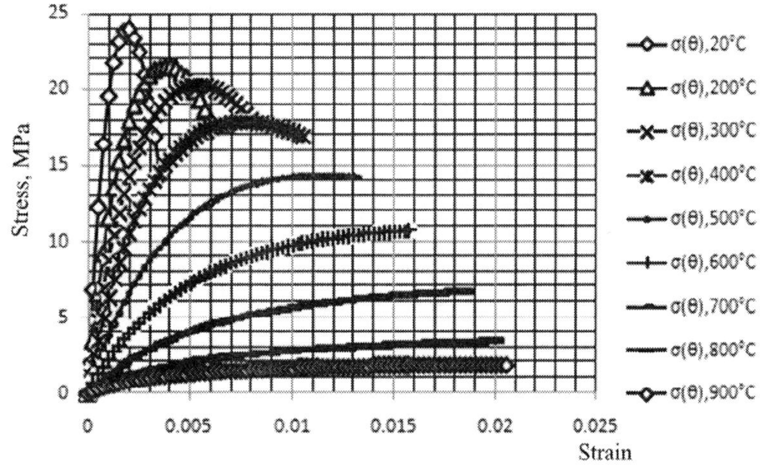

Fig. 3. Diagram "σ_c - ε_c" for concrete class C 16/20 at elevated temperature.

4 Conclusions

The defects of strength and deformation properties of concrete at higher temperature given in EN 1994-1-2:2005 Eurocode were found, which were reflected in the national standard DSTU-N-P B V.2.6-159:2010 in the course of harmonization. A methodology was developed for determining revised "$\sigma_{c,\theta}$ - $\varepsilon_{c,\theta}$" charts for the design of composite steel and concrete structures during a fire. Diagram parameter values were adjusted for concretes based on silicate filler at high temperatures, the mathematical stress-strain model of the concrete under compression and high temperatures was specified.

References

1. *DBN V.2.6-160:2010 Composite steel and concrete structures. Basic provisions.* Minregionbud of Ukraine, Kiev (2011)

2. *DSTU-N-P B V.2.6-159:2010 Structures of buildings and facilities. Design of composite steel and concrete structures. Part 1-2. General provisions. Structural fire design. (EN 1994-1-2:2005, Eurocode 4. MOD)*, Kiev (2010)
3. *EN 1992-1-2: 2004 Eurocode 2: Design of concrete structures – Part 1-2: General rules - Structural fire design*, Brussels (2004)
4. *ENV 1992-1-2:1995 Eurocode 2: Design of Concrete Structures - Part 1-2: General Rules - Structural Fire Design*, Brussels (1995)
5. *DSTU-N B EN 1992-1-2:2012. Eurocode 2: Design of concrete structures. Part 1-2: General rules. Structural fire design. (EN 1992-1-2:2004, IDT).* Minregionbud of Ukraine, Kiev (2012)
6. S.L. Fomin, A.I. Davidenko, V.G. Poklonsky, News of Odessa State Academy of construction and architecture, Odessa, (2012)
7. *DBN V.2.6-98:2009 Structures of buildings and facilities. Concrete and reinforced concrete structures. Basic provisions.* Ministry of regional development and construction of Ukraine, Kiev (2011)
8. EN 1992-1-1:2005 *Eurocode 2: Design of reinforced concrete structures - Part 1-1: General norms and rules for structures*, Brussels (2005)
9. S.L. Fomin, Sh. Kasem, V.G. Poklonskiy, Science bulletin of construction, **73**, 427 (2013)
10. A.N. Bambura, *Materials of the First Ukrainian scientific - technical conference, Scientific - practical problems of modern concrete*, Kiev, (1996)

Features of performance of concrete and reinforced concrete structures of industrial facilities operated under impact of increased and high temperatures

Stanislav Fomin[1*], *Iryna* Plakhotnikova[1], and *Serhii* Butenko[1]

[1]Department of reinforced concrete and stone structures, Kharkiv National University of Civil Engineering and Architecture, 61002, Kharkiv, 40, Sumskaya Str., Ukraine

Abstract. Currently, the Normative base of Ukraine has moved to European standards. The process of harmonization with the Eurocodes was completed, new national normative documents of DBN and DSTU were developed and put into effect. At the same time it was turned out that the big section "Concrete and reinforced concrete structures intended for work in conditions of increased and high temperatures" is absent in the Eurocodes. By order of the Ministry of Regional Development, Construction and Housing and Communal Services of Ukraine, the development of a normative act is provided for the replacement of SNiP 2.03.04-84 with DBN with a change in status. Heat-resistant concretes in the elements of thermal aggregates are used in the iron and steel industry (blast furnaces, air heaters, coke batteries, etc.), non-ferrous metallurgy (graphite furnaces, fluidized bed furnaces, etc.), oil refining and petrochemical industry (tubular furnaces, vertical sectional furnaces, etc.), in the industry of building materials, and in various sectors of industry. These norms and rules apply to the design of concrete and reinforced concrete structures designed to operate under conditions of systematic exposure to increased (50 to 200 °C inclusive) and high (above 200 °C) process temperatures.

1 Analysis of recent studies

When developing a normative act for the replacement of SNiP 2.03.04-84 with DBN with a change in status, it is advisable to use the Book of Rules of SP 27.13330.2011 "Concrete and reinforced concrete structures intended for operation under increased and high temperatures."

This set of rules is built on the basis of numerous theoretical and experimental studies of K.D. Nekrasov [1], V.I. Belskiy [2], A.B. Toturbiev [3], G.F. Gitman, T.N. Malkina, A.F. Milovanov [4, 5], A.E. Desov [6], R. Von der Heyden, P. Fritsch, G. Krishtanits, R. Rumpl [7], and others.

*Corresponding author: sfomin@ukr.net

The rules were developed by NIIZhB named after A.A. Gvozdev - the institute of JSC "Research and Development Center "Stroitelstvo": the head - Grand PhD in Engineering Sciences, prof. A.F. Milovanov Executors: Grand PhDs in Engineering Sciences, prof. A.P. Krichevskiy, S.L. Fomin; PhDs in Engineering Sciences V. N. Goryachev, N.P. Zhdanova, I.M. Zaslavskiy, V.N. Milonov, V.G. Petrov-Denisov, V. N. Samoylenko, V.V. Solomonov, I.S. Kuznetsova, engineers E.M. Bolnykh, V.A. Tarasov, with the participation of LLC "UralNIIstrom" (PhD in Engineering Sciences R.Ya. Akhtyamov) [8].

The papers on the coke-chemical industry for the foundations of coke batteries are also evidence of the need to create DBN. In papers [9,10], we obtained temperature fields in reinforced concrete structures, found new features of the stressed-deformed state of reinforced concrete monolithic structures of coke furnace batteries, and obtained results of experimental studies that revealed a number of new features of the work of reinforced concrete structures and allowed to improve the nature of reinforcement, to prove the feasibility of constructive solutions of foundations of heat-resistant concrete without lining of breaching.

2 The results of the theoretical studies performed

The method of calculation of reinforced concrete elements operating under conditions of the technological environment includes the evaluation of heat and moisture transfer, free temperature and humidity deformations; creep, shrinkage, full diagrams of "σ–ε" concrete and reinforcement, stress and strain state of reinforced concrete elements.

The process of heat spreading in concrete can be approximately described by a scheme, according to which concrete is divided by an evaporation surface with a temperature of 100 °C into two zones: 1st - dry and 2nd - wet, the sizes of which continuously change with increase of temperature. In this case, the thermophysical characteristics of each of the zones are different functions of temperature.

The determination of the temperature field of such a two-zone environment must be carried out taking into account the variable position of the interface of the zones, as well as the continuous absorption of latent heat of vaporization in the zone of separation (in analogy with the Stefan problem):

for dry zone

$$\rho C(t)\frac{\partial t_1}{\partial \tau}+\rho\frac{\partial C(t)}{\partial \tau}t_1=\nabla[\lambda(t)\nabla t_1] \tag{1}$$

for wet zone

$$\rho C(t)\frac{\partial t_2}{\partial \tau}+\rho\frac{\partial C(t)}{\partial \tau}t_2=\nabla[\lambda(t)\nabla t_2] \tag{2}$$

At the interface between zones, the following condition must be met:

$$-\lambda_1\nabla t_1(\xi,\tau)+\lambda_2\nabla t_2(\xi,\tau)=rW\rho\frac{\partial \xi}{\partial \tau} \tag{3}$$

The impact of the technological environment is also manifested in the danger of brittle fracture of concrete [11]. The explosive brittle failure should be excluded, otherwise its impact on operational requirements should be taken into account (Fig. 1).

The explosive brittle failure is unlikely if the concrete moisture content is less than k%. The recommended value of k = 3%. If the moisture value is more than k%, more accurate assessment of the moisture content, type of aggregate, water permeability of concrete and heating conditions should be carried out.

During high-temperature intensive heating, a gradient of the total pressure of moist air arises inside the wet concrete, resulting in an additional transfer of moisture and heat due to the presence of hydrodynamic (filtration) movement of vapour and liquid, while the diffusion transfer is negligibly small.

The mathematical model can be represented in the form of a system of nonlinear differential equations describing the development of the temperature and pressure field with internal sources and drains:

$$C_t \rho \frac{\partial t}{\partial \tau} = \nabla[\lambda_t \nabla t] + r_t \rho B_t \frac{\partial t}{\partial \tau} \quad (4)$$

$$C_p \rho \frac{\partial p}{\partial \tau} = \nabla[\lambda_p \nabla p] - \rho B_t \frac{\partial t}{\partial \tau} \quad (5)$$

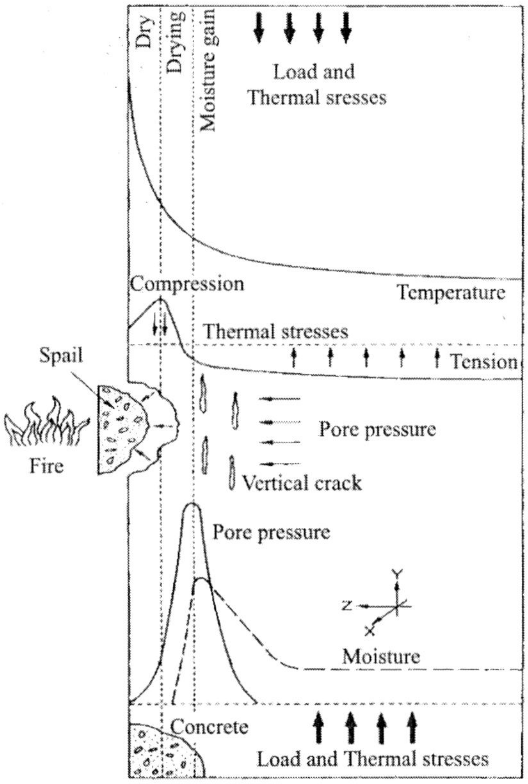

Fig. 1. The explosive peeling of concrete under the influence of pore pressures and thermal stresses [11].

$$\frac{\partial W}{\partial \tau} = B_t \frac{\partial t}{\partial \tau} \quad (6)$$

A modification of the method of V.M. Roitman [12-14] was proposed for determination of the phase transition intensity, the following formula was experimentally obtained:

$$B_t = \frac{\partial W}{\partial t} = -2698 \cdot 10^{-4} + 4.72 \cdot 10^{-6} t - 1.086075 \cdot 10^{-6} t^2 + \\ +7.70504 \cdot 10^{-9} t^3 - 1.722365 \cdot 10^{-11} t^4 \quad (7)$$

The introduction of the characteristic of the intensity of phase transformations makes it possible to combine terms in the equations (4) - (6) with time derivatives, introducing the concept of equivalent heat capacity and the vapour capacity by writing them as follows:

$$C_{red,t}(t)\rho \frac{\partial t}{\partial \tau} = \nabla[\lambda_t \nabla t] \quad (8)$$

$$C_{red,p}(t)\rho \frac{\partial p}{\partial \tau} = \nabla[\lambda_p \nabla p] \quad (9)$$

$$C_{red,t} = C_t - r_t B_t \quad (10)$$

$$C_{red,p} = C_p - B_t \quad (11)$$

With this approach, the system of equations (4) ... (6) breaks up into separate differential equations, each of which shall be solved successively: first, the heat conductivity equation (8), then, using the rate of change of temperature in the elementary volume and the intensity of phase transitions known from the heat conductivity equations, equation of the field of excess pressures (9) and, finally, the equation of moisture conductivity (6).

The coefficients of thermal conductivity and specific heat as a function of temperature and humidity are proposed to be determined by the formulas:

$$\lambda = 1.37(1 + 0.175W - 0.016W^2)(1 - 0.000277t) \quad (12)$$

$$C = 0.83(1 + 0.077W - 0.0007W^2)(1 + 0.001t) \quad (13)$$

3 A numerical method for solving a system of non-linear differential equations of heat-moisture conductivity using a conservative finite-difference scheme

A numerical solution of equations in partial derivatives is considered in the papers of R.D. Rikhtmayer [15], R.D. Richtmyer and K. Morton [16], A.R. Mitchell [17], L.A. Kozdoba [18].

The building of a difference scheme can be regarded as a replacement for a continuous environment described by differential equations with some discrete analogue of it. At finite steps of the grid, the discrete model can differ markedly from the continuous medium. This difference generates parasitic effects of difference origin, which reduce the value of the difference solution. Therefore, it is important to build such schemes that retain their qualities on the grids. The appearance of oscillations of the increasing amplitude caused by superposition of perturbations (perturbations can be generated either by machine rounding errors or transverse motions in a real two-dimensional problem) on the solution due to a large time step is called dynamic instability, which can be eliminated by decreasing the time step, making it smaller than some "critical step in time." The appearance of a monotonically increasing error is called a static instability, which can be eliminated only by a transition to some other finite-difference scheme.

The differential equations of mathematical physics - is a mathematical expression of the basic laws of conservation (mass, momentum, and energy). Therefore, it is reasonable to build a difference scheme so that analogues of these laws are also fulfilled in it. The schemes of this type are called conservative.

A conservative difference scheme should reflect the basic properties of a continuous medium, i.e. the scheme should provide, first of all, the fulfilment of difference analogues of the basic conservation laws [19-20].

The conservative form of the equations of motion of an inviscid compressible fluid was derived by R. Kurant and K.O. Friedrichs. Its practical application was first realized by Lax, who transformed the ordinary equations of hydrodynamics, in which the dependent variables were velocity, density, and temperature, into a system of equations, in which the quantities of motion, density and the specific internal energy of breaking serve as dependent variables. This system of equations reflects the essence of physical conservation laws and allows preserving integral flow characteristics in a finite-difference scheme. If the traditional differential equations are transformed so that the conservative quantities become the sought variables, then the use of conservative finite-difference schemes ensures the conservation of mass, momentum, and energy.

The experience shows that conservative schemes give more accurate results. N.E. Sairus and R. E. Falton found that for elliptic equations, a conservative scheme yields more accurate results than a non-conservative one. A finite-difference method is a conservative method, if it ensures the fulfilment of certain integral conservation laws that are valid for the original differential equations.

A.N. Tikhonov and A.A Samarskiy proposed an integro-interpolation method for building conservative difference schemes, and an example was built, where a non-conservative difference scheme that provides the second order of accuracy in the class of sufficiently smooth coefficients diverges in the class of discontinuous coefficients.

Based on the provisions of the theory of moisture conductivity by V.N. Bogoslovskiy [21-23], a modified system of differential equations of heat and moisture transfer in concrete was proposed:

$$\rho\eta(\theta,t)\frac{\partial\theta}{\partial\tau}+\rho\frac{\partial[\eta(\theta,t)]}{\partial\tau}\theta=\nabla[\kappa(\theta,t)\nabla\theta] \tag{14}$$

$$\rho C(\theta,t)\frac{\partial t}{\partial\tau}+\rho\frac{\partial C(\theta,t)}{\partial\tau}t=\nabla[\lambda(\theta,t)\nabla t] \tag{15}$$

Equations (14) - (15) with boundary conditions compiled from data on the change in temperature and the moisture potential of the environment completely determine the problem of calculation of temperature and humidity fields in cross-sections of reinforced concrete structures.

We considered the building of a conservative difference scheme for equations (14) - (15), which is represented in the form

$$\frac{\partial t}{\partial\tau}=a\nabla^2 t \tag{16}$$

$$\frac{\partial W}{\partial\tau}=\frac{1}{\rho}\nabla[\kappa(t,W)\nabla\theta] \tag{17}$$

In equation (4), the coefficient of moisture conductivity k is considered as a given moisture function W and temperature t.

The computer programs have been developed for the calculation of the models presented.

Conclusions

The study of the concrete and reinforced concrete structures of industrial constructions operating under high and high temperatures was performed to create a new national normative act based on the latest achievements of science, engineering and technology, advanced domestic and foreign experience of designing and building and using the accumulated theoretical and experimental base of the SNiP 2.03.04-84.

This paper considers the impact of the technological environment on the physical and technical characteristics of building structures, including the emergence of the danger of brittle fracture of concrete.

During intensive high-temperature heating, a gradient of the total pressure of moist air arises inside the wet concrete, resulting in an additional transfer of moisture and heat due to the presence of hydrodynamic (filtration) movement of vapour and liquid, while the diffusion transfer is negligibly small.

The mathematical model was built in the form of a system of nonlinear differential equations describing the development of the temperature and pressure field with internal sources and drains. A conservative difference scheme was used, which should reflect the basic properties of a continuous medium, i.e. the scheme should provide, first of all, the fulfilment of difference analogues of the basic conservation laws. Based on the provisions of the theory of moisture conductivity by V.N. Bogoslovskiy, a modified system of differential equations of heat and moisture transfer in concrete was proposed. We considered the building of a conservative difference scheme for equations of heat and mass transfer, and the computer programs was developed.

References

1. K.D. Nekrasov, *Refractory concretes, their properties and applications*, Stroyizdat, Moscow (1949)
2. V.I. Belskiy and others. *Construction of industrial furnaces and pipes of heat-resistant concrete*, Stroyizdat, Moscow (1962)
3. A.B. Toturbiev, Concrete and reinforced concrete, **4**, 10 (2011)
4. G.F. Gitman, T.N. Malkina, A.F. Milovanov, *Heat-resistant concrete and reinforced concrete and areas of their effective use*, VIIGH, Volgograd (1969)
5. A.F. Milovanov, *Heat resisting reinforced concrete*, Stroyizdat, Moscow (1963)
6. A.E. Desov, K.D. Nekrasov, A.F. Milovanov, *Concretes for nuclear reactors, Cube and prismatic strength of concrete at high temperatures*, Detroit, II (1972)
7. R. Von Der Heide, P. Fritsch, R. Krischyanits, P. Rampl, Cement and its application, **1**, 200 (2012)
8. *SP 27.13330.2011. Code of rules Concrete and reinforced concrete structures intended for work in conditions of increased and high temperatures. Updated version of SNiP 2.03.04-84*, Moscow (2011)
9. B.A. Krylov, *Studies in the field of heat-resistant concrete*, Stroyizdat, Moscow (1981)
10. A.M. Kravchenko, L.N. Borisova, *Handbook of the coke-chemist*, Kharkov, (2016)
11. G.A. Khoury, C.E. Majjrana, *Effect heat on concrete*, Imperial College, London (2006)

12. V.M. Roitman, A.G. Apostolov, T.S. Simonova, Fire resistance of building structures, VNIIPO, Moscow, **1**, 220 (1973)
13. V.M. Roitman, A.T. Apostolov, T.S. Simonova. Fire resistance of building structures, VNIIPO, Moscow, **1**, 130 (1973)
14. V.M. Roitman, T.N. Zyrina, Fire resistance of building structures, VNIIPO, Moscow, **58** (1974)
15. R. Richthmaier, *Difference methods for solving boundary value problems*, Mir, Moscow (1970)
16. R. Richthmyer, K. Morton, *Difference methods for solving boundary value problems*, Mir, Moscow (1972)
17. A.R. Mitchell, *Computational methods in partial differential equations*, J. Wiley and Sons Ltd, (1969)
18. L.A. Kozdoba, *Solutions of nonlinear heat conduction problems*, Naukova Dumka, Kiev (1976)
19. A.N. Tikhonov, A.A. Samarskiy, On the convergence of difference schemes in the class of discontinuous coefficients, DAN USSR, **3** (1959)
20. A.N. Tikhonov, A.A. Samarskiy, ZhVM MF, **1**, 562 (1961)
21. V.N. Bogoslovskiy, Construction *Thermophysics*, High School, Moscow (1970)
22. V.N. Bogoslovsky, IFZh, **8**, 31 (1965)
23. V.N. Bogoslovsky, *Building Thermophysics*, High School, Moscow (1982)

Estimate of technical and economic benefits of a new space composite structure

Grygorii Gasii[1,*], *Olena* Hasii[2], and *Oleg* Zabolotskyi[3]

[1]Poltava National Technical Yuri Kondratyuk University, 36011, Poltava, Pershotravnevyi Avenue, 24, Ukraine
[2]Poltava University of Economics and Trade, 36014, Poltava, Koval Str., 3, Ukraine
[3]MTS Company LTD, Almaty, Kazakhstan

Abstract. There is a problem of excessive laboriousness and materials consumption resulting from irrational using of materials in construction. This situation is caused by inconsistencies existing structural concepts to modern requirements of construction industry. It has a direct impact on the overall implementation cost of the project. That is why there is a need for a new structure, which makes it possible to save materials and reduce a construction complexity. Such structure is the new space composite structure. There is summary information about the estimate of technical and economic benefits of a new space composite structure in the paper. Task of paper is arrangement information about estimate of technical and economic benefits of a new space composite structure. Based on the previous research results, there is the optimum ratio for the structure in the paper; also, a relationship between the depth of the module and the span for the new space composite structure were defined. The estimate of both the technical-economic parameters and the advantages of the new space composite structure were described. The estimation results show the effectiveness of the new space composite structure to 10%-37% compared to traditional reinforced concrete structures.

1 Introduction

Improvement of construction sector is accompanied by regular introduction new designs. An important factor in developing constructive solutions is the use of reliable, advanced and lightweight materials. The researchers focus mainly on the study of the stress-strain state of beams, columns, slabs, bridge structures, structures for metallurgical and mining industry, etc. [1-3].

The study of new space composite structures is the actual problem because use such structures in the building construction would save labour and materials. The new space composite structure is a bearing system that consists of concrete slab and space lattice composed of steel tubes or other cross section elements. The distinctive feature of the new space composite structure is a flexible lower belt made of cable or other flexible elements,

* Corresponding author: grigoriigm@gmail.com

which are only in tensile. The new space composite structure is collected of the individual lightweight modules [4].

Manufacturing technology is divided into two independent processes: the production of steel lattice and the formation of the slab. Another distinctive feature of the new space composite structures is their versatility by changing the lengths of the bottom chord can be given different curvatures of structure, buildings or structures any complexity and shape.

2 Literature review

Analysis of recent papers and publications has shown that composite slabs or shells frequently stand out from other composite steel and concrete structures, specifically combining concrete slabs (top chord) and steel rod elements (steel lattice). In such constructions, slabs can be unloaded by means of steel rods. A study of the strength and deformability of the steel and concrete composite structure that consists of steel T-section beams and concrete monolithic slab which are combined with anchors [5]. In [6] a structure was proposed that consists of modular steel elements, top and bottom chords and reinforced concrete slab without anchors.

Considering the above items, the idea to combine the slabs and rods for joint performance in a structure by the new way is the unique and distinguishes the proposed structure among existing as a new type of structure. The concept of the proposed solution is a synthesis of an experience and new developments that consist of the modular elements, the purpose of which is to decrease complexity and difficulty of manufacturing technology, assembly and installation [7].

Efficiency of the constructive solution of spatial structures and their elements were investigated and confirmed earlier [8].

An analysis of previous works shows there is no summary information about the estimate of technical and economic benefits of the new space composite structure. The objective of this article is arrangement information about the estimate of technical and economic benefits of the new space composite structure.

3 Results and Discussions

As it is well known, in the design of building structures often face the problem of finding certain geometric parameters of a structure or its elements, which must satisfy a number of conditions to ensure the normal functioning of the building or structure. The importance of solving this task consist in defining the right geometric parameters, which help to reduce material consumption while ensuring rigidity and reliability of structures with the same bearing capacity comparable options [9]. In most cases, to obtain optimal geometric parameters of structures, it is necessary to use high-strength materials [10].

Solving the problem of choosing the geometric parameters is especially important for the new space composite structure because the lower belt of the structure made of flexible rods, which by their nature and the essence are not designed to in compression. Thus, there is no need to use rigid and massive types of a metal profile to provide the design load-carrying capacity and rigidity of the structure. This feature of the new space composite structure very closely linked to the ability to use high-strength steels to manufacturing elements of the bottom chord, because it is known that for stretched elements effectively apply high-strength steel.

Regarding the possibility of application of the optimization methods, the essence of which is to changing the structure geometry or determining individual geometric dimensions for each item, is unacceptable for the new space composite structure, as it

DOI: 10.1051/matecconf/201711602014

contradicts the essence of the structure, it is using modular elements. Finding the optimal height to span ratio of the module elements is another issue that is closely related to designing of the new space composite structure.

Based on the research results [11] that present the optimum ratio for the previously mentioned parameters for structures, which composed of steel trusses and concrete slabs, were defined relationship between the depth module and the span for the new space composite structure (Fig. 1).

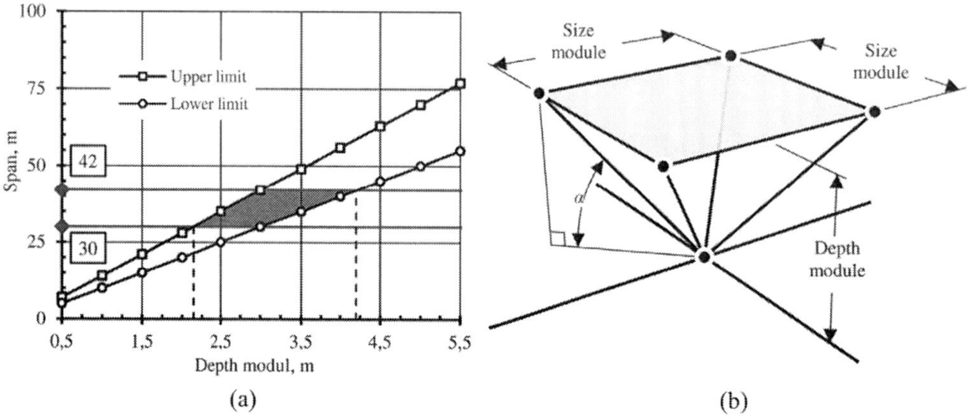

Fig. 1. The depth–span curve for the new space composite structure (a) and space module for the new space composite structure (b).

Knowing the optimum depth of the module can determine the size of the module. However, it should be noted that determination the optimal height to span ratio is the most important for a beam type of the space composite structure that has a small curvature. For the arch type structure, the main issue is to determine the geometrical parameters of curvature and boundary conditions (Fig. 2), which ensure efforts distribution in structure elements so that the top chord will be only in compression and the bottom chord in tension.

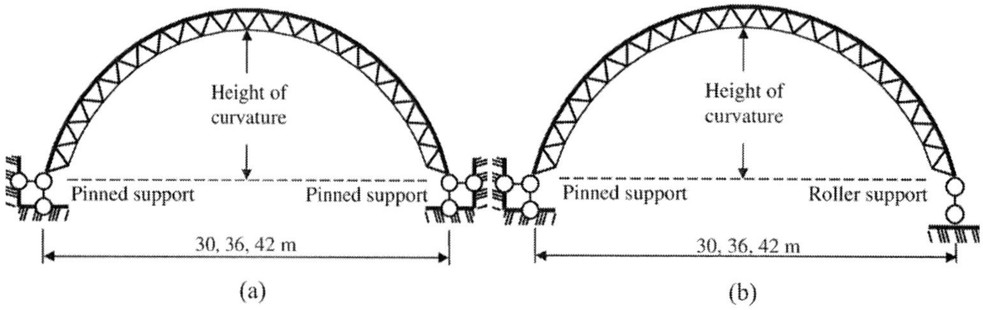

Fig. 2. Types of support of the new space composite structure: pinned–pinned support (a); pinned–roller support (b).

In [12] described the results of research, the task of which was to determine the geometrical parameters of curvature of the new space composite structure with pinned-pinned support (Fig. 2, a). The results of the research have shown the optimal spans of the space composite structure are distances 30, 36 and 42 m. The optimal height of curvature for the spanning structure 30 m is between 1400 and 1450 mm; the optimal height of curvature for the spanning structure 36 m is between 1460 and 1570 mm; the optimal height of curvature for the spanning structure 42 m is between 1580 and 2100 mm. For the new space composite structure with pinned-roller support (Fig. 2, b) the optimal height of curvature may be 10 times higher than in the previous case. Thus, the pinned-roller support gives more opportunities to build different structures. This has been confirmed by experimental research, which was conducted on a full-scale test of the new space composite structure [13].

There is a need to perform the estimate of the technical and economic parameters and to estimate the advantages the new space composite structure compared to existing analogues to determine the field of application of the new space composite structure. Such estimate of both the technical-economic parameters and the advantages of the new space composite structure was obtained in [14]. The estimation results show the effectiveness of the new space composite structure to 10-37% compared to traditional reinforced concrete structures (Fig. 3).

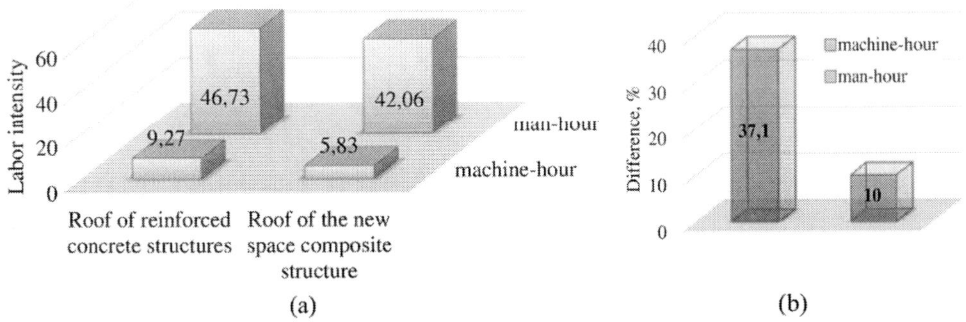

Fig. 3. The technical and economic benefits of a new space composite structure [14]: labor intensity (a); the difference (b).

Research confirms the effectiveness of constructive solutions of the new space composite structure because labour costs of installation and the consumption of materials are reducing compared with analogues.

4 Conclusion

The new space composite structure is the new kind of a structure for long-span systems and shells. The main feature of the structure consists in the structural concept. The new space composite structure has significant advantages; in particular, the structure is lighter and has lower complexity in both manufacturing and assembly than analogues. The new space composite structure consists of the bottom chord; and space modules, which combine both a reinforced concrete slab and a lattice made of a steel tube. This structural concept makes it possible to save materials due to its rational using.

Calculations of technical and economic indicators were solved. These data show the effectiveness and feasibility of the new space composite structure of the different designation. The studies have shown the use of the new space composite structure allows reducing the weight by 32-72%; the man-hour by 10% and the machine-hour by 37,1% in comparison with existing analogues [14].

The new space composite structure is both reliable and efficient in exploitation it has been confirmed by both analytical studies and numerically investigation of the stress-strain state; structural concept allows saving materials it has been confirmed by both experimental and theoretical evidence.

In conclusion, it should be noted that the new space composite structure has different types and shapes; it allows using the structure successfully in construction of different buildings or structures.

References

1. G. M. Gasii, Metallurgical and Mining Industry, **4**, 23–25, (2014)
2. L. I. Storozhenko, H. M. Hasii, Naukovyi Visnyk Natsionalnoho Hirnychoho Universytetu, **4**, 28–34, (2015)
3. L. I. Storozhenko, G. M. Gasii, Metallurgical and Mining Industry, **6**, 40–42, (2014)
4. G. M. Gasii, Bulletin of Dnipropetrovsk National University of Railway Transport, **6**, 158–165, (2016)
5. Yu. Vibranets, Yu. Ivanik, Visnik Lvivskogo natsionalnogo agrarnogo universitetu. Seriya: Arhitektura i silskogospodarske budivnitstvo, **16**, 88–99, (2015)
6. S. M. Krasnov, K. S. Krasnova, M. A. Borisenko, V. Yu. Romenskiy, Resursoekonomni materiali, konstruktsiyi, budivli ta sporudi, **25**, 555–563, (2013)
7. G. M. Gasii, Construction, materials science, mechanical engineering, **87**, 48–53, (2016)
8. D. J. Oehlers, M. A. Bradford, *Composite Steel and Concrete Structures: Fundamental Behavior*, Elsevier, (2013)
9. M. Kristo, M. Heinisuo, Engineering Structures, **79**, 354–364, (2014)
10. T. Tiainen, K. Mela, T. Jokinen, M. Heinisuo, High strength steel in tubular trusses, *Proceedings of the METNET Seminar 2013 in Lulea*, 56–59, (2013)
11. T. T. Lan, *Space Frame Structures*, CRC Press, Boca Raton, FL, (1999)
12. L. I. Storozhenko, G. M. Gasii, Resursoekonomni materials, structures, buildings: Coll. Science. pr., **31**, 511–516, (2015)
13. G. M. Gasii, Academic journal. Industrial Machine Building, Civil Engineering, **3**, 47–51, (2014)
14. G. M. Gasii, Academic Journal of Lviv Polytechnic National University. Series "Theory and Building practice", **844**, 260–265, (2016)

Analysis of influence of the geometric parameters of mast systems on the natural frequency and vibration mode spectrum

Vladimir Grinyov[1], *Tatyana* Alyoshechkina[1*], *Vitaliy* Vynogradov[1], and *Ekaterina* Perepelytsa[1]

[1]Kharkiv National University of Civil Engineering and Architecture, Structural Construction Mechanics 40 Symskaya Street, Kharkov, Ukraine, 61002

Abstract. Natural frequencies and eigenmodes of oscillations are the determining values in the study of dynamic processes in many constructions, including mast systems when they are affected by time-varying loads: wind, seismic, etc. [7]. The article suggests a method of forming, by means of a purposeful variation of the geometrical parameters of the mast, the spectrum of natural frequencies of bending vibrations. The basis is the necessary optimality conditions for one-dimensional models in the form of the maximum principle [1, 3, 8]. The analysis of the sensitivity of the spectrum [9] to small changes in geometric parameters that inevitably arise during the operation of the masts is carried out. An algorithm for using functional derivatives for finding optimal solutions is constructed.

All calculations in the paper are carried out using the example of a mobile phone mast $l = 60$m high with six leveled guy-rope attachment (the diagram is shown in Fig. 1).

The sections are joined together by flanges. The booms are represented by tubular elements having a fixed inner and outer diameter within a section. We will assume that the external diameter remains constant from section to section, and the internal diameter can vary according to the data given in Table 1.

Further, we will assume that variable function is the one determining the distributions of the total area of the booms along the height of the mast *F(z)*. In this case, we will replace the discrete character of variation by continuous variation at a certain stage, i.e. we assume that continuous *F(z)* varies subject to the following constraints

$$F_1 \le F(z) \le F_2, \tag{1}$$

where F_1, F_2 correspond to grades 1 and 7 in Table 1.

Array for the section shown in Fig. 2 consists of round bars with the diameter ∅1.4 cm. In this case, the array remains the same for the different sections.

* Corresponding author: atn4042@gmail.com

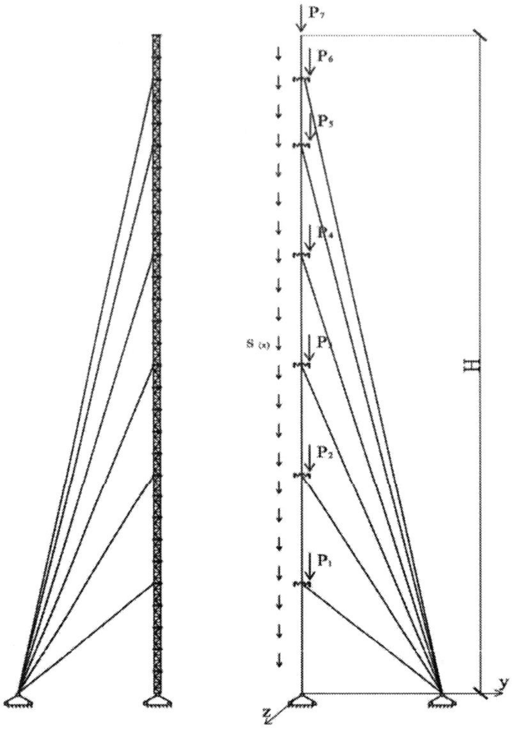

Fig. 1. General view of mast design diagram (beam model).

P1 – P7 – axial concentrated forces arising due to the effect of guy ropes and weight of the equipment on the mast;
s(x) – longitudinal load determined by the mast's own weight.

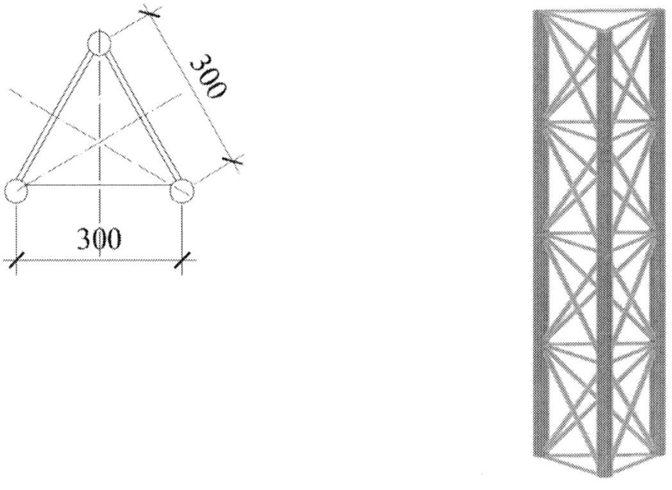

Fig. 2. General view of sections and cross-sectional diagram.

The moment of inertia of cross section during plane bend *zx* will be denoted $I(z)$, while the moment of inertia will be the function of the area within the constraints (1). It can be assumed from the geometry of cross section (Fig. 2) that $I=\alpha F$. Table 1 shows that we can take $\alpha=152.9$.

Table 1. Possible variants of booms' sections.

№	1	2	3	4	5	6	7
$\varnothing$, *cm*	4.8x0.2	4.8x0.22	4.8x0.25	4.8x0.28	4.8x0.3	4.8x0.32	4.8x0.35
F, cm^2	8.670	9.496	10.721	11.928	12.723	13.511	14.679
Iz, cm^4	1325.6	1452.5	1638.3	1822.4	1943.7	2063.8	2241.7

Modulus of elasticity $E = 2\cdot10^6\ \frac{\text{kg}}{\text{cm}^2}$, specific gravity $\gamma = 7.8\cdot10^{-3}\ \frac{\text{kg}}{\text{cm}^3}$, density of the material $\mu = 7.8\cdot10^{-6}\ \frac{\text{kg}\cdot\text{sek}}{\text{cm}^2}$ are the same for booms and arrays. Tie-rope (guy rope) is a regular-lay rope of type T with the diameter of $\varnothing$ 1.4 cm (GOST 3063-80). Rope pre-stressing force is the same for all levels (200 kg).

Boundary value problem for bending vibrations natural mode will be written in the form [2].

For longitudinal deformation differential equations are used

$$\frac{dw}{dx} = \frac{N}{EF};\quad \frac{dN}{dx} = -s; \tag{2}$$

with boundary conditions

$$w(0) = 0;\ N(l) = P_7; \tag{3}$$

and the conditions at the points of application of longitudinal concentrated forces generated by the tension of guy ropes,

$$w(z_i^-) = w(z_i^+);\quad N(z_i^-) + P_i = N(z_i^+);\quad i = \overline{1,6} \tag{4}$$

In these expressions z_i^-, z_i^+ - is the value of coordinate from below and from the top of the point z_i, and functions *w(z), N(z)* represent movements and axial forces. Distributed axial load has the form

$$s(x) = -\gamma(\text{F} + F_d). \tag{5}$$

The value F_d is obtained by dividing the total volume of the array by the length of the section. In our case it is F_d = 8.3 cm^2. Flexural deformations of the mast can be described by the system of differential equations

$$\frac{du}{dz} = \varphi;\quad \frac{d\varphi}{dz} = -\frac{M}{EI};\quad \frac{d\varphi}{dz} = -\frac{M}{EI};\quad \frac{dM}{dz} = Q - N\varphi;$$

$$\frac{dQ}{dz} = -p^2\mu(F + F_d)u; \tag{6}$$

with boundary conditions

$$u(0) = M(0) = 0;\quad M(H) = 0;\quad Q(H) = -p^2 m_H u(l), \tag{7}$$

and the conditions at the fixing points of guy ropes

$$u(z_i^-) = u(z_i^+);\quad \varphi(z_i^-) = \varphi(z_i^+);\quad M(z_i^-) = M(z_i^+);$$

$$\text{Q}(z_i^-) + \text{c}_i\, u(z_i) = Q(z_i^+);\quad i = \overline{1,6} \tag{8}$$

In conditions (7) value m_H represents the mass of equipment on the upper platform of the mast, p is the natural frequency. Rigidities c_i (mounting rigidities) are characterized by the parameters of the guy rope unit and were counted according to the ratios of [5].

For any function $F(x)$, subject to constraint (1), the boundary value problem (6), (7), (8) taking into account relations (2), (3), (4) determines the spectrum of natural frequencies $p_1 < p_2 < p_3 < \ldots$ and corresponding eigenfunctions $u_i(z)$ $(i=1,2,3,\ldots)$. Let the oscillation form $u(x)$ (or rather the set of functions $u(z)$, $\varphi(z)$, $M(z)$, $Q(z)$) belong to the set of eigenfunctions corresponding to the natural frequency p with a given number. The identification of the natural frequency number in case of simple (non-multiple) frequencies is carried out during their numerical determination. Let us pose the problem of finding the function $F(x)$ within the limits of constraints (1) that delivers the extreme value to the natural frequency p within the above mentioned variety of natural modes, that is, let consider the problem of minimizing the functional

$$J(F) = kp^2 = \min \tag{9}$$

For $k=+1(-1)$ this corresponds to minimization (maximization) of the natural frequency. The natural frequency can be represented using the Rayleigh formula [3],

$$p^2 = \int_0^{p+} \frac{M^2}{EI} dz \Big/ \int_0^{H} \mu (F + F_d) u^2 dz \tag{10}$$

The necessary optimality conditions in the form of the maximum principle [8] for the functional (9) taking into account the expression (10) are considered in detail in the monograph [3]. In view of the important property of self-adjointness, the necessary conditions will be represented by the boundary value problem (2)–(4), closed on the optimal $F(x)$ by the Hamiltonian maximum condition

$$H = k \cdot \left[\frac{M^2}{E\alpha F} + p^2 \mu (F + F_d) u^2 \right]. \tag{11}$$

Taking into account that when maximizing the Hamiltonian (11) values M, u are considered as parameters, we can obtain the formulas [3]

for $k=-1$

$$F = \begin{cases} F_2, & F_* \geq F_2; \\ F_*, & F_1 \leq F_* \leq F_2; \\ F_1, & F_* \leq F_1; \end{cases} \tag{12}$$

where $F_* = \dfrac{|M|}{p|u|\sqrt{\alpha \mu E}}$ is found from the factor $H(F)$ for $k=-1$ upward-convex function $\left(\dfrac{\partial^2 H}{\partial F^2} \leq 0 \right)$

for $k=1$

$$F = \begin{cases} F_1, & H(F_1) \geq H(F_2); \\ F_2, & H(F_1) < H(F_2). \end{cases} \tag{13}$$

It is the consequence of the fact that Hamiltonian $H(F)$ is a downward-convex function $\left(\frac{\partial^2 H}{\partial F^2} \geq 0\right)$ and the optimal solution consists of only sections of the boundaries (1).

The relations (12), (13) which close the original boundary value problem make it nonlinear. The approach described in [3], based on the successive approximation of the optimal solution on the sequence of varying functions, will be used in what follows. An important role is played by sensitivity analysis, which consists of finding the functional derivatives $\frac{\partial J}{\partial F}$. Using the results presented in [9], we can write

$$\frac{\partial J}{\partial F} = \frac{\partial H}{\partial F} = k\left[-\frac{M^2}{\alpha E F^2} + p^2 \mu u^2\right], \tag{14}$$

which determines sensitivity functions for the problem under consideration. Fig. 3 shows the vibration modes and sensitivity functions for the first and second eigenfrequencies in the case $F(x)=F_2$. For function $\delta F(z)$ that determine the necessary changes in the cross-sectional area of the mast within the constraints (1), changes of the natural frequency will be determined by the integral

$$\delta J = k\int_0^l \left(\frac{\partial H}{\partial F}\delta F\right)dz = k\int_0^l \left[-\frac{M^2}{\alpha E F^2} + p^2 \mu u^2\right]\delta F dz \tag{15}$$

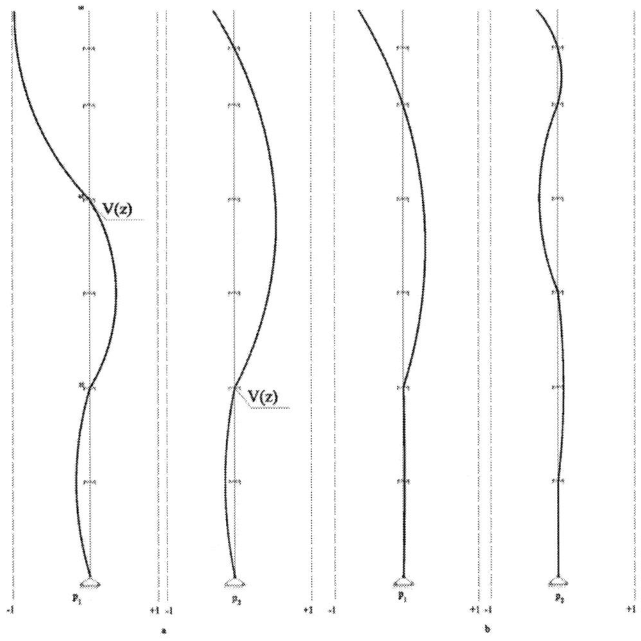

Fig.3 Vibration modes *a* and sensitivity functions *b* for first p_1 and second p_2 natural frequencies.

Variations of the area will be stronger in those sections where the magnitude of derivative is greater and where there is more sensitivity. It should be noted that for the case of maximizing the natural frequency, according to formula (12), the derivative (14) becomes zero in the sections of the optimal solution F^* within the interval $[F_1, F_2]$. For other laws $F(x)$ shows the trend of frequency variation.

The solution of the boundary value problem for the eigenvalue was carried out on the basis of the initial parameter method when using the main program modules from paper [2]. Laws of variation of cross-sectional area $F(z)$ were recalculated in accordance with the dependence

$$F_{i+1}(z) = F_i(z) + k\left[-\frac{M_i^2(z)}{\alpha EF_i^2(z)} + p^i \mu u_i^2(z)\right]\Delta F \qquad (16)$$

Moreover, the new approximation $F_{i+1}(z)$ must be subject to constraints (1). If at any point $F_{i+1}(z) > F_2(< F_1)$, $F_{i+1}(z) = F_2(= F_1)$ was accepted. Step size ΔF was constant based on the height of the mast; in particular, in the conducted calculations $\Delta F=(F_2 - F_1)/3$ was adopted. Figure 4 shows the closed curves in the volume coordinates V – natural frequencies p (for second and first natural frequency).

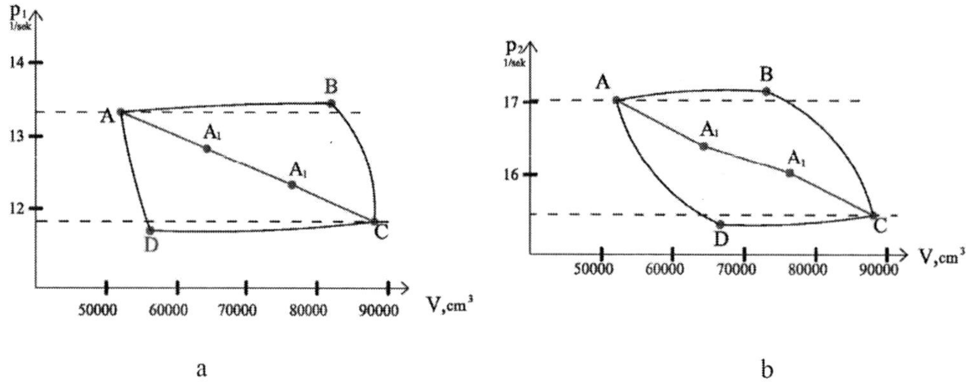

Fig. 4. Zones of possible change of the first (a) and second (b) natural frequencies.

Curves *AA'C* correspond to the frequency dependences on the volume for the masts with constant cross-section along the height. It is characteristic that for cross-sectional area (point A) which is minimal within the limits of constraints (1), there is a larger frequency value than for the maximum area value (point C). Due to the fact that natural frequencies are determined by the ratio of stiffness and inertia characteristics, it can be concluded from this ratio that the increase of the cross-sectional area increases the inertial characteristics to a greater extent. This circumstance occurs both for the first and second natural frequencies.

Figure 5 shows the distributions $F(x)$, corresponding to frequencies p_1^+, p_1^- and p_2^+, p_2^-.

Dashed dotted line indicates the piecewise approximation of the optimal solution in accordance with the discrete variation scheme given in Table 1. For specific z area value within the section is taken equal to the closest of the discrete sets indicated in the table. Such a discretization of optimal distributions $F(z)$ practically does not affect the values of the natural frequencies.

The above relations are generalized for the case of the hill-climbing problem for the natural frequency with a given number provided that the total volume of variable booms is fixed,

$$V = \int_0^l F(z)dz = const \qquad (17)$$

For such a task, the quality functional (9) is modified

$$J(F) = kp^2 + k_0 \int_0^l F dz \,, \tag{18}$$

where k_0 is the indefinite Lagrange multiplier to account for the condition of constancy of the volume (17). Hamiltonian in this case takes the form:

$$H = k \cdot \left[\frac{M^2}{E\alpha F} + p^2 \mu (F + F_d) u^2 \right] + k_0 F \tag{19}$$

For a given optimization problem which is designated as suggested in [3], via $V \to p$, the sensitivity function will have the form

$$\frac{\partial J}{\partial F} = k \left[-\frac{M^2}{\alpha E F^2} + p^2 \mu u^2 \right] + k_0 \,, \tag{20}$$

A similar construction is used for the dual problem $V \to p$ designated in accordance with [3] as the problem of finding *F(x)*, at delivers within the limits of constraints (1) the extreme value to the material volume at a fixed value of the corresponding natural frequency. The solution of these problems forms a closed curve in the coordinates *V, p*. The examples of such curves for the first and second natural frequencies of flexural vibrations are shown in Fig. 4. For any fixed volume $V \in [V_1, V_2]$ the upper sections of *ABC* within the framework of the problem $V \to p$ correspond to the maximum natural frequencies, and the lower sections of *ADC* – to the minimum natural frequencies. For a fixed value of the frequency $p \in [p^-, p^+]$ the left *DAB* sections correspond to the minimum volume values and the right *BCD* sections – to the maximum volume values.

The points of these closed curves were constructed by solving problems $V \to p$ based on the formulas analogous to (16):

$$F_{i+1}(z) = F_i(z) + \left[k \left[-\frac{M_i^2(z)}{\alpha E F_i^2(z)} + (p^i)^2 \mu u^2_i(z) \right] + \mathrm{k}_0 \right] \Delta F \tag{21}$$

Values k_0, k are normalized as $|k| \le 1$, $|k_0| \le 1$. This is achieved at each step by dividing them by $\max(|k|, |k_0|)$. The issue of fulfilling the conditions (17) at each step is solved by choosing the value k_0 in (21). It should be noted that for $p \to V$ such a problem is difficult to solve.

During normalization the value k in (21) changes and allows us to construct the necessary approximations to the optimal solution.

As can be seen from the nature of the closed curve, in *AB* section for the frequency p_1 the value of the maximum frequency depends little on the volume. The same property is possessed by the minimum frequency for the *DC* section. This takes place because of the peculiar relationship between the stiffness and inertia characteristics of this variant of design.

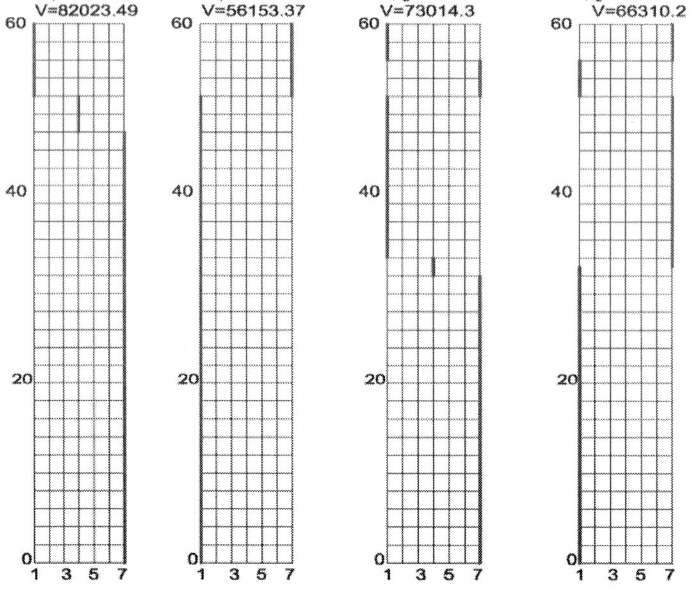

Fig. 5. Distribution of the cross-sectional area of the area for frequencies p_1^+, p_1^- and p_2^+, p_2^-.

Closed *ABCD* curves determine the reserves of this design from the point of view of possible changes in the first and second natural frequencies of bending vibrations and the volume (weight) of booms material.

In order to compare the results of calculations of natural frequencies with other computational models, the first and second natural frequencies of bending vibrations were calculated by the finite element method in the SCAD Office complex. When constructing the computational model, the scheme shown in Fig. 2 was used. The errors for the points of the curves AA_1 A_2C (booms area which is constant along the height of the mast) do not exceed 10%. It should be noted that consideration of the above type optimization problems based on finite element models is very difficult. It is not possible to formulate the necessary optimality conditions analogous to the maximum principle [8]. Optimization of variables based on discrete set is possible, but it entails significant computational difficulties.

Conclusions

The approach to the problem of controlling the natural frequency spectrum of bending vibrations of masts on guy ropes was proposed. The numerical control algorithm is based on the necessary optimality conditions in the form of the maximum principle for beam models. The option of varying the area of the booms along the height of the mast was proposed. The sensitivity analysis for the first and second natural frequencies was carried out and its use for constructing an effective computational process was shown. The possibilities of extending the developed approach to more general problems of spectrum management were shown.

References

1. V.B. Grinyov, Scientific Herald building, Kharkov, HDTUBA HOTV OMB, **62,** 88 – 96 (2011)

2. V.B. Grinev, T.N. Alyoshechkina, *Calculation of one-dimensional structural elements*, Kharkov, 177 (2013)
3. V.B. Grinev, A.P. Filipppov, *Optimization of rods from the spectrum of natural frequencies*, Kiev, (1979)
4. V.B. Grinev, Yu.N. Yarovoy, T.N. Alyoshechkina, V.V. Vinogradov, V.A. Levada, *Proceedings of the IV International Scientific and Technical Internet Conference*, Kharkov, 138-142 (2014)
5. A.Ya. Driving, *Stability of masts on braces*, Moscow, (1964)
6. A.V. Perelmuter. SCAD Office. Calculation of masts on the braces, Kiev, Publishing house SCAD SOFT, (2004)
7. A.V. Perelmuter, V.I. Slivker, *Calculation models of structures and the possibility of their analysis*, Moscow: Publishing house SCAD SOFT, (2011)
8. L.S. Pontryagin, V.G. Boltyanskii, R.V. Gamkrelidze, E.F. Mishchenko. *Mathematical theory of optimal processes*, Moscow, (1969)
9. E. Hog, K. Choi, V. Komkov *Analysis of sensitivity in the design of structures*, Moscow, (1988)

Rigidity of reinforced concrete structures in the presence of different cracks

Igor Iakovenko[1*], *Vladimyr* Kolchunov[1], and *Ianina* Lymar[2]

[1]National Aviation University, 03058, Cosmonaut Komarov Avenu, 1, Kyiv, Ukraine
[2]Limited Liability Company "V. Shimanovsky UkrRDISteelconstruction", 02660, Osvoboditelei Avenu, 1, Kyiv, Ukraine

Abstract. It is proposed a method for rigidity calculating of reinforced concrete structures in the presence of cracks, suitable for rod and flat-strained concrete composite structures. It is based on the operating conditions and includes a new, more complete classification of the various cracks, models of a special crack, the calculation of the two-console model; a special cantilever model to determine the parameters of the joint between the concrete; calculation model of the block with the working section at the beginning and end of the crack to determine the horizontal (vertical) projections of various cracks with the involvement of analytical relationships. They are based on the extremum of a function of many variables and Lagrange multipliers, as well as attracting level model of multi-level development of the various cracks, which allow to find the distance between the cracks and width of their disclosure, with considering the effect of discontinuities. This effect can greatly simplify the process of determining the rigidity of reinforced concrete structures (including composite ones), despite the complexity and diversity of the crack pattern.

1 Review of recent studies and publications

The task of constructing a method for calculating the rigidity of reinforced concrete structures (including composite ones) is already very complicated, because the basic hypothesis of the solid deformable body mechanics (the continuity hypothesis) is not applicable here – the continuity is disturbed by the presence of macrocracks, which greatly reduces their rigidity [1–3].

The analysis shows [1–3], that the solution of the indicated problem is highly disputable and multi-vector. Here, quite complex adjacent tasks of determining the pattern of formation, development and crack opening are included [4]. For example, the values of the crack width opening are measured with a microscope in experiments, and the calculation is carried out with the aid of a rather rough calculating apparatus. It is not necessary to expect the coincidence of experimental and theoretical data in which case. The distances between adjacent cracks are not a functional one, but a discrete parameter, and fracture levels, as shown by experimental studies, can be not one but several [1, 4, 5]. However, despite the polarity of opinions, points of their contact have been found. So, it is generally accepted

* Corresponding author: i2103@ukr.net

that the calculated dependencies should have an engineering-visual form, a clear physical meaning that allows them to be used more efficiently and creatively in engineering practice, which requires the solution of far from trivial problems of an applied nature.

On the other hand, it is hardly necessary to resort to categorical formulations, for example, with a complete denial of the use of integro-differential equations, which lead, as a rule, to engineering-immense results. This occurs only in the case of formalized approximate solutions of such equations. Here, reaching the level of their analytical solvability is important to be able to obtain very “elegant” solutions.

2 Basic material and results

It is proposed a new classification of basic cracks in reinforced concrete structures (including composite ones) on the basis of an in-depth study of the nature and nature of the crack formation [1, 5]. It is based on the geometric, force (deformation) and interspersed concentration of the stress-strain state, with the corresponding sources-concentrators.

The proposed calculation methodology is based on the following working assumptions:

1. The formation of the subsequent level of cracks occurs after the concrete fibers stretched by the fibers have been reached along the axis of the transverse (longitudinal) reinforcement of limiting deformations $\varepsilon_{bt,u}$. There may be several levels of cracking. The development of various cracks occurs along trajectories approaching straight lines in the reinforced concrete structures (including composite ones).

2. The opening of a crack is considered as the accumulation of relative mutual displacements of reinforcement and concrete in areas located on both sides of the fracture. The effect associated with violation of the continuity of concrete is taken into account [1] in this case.

3. Classification of basic cracks is introduced. In the reinforced concrete structures (including composite ones), the following basic cracks can occur: 1) cracks developing to zones or from zones of geometric concentration of stress-strain state (in places where cross-sectional dimensions change, input concentration angles, concentration from non-circular holes, etc.); 2) cracks developing to zones or from zones of concentration of force and deformation loading (the location of support reactions and concentrated forces, the place of change in the intensity of loading along the contour of the structure, the place of deformation loading from subsidence, the type of loading is of particular importance, bending, shear, compression-tension); 3) longitudinal cracks developing in zones of inter-medial strain concentration (in the seams between concrete in reinforced concrete composite structures, along longitudinal reinforcement in anchoring zones, etc.). In the building practice (for example, in complex engineering-geological conditions), the most frequent schemes are from force and deformation loading, which, as a rule, cause the imposition of various cracks [6].

4. The actual crack is replaced by the proposed model (Fig. 1) by means of the finite element (FE) casing in the PC "Lira-CAD" [7, 8] and the deformation action $\Delta=a_{crc,j}$ along the crack profile (the discontinuity effect [1] is introduced by the actual crack profile, and with the help of the formula for determining the crack opening width).

5. A special design two-element cantilever model is used to compute (Fig. 2). The sequential iterative analysis is performed with the help of the stress-strain state of flat cantilever elements. They are located along the fracture path are realized by means of a "shear" and deformation effect that takes into account the discontinuity effect of concrete [7, 8].

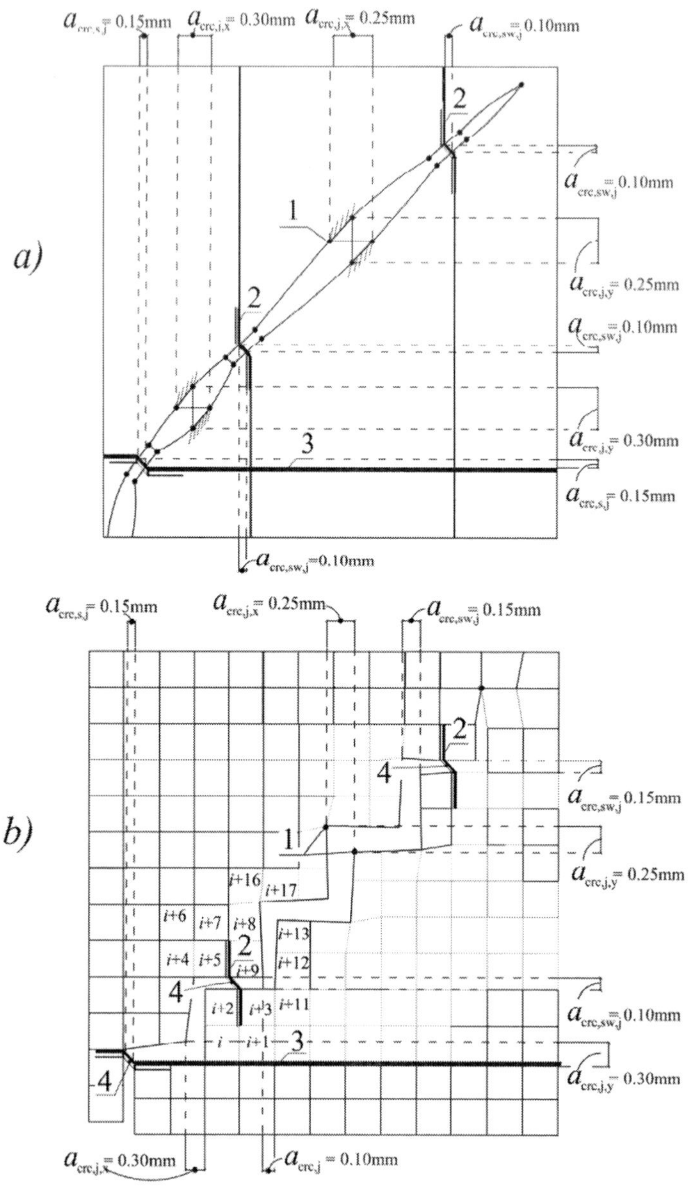

Fig. 1. The proposed model of a crack: *a* – real crack; *b* – the crack modeled by means of FE embroidering and deformation effect $\Delta = a_{crc,j}$; 1 – crack; 2 - transverse reinforcement and its simulation with 201 FE; 3 – longitudinal reinforcement and its simulation with 201 FE; 4 – possible closure of the fracture and its simulation using 255 FE

6. A special design two-element cantilever model is used to calculate the composite structures [9], which makes it possible to determine the parameters of the seam ($G(\lambda)$, $E(\lambda)$, $\mu(\lambda)$ in two rows of small size FE's adjacent to the seam) between the concrete).

It is based on the experiments of shearing the seam of prisms with different concretes, including reinforcing rods in the seam between concrete. At the same time, the capabilities of Lira-SAPR software are used to determine the stress-strain state of the FE (including those adjacent to the seam).

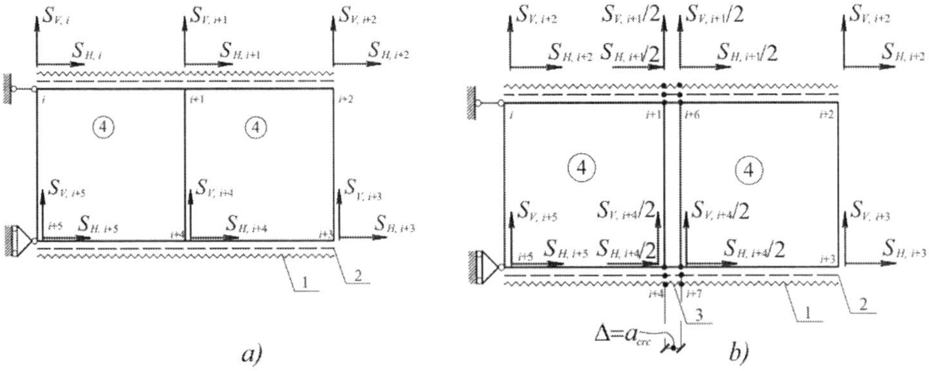

Fig. 2. The construction of a two-element model: *a* – the design scheme of a two-element model without embroidering; *b* – the design scheme of the two-element model after embroidering; 1 – 255 FE before embroidering; 2 – 201 FE; 3 – 255 FE after embroidering; 4 - 233 FE

7. The projections of various cracks on the horizontal (vertical) are sought on the basis of the block model with the calculated cross-sections I-I and II-II (Fig. 3), passing through

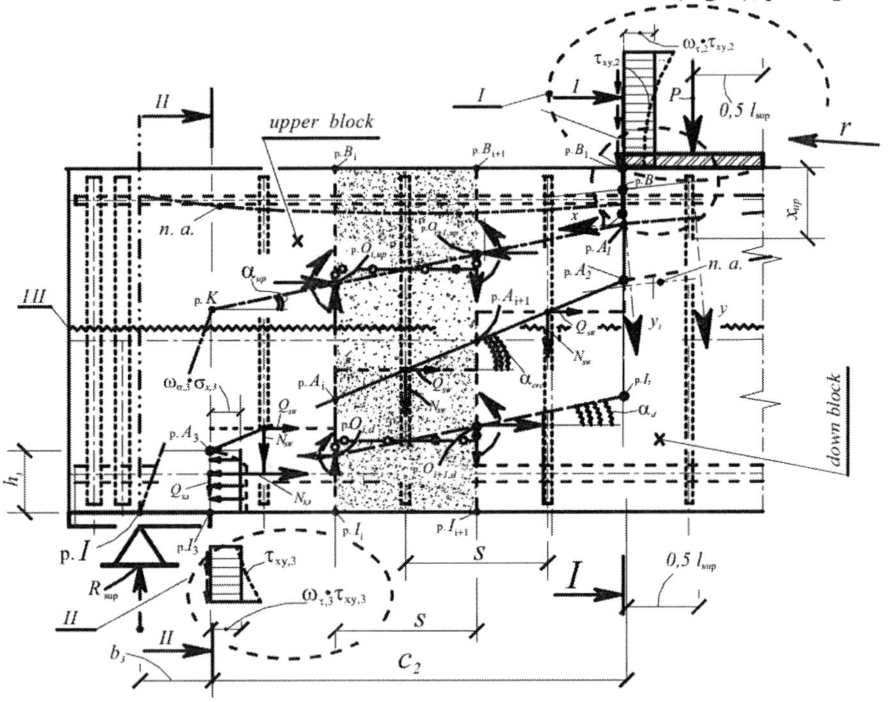

Fig. 3. The calculation scheme for blocks with cross sections I-I and II-II to determine the projection of an inclined crack in a plane-stressed reinforced concrete structure (including composite ones): I and II – diagrams of tangential stresses in sections I-I and II-II, respectively; III – the seam joint between concrete

the beginning and end of the crack (refined in the iteration process). One of these sections, as a rule, is attached to the largest force – the support reaction R_{sup} or goes to one of the faces of the structure with the use of analytic dependences [4], based on the extremum of the function of several variables and Lagrange multipliers.

8. A multi-level process of development of various cracks takes place in plane-stressed reinforced concrete structures (including composite ones) [4], which is determined by special design models (Fig. 4) of their level appearance. It allows finding the distances

between cracks and the width of their opening, taking into account the effect of discontinuity.

In this case,

$$\varepsilon_{\mathrm{bt}}(y)=\varepsilon_{\mathrm{sw}}\cdot E_{\mathrm{sw}}\cdot A_{\mathrm{sw}}\cdot\frac{1}{D_{13}}-\varepsilon_{\mathrm{sw}}(y)\cdot E_{\mathrm{sw}}\cdot A_{\mathrm{sw}}\cdot\frac{1}{D_{13}}+\frac{D_{14}}{D_{13}}\cdot y+\frac{D_{15}}{D_{13}}, \tag{1}$$

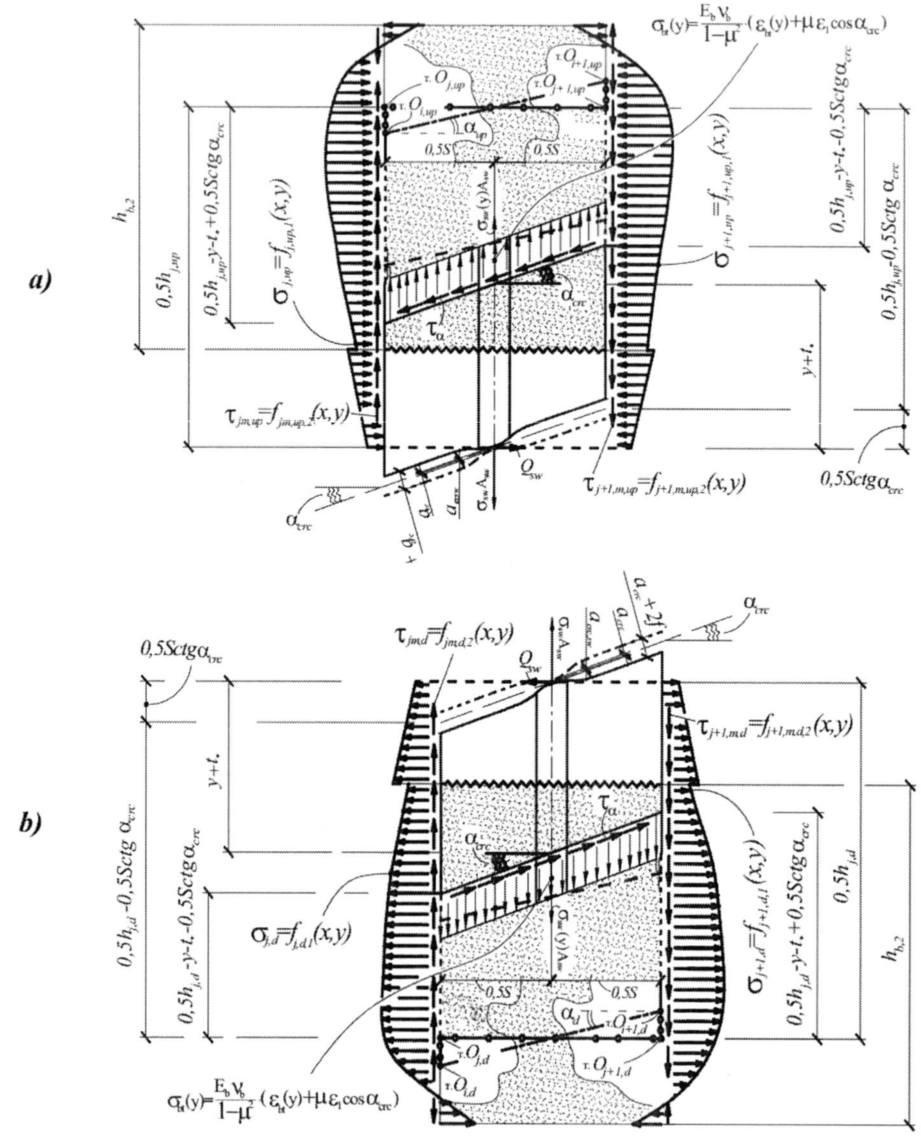

Fig. 4. The level model for determining the deformations of tension concrete $\varepsilon_{bt}(y)$ along the axis of transverse reinforcement of the *i*-th level of formation various cracks and the distance between them in a reinforced concrete composite structure: *a* – first case; *b* – second case

Careful analysis [4, 10, 11] of the pricked cracks and cracks adjacent to the concentrated load, makes possible to reveal a "fan" of inclined cracks adjacent to the concentrated load and to the support (Fig. 5).

The projection of an inclined crack, whose opening width has the maximum value, is determined from the condition of an extremum of functions $F_1=f(q_{sw}, x_B, \sigma_s, x, \sigma_b, \sigma_{s,1}, \sigma_{b,1}, C_2, \lambda_1, \lambda_2, \lambda_3, \lambda_4, \lambda_5, \lambda_6, \lambda_7)$ of several variables and $F_2 = f(q_{sw}, x_{B,2}, \sigma_{s,3}, C_2, \lambda_1, \lambda_2, \lambda_3)$, respectively, and the resulting conditions for the equality of partial derivatives to zero:

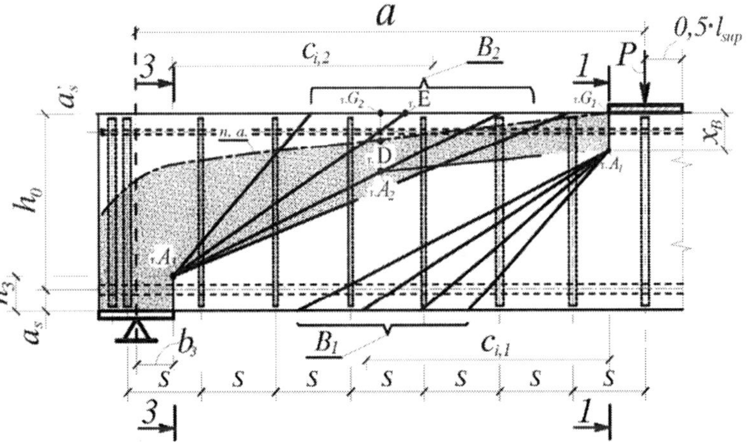

Fig. 5. To analyze the prismatic cracks and cracks adjacent to the concentrated load under flexural loading: B_1 – "fan" of inclined cracks adjacent to the concentrated load; B_2 – "fan" of possible inclined cracks adjacent to the support

$$\left.\begin{array}{l} \dfrac{\partial f}{\partial x_1} + \lambda_1 \dfrac{\partial \varphi_1}{\partial x_1} + \lambda_2 \dfrac{\partial \varphi_2}{\partial x_1} + \ldots + \lambda_m \dfrac{\partial \varphi_m}{\partial x_1} = 0 \\ \dfrac{\partial f}{\partial x_2} + \lambda_1 \dfrac{\partial \varphi_1}{\partial x_2} + \lambda_2 \dfrac{\partial \varphi_2}{\partial x_2} + \ldots + \lambda_m \dfrac{\partial \varphi_m}{\partial x_2} = 0 \\ \ldots\ldots\ldots\ldots\ldots\ldots\ldots\ldots\ldots\ldots \\ \dfrac{\partial f}{\partial x_n} + \lambda_1 \dfrac{\partial \varphi_1}{\partial x_n} + \lambda_2 \dfrac{\partial \varphi_2}{\partial x_n} + \ldots + \lambda_m \dfrac{\partial \varphi_m}{\partial x_n} = 0 \end{array}\right\} . \quad (2)$$

In this case, having stresses in the longitudinal reinforcement at the cross-section with a dangerous inclined crack (according to the criterion of the maximum opening width at the level of the longitudinal reinforcement axis), it is possible to determine the functional distance between different cracks along the axis of the longitudinal reinforcement, and also by arranging the stresses in the transverse reinforcement at the cross-section with a dangerous inclined crack (according to the criterion of the maximum opening width at the level of the axis of the transverse reinforcement) along the axes of the transverse reinforcement. As a rule, second-level design schemes are used (Fig. 4, 6).

Then the distance between cracks is calculated by the formula below:

$$l_{crc} = \frac{2(\ln B_4 - Bt_*)}{-B}, \quad (3)$$

where parametrs

$$B_4 = 1 + \frac{\sigma_{bt,c}}{(K-1)B_{3,*}\nu_b E_b} + \frac{\varepsilon_{bt,u}}{B_{3,*}(K-1)} . \quad (4)$$

$$B_3 = \varepsilon_s + \frac{\Delta T}{E_s A_s} - \frac{\sigma_{\mathrm{bt,c}}}{\nu_b E_b} - B_2\,; \qquad B_2 = \frac{\delta Q}{t\,B}, \tag{5}$$

B is the parameter of bond reinforcement with concrete, is taken equal to

$$B = \frac{S_S \cdot G}{K \cdot A_S \cdot E_S}\,; \tag{6}$$

$$\frac{1}{K} = 1 + 2\mu\alpha\,, \tag{7}$$

where

$$t = \frac{A_S E_b}{2\mu}, \qquad \alpha = \frac{E_S}{E_b}, \qquad \mu = \frac{A_S}{b \cdot h_0}. \tag{8}$$

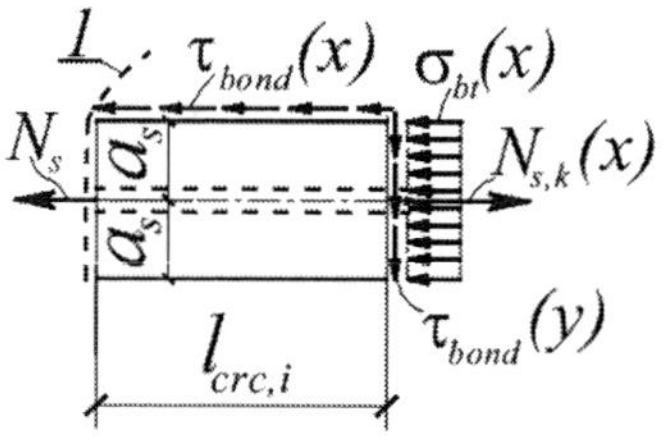

Fig. 6. The calculation scheme for determining the deformations of tensioned concrete $\varepsilon_{bt}(x)$ along the longitudinal reinforcement axis for the i-th level of the of inclined cracks formation of the first, second or third types in reinforced concrete structures (including composite)

It should be noted that the field of real numbers by the parameter following restrictions.

$$0 < B_4 < e^{B \cdot t_b}, \tag{9}$$

Comparing the functional and level values of l_{crc}, an analysis is performed of the possible realization of the appearance of subsequent fracture levels.

Analysis of formula (6) shows that an increase in deformations in the reinforcement with increasing load causes a decrease in the distance between the cracks. At the same time, the appearance of a new level of crack formation corresponds to the level of loading at which the following inequality is observed.

$$l_{\mathrm{crc,i}} \le \eta \cdot l_{\mathrm{crc,i-1}}, \tag{10}$$

where η is determined from the following relationships (Fig. 7) to the right of the dangerous inclined crack:

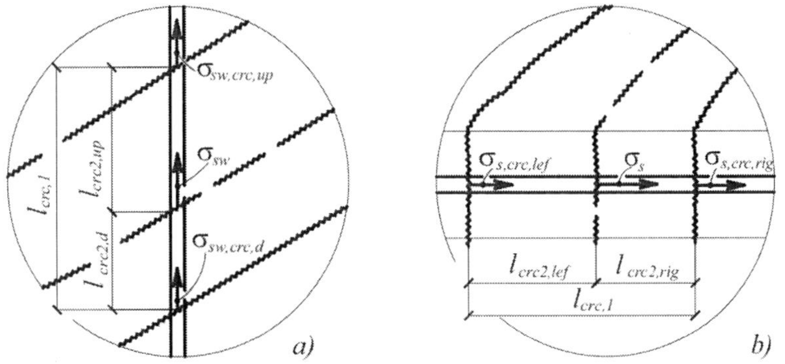

Fig. 7. The location of the adjacent crack of the next level between the two cracks of the previous level: *a* – along the axis of the transverse reinforcement; *b* – along the axis of the longitudinal reinforcement

$$\frac{\sigma_{sw,crc,d}}{\sigma_{sw,crc,up}} = \frac{l_{crc,2,up}}{l_{crc,2,d}} = \eta ; \tag{11}$$

$$l_{crc,2,up} + l_{crc,2,d} = l_{crc,1} . \tag{12}$$

Along the longitudinal reinforcement, such relations will take the form (Fig. 7, *b*):

$$\frac{\sigma_{sw,crc,rig}}{\sigma_{sw,crc,lef}} = \frac{l_{crc,2,lef}}{l_{crc,2,rig}} = \eta , \tag{13}$$

$$l_{crc,2,lef} + l_{crc,2,rig} = l_{crc,1} . \tag{14}$$

Thus, the cracking continues until the moment of destruction is started. In this case, not one is allocated (as is customary in a number of known techniques), but several levels of crack formation:

$$\left.\begin{array}{r} l_{crc} > l_{crc,1} - no \;\; cracks \\ l_{crc,1} \gg l_{crc} > l_{crc,2} \;\; - first \;\; level \\ l_{crc,2} \geq l_{crc} > l_{crc,3} - \sec ond \;\; level \\ \ldots \;\; \ldots \;\; \ldots \\ l_{crc} \geq 6 \cdot t_{*} - last \;\; level \end{array}\right\} . \tag{15}$$

Having a cracking level along the longitudinal and transverse reinforcement of a flat-stressed reinforced concrete structures (including composite), a “fan” of various cracks is built up adjacent to the concentrated force and to the support (Fig. 5). The degree of realization of cracks (whether these cracks intersect the lateral reinforcement, or it will be crossed only by a dangerous inclined crack) are determined from consideration of the stress-strain state along the clamps on the basis of the design scheme, the subsequent level shown in Fig. 4, which "closes" to a multi-iterative process, provided using Lira-CAD software.

The distance between the cracks is the most important parameter necessary for determining the crack opening width in reinforced concrete structures. From physical considerations, it is logical to use the level (discrete) value in the formula for determining the crack opening width in accordance with inequalities (10) and (14).

Having the stresses in the transverse reinforcement or stresses in the longitudinal reinforcement in the section with the y coordinate, it is possible to find the relative mutual displacements of the reinforcement and concrete:

$$\varepsilon_g(y) = \varepsilon_{sw}(y) - \varepsilon_{bt}(y), \tag{16}$$

where, using [1, 5]

$$\varepsilon_{sw}(y) = \varepsilon_{sw} + \frac{\Delta T}{E_{sw}A_{sw}} - \frac{S}{A_{sw}E_{sw}}\int_0^y \tau(y)dy. \tag{17}$$

Here S_s is the perimeter of the reinforcement cross-section; ε_s – deformation of the reinforcement in the crack; and ΔT – resulting conditional tangential stresses at local zone.

The nature of the diagram $\varepsilon_{sw}(y)$, performed with the involvement of the experiments of other authors, shows that, at a certain strain load, the deformations on the areas adjacent to the cracks begin to decrease and even change sign, while deformations in the middle of the section between the cracks continue to increase until at this point new crack. Analysis of the nature of the diagram $\varepsilon_{bt}(y)$ shows the necessity (Fig. 8) of taking into account the deformation effect in crack [1, 4].

Substituting $\varepsilon_{bt}(y)$ into the formula (16) and performing the corresponding algebraic transformations, obtain:

$$\varepsilon_g(y) = \left(\varepsilon_{sw} + \frac{\Delta T}{E_{sw}A_{sw}} - \frac{\sigma_{bt,c}}{\nu_b E_b} - \frac{D_{14}}{D_{13}\cdot B}\right)e^{-By} - \frac{D_{14}}{D_{13}\cdot B}. \tag{18}$$

Here $\frac{\sigma_{bt,c}}{\nu_b E_b}$ corresponds to the deformation of concrete $\varepsilon_{bt}(y)$ in a section located at a distance from t_* from the section with a crack (Fig. 8). In this case, the values $\sigma_{bt,c}$ are taken with the minus sign here and in all the below formulas.

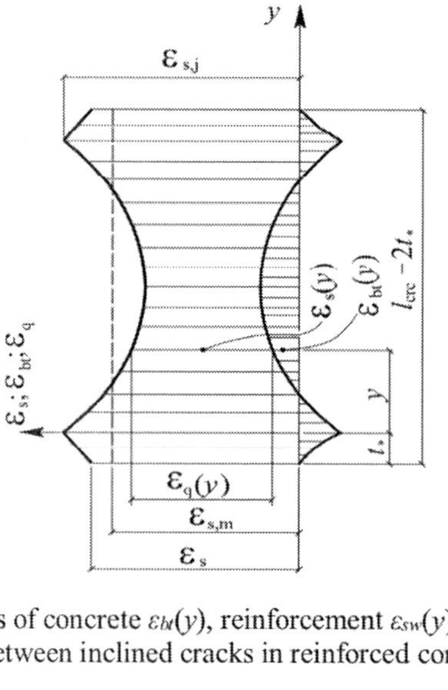

Fig. 8. Diagrams of deformations of concrete $\varepsilon_{bt}(y)$, reinforcement $\varepsilon_{sw}(y)$ and their relative mutual displacements $\varepsilon_g(y)$ in the area between inclined cracks in reinforced concrete structures

Thus, the problem by definition a_{crc} is reduced to finding the relative mutual displacements $\varepsilon_g(y)$ of reinforcement and concrete in different sections between the cracks:

$$a_{crc} = 2\int_0^{t_*} \varepsilon_g(y_1)dy_1 + \int_0^{\eta \cdot l_{crc}} \varepsilon_g(y)dy + \int_{\eta \cdot l_{crc}}^{l_{crc}} \varepsilon_g(y)dy. \tag{19}$$

After integration and some simplifications, we get:

$$a_{crc} = -\frac{2\Delta T}{G} - \frac{2B_{a2}}{B} - \frac{2B_2}{B}\ln\left(1 + \frac{B_{a2} \cdot A_{sw}E_{sw}}{q_{sw}S + B_{a,1}A_{sw}E_{sw}}\right). \tag{20}$$

The distance between the cracks (3), (10), (14), (15) is the most important parameter necessary for determining the resistance of tensioned concrete between cracks.
Integration ε_s within the distance between cracks, makes it possible to obtain the value of the coefficient Ψ_s of stretched concrete work account of according to the formula [12]:

$$\psi_s = \frac{2 \cdot K \cdot B_3}{\varepsilon_s \cdot l_{crc} \cdot B} \cdot \left[1 - e^{-B\cdot(0.5l_{crc} - t_*)}\right] + \frac{2}{\varepsilon_s \cdot l_{crc}}\left(\varepsilon_s + \frac{\Delta T}{E_s A_s} - K \cdot B_3\right) \cdot (0,5l_{crc} - t_*) +$$

$$+ \frac{\delta Q \cdot K}{\varepsilon_s \cdot l_{crc} \cdot B \cdot t_*}(0,5l_{crc} - t_*)^2 + \frac{2 \cdot \varepsilon_s \cdot E_s \cdot A_s + \Delta T}{\varepsilon_s \cdot l_{crc} \cdot E_s \cdot A_s} \cdot t_*. \tag{21}$$

Obtained parameters according to (3), (10), (14), (15), (20), (21) account the effect of a number of ***important factors***, such as deformation of reinforcement in the cross section with a crack, parameter of bond reinforcement with the concrete, geometrical sectional characteristics of concrete and reinforcement, deplanation on the surface of concrete structures with respect to the radius of the boundary layer, the deformation effect (appearing in reinforced concrete member after discontinuity) centered relative conditional mutual displacements concrete and reinforcement and, where necessary, the effect of transverse force.

3 Conclusions

On the basis of the proposed design model for the width of the opening of inclined cracks in reinforced concrete structures (including composite ones), it was also possible to refine the parameters of the stress-strain state of reinforced concrete structures in the presence of various cracks, taking into account the discontinuity effect [1]. This served as the basis for the development of a method for calculating the stiffness in reinforced concrete structures (including composite ones), which is reduced to the following algorithm, considered in two versions with the use of the PC Lira-CAD software complex.

The first variant is performed without changing the given order and the flat FE numbers. In FE, adjacent to implicit cracks, their thickness decreases. The operation of each pair of CE's is calculated twice using a two-element cantilever model (TECM) (Fig. 2): before embroidering the FE (W_1) and after embroidering the FE from the crack opening taking into account the effect of discontinuity (W_2) (the field reinforcement is replaced by two fibreboard FE). Efforts in the TECM nodes are determined from the nonlinear calculation of the entire plane-stressed structure. The displacement of the nodes is determined from the calculation of the TECM with the forces exerted in the nodes (Fig. 2). As a result, the new thickness of FE's adjacent to the crack is determined by the formula:

$$b_i = \frac{W_1}{W_i} \cdot b_1 . \tag{22}$$

The calculation algorithm assumes the presence of an iterative process, which is regulated by the accuracy achieved in the thickness of the specified FE.

In the second variant, the rigidity of reinforced concrete structures is determined using by special method of modeling the sheer cracks – gaps, when they are opened and closed Fig. 1, *b*, taking into account the effect of discontinuity and the incompatibility of deformations of reinforcement and concrete. Reinforcing rods are modeled by an additional 201 FE, and a possible closure of a crack, – concrete 255 FE.

Thus, a method is proposed for calculating the rigidity of reinforced concrete structures in the presence of cracks, suitable for rod and flat-concrete reinforced concrete structures (including composite ones), based on working assumptions on a new, most complete classification of various cracks; special model of cracks; two-element cantilever (console) model; the special console model for determining the parameters of the seam between concrete (in the case of a composite structure); block calculation model with working sections, passing through the beginning and end of the crack to determine the projections of various cracks to the horizontal (vertical) with the use of analytical dependencies based on the extremum of the function of many variables and Lagrange multipliers, and also involving level models of the multilevel development of various cracks that allow to find the distances between the cracks and the width of their opening, with the discontinuity effect. It allows to substantially simplify the process of determining the hard of reinforced concrete structures, despite the complexity and diversity of the pattern of cracking.

References

1. A.B. Golyshev, V.I. Kolchunov. *The resistance of reinforced concrete*, 432 p. (Osnova, Kyiv, 2009)
2. Li Yongzhen. *Final Report Predicting of the Stiffness of Cracked Reinforced Concrete Structures*, 147 p. (2010)
3. P. N. Saavedra, L. A. Cuitino. Crack detection and vibration behaviour of cracked beam, Computers and Structures, **79**, pp. 1451 – 1459 (2001)
4. H.Z. Bashirov, V.I. Kolchunov, V.S. Fedorov, I.A. Iakovenko. *Reinforced concrete structures of buildings and structures*, 248 p. (ASV, Moskow, 2017)
5. T.V. Tugay. The main results of experimental studies of flat-strained stiffness of reinforced concrete composite structures, *Resource-economical materials, structures, buildings*, **29**, pp. 369-375 (Rivne, 2014).
6. A.B. Golyshev, V.I. Kolchunov, I.A. Iakovenko. *Resistance of reinforced concrete structures, buildings and structures erected in complex engineering-geological conditions*, 371 p. (Talcom, Kyiv, 2015)
7. V.I. Kolchunov, N.G. Maryenkov, T.V. Tugai, I.A. Iakovenko. Resistance of plane-stressed walls to a skew in the presence of a diagonal crack with the use of the Lira-Pro software complex. *Academic journal. Industrial Machine Building, Civil Engineering Poltava National Technical Yuri Kondratyuk University*, **1 (43)** (Poltava, 2015)
8. V.I. Kolchunov, T.V. Tugai, I.A. Iakovenko, N.G. Maryenkov. Calculation of the stiffness of plane-stressed walls with the use of PC "Lira-Pro" by the method of new norms. *City and territorial planning*, **53**, pp. 209-221.
9. V.I. Kolchunov, N.G. Maryenkov, T.V. Tugai, I.A. Iakovenko. Method for determination of stiffness of flat-stressed and rod-shaped reinforced concrete composite structures under seismic influences // Industrial and civil construction, **№2**, pp. 12-15 (Moscow, 2014)

10. P. Rosko and A. Bekw. Reinforced concrete shear wall: structural element – finite element. Vienna University of Technology. Center of Mechanics and Structural, Dynamics Institute for Building Construction and Technology, 7p (Vienna, 2014).
11. Yo. Yamamoto, H. Nakamura, I. Kuroda and N. Furuya. Simulation of crack propagation in rc shear wall using a 3d rigid-body-spring model with random geometry. *VIII International Conference on Fracture Mechanics of Concrete and Concrete Structures*, 12 p. (Toledo – Spain, 2013)
12. I.A. Iakovenko, G.K. Bidzhosyan. New dependencies for estimating the resistance of stretched concrete between cracks in compound eccentrically compressed reinforced concrete structures. *Bulleten of Pridniprovsk State Academy of Construction and Architecture*, **9**, pp. 9-15 (Dnipropetrovsk, 2012).

Use of information technologies for energetic portrait construction of cylindrical reinforced concrete shells

Oleg Kalmykov[1], *Ludmila* Gaponova[1], *Petro* Reznik[1*], *and Sergey* Grebenchuk[1]

[1]O.M.Beketov National University of Urban Economy in Kharkiv, Department of Building Constructions, 61002, Kharkiv, Marshal Bazhanov str., 17 Ukraine

Abstract. The analysis of strain-stress state of new type of architectural and construction system 'Monofant' was examined. The analysis of the advanced graphic, computing software packages was carried out. The possibility of joint applying these packages to the problem of sustainable conjunction study of a rational combination of the geometric parameters of the design-built system "Monofant" was analyzed. From the constructive point of view, the search for the structural element shape that provides a minimum material consumption under desired conditions is of some interest. The approach based on the energy criterion of rationalization was adopted to solve this problem. Fundamentally, new opportunities in the field of building structures optimization are offered with introduction of visual programming complexes adopted for designers (Grasshopper, Dynamo). Applying the described approach to the problem of rationalizing of the constructive system "Monofant" offers the opportunity of constructing, calculating, analyzing and rationalizing of construction that has complex external and internal geometry. An illustration of a possible approach is given in a specific numerical example.

Introduction

The development of modern approaches to building construction design leads to the introduction of new systems that differ from the known solutions, generally, by more sustainable use of materials. Thus, as part of the research of the Building Structures Department of O.M.Beketov Kharkiv National University of Urban Economy under the guidance of prof. V.S. Shmukler, the design-built system "Monofant" (abbreviation "monolithic fantasy") was developed. The system was created with the comprehensive approach: technological solutions were adapted; sustainable materials and their contents were determined; and the calculation methodology was elaboarted [1]. The main idea of this system is creation of load-carrying elements made of monolithic reinforced concrete with an predefined unrestricted external shape and a complex intrinsic geometry that is provided by volume inserts storage in the reinforced concrete element body (usually from foam

*Corresponding author: engipr@gmail.com

polystyrene). It has been noted that this development provides concrete savings of up to 40% and was implemented in a number of Kharkiv facilities.

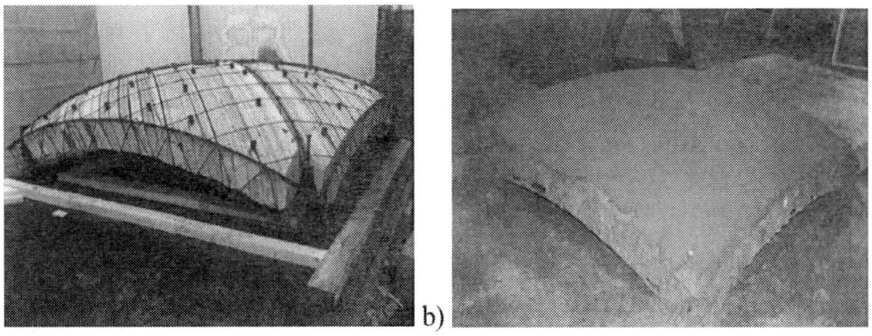

a) b)

Fig.1. General view of the shell: a) prefabricated reinforcement cage with non-removable insets; b) end product.

Relevance of the issue

From the constructive point of view, the search for the structural element shape that provides a minimum material consumption under desired conditions is of some interest [2]. The approach based on the energy criterion of rationalization was adopted to solve this problem. It is considered that of all geometric parameters correlations describing the external geometry of the system, the potential energy of deformation (PED) will adopt the extreme value at their rational combination. The implementation of the approach includes determination of interrelations between the energy magnitude of the system deformation and one or more geometric parameters. Taking into account the complexity of the analyzed system configurations, forming up of such an analytical interrelation is practically impossible. This problem can be expressed numerically with the use of computing software packages. However, the operation presumes the construction of a significant number of models, that to some extent complicates the process [3, 4]. At the same time, when it comes to constructing a finite-element model of a curvilinear shell-type system, which also has internal cavities, it is insufficient to have an interface and a graphical editor of computer complexes.

In the related research fields, for example, mechanical engineering, a similar problem occurred when the first computer complexes appeared based on the finite elements method. However, today, the analysis of the strength characteristics of parts and mechanisms is performed by importing complex geometric shapes from graphic software complexes (Inventor, SolidWorks) into calculated (Ansys, Robot) without problems [5,6]. Along with this, the rapidly growing progress in the information technologies offers fundamentally new opportunities in calculation, analysis, and, consequently, building structures optimization. The solution of a specific layer of the optimization problem is impossible without the use of specialized software, which, in its turn, is developed and upgraded taking into account the requirements and requests based on the result of development of the construction design.

Fundamentally, new opportunities in the field of building structures optimization are offered with introduction of visual programming complexes adopted for designers (Grasshopper, Dynamo). An example of this can be the article [7], where the problem of rational topology determining of the tubular bridge shell was solved using the approach of interrelation of a number of software complexes. In the work [8], the principles of the design processes optimizing, based on the interrelation of computer and graphic software packages are demonstrated using the example of trussed systems.

Objective of the study

Applying the described approach to the problem of rationalizing of the constructive system "Monofant" offers the opportunity of constructing, calculating, analyzing and rationalizing of construction that has complex external and internal geometry. As an example, a simplified analysis of the external parameters of a cylindrical reinforced concrete shell with internal cavities will be performed (Fig.2). Two types of shells will be considered: long (B/L =2) and short (B/L=0.5). The variable parameter is the lifting arrow (X [0; 9]). All other parameters will be accepted as constant (Table 1).

Table 1. The adopted geometric parameters of the cylindrical shell.

Constant parameters	
band thickness	50 mm
fin thickness	150 mm
fin pitch longitudinally	1000 mm
fin pitch transversally	1000 mm
contour interval	300 mm
span, L	18000 mm
width, B (short shell)	9000 mm
width, B (long shell)	36000 mm

Variable parameters		
	x	R
1	0	∞
2	1125	36562.5
3	2250	19125
4	3375	13687.5
5	4500	11250
6	5625	10012.5
7	6750	9375
8	9000	9000

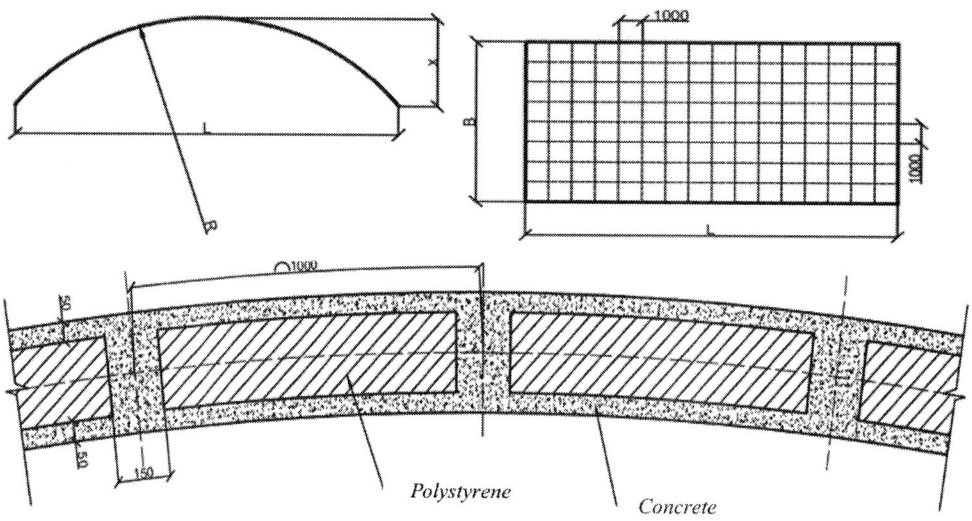

Fig. 2. Diagram of the analyzed shell.

Initially, the algorithm is written in Dynamo software package (Fig. 3). The written script allows to create automatically a 3D model of the shell by indicating its geometric parameters. Further, the script is loaded into Inventor software package (Fig. 4,a). Nastran software package is installed as an adjustment to the software package Inventor. Choosing the appropriate tab, it becomes possible to convert the created 3D model into a calculation scheme (finite element model, Fig. 4,b). Rigidity, support and load are assigned. In this case:

a load of 200 kg/m^2 is applied to the upper surface, the dead load is not taken into account; the rigidity is assigned automatically by setting the class of concrete (C25/30); the buttress is taken along the longitudinal ribs on one side hinged-fixed, on the other hand they are movable. After performing the described procedure, a calculation is performed, the result of which is the density distribution field of the electrical submersible motor, as well as the total value of the electrical submersible motor of the system (Fig.5). The described procedure is cyclic. On each subsequent cycle, the shell lifting arrow changes.

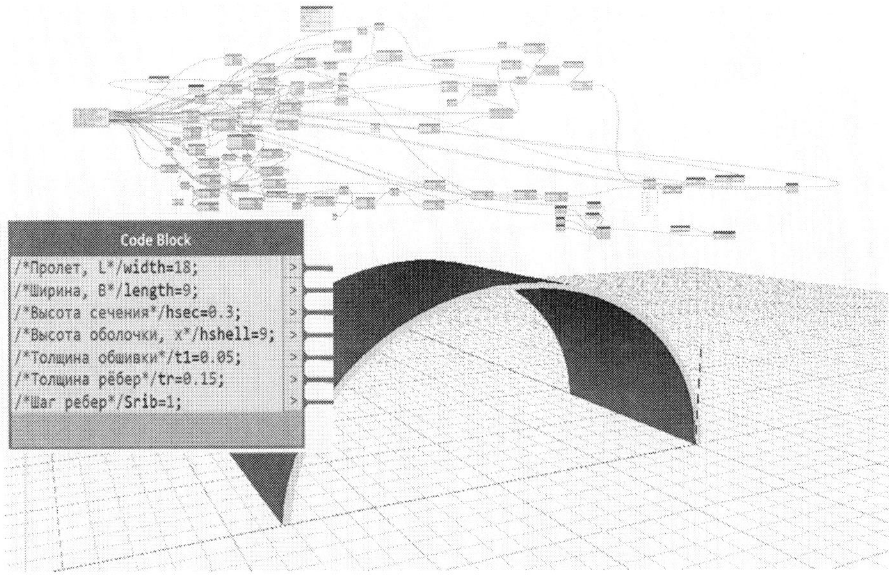

Fig. 3. Dynamo software package script for the shell topology forming.

a) b)

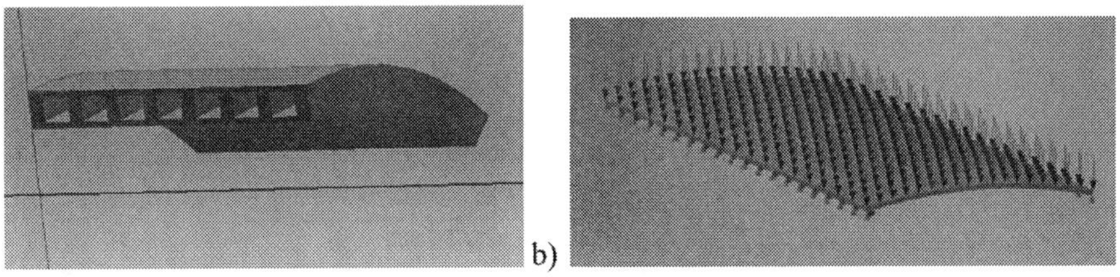

Fig. 4. General view of the shell topology in the software package: a) Inventor; b) Nastran [12].

a) b)

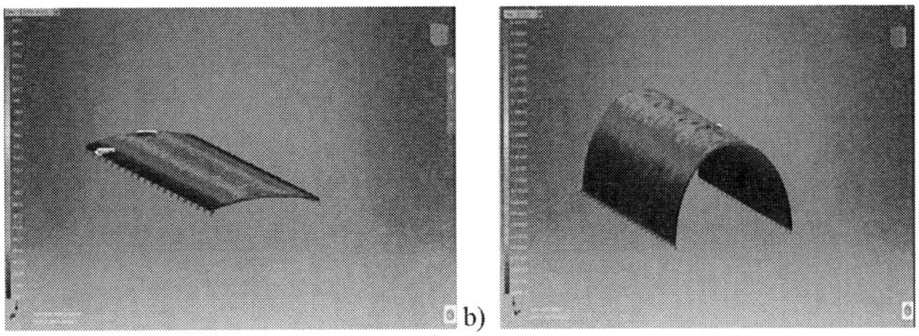

Fig. 5. Field of energy deformation distribution: a) X=1.125m b) X=9m.

 DOI: 10.1051/matecconf/201711602017

As a result, the values of the electrical submersible motor for the shells were determined. Also, during the calculation, the volume of each shell was recorded (Table 2). The ratio of the structure volume to the electrical submersible motor, in this case, determines the degree of the system material tension, its dependence on the lift arrows is shown in Fig. 6.

Table 2. Results of calculation.

X, m	*B/L*=0.5			*B/L*=2		
	Energy *U*, relative unit	Volume *V*, m^3	*U/V*	Energy *U*, relative unit.	Volume *V*, m^3	*U/V*
0	1.171204	25.7	4.557214	4.569075	101.5	4.501552
1.125	1.17504	25.9	4.536834	4.58597	102.3	4.482864
2.25	1.212534	26.9	4.507561	4.745556	106.3	4.464305
3.375	1.281871	28.1	4.561819	5.027976	110.8	4.537884
4.5	1.373526	29.9	4.593732	5.406874	118.1	4.578217
5.625	1.4562	31.8	4.579245	5.729	125.4	4.568581
6.75	1.54325	34.3	4.499271	6.0452	135.5	4.461402
9	1.631011	40.2	4.057241	6.416271	158.7	4.043019

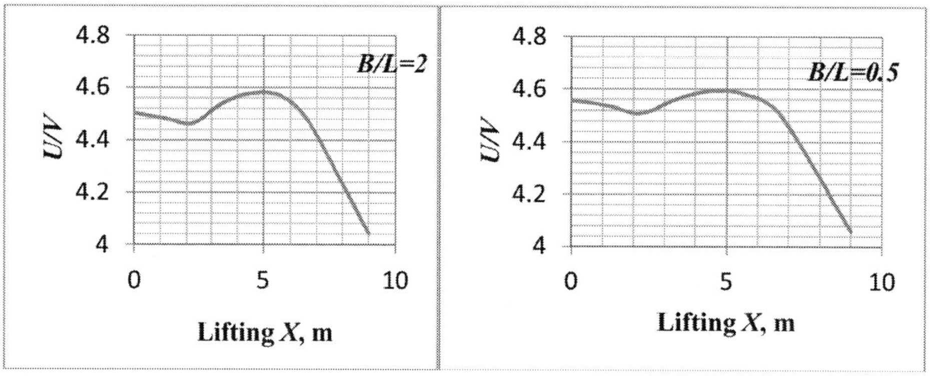

Fig.6. Dependence of the shell lifting arrow on the ratio of the system deformation energy to the volume.

From these diagrams it follows that for the considered standard sizes of cylindrical reinforced concrete shells, the function acquires an extreme value of the corresponding lifting arrow $X\approx 5$m. In this formulation of the problem, denoted value is considered as sustainable.

Conclusion

Application of the corresponding approach to the problem of finding sustainable shapes of the construction system "Monofant" offers significant opportunities in this field. It is noted that the described formulation shows the possibility of solving similar problems, once the external parameter is considered. The search for sustainable combinations of internal parameters of similar systems is of certain interest in further research.

References

1. V. Shmukler, V. Babayev, Bugayevskiy S., Berezhna K., Kariakin I., Kondrashchenko V., Seirski I. *Framework Building "MONOFANT"* 89464 Ukraine

2. Shmukler V., Klymov Y., Burak M. Frameworks of Hollow Core Types. (Kharkiv: Golden Pages, 2008)
3. Li Xue-ping, Zhao Lian-yu, Liu Zheng-zhong, *MATEC Web of Conferences* 95 (2017)
4. E. Carrera. Arch. Comput. Meth. Engng,. 9**(2)** 87-140, (2002)
5. M. Yaghoubi, S.S. Mohtasebi, A. Jafari, H. Khaleghi. Design and manufacture of a new and simple mechanism for transmission of power between crossover shafts up to 135 degrees for farm. (XVIIth World Congress of the International Commission of Agricultural and Biosystems Engineering Hosted by the Canadian Society for Bioengineering Québec City, Canada, 2010)
6. Bankole I. Oladapo, Balogun A. Vincent, Augustine O. Oke, Esoso A. Agbor. Design And Finite Element Analysis On Car Seat Height Screw Adjuster Using Autodesk Inventor (Dept. Of Mechanical & Mechatronics Eng,Afe Babalola University, Ado-Ekiti, Nigeria Volume 2 Issue 8, August 2015)
7. Mathis Chlosta Master's thesis. "*Feasibility study on fiber reinforced polymer cylindrical truss bridges for heave trafic*" Netherlands (2012)
8. Mehdi Nourbakhsh Generalizable surrogate models for the improved early stage exploration of structural design alternatives in building construction. (School of Building *Construction Georgia Institute of Technology* 2016).

Study of fire-resistance of reinforced concrete slab of a new type

Oleg Kalmykov[1], *Ludmila* Gaponova[1], *Petro* Reznik[1*], *and Sergey* Grebenchuk

[1]O.M.Beketov National University of Urban Economy in Kharkiv, Department of Building Constructions, 61002, Kharkiv, Marshal Bazhanov str., 17 Ukraine

Abstract. Reinforced concrete structures with complex inner geometry under the effect of high temperatures considering void former materials were examined. The analysis of strain-stress state of new type of architectural and construction system 'Monofant' under the effect of high temperature heating in standard fire mode, considering the change of design pattern was carried out. Numerical study of concrete slab with given reinforcement and complex inner geometry was carried out with use of software packages based on finite element method. Temperature fields throughout the depth of cross section of the slab of new type of architectural and construction system 'Monofant' upon heating in standard fire mode for time interval 0-240 min. were obtained. The carrying capacity of sections exposed to high temperatures was determined by deformation method. Offered the algorithm that considers the transformation of design patterns depending on temperature values and excessive pressure in thermal insulation cavities taking into account influence of deformation fields on temperature distribution.

1 Introduction

The requirements to constructed structures increase continually in the course of development of modern construction technologies that provides for the use of pre-cast and cast-in-place concrete. These requirements include the following: creation of new architectural forms for construction of different civil buildings, lightening of applied structures and improvement of traditional concreting technologies due to use of new materials. One of the systems that meets modern requirements to reinforced concrete structures is the architectural and construction system 'Monofant' [1] (Fig. 1).

The design of the slab is characterized by simple external and complex internal geometry. The main feature of this design is concealing liners of the certain shape made of light and widely used materials, such as polystyrene foam and mineral wool, in slab body.

The efficiency of this structure is also conditioned by section symmetry that determines the ability for random supporting. This results in variableness of bending moment field of cross-section.

[*]Corresponding author: engipr@gmail.com

2 Research Significance

Small number of articles dealing with reinforced concrete structures with complex inner geometry under the effect of high temperatures considering void former materials give rise to the need in either field or numerical studies. The purpose of these studies should be not only determining actual temperature field in structural sections in time intervals that are established by normative documents but the study of the transformation of design diagrams (geometrical nonlinearity) due to the effect of high temperatures and development of general approach for managing these factors. Architectural and construction system 'Monofant' was selected as a study object.

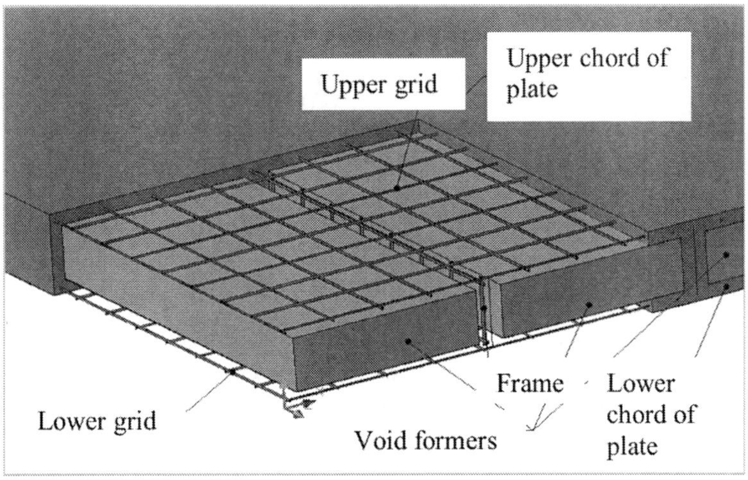

Fig. 1. Design features of the slab of structural system 'Monofant'.

3 Proposed algorithm

Since the inner design/anisotropic structure of slab elements consists of different materials, it is characterized by the increase of non-uniform thermal fluxes. Theoretically, such non-uniformity may be the course of reciprocal influence of temperature on the deformation and vice versa. Therefore, it is reasonable to assess this interaction based on the philosophy of dynamic theory of thermoelasticity [2, 3].

This theory is based on V. Novatsky's assumption that the deformation of a solid body relates to the change of the heat that such body contains and, therefore, the change of temperature distribution of envelope building. Time-dependent temperature field causes the change of deformation field and vice versa. Inner energy of the structure depends on deformations and temperature.

It should be noted that the effect of deformation field on temperature distribution in a solid body is significant except for acute (jump-in) change of temperatures [2, 3]. Nonetheless, considering nonhomogeneous internal structure of formed structure and lack of knowledge about the processes that may take place in normal and, especially, abnormal conditions, it is reasonable to waive conventional suppositions and admissions that prove the separation of equations of Voigt- Jeffreys and Lame.

It is supposed to model 3 design patterns [4] that consider fluctuations of liner - void former (Fig. 2).

Then solution algorithm will include the following operations:

-Task of temperature field T that should meet initial conditions;

- Determination by displacement vector of finite elements method $\{\vec{u}\}^0$;

- Substitution of determined vector $\{\vec{u}\}^0$ and determination of new field T;
- Substitution of temperature field T and determination of the ne displacement vector $\{\vec{u}\}^n$;
- Comparison of original vector $\{\vec{u}\}^0$ and new displacement vector $\{\vec{u}\}^n$;

$$|u_k^0| - |u_k^n| \leq \varepsilon \quad (1)$$

k - 1,2.........N,
N is a number of reference points;
ε is required accuracy.

In case of non-fulfilment (1) the procedure stops. Otherwise, $\{u\}^0 = \{u\}^n$ is assigned and the entire procedure repeats ones again until fulfilment (1).

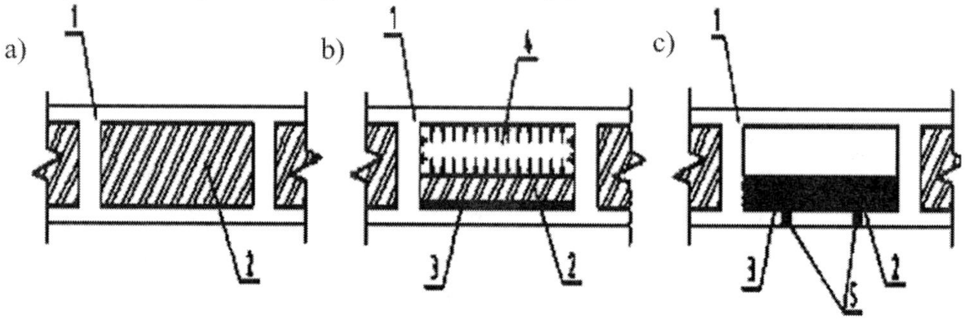

Fig. 2. Design patterns that consider fluctuations liner - void former: a) for $T \leq T^*$ (T^*- is fusion temperature); b) for temperatures $T > T^*$ (before crack formation); c) for temperatures $T > T^*$ (after crack formation) 1 is reinforced concrete housing; 2 is liner - void former; 3 is liquid substation; 4 is gaseous substation with pressure P; 5 is cracks.

4 Numerical study

4.1 The structure of numerical study

Square slab with border size of 4 m made of heavy concrete class C25/30 was chosen as an object of numerical study. Operating temperature in each areas: bottom - 1∅14A400C and top - 1∅8A400C. Depth of cross section of the slab is 300 mm. Arris pitch is 1 m and arris width is 150 mm. The thickness of outer and inner trimming is 150 mm. Total vertical load, including dead weight, is 10 kN/m^2. The procedure of numerical study includes the following stages:

1. Determination of temperature field in floor slab caused by high temperatures during bottom-up heating in standard fire mode;
2. Analysis in finite-element setting of slab flat sections in different time intervals with creation of vertical loads, applied temperature and excessive pressure due to melting of polystyrene foam provided that such event takes place in considered moment of time to determine transformation moment of design pattern;
3. Determination of effective bending thickness of floor slab and determination of resulting forces caused by the temperature in set moment of time and vertical load;
4. Determination of carrying capacity of the cross section of the slam that is subject to load by deformation method DSTU 2.6-156:2010 [5] considering requirements of EN 1992-1-2 [6] and comparison with resulting forces.

FEM-based software packages were used for creation of numerical study models. The configuration and sizes of the elements in these models are established on the grounds of test tasks by checking the convergence of finite elements method and evaluation of the accuracy of obtained results.

4.2 Temperature fields

Temperature fields on cross section of floor slab upon one-sided heating in standard fire mode were obtained by means of Ansys software package [7] at volumetric FE model for heating time from 0 min. to 240 min. Floor slabs of dwelling and administrative buildings are required to be fire-resistant R60. Therefore, temperature fields for the period of time of not less than 60 min. (Fig. 3) were used for further analysis.

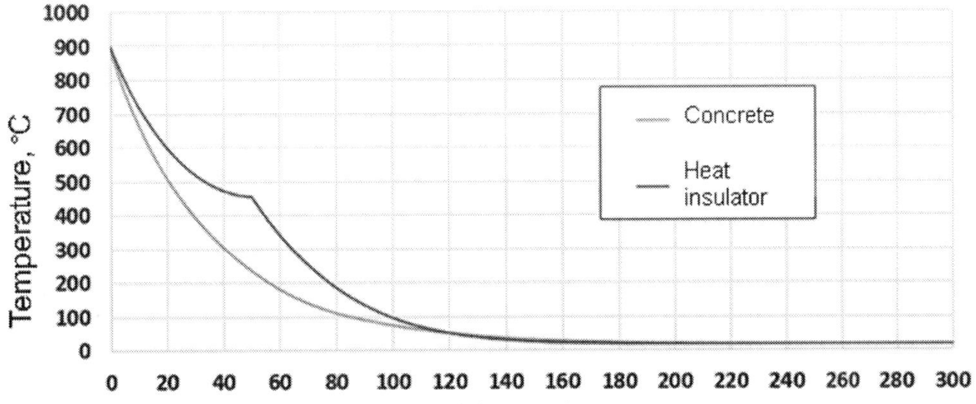

Fig. 3. Temperature distribution along the depth of cross section of floor slab in 60 min. upon heating in the standard fire mode.

4.3 Excessive pressure upon heating

To prevent the effects of excessive pressure at the insulation bottom upon heating, we have created 12 plane models, FE hull of zero gauss's curvature, by means of Lira-SAPR software package [8]. Each model reflected the cross section of floor slab in definite period of time (6-50 min., in increment of 5 min.). The temperature was set as a load in each FE model according to previously received temperature fields. Moreover, full vertical load was applied to all models at those moments of time when insulation temperature reached melting values [2]. Excessive pressure caused by insulation melting was applied. Calculation results have shown that excessive pressure in slab cavity was so high that the tensile strength for a concrete C25/30 reached threshold values after 30 min. (f_{ctm}=2.6 MPa) (figure 4).

4.4 Effective thickness of floor slab

To obtain final forces from joint effect of vertical loads and temperature, we also considered FE model of the slab made of hulls of zero Gaussian curvature. Effective thickness of the slab was determined by the formula (2). Effective thickness was 27.4 cm before the effect of excessive pressure caused by insulation melting and it was 21.5 after damage of the integrity of bottom trimming.

$$\frac{b_{eff}\Delta}{12} = I_x \Rightarrow \Delta = \sqrt[3]{\frac{12 I_x}{b_{eff}}}, \quad (2)$$

where
b_{eff} is an effective thickness of profile section;
I_x is preliminary calculated second area moment
Δ is an effective thickness of the slab.

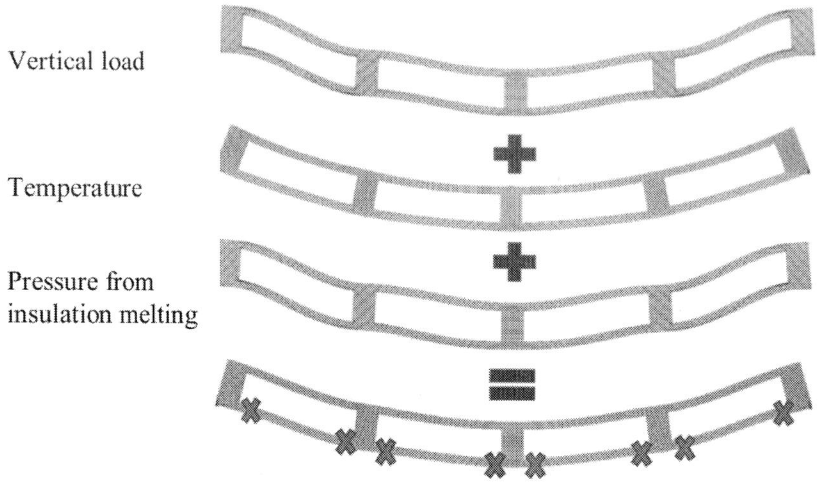

Fig. 4. Feature of deformation of cross section of floor slab after 30 min. of standard fire.

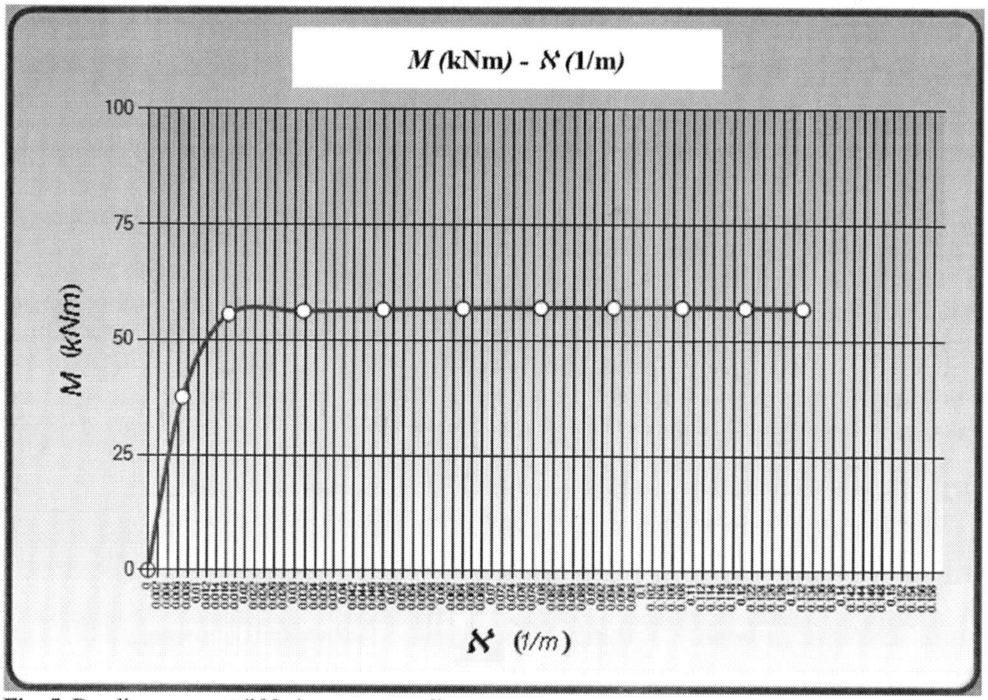

Fig. 5. Bending moment (kNm) – curvature diagram.

4.5 Carrying capacity of the section based on deformation method

The carrying capacity of sections exposed to high temperatures was determined by deformation method described in [5]. The temperature effect was traced by the method of isothermal curve 500°C [6]. As a result, we have obtained bending moment-curvature diagram (figure 5) for effective thickness. Numerical experiment of considered floor slab of architectural and construction system 'Monofant' has shown the correspondence to fire resistance R60. The deformation of design section does not have significant effect. The reduction of carrying capacity is 17% in comparison with non-heated slab.

5 Conclusions

- We offer the algorithm that considers the transformation of design patterns depending on temperature values and excessive pressure in thermal insulation cavities.
- We have obtained fields throughout the depth of cross section of the slab of new type of architectural and construction system 'Monofant' upon heating in standard fire mode for time interval 0-240 min.
- We have carried out the analysis of strain-stress state of new type of architectural and construction system 'Monofant' under the effect of high temperature heating in standard fire mode, considering the change of design pattern. It was disclosed that this slab corresponds to fire resistance rating R60.

References

1. V. Shmukler, V. Babayev, S. Bugayevskiy, K. Berezhna, I. Kariakin, V. Kondrashchenko, I. Seirski *Framework Building "MONOFANT"* 89464 Ukraine
2. A.I. Lurie *Theory of elasticity* (Moscow, Science, 1970)
3. L.D. Landau, E.M. Lifshits, *Elastic theory* (Moscow, Science, 1987)
4. V.S. Shmukler, Yu.A. Klimov, N.P. Burak *Frame systems of reduced-weight type* (Kharkov, Golden Pages, 2008)
5. *DSTU 2.6-156:2010: Designs of buildings and structures. Concrete and reinforced concrete structures of heavy concrete.* (Ministry of Regional Construction of Ukraine, 2011)
6. *EN 1992-1-2:2004 Eurocode 2: Design of concrete structures – Part 1-2: General rules – Structural fire design* (European committee for standardization, Brussels, 2004)
7. http://www.ansys.com.
8. A.S. Gorodetsky, I.D. Evzerov, E.B. Strelets-Sagittarius *Finite elements method: Theory of numerical implementation.* (Kiev: Fact, 1997)

The building of the distribution function of the normative generalized power influence on brick buildings

Oxana Kichaeva*

O.M. Beketov National University of Urban Economy in Kharkiv, 61002 Kharkiv, Ukraine

Abstract. Loads and impacts which have an influence on the performance of the structure have high statistical distribution, being random variables or random processes. The influence on frameless brick buildings and buildings with a mixed frame from the following types of loads and impacts is considered: load from own weight of structures, load from people and equipment (payload), snow load, wind load, external temperature effects and impacts due to ground motion (settlement). The types of buildings were considered: frameless buildings and buildings with a mixed frame. The article proposes an algorithm for constructing the distribution function of the generalized force effect on brick buildings and structures. Obtaining such a function is one of the stages of the critical normative risk (probability) of failure of brick structures of buildings. It is determined that the distribution function of the generalized force action obeys the normal distribution law of Gauss, for which the values of the mathematical expectation and the coefficient of variation are obtained.

1 Introduction

In the continuous operation, brick buildings and structures undergo simultaneous effect of loads and impacts of different types and intensity. The combination of impacts leads to reduction of the durability of buildings and structures. Considering the reliability of the entire "base – foundation – structure" system, it can be concluded that its level is influenced by the following parameters: 1) external loads and impacts, which include impacts from engineering-geological processes that occur at the base; 2) Strength and deformability of structural elements of a building (structure); 3) strength and deformability of the base. These parameters are not constant in time. So, the probabilistic approach to determining the level of reliability is due to the fact that all characteristics of the strength and deformation of the "base – foundation – structure" system, as well as all loads and impacts on the system are random variables or random processes.

In this article, an approach to the structure of the distribution function of the normative generalized force action on buildings and structures made of bricks is presented. This

* Corresponding author: o_kichaeva@ukr.net

function is one of the stages of determining the value of the critical normative risk (probability) of failure of brick buildings.

2 Discussion

When assessing the reliability of brick buildings, the loads and impacts, regulated by existing engineering standards, must be taken into account. According to the type of loads, these classes of loads and impacts are distinguished:

1. The own weight of structures F_G.
2. The load from people and equipment (payload) F_q.
3. Climatic impacts (snow load F_S, wind load F_W, external temperature effects F_T).
4. Impacts due to ground motion (settlement) F_D.

Loads depending on the variability in the time of impact are divided into constants and variables, depending on the causes of the load - the main and episodic.

The above loads and impacts are considered in this paper. Characteristic values of random natural influences (wind, snow, temperature) are often assigned as values exceeding 1 time in T_c years. Then the possibility of occurrence of impact with an average period of recurrence T_c years for one arbitrary taken year is $1/T_c$, and the probability that an impact will never occur over the period of maintenance T years, is calculated by the formula:

$$P = (1 - 1/T_c)^T. \quad (1)$$

By Norms of Ukrainian engineering it is established that the values of some of the considered loads and impacts (wind, temperature, snow and useful) can be exceeded once in 50 years. Thus, the annual probability of exceeding of these loads is:

$$p_w = p_T = p_s = p_q = 0.02 \ 1/year. \quad (2)$$

The probability of either impact will never occur over the period of maintenance T years (for a period of T we will adopt the reference value of the life of residential and public buildings), equal to 100 years, will be:

$$P = (1 - 1/50)^{100} = 0.133. \quad (3)$$

When performing calculations for assessing the reliability of buildings made of bricks in conditions of an unevenly deformable base due to the normative method there should be considered calculated cases for the main combination of loads and impacts. In the calculated case of the basic combination, loads and impacts with normative excess probabilities corresponding to this engineering case are taken into account.

Loads influencing the structure and other external impacts are the most uncertain values and have high statistical distribution. The structure is usually effected by several loads, each of which has its own characteristics. In order to lead all these external influences to one generalized load (impact), it is necessary to understand the specific characteristics of each impact and their combinations. To determine the stock coefficient, it is desirable that the entire set of loads, randomly varied in time, is led to one random variable F, which can be inserted into the calculation formulas.

Loads and impacts on bricks buildings and structures can be regarded as independent random variables. When performing probabilistic calculations to determine the risk of the emergence of one of the critical conditions of buildings and structures made of bricks, the laws of distribution of these loads and the impacts with respect to their variability should be known. On the basis of the laws of distribution of each impacts, the distribution law of the generalized force action F can be obtained, which can be used to perform calculations.

The following loads and impacts have influence on brick buildings and structures, such as payload on the overlap F_Q, snow load F_S, wind load F_W, temperature impacts F_T, impacts of the uneven settlement F_D, can be considered independent random variables. The own weight F_G of the structure can be regarded in some cases as a deterministic value due to its small variability, and in some cases as a random variable.

Below is an approach for determining the parameters of the distribution laws of loads and impacts on brick buildings and structures, as well as the parameters of the distribution law of generalized force impact. Thus, it is possible to be confined to values of standard factors and values of standard probabilities of excess of loads and impacts.

The distribution functions of random loads and impacts can be adopted in accordance with the following distribution laws:

– own weight F_G of the structure is the normal distribution law, which is characterized by the mathematical expectation μ_G and the mean squared deviation σ_G; we will also note that the value of the structures' own weight can be assumed deterministic because of its small variability;

– payload on overlaps F_Q - double exponential distribution law of Gumbel, which is characterized by mathematical expectation μ_Q and the mean squared deviation σ_Q;

– snow load F_S is also the distribution law of Gumbel, which is characterized by the mathematical expectation μ_S and the mean squared deviation σ_S;

– wind load F_W - Gumbel distribution law, which is characterized by the mathematical expectation μ_W and the mean squared deviation σ_W;

– the temperature impacts F_T can be taken in accordance with the normal distribution law, which is characterized by the mathematical expectation μ_T and the mean squared deviation σ_T;

– the impact of uneven settlement F_D can also be taken in accordance with the normal distribution law, which is characterized by the mathematical expectation μ_D and the mean squared deviation σ_D.

According to EN 1990 [1] and DBN B.1.2-2:2006 [2], the calculated combinations should include loads that most adversely affect the structures (bases) from the point of view of the critical state. According to EN 1990, the combinations of impacts for permanent or transitional calculated situations (basic combinations) are generally described as follows:

$$E_d = \gamma_{Sd} E\{\gamma_{G,j} G_{k,j};\ \gamma_P P;\ \gamma_{Q,1} Q_{k,1};\ \gamma_{Q,i}\, \psi_{0,i}\, Q_{k,i}\} \quad j \geq 1,\ i > 1. \tag{4}$$

The combination of impacts must include:

- the calculated value of the leading (dominant) time impact, and
- the calculated combination of accompanying temporal effects.

According to DBN B.1.2-2:2006, combinations of two types – basic and emergency – can be used in structure calculations. To test the critical states of the first group, basic combinations are used, which include constant loads with the critical calculated values, as well as critical calculated, cyclic, or quasi-constant values of variable loads. To test the marginal states of the second group, basic combinations are used, including constant loads with operational values, as well as operational design, cyclic or quasi-permanent values of variable loads. In emergency combination, in addition to permanent and variable loads, there can be one episodic value.

For basic combinations involving constant and not less than two variable loads, the latter are adopted with a combination coefficient $\psi_1 = 0.95$ for long loads and $\psi_2 = 0.90$ for short-term loads. Then the basic combination of loads can be written in the notations of EN

1990 for the combination of the constant G and the two variable loads $Q_{k,1}$ and $Q_{k,2}$, while for them the values are taken for both short-term loads:

$$C_m = \gamma_G^* G_k^* + \gamma_{Q,1}^* \psi_{t,1} Q_{k,1}^* + \gamma_{Q,2}^* \psi_{t,2} Q_{k,2}^*. \tag{5}$$

The symbols G_k^*, $Q_{k,1}^*$ mean the normative values of the constant and variable loads. The coefficient of reliability γ_G^* (in DBN - γ_{fm}) for a constant load, which is meant to be the weight of structures and soils, is adopted depending on the type of material and varies within the range of 1.05 ... 1.3. The reliability coefficient for a variable load is adopted depending on its type. The load reliability coefficient γ_{fm} for evenly distributed loads should be taken: 1.3 – at a characteristic value of less than 2.0 kPa (200 kgf/m^2); 1.2 – with a characteristic value of 2.0 kPa (200 kgf/m^2) and more. The reliability coefficient of responsibility (responsibility coefficient) γ_n is determined depending on the class of consequences (responsibility) of the object depending on the calculated situation and the group of critical states.

The active loads and impacts on the "base – foundation – structure" system for the problems considered in this paper can be summarized as follows (Table 1). The data given in Table 1, are obtained both from the results of the normative literature and authors dealing with this subject, and from the author's own data.

Table 1. Statistical parameters of load distributions.

Type of load	The designation	Units	Average value	Variation coefficient V	The distribution law
Constant load	G	kN	G	0.1	Normal
Payload (50 years)	Q	kN/m^2	$0.6Q$	0.35	Gumbel
Snow load (50 years)	S	kN/m^2	$0.7S_0$	0.5	Gumbel
Wind load	W	kN/m^2	$0.75W_0$	0.35	Gumbel
Temperature impact	T	C^o	T	0.15	Normal
Uneven settlement impact (effort in the foundation)	D	kPa	P	0.59	Normal

The distribution function of the generalized force action obeys the normal distribution law of Gauss with the mathematical expectation of μ_F and the variation coefficient V_F. This statement can be explained by the fact that the generalized force action, considered as a random variable, is the sum of the loads and impacts on the structure, and the loads and impacts are independent random variables. In accordance with the central critical theorem of probability theory, when summing a sufficiently large number of independent random variables, the distribution of their sum is close to normal.

3 Results

Knowing the values of the variation coefficients of various loads, we can determine the value of the variation coefficient of the generalized force action for both the sum of the random variables, assuming that the values of the generalized mathematical expectation μ_Σ and the mean squared deviation σ_Σ are known.

In this case, μ_Σ is the total summary value of the mathematical expectation from various impacts, equal to:

$$\mu_\Sigma = \mu_G + \mu_Q + \mu_S + \mu_W + \mu_D. \tag{6}$$

The mean squared deviation:

$$\sigma_\Sigma = \sqrt{\sigma_G^2 + \sigma_Q^2 + \sigma_S^2 + \sigma_W^2 + \sigma_D^2}\ . \tag{7}$$

The contribution of climatic temperature loads for the entire building (but not for a separate structure) can be neglected because of its small contribution to the generalized force impact.

Then, according to the theory of probability:

$$V_\Sigma = \frac{\mu_\Sigma}{\sigma_\Sigma}, \tag{8}$$

$$V_\Sigma = \sqrt{V_G^2 + V_Q^2 + V_S^2 + V_W^2 + V_D^2}\ . \tag{9}$$

Figure 1 shows the distribution functions of the impacts distributed according to the normal distribution law with mathematical expectation equal to 1 and the corresponding variation coefficients for each type of load. The graphs were obtained using the MathCAD program.

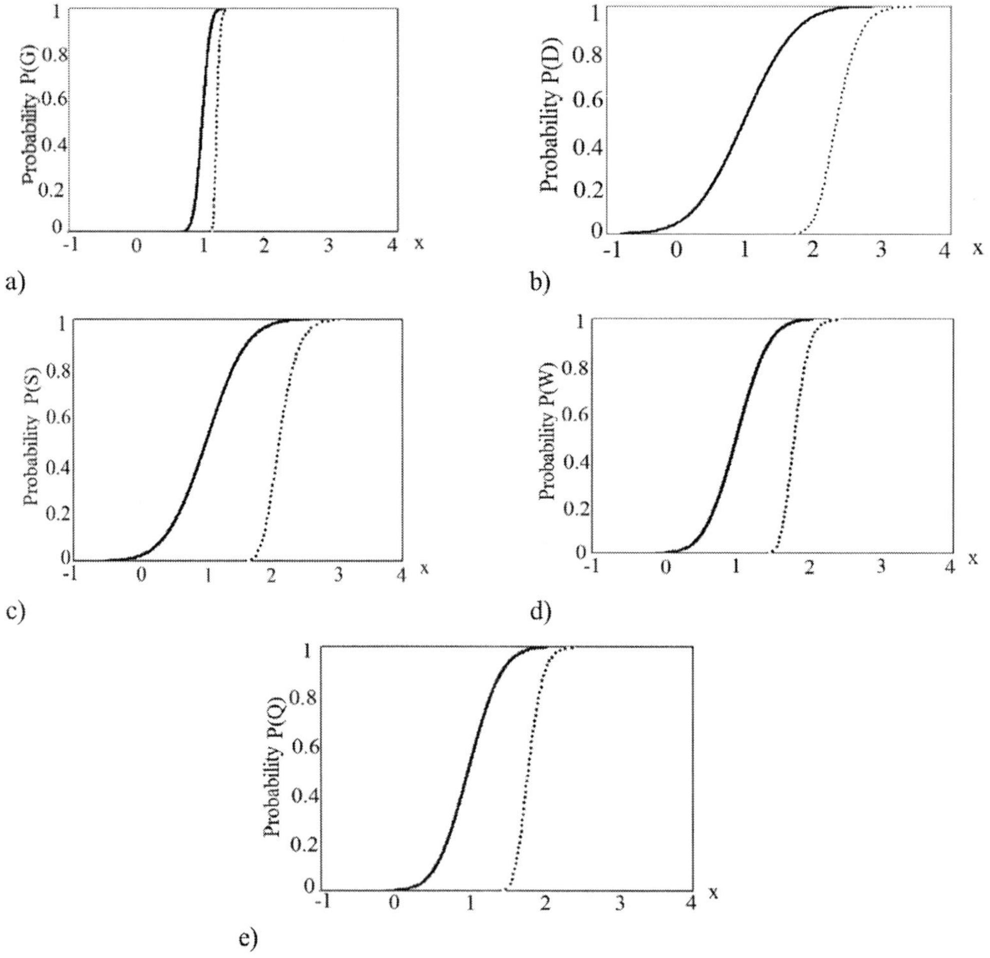

Fig. 1. Graphs of the distribution function of force impacts (x): a) from own weight; b) from the influence of uneven settlement; c) from the snow load; d) from the wind load; e) from the payload.

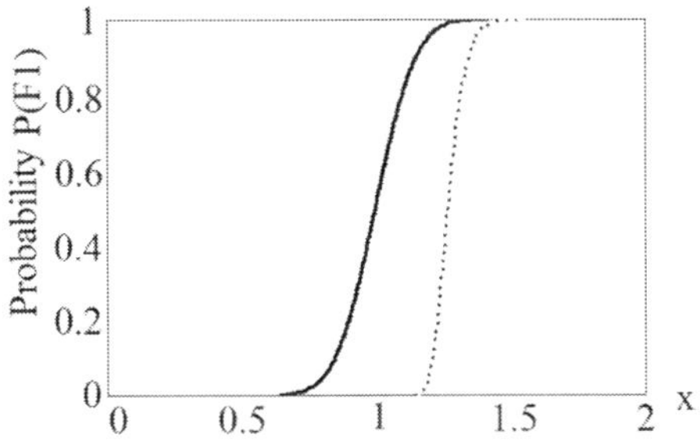

a)

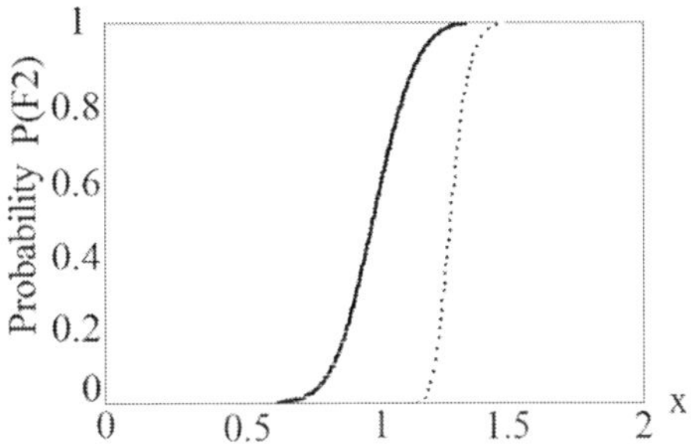

b)

Fig. 2. Graphs of distribution functions of the generalized force (x): a) for case 1 with $V_{\Sigma 1}= 0.119$; b) for case 2 with $V_{\Sigma 2} = 0.122$.

The following variants were considered for determining V_{Σ}.

1) Frameless buildings with a cargo area of A = 3 m^2, 6 m^2 и 9 m^2 with the payload value q = 150 kgf /m^2, 200 kgf /m^2, 300 kgf /m^2, 5th snow area accepted [2], 2nd wind zone, the value of loads on foundation with its vertical displacement is accepted, with the value of deformations up to 5 cm inclusive, the soil deformation modulus E = 8 MPa, 10 MPa, 12 MPa, 14 MPa, 16 MPa, Poisson's coefficient is 0.2, 0.3, 0.4, the square of the base of foundation varied from 1 m^2 to 2 m^2.

2) Buildings with a mixed frame. Parameters: A = 16 m^2 with the payload value q=150 kgf/m^2, 200 kgf/m^2, 300 kgf/m^2, 5th snow area accepted, 2nd wind zone [2], the value of loads on foundation with its vertical displacement is accepted, with the value of deformations up to 5 cm inclusive, the soil deformation modulus E = 8 MPa, 10 MPa, 12 MPa, 14 MPa, 16 MPa, Poisson's coefficient is 0.2, 0.3, 0.4, the square of the base of foundation varied from 1 m^2 to 4 m^2.

On figure 2 there are graphs of the distribution functions of the generalized force impact: a) for case 1 with $V_{\Sigma 1} = 0.119$; B) for case 2 with $V_{\Sigma 2} = 0.122$.

On figures 1, 2 the dashed lines show the distribution functions which represent the annual probabilities which correspond to the values of F, and the solid lines show the distribution functions of the impacts for T = 100 years. The value of the variation coefficient of the generalized force impact, calculated by the formula (9), equals $V_{\Sigma 1} = 0.119$ and $V_{\Sigma 2} = 0.122$.

4 Conclusions

Thus, the function of the distribution of the normative generalized force impact on buildings and structures made of bricks is constructed. With the help of this function, it is possible to determine the values of the critical normative risk (probability) of failure of brick buildings.

References

1. EN 1990:2002, *Eurocode: Basis of Structural Design* (Brussels, 2002)
2. DBN B.1.2-2:2006, *Load and impact* (Kiev, 2006)

DOI: 10.1051/matecconf/201711602020

Calculation methodology of reinforced concrete elements based on calculated resistance of reinforced concrete

Dmitriy Kochkarev [1*], and *Tatyana* Galinska[2]

[1]National University of Water and Environmental Engineering, Department Urban building and development, 33028 Rivne, Soborna St.,11, Ukraine
[2]Poltava National Technical Yuri Kondratyuk University, Department Architecture and urban Development, 36011 Poltava, Pershotravnevyi avenue, 24, Ukraine

Abstract. Calculation methodology of reinforced concrete elements based on the calculated resistance of reinforced concrete is presented. The basic dependence which allows setting the strength of bending sections and non-central compressed elements is obtained. The proposed method for calculating reinforced concrete elements is based on the use of nonlinear diagrams of material deformation, the hypothesis of flat sections and deformation criteria for the destruction of materials. The basic equations of strength are reduced to dimensionless quantities and are tabulated. When compiling the tables, the formula proposed in Eurocode 2 was adopted as the diagram of concrete deformation, and for the reinforcement two linear Prandtl diagram was used. The calculated formulas of the proposed method fully correspond to the formulas of the classical resistance of materials, and make it possible to solve the most frequently encountered problems in the practice of modern construction. The reliability of the dependencies is experimentally confirmed. There are calculation examples of bending and non-central compressed elements by the developed methodology.

1 Review of recent studies and publications

Design of reinforced concrete elements based on nonlinear dependencies are listed in the standards of many countries [1, 2, 3], has set a question about their practical calculation for most of the engineers. This is because the usage of the formulas of these rules without a computer [1, 2, 3] is practically impossible. It is generally accepted that any calculation by using computers should be checked or evaluated through classical and practical techniques. Creating such a technique would allow making an assessment, verification and analysis conducted computer calculations. Simplified and approximate methods [4, 5, 6, 7] not correspond to exact solution that is obtained by nonlinear dependencies in most of cases.

Consider the calculation of resistibility of sections of flexural reinforced concrete elements with a single reinforcement. Carrying force of these elements must be determined by following conditions:

* Corresponding author: dim7@ukr.net

1) achievement of extremum of function of carrying force $dM / d\varepsilon_c = 0$ in limits $\varepsilon_{c1} - \varepsilon_{cu}$ and occurrence of fluidity in the stretched reinforcement;
2) achievement of boundary deformation of concrete compressed brink ε_{cu} and limits of fluidity in the stretched reinforcement in the absence of extremum of function of carrying force;
3) occurrence of extremum of function of carrying force $dM / d\varepsilon_c = 0$ in limits $\varepsilon_{cl} - \varepsilon_{cu}$ without achievement fluidity of reinforcement;
4) achievement of boundary deformation of concrete ε_{cu} without achievement fluidity of reinforcement and in the absence of extremum of function of carrying force in limits $\varepsilon_{cl} - \varepsilon_{cu}$;
5) achievement deformations in the stretched reinforcement with value of ε_{cu}.

2 Basic material and results

Calculation of resistibility performed by using the equation of equilibrium of external forces and internal forces, deformation diagrams of concrete and reinforcement and the function of changes of deformation by height section. As a function of diagrams of concretes deformation should be taken like function of stresses in the concrete which would correspond to the conditions of nonlinear deformation of concrete. The following dependences are proposed for calculations of nonlinear structures in the standards [1]: formula of rules Eurocode 2 (3.4), fifth degree polynomial (3.5), two- and three a linear dependence between the stresses and deformations. The dependence of stress-deformation for reinforcement is taken as two linear Prandtl diagrams. The distribution of deformation by height section at the moment of the destruction taking in a linear dependence like:

$$\varepsilon = \frac{1}{r} x \ or \ x = \frac{\varepsilon}{1/r}, \ or \ \frac{1}{r} = \frac{\varepsilon}{x}, \quad (1)$$

where ε – relative deformation of the material at a distance x from neutral line, $1/r$ – curvature of element in section.

In most cases, diagrams of concrete deformation are function of related strength and deformation characteristics: the calculated resistance of concrete to axial junction f_c, concrete deformation module E_c and limits of concrete deformation ε_{cl}, ε_{cu}. This can be expressed in the following functional dependence

$$\sigma_c = f(f_c, E_c, \varepsilon_{c1}, \varepsilon_{cu}). \quad (2)$$

There are many formulas that associate deformation and strength characteristics of concrete. The following expressions can be written by its summarizing

$$E_c = f(f_c),\ \varepsilon_{c1} = f(f_c),\ \varepsilon_{cu} = f(f_c). \quad (3)$$

With taking into account (3) the final dependence (2) will take the following form

$$\sigma_c = f(f_c). \quad (4)$$

The function laid down in the current norms [1, 2] will be examined for further research. It is also called Eurocode function

$$\sigma_c = \frac{E_c \varepsilon_c - f_{cm}\left(\frac{\varepsilon_c}{\varepsilon_{c1}}\right)^2}{1+\left(\frac{E_c \varepsilon_{c1}}{f_{cm}} - 2\right)\frac{\varepsilon_c}{\varepsilon_{c1}}} \; or \; \frac{\sigma_c}{f_c} = \frac{k\eta - \eta^2}{1+(k-2)\eta}, \tag{5}$$

where $\eta = \varepsilon_c / \varepsilon_{c1}$, $k = 1.05 E_c \varepsilon_{c1} / f_c$.

Here is the expression (4) for the function (5)

$$\sigma_c = f_c \times f(k,\eta) = f_c \times f(f_c). \tag{6}$$

The bending reinforced concrete elements at a single reinforcement are proposed to consider. After putting equilibrium equation and conducting simple transformation with considering the hypothesis of flat sections we will get:

- for non-overreinforced

$$\left(\frac{\int_0^{\varepsilon_c} \sigma_c \varepsilon_c d\varepsilon_c}{\left(\int_0^{\varepsilon_c} \sigma_c d\varepsilon_c \right)^2} - \frac{\varepsilon_c}{\int_0^{\varepsilon_c} \sigma_c d\varepsilon_c} \right) \rho_f^{\,2} f_{yd}^2 + \rho_f f_{yd} = \frac{M_{Ed}}{bd^2}; \tag{7}$$

- for overreinforced

$$\left(\frac{\int_0^{\varepsilon_c} \sigma_c \varepsilon_c d\varepsilon_c}{\left(\int_0^{\varepsilon_c} \sigma_c d\varepsilon_c \right)^2} - \frac{\varepsilon_c}{\int_0^{\varepsilon_c} \sigma_c d\varepsilon_c} \right) \rho_f^2 E_s^2 \varepsilon_c^2 + \rho_f E_s \varepsilon_c = \frac{M_{Ed}}{bd^2}. \tag{8}$$

Both sides of equation (7) i (8) are divided into f_c and taking into account (6), we will get:

- for non-overreinforced

$$\left(\frac{\int_0^{\varepsilon_c} f(f_c) \varepsilon_c d\varepsilon_c}{\left(\int_0^{\varepsilon_c} f(f_c) d\varepsilon_c \right)^2} - \frac{\varepsilon_c}{\int_0^{\varepsilon_c} f(f_c) d\varepsilon_c} \right) \frac{\rho_f^{\,2} f_{yd}^2}{f_c^2} + \frac{\rho_f f_{yd}}{f_c} = \frac{M_{Ed}}{f_c bd^2}; \tag{9}$$

- for overreinforced

$$\left(\frac{\int_0^{\varepsilon_c} f(f_c) \varepsilon_c d\varepsilon_c}{\left(\int_0^{\varepsilon_c} f(f_c) d\varepsilon_c \right)^2} - \frac{\varepsilon_c}{\int_0^{\varepsilon_c} f(f_c) d\varepsilon_c} \right) \frac{\rho_f^2 f_{yd}^2 E_s^2 \varepsilon_c^2}{f_c^2 f_{yd}^2} + \frac{\rho_f f_{yd} E_s \varepsilon_c}{f_{yd} f_c} = \frac{M_{Ed}}{f_c bd^2}. \tag{10}$$

The following notations are introduced

$$\omega = \frac{\rho_f f_{yd}}{f_{cd}}, \ \eta_s = \frac{E_s \varepsilon_c}{f_{yd}}. \tag{11}$$

The name parameter ω is mechanical reinforcement coefficient [4, 5]. The formulas (9) and (10) with introduced notations (11) will take the following form:
- for non-overreinforced

$$\left(\frac{\int_0^{\varepsilon_c} f(f_c)\varepsilon_c d\varepsilon_c}{\left(\int_0^{\varepsilon_c} f(f_c) d\varepsilon_c \right)^2} - \frac{\varepsilon_c}{\int_0^{\varepsilon_c} f(f_c) d\varepsilon_c} \right) \omega^2 + \omega = \frac{M_{Ed}}{f_c b d^2}; \tag{12}$$

- for overreinforced

$$\left(\frac{\int_0^{\varepsilon_c} f(f_c)\varepsilon_c d\varepsilon_c}{\left(\int_0^{\varepsilon_c} f(f_c) d\varepsilon_c \right)^2} - \frac{\varepsilon_c}{\int_0^{\varepsilon_c} f(f_c) d\varepsilon_c} \right) \omega^2 \eta_s^2 + \omega \eta_s = \frac{M_{Ed}}{f_c b d^2}. \tag{13}$$

The notation is introduced to both equations

$$k_z = f(f_c, \omega) = \frac{6 M_{Ed}}{f_c b d^2} = \frac{M_{Ed}}{f_c W_c}. \tag{14}$$

Thereby
- for non-overreinforced

$$k_z = \frac{1}{6} \left(\left(\frac{\int_0^{\varepsilon_c} f(f_c)\varepsilon_c d\varepsilon_c}{\left(\int_0^{\varepsilon_c} f(f_c) d\varepsilon_c \right)^2} - \frac{\varepsilon_c}{\int_0^{\varepsilon_c} f(f_c) d\varepsilon_c} \right) \omega^2 + \omega \right); \tag{15}$$

- for overreinforced

$$k_z = \frac{1}{6} \left(\left(\frac{\int_0^{\varepsilon_c} f(f_c)\varepsilon_c d\varepsilon_c}{\left(\int_0^{\varepsilon_c} f(f_c) d\varepsilon_c \right)^2} - \frac{\varepsilon_c}{\int_0^{\varepsilon_c} f(f_c) d\varepsilon_c} \right) \omega^2 \eta^2 + \omega \eta_s \right). \tag{16}$$

Connection between the calculated resistance of reinforced concrete and characteristics that derived above is following

$$f_{zM} = k_z f_c. \tag{17}$$

Introduced parameter k_z generally depends on the mechanical reinforcement coefficient ϖ and deformation characteristics of concrete. There are more details about the impact of deformation characteristics of concrete on bearing capacity of bending elements. Here is a submission of parameter dependence schedule k_z depending on the mechanical reinforcement coefficient ϖ for different kinds of concrete (Fig.1). The impact of deformation characteristics of concrete (parameters $\eta = \varepsilon_c / \varepsilon_{c1}, k = 1.05 E_c \varepsilon_{c1} / f_c$ on bearing capacity of bending reinforced concrete elements for considered kinds of concrete varies with differences within 10% (Fig. 2). This error is completely allowed by normative coefficients of variation of strength for such elements. This makes possible to offer slightly simplified force model of reinforced concrete elements calculation. The impact of deformation characteristics of strength sections is ignored and common functional dependence is taken $k_z = f(\varpi)$ for all kinds of concrete. This dependence is true not only for different kinds of concrete and reinforcement and even for different duration of the load. Formulas for non-central compression can be derived by similar arguments. Dependence of parameter k_z from the mechanical reinforcement coefficient ϖ is presented in tables.

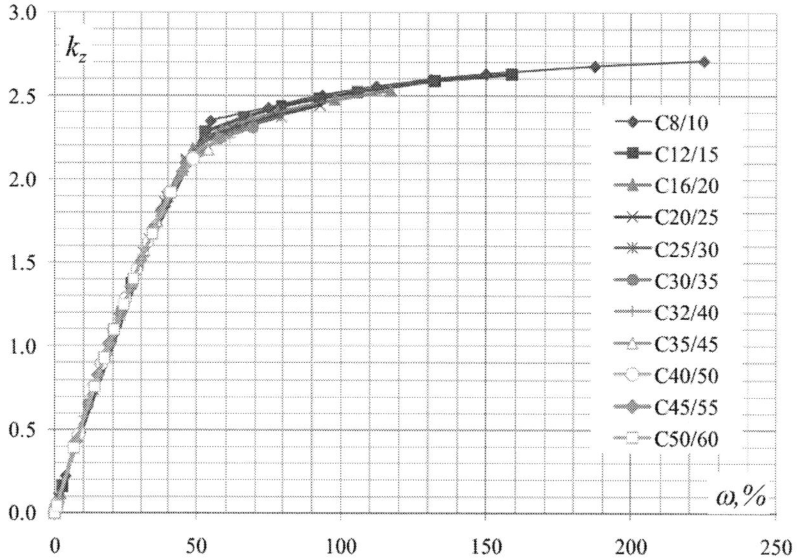

Fig.1. Parameter k_z of reinforced concrete depending on the mechanical reinforcement coefficient at single reinforcement $f_{yd} = 365$ MPa.

The strength conditions by a modified method of calculation resistances are formulated:
- bending elements

$$M_{Ed} = W_c f_{zM}. \tag{18}$$

where f_{zM} – estimated resistance of reinforced concrete on bend, which depends on kinds of concrete and reinforcement, section shapes and percent of reinforcement section that defined by the expression (17). Parameter k_z in expression (7) is calculated by table (Table 1, 2, 3) depending on the mechanical reinforcement coefficient ϖ.
- non-central compressed

$$N_{Ed} = A_c f_{zN}. \tag{19}$$

where f_{zN} – estimated resistance of reinforced concrete on bend, which depends on kinds of concrete and reinforcement, section shapes and percent of reinforcement section, relative initial eccentricity of the force application e_0/d, that defined by the expression $f_{zN} = k_z f_c$, Parameter k_z is also calculated by table (Table 4) depending on the mechanical reinforcement coefficient ϖ; A_c – working sectional area of concrete: for rectangular elements $A_c = bd$; for round ones $A_c = \pi d^2/4$.

The conditions (18, 19) satisfy the conditions of bending element's strength of material's classic resistance and allow solving a number of problems, including:
verification of element's strength; determination of bearing capacity of reinforced concrete elements according defined reinforcement; selection of a reinforcement section of the reinforced concrete element; setting the required sizes of cross-section according defined reinforcement; variant design of bending reinforced concrete elements.

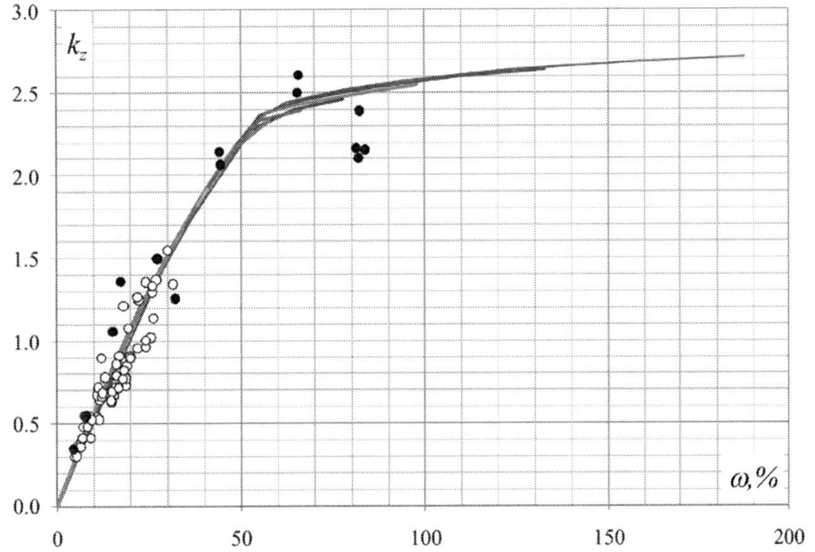

Fig. 2. Experimental verification of dependence $\varpi = f(k_z)$

Table 1. Dependence of $\varpi - k_z$ by the flat bend for a rectangular section with single reinforcement.

ϖ	0	0.1	0.15	0.2	0.25	0.3	0.35	0.4	0.45	0.5	0.6	0.7	1	2	3
k_z	0	0.568	0.828	1.071	1.299	1.511	1.706	1.885	2.028	2.07	2.14	2.195	2.31	2.476	2.542

Table 2. Dependence of $\varpi - k_z$ by the flat bend for a rectangular section with symmetrical reinforcement.

ϖ	0	0.1	0.15	0.2	0.25	0.3	0.35	0.4	0.45	0.5	0.6	0.7	1	2	3
k_z	0	0.292	0.429	0.565	0.700	0.834	0.967	1.100	1.232	1.36	1.63	1.888	2.66	4.85	6.671

Table 3. Dependence of $\varpi - k_z$ by the band for round section with symmetrical reinforcement.

ϖ	0	0.1	0.15	0.2	0.25	0.3	0.35	0.4	0.45	0.5	0.6	0.7	1	2	3
k_z	0	0.368	0.523	0.673	0.821	0.967	1.11	1.253	1.394	1.53	1.81	2.072	2.78	5.06	7.32

Table 4. Parameter values k_z for non-central compressed elements of rectangular cross-section with symmetrical reinforcement, $\lambda \leq 4$ ($\lambda \leq 0$).

Mechanical reinforcement coefficient ϖ	Relative initial eccentricity of longitudinal force application e_0/d									
	0.01	0.15	0.30	0.65	1.00	1.50	2.00	3.00	4.00	5.00
0.10	1.159	0.817	0.575	0.206	0.090	0.047	0.032	0.019	0.014	0.011
0.15	1.209	0.875	0.640	0.272	0.131	0.070	0.047	0.029	0.020	0.016
0.20	1.259	0.918	0.682	0.326	0.169	0.092	0.062	0.038	0.027	0.021
0.25	1.309	0.959	0.721	0.375	0.204	0.114	0.077	0.047	0.034	0.026
0.30	1.359	0.999	0.758	0.420	0.238	0.136	0.092	0.056	0.040	0.031
0.35	1.409	1.040	0.794	0.461	0.270	0.157	0.107	0.065	0.047	0.037
0.40	1.458	1.077	0.829	0.499	0.301	0.177	0.123	0.075	0.054	0.042
0.45	1.509	1.113	0.864	0.535	0.330	0.197	0.137	0.084	0.060	0.047
0.50	1.558	1.148	0.897	0.568	0.359	0.217	0.152	0.093	0.067	0.052
0.60	1.658	1.216	0.964	0.619	0.413	0.256	0.180	0.111	0.080	0.062
0.70	1.758	1.284	1.029	0.667	0.464	0.293	0.209	0.130	0.093	0.073
1.00	2.058	1.544	1.221	0.804	0.597	0.400	0.290	0.183	0.133	0.104
2.00	3.058	2.297	1.340	1.222	0.916	0.676	0.535	0.354	0.261	0.205
3.00	4.057	3.068	1.462	1.665	1.256	0.930	0.738	0.515	0.384	0.305

3 Examples of application of the proposed calculation method

3.1 Example 1

Reinforced concrete beams has a working section $b \times d = 20 \times 55$ cm. Determine reinforcement for the perception estimated moment $M_{Ed} = 242$ kNm. Reinforcement grade is accepted A500C ($f_{yd} = 435$ MPa), concrete grade C20/25 ($f_{cd} = 14.5$ MPa).

Solution. The estimated resistance of reinforced concrete that provides beam's strength and parameter k_z are calculated

$$f_{zM} = \frac{6M_{Ed}}{bd^2} = \frac{6 \times 242 \times 10^3}{20 \times 55^2} = 24.0 \text{ MPa}; k_z = \frac{f_{zM}}{f_{cd}} = \frac{24.0}{14.5} = 1.655.$$

According to Table 1 $\varpi = 0.337$ depending on $k_z = 1.655$.

The required contents of reinforcement

$$\rho_f = \frac{f_{cd}}{f_{yd}} \omega = \frac{14.5}{435} 0.337 = 0.011.$$

The required sectional area of reinforcement

$$A_s = \rho_f \times b \times d = 0.011 \times 20 \times 55 = 12.25 \text{ cm}^2.$$

3.2 Example 2

The longitudinal strength $N_{Ed} = 2000$ kN and bending moment $M_{Ed} = 120$ kNm on the reinforced concrete elements with section $b \times d = 30 \times 40$ cm length $l_0 = 200$ cm, made of kind of concrete C20/25 ($f_{cd} = 14.5$ MPa). Determine the required sectional area of steel reinforcement of grade A400C ($f_{yd} = 365$ MPa) with symmetrical reinforcement.

Solution. The relative initial eccentricity and element's flexibility are determined

$$\frac{e_0}{d} = \frac{M_{Ed}}{N_{Ed} d} = \frac{120 \times 10^2}{2000 \times 40} = 0.15; \lambda = l_0/d = 200/40 = 5.$$

The element will be considered as rigid according to flexibility $\lambda = 5$. The required estimated resistance of reinforced concrete in compression and parameter k_z are calculated

$$f_{zN} = \frac{N_{Ed}}{bd} = \frac{2000}{30 \times 40} \times 10 = 16.67 \text{ MPa}; k_z = \frac{f_{zN}}{f_{cd}} = \frac{16.67}{14.5} = 1.150.$$

According to Table 4, using interpolation on k_z and relative initial eccentricity e_0 / d, $\varpi = 0.503$ is defined.

Reinforcement coefficient is calculated

$$\rho_f = \frac{f_{cd}}{f_{yd}} \omega = \frac{14.5}{365} 0.503 = 0.02.$$

The required sectional area of reinforcement

$$A_s + A_{sc} = \rho_f \times b \times d = 0.02 \times 30 \times 40 = 24.0 \text{ cm}^2.$$

4 Conclusions

The methodology of practical calculation of strength of reinforced concrete elements based on deformation model is proposed. The tables are compiled based on this methodology allowing to quickly and easily perform calculations of strength of overreinforced and non-overreinforced beams by deformation methodology without using computer programs. To simplify the work of designers, engineers-practitioners and students of construction specialties it is appropriate to include the table into design standards of concrete elements.

References

1. *DBN V.2.6-98:2009. Konstruktsiyi budynkiv i sporud. Betonni ta zalizobetonni konstruktsiyi. Osnovni polozhennya.*, Minrehionbud, Kyiv, (2011).
2. *SP 52-101-2003. Betonnyie i zhelezobetonnyie konstruktsii bez predvaritelnogo napryazheniya armaturyi*, Gosstroy Rossii, Mockow, (2003).

3. *ENV 1992-1. Eurocode 2. Design of concrete structure. Part 1, General rules and rules for buildings*, GEN, (1993).
4. J. K. Wight, J.G MacGregor . *Reinforced Concrete: mechanics and design*, New Jersey: Upper Saddle River, (2009).
5. J. C. McCormac, J. K. Nelson. *Design of Reinforced Concrete. Hoboken*, NJ: John Wiley & Sons, (2006).
6. V.B. Filatov, A.A. Suvorov. **153**, p. 144–150, (2016).
7. D. A. Panfilov, A. A. Pischulev, V. V. Romanchkov. **153**, p. 531-536, (2016).

Experimental studies of fiber concrete creep

Irina Korneeva[1,*], *Stepan* Neutov[1], and *Mykola* Suriyaninov[1]

[1]Odessa State Academy of Civil Engineering and Architecture, Structural Mechanics Department, Didrikson str., 4, Odessa 65029, Ukraine

Abstract. The results of two-stage experimental studies of the strength and deformation characteristics of fibrous concrete reinforced with steel fiber. In the experiments we used steel fiber with bent ends, which practically does not form "hedgehogs", which allows to achieve an even distribution of the fiber by volume. At the first stage, the cube and prismatic strength, deformability at central compression, a number of special characteristics are determined: water absorption, frost resistance, abrasion; the optimal percentage of fiber reinforcement and the maximum size of the coarse aggregate fraction were selected. Fiber reinforcement led to an increase in the strength of concrete at compression by 1,35 times and an increase in the tensile strength at bending by 3,4 times. At the second stage, the creep of fibrous concrete and plain concrete of similar composition at different stress levels was researched. Creep curves are plotted. It is shown that the use of fiber reinforcement leads to a decrease in creep strain by 21 to 30 percent, depending on the stress level.

1 Introduction

In recent decades, the distinctive feature of the development of industrial production is the use of new materials, and in particular, composites. In construction, this material is fiber concrete. The appeal of fiber concrete is due to the fact that it has higher strength and deformation characteristics than ordinary concrete. This makes it possible to effectively use fiber concrete in structures that are used in difficult conditions – bridges, tunnels, airfields, offshore structures, etc.

The use of fiber-reinforced concrete in fiber-reinforced concrete structures requires knowledge of a wide range of its physic-mechanical characteristics obtained as a result of laboratory tests. Many specialists were engaged in such studies [1-3], however, analysis of data on the effect of disperse (fibrous) reinforcement on the strength properties of fiber-reinforced concrete indicates their ambiguity. This difference can be explained by a number of objective reasons - the composition of the matrix, the type of fiber and the percentage of fiber reinforcement, the size of the samples, the conditions of the experiment, etc.

Nevertheless, the strength characteristics of fiber-reinforced concrete are now sufficiently well studied both theoretically and experimentally [4,5]. The dependence of the strength on such parameters as the percentage of reinforcement, the length, diameter and shape of the fibers was studied, the effect of the fiber orientation, the strength of the matrix

* Corresponding author: sng@ogasa.org.ua

material, and the composition of the concrete was taken into account. On the basis of the work done, regulatory documents have been created for the calculation of steel fiber-reinforced concrete structures [6-8].

Deformation characteristics, especially creep and shrinkage under long-term load action, have been studied much worse.

2 Results

In order to fill the existing gap, we conducted experimental studies, during which we studied the work of fiber concrete with a long-term load action.

The test program provides for two phases. At the first stage, cubical and prismatic strength, deformability under central compression, a number of special characteristics (water absorption, frost resistance, etc.) were determined. At the same time, the percentage of fibrous reinforcement varied, which was 0,5%, 1,0% and 1,5%, and the maximum size of a large aggregate (granite crushed stone) - with the size of the fraction ≤ 10 mm in the first series of tests and ≤ 20 mm in the second.

In the experiments we used steel fiber with bent ends (Fig. 1), produced by PJSC "PA" Stalkanat-Silur" in accordance with the European standard EN 14889-1: 2006 [9].

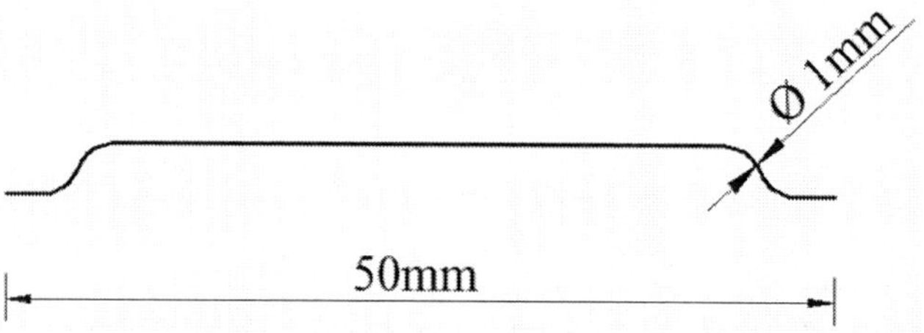

Fig. 1. Fiber with bent ends.

The fiber is made of wire with a temporary resistance of 1150 MPa (1 class), 1335 MPa (2 class) and 1550 MPa (Grade 3).

World experience of practical application indicates that the fiber with folded ends practically does not form "hedgehogs", which allows to achieve its uniform distribution by volume.

In the studies carried out, class 2 fibers with a diameter of 1 mm were used. The main characteristics of the fiber are given in Table 1.

Table 1. Main characteristics of fiber before.

Type	Length, mm	Diameter, mm	σ_u, MPa	E, MPa	μ
HE50	50	1.0	1335	200000	0.3

For the production of the matrix, cement 400 was used, washed river sand. The water-cement ratio was 0.449.

In order to compare the properties of concrete and steel-fiber-concrete, all types of tests were carried out on samples made of both materials.

Table 2 shows the parameters of the samples used and the types of tests performed.

The conducted studies showed that, in comparison with ordinary concrete, the structure of fiber concrete is characterized by a smaller number of large pores and a higher uniformity. As is known, contact zones of the interphase layer with improved mechanical properties are formed in conventional concrete. In fiber-reinforced concrete these zones were 5 to 6 times larger, and their total number increased. Obviously, this leads to an increase in the physical and mechanical characteristics of the material as a whole. In addition, the steel fiber arbitrarily oriented in the sample redistributes the gradients of stresses arising under the load, resulting in a decrease in stress concentration.

Table 2. Samples and types of tests.

Type of sample	Material	Sizes, mm	Determined characteristics
Cube	Concrete	100x100x100	Cubic strength, frost resistance, abrasion, water absorption, stretching during splitting
Prism	Concrete	100x100x400	Prismatic strength, Poisson's ratio, modulus of elasticity, crack resistance in bending
Cube	SFC	100x100x100	Cubic strength, frost resistance, abrasion, water absorption, stretching during splitting
Prism	SFC	100x100x400	Prismatic strength, Poisson's ratio, modulus of elasticity, crack resistance in bending tensile strength in bending

The tests of cubes and prisms were carried out according to the standard procedure in accordance with the current recommendations. Before installing the prism in the universal test machine (press) or the stand (depending on the type of test), strain gauges were attached to its face, and one hour-type indicator with a fission value of 0.001 mm and a base of 250 mm was attached to it. The prisms were loaded with steps: first three steps were applied with load of $0.1F_{max}$, next with load of $0.2F_{max}$ With an exposure at each stage of 5-8 minutes with a total test duration of at least 45 minutes. In the intervals between the loading stages using the Brunell tube, the cracking process was monitored on the prism surfaces. At a load equal to $0.85F_{max}$, the measuring instruments were dismantled and the prisms were brought to destruction. Thus, as a result of the tests, data were obtained that described only the ascending branch of the "strain-load" diagram.

The qualitative form of the "strain-load" diagram is shown in Fig.2.

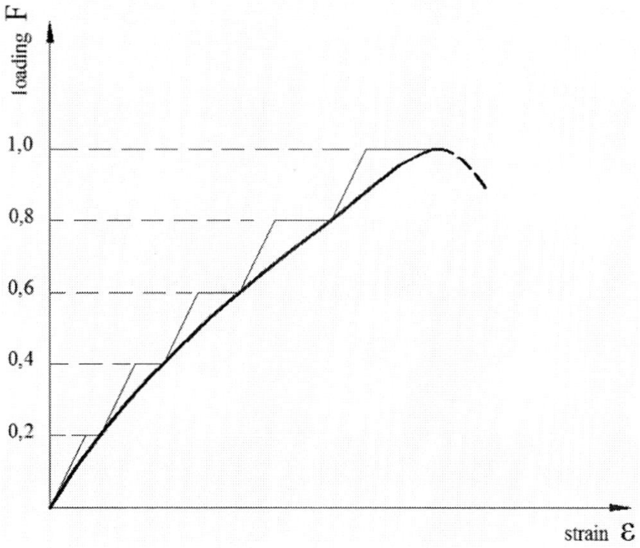

Fig. 2. "Strain-load" diagram.

The destruction of the samples was preceded by the formation of cracks along the axis of action of the compressive load. At the same time, the destruction of samples with 0.5% fibrous reinforcement, as well as of ordinary concrete, was clearly frail and accompanied by their complete separation into separate blocks. At 1.0% and 1.5% fibrous reinforcement of frail fracture and sample separation were not observed.

Based on the test results, complete longitudinal strain diagrams $\sigma - \varepsilon$ for 0.5%, 1.0% and 1.5% fiber reinforcement for two matrix variants are formulated.

The results of the first stage of the tests showed that the optimal characteristics of the fiber concrete mixture is a matrix with a large aggregate of the fraction mm at 1.0% fiber reinforcement. Therefore, to study the creep in the second stage of the tests, there were made samples-prisms of 100x100x400 mm in size from fiber concrete and ordinary concrete of the same composition (Fig. 3), which were tested on special test benches designed to study creep (Fig. 4).

Fig. 3. Samples for testing.

Fig. 4. Test benches for creep and long-term strength testing.

Before the beginning of the main tests, a number of auxiliary (preliminary) experiments were carried out, the results of which determined the levels of prolonged loading – 0.3; 0.4; 0.5; 0.67 and 0.8 from the prismatic strength. During long tests, after reaching a predetermined loading level, the load was fixed, and with the help of a spring cassette and a jack was maintained throughout the experiment (180 days).

In order to compare the results of experimental studies on each stand, two prisms were located in one power row: one of ordinary concrete, and the second of fiber concrete. This loading scheme makes it possible to apply the same load level to the two samples and maintain it throughout the experiment.

The creep curves formulated by mathematical processing of the test results are shown in Fig. 5.

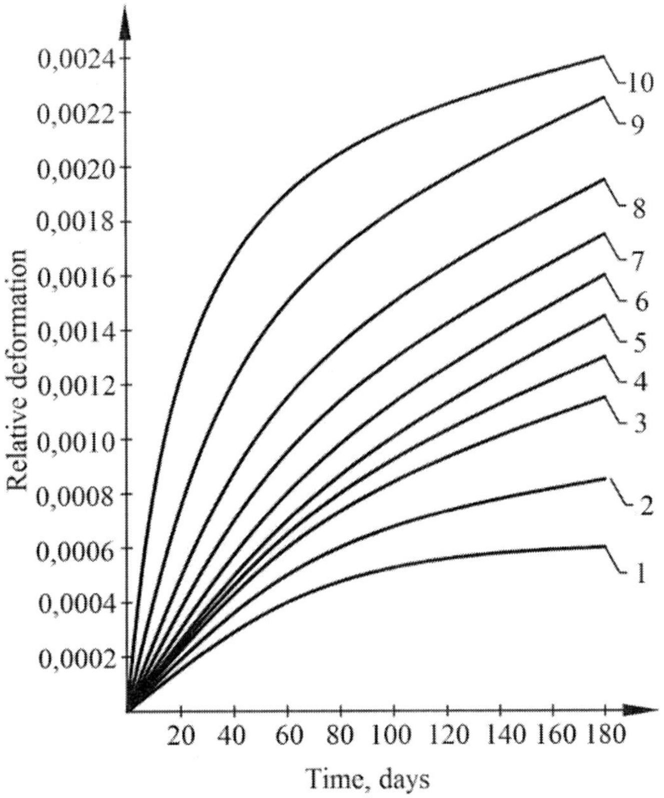

Fig. 5. Creep curves: 1, 3, 5, 7, 9 — fiber concrete; 2, 4, 6, 8, 10 — ordinary concrete at 0.3*R*; 0.4*R*; 0.5*R*; 0.67*R*; 0.8*R*.

3 Conclusions

The destruction of samples from 0.5% fiber reinforcement is fragile, which is not observed at 1.0% and 1.5% fiber reinforcement. Optimum characteristics of the fiber-reinforced concrete mixture is a matrix with a large aggregate of the fraction ≤ 10 mm at 1.0% fiber reinforcement. Fiber reinforcement led to an increase in the strength of concrete for compression by 1.35 times and an increase in the tensile strength by bending by 3.4 times. In this case, the creep strains decrease by 21-30 percent, depending on the stress level.

References

1. V.B. Aronchik, *Thesis for Doctor of Philosophy*. (Riga, 1982)
2. F.N. Rabinovich, *Composites based on dispersed reinforced concrete. Issues of theory and design, technology* (Moscow, 2006)
3. K.V. Talantova, Polzunovsky vestnik, **1**, 194-199 (2011)
4. N.P. Bleschik, Problems of modern concrete and reinforced concrete: materials of the III International Symposium, **2**, 80-113 (2011)
5. A.M. Bragov, Vestnik of Lobachevsky University of Nizhni Novgorod, **4**, 123-129 (2001)
6. *EN 14845-2:2006 Test methods for fibers in concrete. Effect on concrete*, Brussels (2006)
7. *SP 52-104-2006 Steel-fiber-concrete structures*, Moscow (2006)
8. *DSTU-N-B V.2.6-78:2009 Guidelines for the design and manufacture of steel-fiber-concrete structures*, Kiev (2009)
9. *BS EN 14889-1:2006 Fibres for concrete. Steel fibres. Definitions, specifications and conformity*, London (2006)

Exact solution of the differential equation of transverse oscillations of the rod taking into account own weight

Yurii Krutii[1,*], *Mykola* Suriyaninov[1], and *Victor* Vandynskyi[1]

[1]Odessa State Academy of Civil Engineering and Architecture, Structural Mechanics, Department Didrikson str., 4, Odessa, 65029, Ukraine

Abstract. The free transverse oscillations of the rod of uniform cross section taking into account its own weight are considered in this work. The appropriate partial differential equation of transverse oscillations of a rod was reduced to two ordinary differential equations for the time function and the amplitude function of deflections. Concurrent with the differential equations for the amplitude state, the equivalent system of differential equation is considered. In total, the exact solution of the initial partial differential equation of transverse oscillations of a rod, expressed in non-dimensional fundamental functions and initial parameters is attained. The method of power series was used for the construction of fundamental functions. Due to the exact solution, the formulas in an explicit form for dynamic variables of the state of a rode – deflections, angular displacement, bending moment and transverse force – were defined. The analytical form for equation of free oscillation frequency is defined. That has limited the finding of frequency to definition the unknown non-dimensional parameter through the frequency equation. As a result, the presence of derivative exact solutions provides the possibility to investigate the free oscillations of rod with various types of boundary conditions.

1 Introduction

It is generally known, that oscillation is the most widespread type of motion. There is no any branch of technology without phenomena where the vibrations take place. At this point, the most important characteristics of oscillatory system are natural frequencies and fundamental mode shapes. Problem of characterization is the actual scientific and practical question, which often reduces to the problem of the solving of appropriate differential equation of motion.

Among others, the important question is the investigation of bending vibration in various buildings and constructions taking into account the effect of axial force. Unfortunately, in the most part of the publication [1–5], where the oscillations of these

* Corresponding author: yurii.krutii@gmail.com

constructions are considered, an axial force is accepted constant from any height for simplification.

However, in the real structures the axial forces in different cross sections have different values. For example, columns in industrial buildings have different compression ratio at different stories due to concentrated load at floor level. The industrial high-rise buildings like smokestacks, water towers, multipurpose steel towers, used for electric-power transmission line, wind generators supports, antennas of various constructions may be added to this example.

One of the most widespread design models for investigation of transverse oscillations of mentioned buildings is the uniform cross-section rod under the action of building weight in the capacity of variable axial force. The math model of this physical phenomenon is the differential variable-coefficient equation [3, 6, 7], the exact solution is still unknown. So the investigations are usually made by variational methods. Because of this, the investigations are usually made by variational methods. However, it is evident that the most full and qualitative evaluation of mechanical system can be gained only on the basis of the exact solution of differential equation. As a matter of fact, the focus of this paper is the definition of this solution.

2 Results

2.1 General symbols and equations

Let's analyze the free bending vibrations of the rod with uniform flexural rigidity and intensity of the distributed mass taking into account own weight (Fig.1, Fig.2).

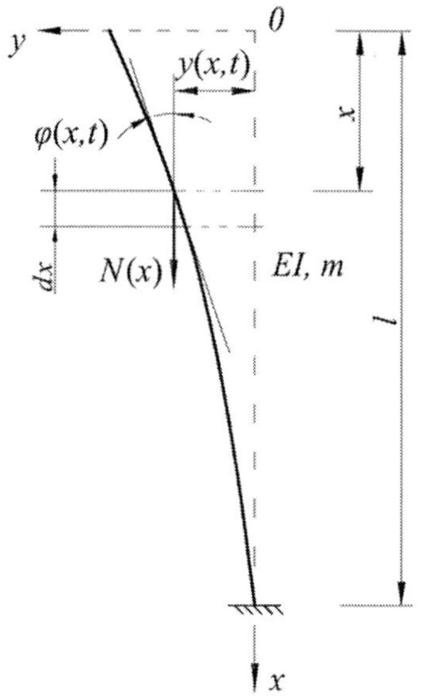

Fig. 1. The scheme of free transverse oscillations of a straight rod.

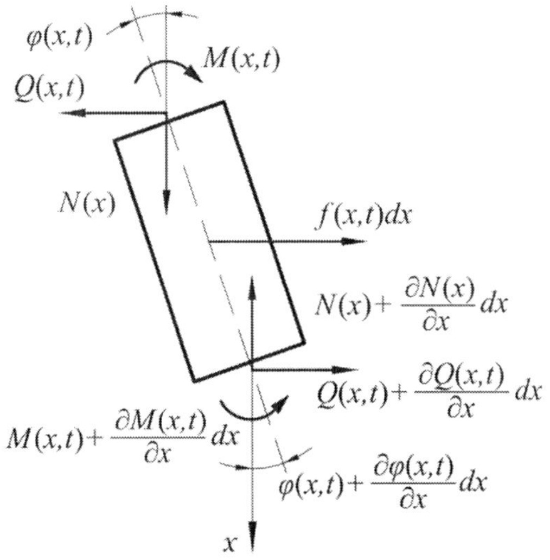

Fig. 2. The scheme of the forces acting upon the element of a rod.

List of symbols:
EI – the flexural rigidity of a rod;
m – the intensity of the distributed mass (own weight) of the rod;
$N(x) = qx$ – the variable axial (compressive) force, where q – weight per unit length of beam;
$y(x,t)$ – the cross motion of the axis point of the rod with coordinate x at time t (dynamic deflection);
$\varphi(x,t)$ – the dynamic angular displacement;
$M(x,t)$ – the dynamic bending moment;
$Q(x,t)$ – the dynamic transverse force;
$f(x,t)$ – the intensity of inertial forces that appear during oscillation (D'Alembert force).

However, we should note, that all the following formulas are valid for any boundary conditions at the ends of the rod.

It is known [3, 6], that the equation of free transverse oscillations taking into account own weight is written as:

$$EI\frac{\partial^4 y}{\partial x^4}+q\frac{\partial}{\partial x}\left(x\frac{\partial y}{\partial x}\right)+m\frac{\partial^2 y}{\partial t^2}=0\,. \tag{1}$$

This equation is valid for a model where the longitudinal displacement of cross sections and their twists and shifts are decided to ignore.

Defined dynamic deflection $y(x,t)$ from Eq.(1), other dynamic parameters of the rod state are given according to the known formulas [3, 6]:

$$\varphi(x,t)=\frac{\partial y}{\partial x};\;\; M(x,t)=-EI\frac{\partial^2 y}{\partial x^2};\; Q(x,t)=\frac{\partial M}{\partial x}-qx\frac{\partial y}{\partial x}. \tag{2}$$

Using Fourier method, the solution of partial derivative equation (1) is given as

$$y(x,t)=v(x)T(t)\,, \tag{3}$$

where $v(x)$ – the amplitude of the transverse displacement, which depends only on variable x; $T(t)$ – the time function t. After the implement (3) in formulas (2), we'll have the similar formulas for the other dynamic parameters:

$$\varphi(x,t)=\varphi(x)T(t);\ \ M(x,t)=M(x)T(t);\ \ Q(x,t)=Q(x)T(t)\,, \tag{4}$$

where $\varphi(x), M(x), Q(x)$ – amplitude functions, which are linked by the equalities

$$\varphi(x)=v'(x);\ \ M(x)=-EIv''(x);\ Q(x)=M'(x)-qx\,v'(x). \tag{5}$$

If we substitute (3) into Eq. (1) and separate the variables there, we will have two differential equations:

$$\ddot{T}(t)+p^2T(t)=0\,; \tag{6}$$

$$EI\,v''''(x)+q(x\,v'(x))'-p^2m\,v(x)=0\,, \tag{7}$$

where p^2 – Fourier method constant.

The solution of the Eq. (6), expressed in terms of parameters of initial motion conditions $T(0), \dot{T}(0)$, is given by

$$T(t)=T(0)\cos pt+\frac{\dot{T}(0)}{p}\sin pt\,.$$

Its analysis shows that motion of mechanical system is oscillatory in nature. The constant is the frequency of free oscillations.

The fundamental mode is the solution of Eq. (7). The main difficulty of the problem is exactly the definition of this solution.

The equivalent system of differential equations is considered along with Eq. (7). The state vector of the rod is used in the function of vector of unknowns. The components of the state vector are amplitude deflections, angular displacement, bending moment and transverse force:

$$\Phi(x)=\begin{pmatrix} v(x)\\ \varphi(x)\\ M(x)\\ Q(x)\end{pmatrix}=\begin{pmatrix} v(x)\\ v'(x)\\ -EI\,v''(x)\\ -EI\,v'''(x)-qx\,v'(x)\end{pmatrix}. \tag{8}$$

Then, writing the collection of formulas (5), (7) in form of matrix we'll have:

$$\frac{d\Phi(x)}{dx}=D(x)\Phi(x)\,. \tag{9}$$

where $D(x)$ – the coefficient matrix of the system, which have the form

$$D(x)=\begin{pmatrix} 0 & 1 & 0 & 0 \\ 0 & 0 & -\frac{1}{EI} & 0 \\ 0 & qx & 0 & 1 \\ -p^2m & 0 & 0 & 0 \end{pmatrix}.$$

2.2 The exact solution of amplitude equation of oscillation. Formulas of state parameters of the rod

For the construction of an exact solution of Eq. (7) we use integration technique proposed and developed in [8].

There are four infinite systems of functions $b_{n,0}(x)$, $b_{n,k}(x)$ $(n=1,2,3,4)(k=1,2,3,...)$, which are decided continuous ones with their derivatives from the first-order to fourth-order inclusive. With the help of these functions and their derivatives, we make the expansion in powers of parameter $P=p^2m$:

$$U_n(x)=b_{n,0}(x)+Pb_{n,1}(x)+P^2b_{n,2}(x)+\cdots; \tag{10}$$

$$U_n^{(v)}(x)=b_{n,0}^{(v)}(x)+Pb_{n,1}^{(v)}(x)+P^2b_{n,2}^{(v)}(x)+\cdots, \tag{11}$$

where (v) – the order of derivate, $v=1,2,3,4$. At the moment we allow, that the ranges (10), (11) are uniform convergent, so it's possible to do term-by-term differentiation.

Following the terminology, accepted in [8], the functions $b_{n,0}(x)$ is called primary, and $b_{n,k}(x)$ $(k=1,2,3,...)$ – generating. We are going to define them on condition that $U_n(x)$ is satisfied for Eq. (7), so

$$EI\,U_n''''(x)+q(xU_n'(x))'-PU_n(x)=0\ (n=1,2,3,4). \tag{12}$$

Substituting the values (10), (11) instead of $U_n(x), U_n'(x), U_n''(x), U_n''''(x)$, after transformations one arrives the equality

$$EI\,b_{n,0}''''(x)+q(xb_{n,0}'(x))'+\sum_{k=1}^{\infty}P^k\,(EI\,b_{n,k}''''(x)+q(x\,b_{n,k}'(x))'-b_{n,k-1}(x))=0. \tag{13}$$

For its satisfaction set all the coefficients at powers of parameter P, beginning from zero power:

$$EI\,b_{n,0}''''(x)+q(x\,b_{n,0}'(x))'=0 \tag{14}$$

$$EI\,b_{n,k}''''(x)+q(x\,b_{n,k}'(x))'=b_{n,k-1}(x)\ \ (k=1,2,3,...) \tag{15}$$

So, the differential equations for primary and generating function definition are attained. Use the power series method for integration this equations, previously accepted next boundary conditions:

$$\begin{pmatrix} b_{1,0}(0) & b_{2,0}(0) & b_{3,0}(0) & b_{4,0}(0) \\ b'_{1,0}(0) & b'_{2,0}(0) & b'_{3,0}(0) & b'_{4,0}(0) \\ b''_{1,0}(0) & b''_{2,0}(0) & b''_{3,0}(0) & b''_{4,0}(0) \\ b'''_{1,0}(0) & b'''_{2,0}(0) & b'''_{3,0}(0) & b'''_{4,0}(0) \end{pmatrix} = \begin{pmatrix} 1 & 0 & 0 & 0 \\ 0 & 1 & 0 & 0 \\ 0 & 0 & 1 & 0 \\ 0 & 0 & 0 & 1 \end{pmatrix} \tag{16}$$

$$b_{n,k}(0) = b'_{n,k}(0) = b''_{n,k}(0) = b'''_{n,k}(0) = 0 \ \ (n = 1,2,3,4)(k = 1,2,3,...). \tag{17}$$

Integrating both parts (14), one arrives at the equation

$$EI\, b'''_{n,0}(x) + qx\, b'_{n,0}(x) = A_n \ \ (n = 1,2,3,4) \tag{18}$$

where A_n – permanent integration. Accepted the equation $x = 0$, define $A_n = EI\, b'''_{n,0}(0)$. Taking into account (16) we have: $A_1 = A_2 = A_3 = 0$; $A_4 = EI$.

Integrating both part (15) and taking into account boundary conditions (17), we will have

$$EI\, b'''_{n,k}(x) + qx\, b'_{n,k}(x) = \int_0^x b_{n,k-1}(x)\,dx \ (k = 1,2,3,...) \tag{19}$$

Solution of Eq. (18), (19) is definite in form of series:

$$b_{n,0}(x) = x^{n-1} \sum_{j=0}^{\infty} C_{n,0,j} x^{3j} \tag{20}$$

$$b_{n,k}(x) = x^{n+4k-1} \sum_{j=0}^{\infty} C_{n,k,j} x^{3j} \ \ (k = 1,2,3,...) \tag{21}$$

According to this choice, the boundary conditions for generating functions (17) are satisfied in advance, and after boundary condition realization for primary function (16), we'll have:

$$C_{n,0,0} = \frac{1}{(n-1)!} \ (n = 1,2,3,4); \ \ C_{1,0,1} = 0 \tag{22}$$

Based on (20), (21), we find:

$$x b'_{n,0}(x) = x^{n-1} \sum_{j=0}^{\infty} e_{n,0,j} C_{n,0,j} x^{3j}; \tag{23}$$

$$\begin{gathered} b'''_{n,0}(x) = x^{n-1} \sum_{j=0}^{\infty} f_{n,0,j} C_{n,0,j} x^{3j-3} = (n-3)(n-2)(n-1) C_{n,0,0} x^{n-4} + \\ + x^{n-1} \sum_{j=0}^{\infty} f_{n,0,j+1} C_{n,0,j+1} x^{3j}; \end{gathered} \tag{24}$$

$$x b'_{n,k}(x) = x^{n+4k-1} \sum_{j=0}^{\infty} e_{n,k,j} C_{n,k,j} x^{3j} \ \ (k = 1,2,3,...); \tag{25}$$

$$b'''_{n,k}(x) = x^{n+4k-1}\sum_{j=0}^{\infty} f_{n,k,j}C_{n,k,j}x^{3j-3} = x^{n+4k-4}f_{n,k,0}C_{n,k,0} + x^{n+4k-1}\sum_{j=0}^{\infty} f_{n,k,j+1}C_{n,k,j+1}x^{3j} \quad (k=1,2,3...); \tag{26}$$

$$\int_0^x b_{n,k-1}(x)\,dx = x^{n+4k-1}\sum_{j=0}^{\infty}\frac{C_{n,k-1,j}}{e_{n,k,j-1}}x^{3j-3} = \frac{C_{n,k-1,0}}{e_{n,k,-1}}x^{n+4k-4} + x^{n+4k-1}\sum_{j=0}^{\infty}\frac{C_{n,k-1,j+1}}{e_{n,k,j}}x^{3j} \quad (k=1,2,3,...), \tag{27}$$

where

$$e_{n,k,j} = n+4k+3j-1\,,\ f_{n,k,j} = (n+4k+3j-3)(n+4k+3j-2)e_{n,k,j}\,.$$

After substitution (23), (24) in Eq. (18), and (25)–(27) in Eq. (19), we arrives at the equalities:

$$(n-3)(n-2)(n-1)C_{n,0,0}x^{n-4} + x^{n-1}\sum_{j=0}^{\infty} f_{n,0,j+1}C_{n,0,j+1}x^{3j} + \frac{q}{EI}x^{n-1}\sum_{j=0}^{\infty} e_{n,0,j}C_{n,0,j}x^{3j} = \frac{A_n}{EI} \quad (n=1,2,3,4); \tag{28}$$

$$\begin{gathered} f_{n,k,0}C_{n,k,0}x^{n+4k-4} + x^{n+4k-1}\sum_{j=0}^{\infty} f_{n,k,j+1}C_{n,k,j+1}x^{3j} + \frac{q}{EI}x^{n+4k-1}\sum_{j=0}^{\infty} e_{n,k,j}C_{n,k,j}x^{3j} = \\ = \frac{1}{EI}\frac{C_{n,k-1,0}}{e_{n,k,-1}}x^{n+4k-4} + \frac{1}{EI}x^{n+4k-1}\sum_{j=0}^{\infty}\frac{C_{n,k-1,j+1}}{e_{n,k,j}}x^{3j} \quad (k=1,2,3,...). \end{gathered} \tag{29}$$

So the coefficients at the same powers are equated, we'll have:

$$C_{n,0,j+1} = -\frac{q}{EI}\frac{e_{n,0,j}}{f_{n,0,j+1}}C_{n,0,j} \quad (j=0,1,2,...) \tag{30}$$

$$C_{n,k,0} = \frac{1}{EI}\frac{C_{n,k-1,0}}{f_{n,k,0}\,e_{n,k,-1}} \quad (k=1,2,3,...) \tag{31}$$

$$C_{n,k,j+1} = \frac{1}{EI\,f_{n,k,j+1}}\left(\frac{C_{n,k-1,j+1}}{e_{n,k,j}} - qe_{n,k,j}C_{n,k,j}\right) \quad (k=1,2,3,...)(j=0,1,2,...). \tag{32}$$

Formulas (30), (31) taking into account (22) are possible to be written in the explicit form. On the whole, the collection (22), (30) will be equal to formulas:

$$C_{n,0,0} = \frac{1}{(n-1)!} \quad (n=1,2,3,4) \tag{33}$$

$$C_{n,0,j+1} = (-1)^{j+1}\left(\frac{q}{EI}\right)^{j+1}\frac{(n-1)(n+2)...(n+3j-1)}{(n+3j+2)!} \quad (j=0,1,2,...), \tag{34}$$

based on formula (31) we define

$$C_{n,k,0} = \frac{1}{(EI)^k} \frac{1}{(n+4k-1)!} \quad (k = 1,2,3,...) \tag{35}$$

Thus, the coefficients of series (20), (21) are totally definite by formulas (33)–(35) and recurrence formula (32). Disadvantage of these formulas is that they are depending on initial dimensional problem parameters EI and q. This disadvantage can be avoided making the next substitution

$$C_{n,k,j} = \frac{(-1)^j}{(EI)^k}\left(\frac{q}{EI}\right)^j c_{n,k,j} \quad (n = 1,2,3,4)(k = 1,2,3,...)(j = 0,1,2,...). \tag{36}$$

Substituted (36) in (32)–(35) and lowered the j index in the finally formulas by one, we'll have next formulas for definition of new non-dimensional coefficients $c_{n,k,j}$:

$$c_{n,0,0} = \frac{1}{(n-1)!} \quad (n = 1,2,3,4); \tag{37}$$

$$c_{n,0,j} = \frac{(n-1)(n+2)...(n+3j-4)}{(n+3j-1)!} \quad (j = 1,2,3,...); \tag{38}$$

$$c_{n,k,0} = \frac{1}{(n+4k-1)!} (k = 1,2,3,...); \tag{39}$$

$$c_{n,k,j} = \frac{1}{f_{n,k,j}}\left(\frac{c_{n,k-1,j}}{e_{n,k,j-1}} + e_{n,k,j-1}\, c_{n,k,j-1}\right) \quad (k = 1,2,3,...)(j = 1,2,3,...). \tag{40}$$

As a result, formulas (20), (21) take the form:

$$b_{1,0}(x) = 1; b_{n,0}(x) = x^{n-1}\left(c_{n,0,0} + \sum_{j=1}^{\infty}(-1)^j\left(\frac{q}{EI}\right)^j c_{n,0,j}\, x^{3j}\right) \quad (n = 2,3,4); \tag{41}$$

$$b_{n,k}(x) = \frac{x^{n+4k-1}}{(EI)^k}\left(c_{n,k,0} + \sum_{j=1}^{\infty}(-1)^j\left(\frac{q}{EI}\right)^j c_{n,k,j}\, x^{3j}\right) \quad (k = 1,2,3,...). \tag{42}$$

So four solutions $U_n(x)$ $(n = 1,2,3,4)$ are defined by formulas (10), (37)–(42) Eq.(15). Accordingly each of these solutions by formula (8) generates the solution of system (9)

$$\Phi_n(x) = \begin{pmatrix} U_n(x) \\ U_n'(x) \\ -EI\,U_n''(x) \\ -EI\,U_n'''(x) - qxU_n'(x) \end{pmatrix} (n = 1,2,3,4) \tag{43}$$

Then the matrix $\Omega(x) = \|\Phi_1(x)\ \Phi_2(x)\ \Phi_3(x)\ \Phi_4(x)\|$, consisted of these vectors, also will satisfy the system (9).

Based on formulas (10), (11) and boundary conditions (17) for vector $\Phi_n(0)$ $(n = 1,2,3,4)$ we'll have

$$\Phi_n(0)=\begin{pmatrix} U_n(0) \\ U'_n(0) \\ -EI\,U''_n(0) \\ -EI\,U'''_n(0) \end{pmatrix}=\begin{pmatrix} b_{n,0}(0) \\ b'_{n,0}(0) \\ -EI\,b''_{n,0}(0) \\ -EI\,b'''_{n,0}(0) \end{pmatrix}.$$

Therefore taking into account (16) we find:

$$\Phi_1(0)=\begin{pmatrix}1\\0\\0\\0\end{pmatrix},\Phi_2(0)=\begin{pmatrix}0\\1\\0\\0\end{pmatrix},\Phi_3(0)=\begin{pmatrix}0\\0\\-EI\\0\end{pmatrix},\Phi_4(0)=\begin{pmatrix}0\\0\\0\\-EI\end{pmatrix},$$

i.e.

$$\Omega(0)=\begin{pmatrix}1&0&0&0\\0&1&0&0\\0&0&-EI&0\\0&0&0&-EI\end{pmatrix}.$$

Once $|\Omega(0)|=(EI)^2\neq 0$, that the vectors (43) and the functions $U_n(x)$ $(n=1,2,3,4)$ are linearly independent[9]. Thus, the matrix $\Omega(x)$ is the fundamental matrix of the system (9), and $U_n(x)$ $(n=1,2,3,4)-$ are the fundamental solutions of Eq. (7).

Multiplied matrix on the right $\Omega(x)$ by the constant matrix $\Omega^{-1}(0)$ we'll get new fundamental matrix of the system (9) $\Lambda(x)=\Omega(x)\Omega^{-1}(0)$, and this

$$\Lambda(0)=\begin{pmatrix}1&0&0&0\\0&1&0&0\\0&0&1&0\\0&0&0&1\end{pmatrix}. \tag{44}$$

Fundamental matrix, which satisfies the condition (44), is uniquely determined and called matrizant [9].

Thus, allowing for the uniform convergence at the beginning (10), (11), we arrive at a conclusion that the matrix $\Lambda(x)$, which consists of the sums of these series $U_n(x), U'_n(x), U''_n(x), U'''_n(x)$ $(n=1,2,3,4)$, is the matrizant of the system (9). On the other hand, in the differential equations theory it is proved that matrizant of equation system with continuous coefficients is always absolutely and uniform convergent matrix series [9]. It follows that because of the uniqueness of matrizant the series (10) and (11) are absolutely and uniform convergent for $\nu=1,2,3$. As for series (11) for $\nu=4$, its uniform convergence follows immediately from identity (12). So, previously accepted designations $U_n^{(\nu)}$ $(\nu=1,2,3,4)$ for series (11) are correct.

General solution of differential equation system with the presence of matrizant is given by known [9] formula $\Phi(x)=\Lambda(x)\Phi(0)$. Writing it in expanded form we get the formulas for amplitude deflections and internal forces:

$$v(x) = v(0)U_1(x) + \varphi(0)U_2(x) - M(0)\frac{1}{EI}U_3(x) - Q(0)\frac{1}{EI}U_4(x); \tag{45}$$

$$\varphi(x) = v(0)U_1'(x) + \varphi(0)U_2'(x) - M(0)\frac{1}{EI}U_3'(x) - Q(0)\frac{1}{EI}U_4'(x); \tag{46}$$

$$M(x) = -v(0)EI\,U_1''(x) - \varphi(0)EI\,U_2''(x) + M(0)U_3''(x) + Q(0)U_4''(x); \tag{47}$$

$$\begin{aligned} Q(x) = &-v(0)(EI\,U_1'''(x) + qxU_1'(x)) - \varphi(0)(EI\,U_2'''(x) + qxU_2'(x)) + \\ &+ M(0)\left(U_3'''(x) + \frac{qx}{EI}U_3'(x)\right) + Q(0)\left(U_4'''(x) + \frac{qx}{EI}U_4'(x)\right). \end{aligned} \tag{48}$$

2.3 Parameters of free transverse oscillation of the rod, expressed in terms of non-dimensional functions

Pass from the fundamental system of solutions Eq. (7) $U_n(x)\ (n = 1,2,3,4)$ to a new fundamental system of non-dimensional functions $X_n(x)\ (n = 1,2,3,4)$. To this end, the non-dimensional multipliers in the formulas of primary and generating functions (41), (42) must be identified. As a result, we'll have:

$$b_{n,0}(x) = l^{n-1}\beta_{n,0}(x);\ \ b_{n,k}(x) = \frac{l^{n+4k-1}}{(EI)^k}\beta_{n,k}(x)\ \ (n = 1,2,3,4)(k = 1,2,3,...), \tag{49}$$

where $\beta_{n,0}(x)$, $\beta_{n,k}(x)\ (n = 1,2,3,4)\ (k = 1,2,3,...)$ – new non-dimensional primary and generating functions, are given by:

$$\beta_{1,0}(x) = 1; \beta_{n,0}(x) = \left(\frac{x}{l}\right)^{n-1}\left(c_{n,0,0} + \sum_{j=1}^{\infty}(-1)^j\alpha^j c_{n,0,j}\left(\frac{x}{l}\right)^{3j}\right)\ \ (n = 2,3,4); \tag{50}$$

$$\beta_{n,k}(x) = \left(\frac{x}{l}\right)^{n+4k-1}\left(c_{n,k,0} + \sum_{j=1}^{\infty}(-1)^j\alpha^j c_{n,k,j}\left(\frac{x}{l}\right)^{3j}\right)\ \ (k = 1,2,3,...), \tag{51}$$

where $\alpha = \dfrac{ql^3}{EI}$ – non-dimensional parameter.

Using (49), we comes to new primary and generating functions in the formula (10) . As a result, we have:

$$U_n(x) = l^{n-1}X_n(x)\ \ (n = 1,2,3,4), \tag{52}$$

where $X_n(x)\ (n = 1,2,3,4)$ – new fundamental functions, are given by:

$$X_n(x) = \beta_{n,0}(x) + K^2\beta_{n,1}(x) + K^4\beta_{n,2}(x) + ...\ \ (n = 1,2,3,4);$$

$$K = p\,l^2\sqrt{\frac{m}{EI}} \tag{53}$$

The zero dimension of functions $\beta_{n,0}(x)$, $\beta_{n,k}(x)$ $(n=1,2,3,4)$ $(k=1,2,3,...)$, and the zero dimension of the parameter K provide evidently dimensionless character of functions $X_n(x)$ $(n=1,2,3,4)$.

Differentiating (52) we'll also have

$$U_n'(x)=l^{n-2}\tilde{X}_n(x)\,,\;U_n''(x)=l^{n-3}\hat{X}_n(x)\,,\;U_n'''(x)=l^{n-4}\hat{X}_n(x)\;(n=1,2,3,4),\qquad(54)$$

where

$$\tilde{X}_n(x)=lX_n'(x),\;\hat{X}_n(x)=l\tilde{X}_n'(x),\;\hat{X}_n(x)=l\hat{X}_n'(x)\quad(n=1,2,3,4).$$

It proves that the functions $\tilde{X}_n(x)$, $\hat{X}_n(x)$, $\hat{X}_n(x)$ $(n=1,2,3,4)$ are expressed in the form of series:

$$\tilde{X}_n(x)=\tilde{\beta}_{n,0}(x)+K^2\tilde{\beta}_{n,1}(x)+K^4\tilde{\beta}_{n,2}(x)+K^6\tilde{\beta}_{n,3}(x)+...;$$

$$\hat{X}_n(x)=\hat{\beta}_{n,0}(x)+K^2\hat{\beta}_{n,1}(x)+K^4\hat{\beta}_{n,2}(x)+K^6\hat{\beta}_{n,3}(x)+...;$$

$$\hat{X}_n(x)=\hat{\beta}_{n,0}(x)+K^2\hat{\beta}_{n,1}(x)+K^4\hat{\beta}_{n,2}(x)+K^6\hat{\beta}_{n,3}(x)+...,$$

where

$$\tilde{\beta}_{n,k}(x)=l\beta'_{n,k}(x),\;\hat{\beta}_{n,k}(x)=l\tilde{\beta}'_{n,k}(x),\hat{\beta}_{n,k}(x)=l\hat{\beta}'_{n,k}(x)\;\;(n=1,2,3,4)\;(k=0,1,2,...).\quad(55)$$

Based on (50), (51), (55), it can be seen that $\tilde{\beta}_{n,k}(x)$, $\hat{\beta}_{n,k}(x)$, $\hat{\beta}_{n,k}(x)$ $(n=1,2,3,4)$ $(k=0,1,2,...)$ functions are non-dimensional. So the functions $\tilde{X}_n(x), \hat{X}_n(x)$, $\hat{X}_n(x)$ $(n=1,2,3,4)$ will also be non-dimensional.

Using (52), (54) we'll get the final formulas for amplitude parameters of free transverse oscillation of the rod, expressed in terms of initial parameters and non-dimensional function:

$$v(x)=v(0)X_1(x)+\varphi(0)lX_2(x)-M(0)\frac{l^2}{EI}X_3(x)-Q(0)\frac{l^3}{EI}X_4(x);\qquad(56)$$

$$\varphi(x)=v(0)\frac{1}{l}\tilde{X}_1(x)+\varphi(0)\tilde{X}_2(x)-M(0)\frac{l}{EI}\tilde{X}_3(x)-Q(0)\frac{l^2}{EI}\tilde{X}_4(x);\qquad(57)$$

$$M(x)=-v(0)\frac{EI}{l^2}\hat{X}_1(x)-\varphi(0)\frac{EI}{l}\hat{X}_2(x)+M(0)\hat{X}_3(x)+Q(0)l\hat{X}_4(x);\qquad(58)$$

$$\begin{aligned}Q(x)=&-v(0)\left(\frac{EI}{l^3}\hat{X}_1(x)+\frac{qx}{l}\tilde{X}_1(x)\right)-\varphi(0)\left(\frac{EI}{l^2}\hat{X}_2(x)+qx\,\tilde{X}_2(x)\right)+\\&+M(0)\left(\frac{1}{l}\hat{X}_3(x)+\frac{ql\,x}{EI}\tilde{X}_3(x)\right)+Q(0)\left(\hat{X}_4(x)+\frac{ql^2x}{EI}\tilde{X}_4(x)\right).\end{aligned}\qquad(59)$$

It is important to note, that dimensional of constant coefficients in non-dimensional functions in the right parts of formulas (56)–(59) are same as the appropriate left parts.

Realized specified boundary condition with the help of formulas (56)–(59), we'll get the frequency equation in terms of unknown non-dimensional parameter K. Found out the frequency equation roots $K_1, K_2, K_3, \ldots$, we'll have the frequency spectrum in according to formula (53)

$$p_j = \frac{K_j}{l^2}\sqrt{\frac{EI}{m}} \; (j = 1,2,3,\ldots).$$

Appropriate fundamental modes are given by formula (56).

3 Conclusions

In this work, the exact solution of differential partial derivative equation of free transverse oscillation of a rod has been attained. Consequently, the formulas for dynamic variables of rode state (deflections, angular displacement, bending moment and transverse force) expressed in terms of non-dimensional fundamental solutions, have been defined. It made possible to attain the analytical form for equation of oscillation frequency.

In prospect, the exact solutions will allow to investigate the free oscillations of the rod taking into account own weight with any boundary conditions on their ends.

References

1. S. Timoshenko, D.H. Young, W. Weaver, Vibration Problems in Engineering. (John Wiley & Sons, New York, 1990)
2. I.M. Babakov, *Oscillation theory* (Moscow, 1965)
3. B. Balachandran, E. Magrab, *Vibrations, second edition* (Cengage Learning, Toronto, 2009)
4. S.S. Rao, *Mechanical vibrations* (Pearson, Prentice Hall, 2004)
5. F. J. Shaker, Lewis Research Center Report NASA TN D-8109 (1975)
6. M. Vasilenko, A. Alekseychuk, *Theory of oscillations and stability of motion* (Kyiv, High School, 2004)
7. S. Kukla, B. Skalmierski, Journal of Theoretical and Applied Mechanics, **31 (2)**, 413–430 (1993)
8. Yu. S. Krutiy, Thesis for Doctor Degree in Technical Sciences (Odessa, OSACEA, 2016)
9. F. R. Gantmacher, *The Theory of Matrices* (AMS Chelsea Publishing: Reprinted by American Mathematical Society, 2000)

Modeling of the stress-strain state of the ground mass contaminated with peracetic acid

Anna Levenko[1]*

[1]Department of Soil Mechanics, Foundations and Engineering Geology, O. M. Beketov National University of Urban Economy in Kharkiv, 17, Marshal Bazhanov Street, Kharkiv, 61002, Ukraine.

Abstract. None of the methods described previously provides a solution to the problem that deals with the SSS evaluation of the ground mass which is under the influence of chemically active substances and, in particular, under the influence of peracetic acid. The stress-strain state of the ground mass contaminated with peracetic acid was estimated. Stresses occurring in the ground mass in the natural state were determined after the entry of acid into it and after the chemical fixation of it with sodium silicate. All the parameters of the stress-strain state of the ground mass were obtained under a number of physical and mechanical conditions. It was determined that following the work on the silicatization of the ground mass contaminated with peracetic acid the quantity of strain decreased by 26.11 to 48.9%. The comparison of the results of stress calculations indicates the stress reduction in the ground mass in 1.8 – 2.6 times after its fixing.

Introduction

Currently, there is a problem resulting from the ground contamination with industrial effluents directly under industrial buildings, which in turn leads to the swelling of ground bases, the shifting of foundations and the destruction of building structures [1, 2, 5].

Therefore, when designing it is necessary to take into account the swelling effect on the entire "ground base–foundation–building" system for more reliable operation of the structure.

The existing standards [3, 4] only regulate the design rules for swelling grounds but do not provide an opportunity to predict the further behavior of the ground base in case of chemical contamination.

In addition, not much is known about the behavior of the swelling grounds after their chemical fixation and the stress-strain state (SSS) of the "fixed ground base–foundation – building" system.

Analysis of recent studies and publications

The solution to the problem taking into account the joint action of the "ground base – foundation – building" system is explained in the paper [9]. It is proposed to represent the

* Corresponding author: levenkoanna@mail.ru

ground mass consisting of the swelling grounds as a linearly deformable medium, to take into account the swelling of the ground base as an additional action that is close in nature to the influence of temperature, and to consider the swelling ground as a material possessing orthotropic properties.

Some authors [8] believe that the finite element method is the most suitable for evaluating the stress-strain state (SSS) of the "ground base – foundation – building" system. As this takes place, it is recommended to simulate separately each of the parts of the system that makes it possible to see changes in the SSS of system and to adjust the decisions taken while performing work.

In papers [6, 7] the methods of accounting for the effect of the ground base swelling on the SSS of the "ground base – foundation – building" system under plane deformation are presented.

Additional pressure caused by the ground swelling as an external load applied to the foundations in the areas of the swelling knoll influence is recommended to be taken into account.

However, none of the methods described previously provides a solution to the problem that deals with the SSS evaluation of the ground mass which is under the influence of chemically active substances and, in particular, under the influence of peracetic acid.

The objectives and tasks of the problem research

The objective of the research is to assess the stress-strain state of a ground mass contaminated with peracetic acid.

The main part of the research

The present research has been implemented as a numerical experiment consisting of a series of calculations for various ground conditions; the four types of ground bases with the four calculated behavioral patterns were under research.

Table 1. Values of characteristics of the deformation modulus and the Poisson's ratio.

Ground type	Poisson's ratio			Deformation modulus, MPa		
	Before fixing while in the natural state	After raising the acid content level	After fixing	Before fixing while in the natural state	After raising the acid content level	After fixing
Fine quartz sand	0.30	0.35	0.27	31.5	21.5–22.03	38.9
Brownish-yellow sandy loam with plastic behavior	0.34	0.39	0.30	22.6	19.9–21.9	42
Heavy clay loam with hard plastic behavior	0.36	0.41	0.30	26.7	16.9–24.8	38
Gray semisolid clay	0.42	0.50	0.36	22.4	18.9–21.9	37.9

Basic hypotheses and assumptions are as follows: the model of the ground base is the model of a linearly deformable half-space.

The specified stiffness characteristics are as follows: E_i – the deformation modulus, v_i – Poisson's ratio, and P_i – the load transferred to the ground base.

Linearly deformable medium is considered; the ground mass with the dimensions of 10×10 m is under the plane deformation. On centre the ground mass perceives a load from the foundation with a width of 1 m (Area 3).

Inside the ground mass there is Area 1 with the dimensions of 5×5 m subjected to the raising of the acid content level and further fixing. The rest of the ground mass is represented by Area 2.

The following calculation schemes are taken into account:

- the calculation scheme 1. Areas 2, 3 are the grounds with normal (natural) characteristics. Area 3 is the C16/20 concrete foundation. The load on Area 3 is the concentrated force of 100 tons;
- the calculation scheme 2. Area 2 is the ground with normal (natural) characteristics. Area 1 is the ground with the raised acid content levels. Area 3 is the C16/20 concrete foundation. The load on Area 3 is the concentrated force of 100 tons;
- the calculation scheme 3. Area 2 is the ground with normal (natural) characteristics. Area 1 is the ground with the raised acid content level characteristics. Area 3 is the C16/20 concrete foundation made from. The load on Area 3 is the concentrated force of 125 tons;
- the calculation scheme 4. Area 2 is the ground with normal (natural) characteristics. Area 2 is the ground with fixed characteristics. Area 3 is the C16/20 concrete foundation. The load on area 3 is the concentrated force of 125 tons.

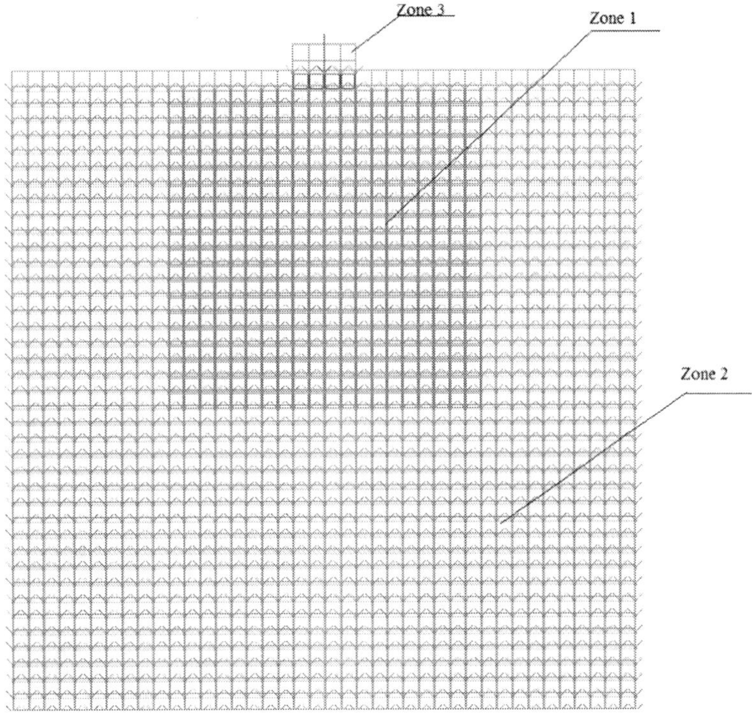

Fig. 1. Finite element model. Area 1 is the ground with the raised acid content level (fixed) characteristics; Area 2 is the ground with natural characteristics; Area 3 is the foundation.

The coordinates of the model points were specified according to the results of the evaluated amount of the calculation scheme deformation.

Modeling was carried out in the CAD Lira. The finite element model (Fig. 1) was formed from FE21 that was a rectangular element of the linearly deformable medium (beam – wall).

The model included 1.686 points and 1.604 elements. The thickness of elements was arbitrarily set as equal to 1 m.

As a result, the characteristics of the stress-strain state of the ground base were obtained under the four different conditions of contamination and loading.

The main criterion for comparison was taken by the values of shifts along the X, Y and Z axes, as well as the values of stresses arising in the ground mass.

The results of the calculation are illustrated by the fields of vertical shifts of the ground mass (Figure 3) and in a tabular form (Table 2).

Table 2. Stresses and vertical shifts of the ground mass.

			Ground type			
			Fine quartz sand	Brownish-yellow sandy loam with plastic behavior	Heavy clay loam with hard plastic behavior	Gray semisolid clay
Vertical shift of the ground mass, mm	*CS-1*		92	128	108	129
	CS -2		116	134	125	138
	CS -3		137	158	148	138
	CS -4		80,3	123	108	117
Stresses in the ground mass, kN/cm²	*CS -1*	σ_x	– 0.0377	– 0.0369	– 0.0366	– 0.355
		σ_z	– 0.101	– 0.101	– 0.101	– 0.1
		τ_{xz}	– 0.0679	– 0.0679	– 0.0678	– 0.0678
	CS -2	σ_x	– 0.37	– 0.036	– 0,357	– 0,0339
		σ_z	– 0,101	– 0,101	– 0.1	– 0.0998
		τ_{xz}	– 0.0679	– 0.0678	– 0.678	– 0.0676
	CS -3	σ_x	– 0.0462	– 0.045	– 0.447	– 0.0424
		σ_z	– 0.126	– 0.0126	– 0.125	– 0.125
		τ_{xz}	– 0.0849	– 0.0847	– 0.847	– 0.845
	CS -4	σ_x	– 0.0373	– 0.0368	– 0.369	– 0.0357
		σ_z	– 0.101	– 0.101	– 0.101	– 0.1
		τ_{xz}	– 0.0832	– 0.0833	– 0.833	– 0.0832

Table 2 also shows the stresses arising in the ground mass for four ground groups under conditions specified by the calculation schemes (CS) CS-1, CS-2, CS-3 and CS-4.

From the results of the calculations, the picture (Fig. 2) of the change in the magnitude of the deformation of the ground mass under the four given conditions is clearly visible. After carrying out works on fixing the ground mass contaminated with solutions of peracetic acid the value of deformations decreases

by 48.9% for grounds with the parameter L <3%,

by 26.11% for grounds with the parameter of L 3 –10%,

by 32% for grounds with the parameter of L 10 – 30%,

by 33.4% for grounds with the parameter of L > 30%.

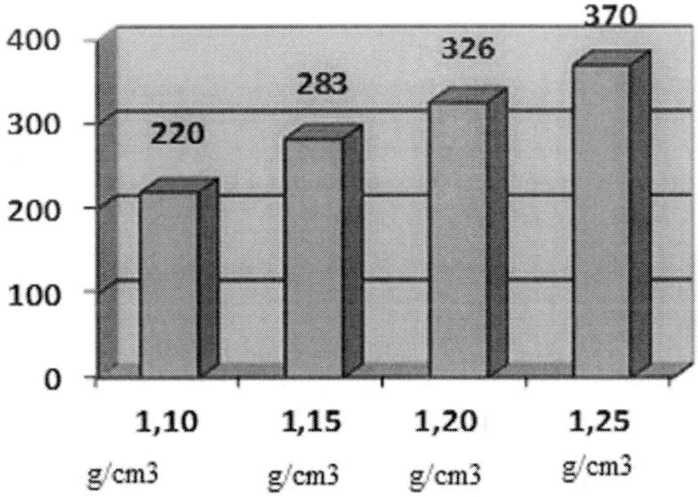

Fig. 2. The graph of the change in the magnitude of deformations of the ground mass under various calculation schemes.

This clearly indicates a qualitative improvement in the state of the ground mass after its fixation according to the recommended formulas. After carrying out works aimed at improving the structural properties of the ground mass contaminated with industrial effluents of peracetic acid the deformations of grounds are almost equal to the deformations in their natural state.

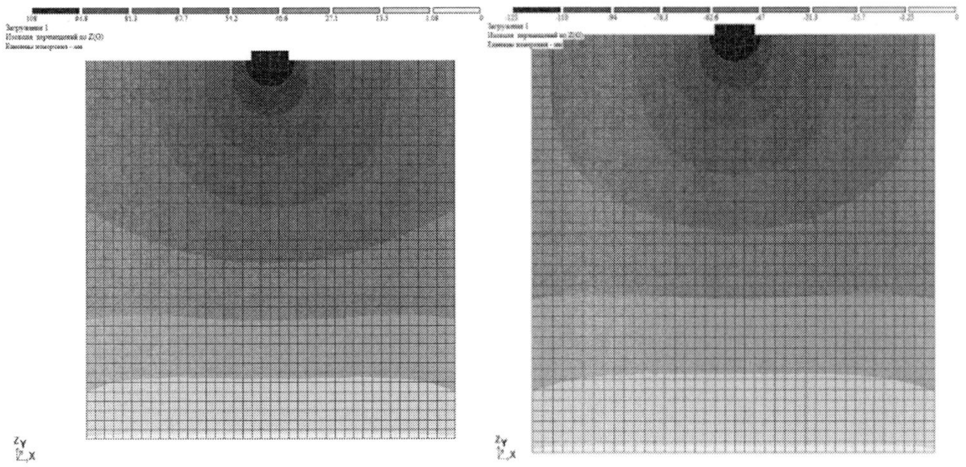

a) b)

Fig. 3. Vertical shifts of the ground mass: a) CS-1; b) CS-2

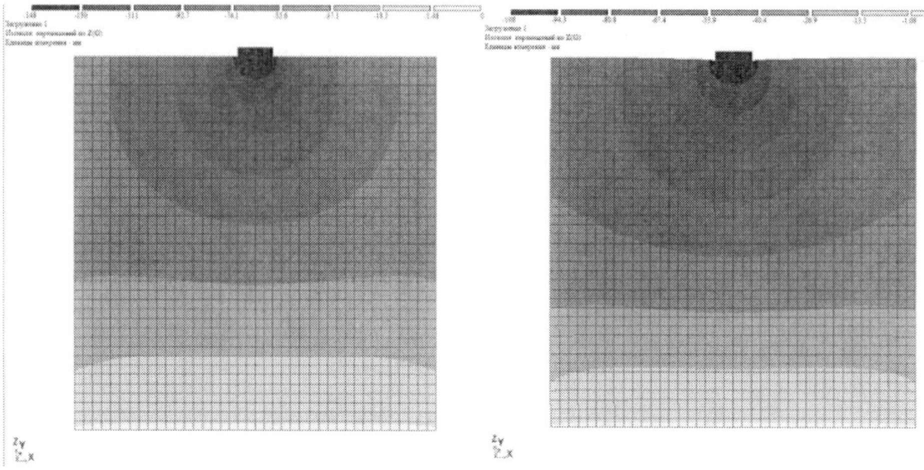

c) d)

Fig. 4. Vertical shifts of the ground mass: c) CS-3; d) CS-4

Conclusion

1. The evaluation of the stress-strain state of the ground mass contaminated with peracetic acid is performed.
2. As a result of the computer modeling of the ground mass contaminated with the solutions of peracetic acid it is determined that after the fixation of grounds with silicization, the value of deformations decreases by 26.11 – 48.9% depending on the ground base type.
3. The comparison of the results of stress calculations indicates the stress reduction in the ground mass in 1.8 – 2.6 times after its fixing.

References

1. G. Radhakrishnan, Dr M. Anjan Kumar. Dr GVR Prasada Raji. American Journal of Engineering Research, **03(04)**, 245-250, (2014)
2. Patil B.M., Patil K.A. Facta Universitatis Series: Architecture and Civil Engineering **11(3)**, 231-236 (2013)
3. *DBN A.2.1-1-2008 Inzhenernyie izyiskaniya dlya stroitelstva* Kiev, (2008)
4. *DSTU B V.2.1-3-96 (GOST 30416-96). Grunti. LaboratornI viprobuvannya. ZagalnI polozhennya* Kiev, (1997)
5. Takahashi A., D. W. H. Fung, R. J. Jarsine. *Proc. 16th Intern. Conf. Soil Mecg. And Geot. Engng.* Osaca, 443-446 (2005)
6. Vinnikov Yu.L. *Matematichne modelyuvannya vzaemodiyi fundamentiv z uschilnenimi osnovami pri yih zvedenni ta nastupniy roboti.* Poltava, PoltNTU (2004)
7. Gabibov F.G. *Problemyi regulirovaniya svoystv strukturno-neustoychivyih glinistyih gruntov v osnovaniyah sooruzheniy.* Baku, «ELM», (1997)
8. Voblyih V.A., Kichaeva O.V. Naukoviy vIsnik budIvnitsta. Kharkiv, HGTUSA, **23**, 55-59 (2003)
9. Hrapatova I.V. svit geotehniki: naukovo-tehnichniy zhurnal, Zaporizhzhya, **2**, 26–29 (2008)

The research of bearing capacity of reinforced concrete beam with use combined experimental-computational method

Olga Nekora[1], *Vitaliy* Slovynsky[1], and *Serhii* Pozdieiev[1*]

[1]Cherkassy Institute of Fire Safety named after Chernobyl Heroes National University of Civil Protection of Ukraine, 18000 Cherkassy, Ukraine

Abstract. The article shows results of behavior research of reinforced concrete beam exposed to fire using standard "time-temperature" curve. Practice of fire resistance estimating for reinforced concrete beams based on experimental and calculation method was developed and tested on the basis of the study results. Block diagram of temperature determining algorithm for nodal points of the cross-section by interpolation of temperature values in measurement points of the cross-section was showed. The fire test of reinforced concrete beams was carried out and a temperature distribution in cross-section of beams was defined based on it by application of interpolation techniques developed. The sequence of procedures of the strength calculation method based on temperature measuring in the inner layers of reinforced concrete beams was demonstrated. Fire resistance of the reinforced concrete beams under question was evaluated based on the interpretation of the data obtained during the fire tests and using the proposed method of strength calculating.

Timeliness

Due to a large volume of construction in our country, complexity and increased risk of accidents for people, who stay in buildings, in case of fire, there are many crucial issues to reduce the risk of injuries and deaths in case of collapse of building structures.

One of the main sources of risk in building structures due to fire heat exposure is low design safe load factor for such extreme conditions. Collapse of structures is of particular danger for people staying in a building for a time longer than the one required for the structure to exhaust its strength resources. Such people are exposed to danger during evacuation, especially when it is difficult, as the personnel of rescue units engaged in localization and liquidation of fire in the building are. Bent structural elements in such circumstances are of the greatest danger of collapse. Reinforced concrete beams are among these elements, as they have the largest size and load.

Taking into account all aforementioned, it can be noted that increasing fire safety of building structures is connected with improving of forecasting methods of reinforced concrete elements including reinforced concrete beams behavior for these conditions based

* Corresponding author: svp_chipbbk@ukr.net

on tests for fire resistance [1 to 3]. Therefore, study of patterns of temperature distribution in cross-sections of reinforced concrete beams during the fire resistance test is a critical task.

Analysis of recent achievements and publications

In the work [4] the approach proposed is lying in testing without application of mechanical stress for further calculating interpretation of temperature measurements by solving a strength task. The strength task can be solved using revised and simplified calculation methods [5], but use of any of them means the need to determine temperature distribution in each point of cross-section. It is necessary to develop a method of interpolation of temperatures in every point of the cross-section of reinforced concrete columns by temperature measuring during the tests carried out in specific measurement points of the inner layers in order to implement the method of data interpretation fire tests.

The objective of the research is formulated in this regard.

The objective is to develop methods of calculating strength on the basis of temperature measuring in the inner layers of reinforced concrete beam during standard fire resistance tests without application of mechanical load.

High-temperature testing of two identical sample beams was carried out within the testing stand, which is shown in Fig. 1, in order to exercise procedures for developed methods of interpolation.

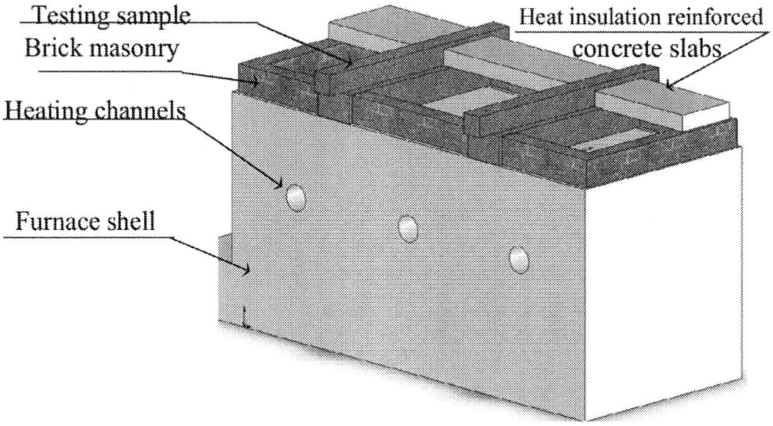

Fig. 1. Structure and appearance of the testing stand.

Sample reinforced concrete beams for testing were the beam elements having folds for placing of reinforced concrete panels into them. Sample beam dimensions are the following: length is 2560 mm; width — 400 mm; height — 450 mm. Fig. 3 shows their cross-sectional geometry and layout of reinforcement bars.

According to the technical documentation a reinforced concrete beam should be made of concrete of C 30/35 (B30) grade. Fig. 3 shows the layout of temperature sensors for temperature measuring in inner layers of the reinforced concrete beams.

Below the results of the beam tests, the results of interpolation, and the results of study of interpolation adequacy are given.

In order to acquire temperature distribution pattern in the reinforced concrete beam cross-section using the acquired data the following algorithm was used, see Fig. 2 [4].

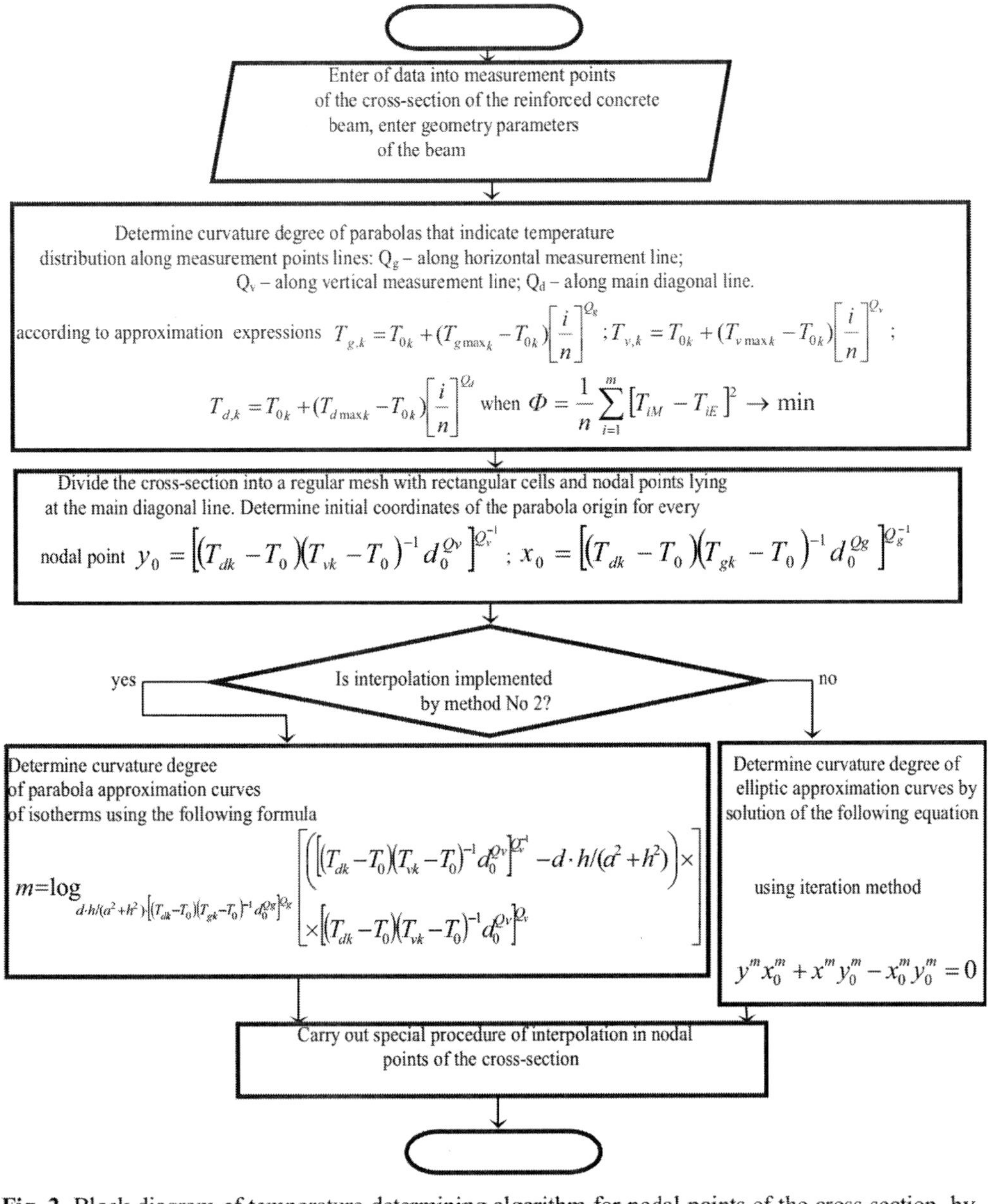

Fig. 2. Block diagram of temperature determining algorithm for nodal points of the cross-section by interpolation of temperature values in measurement points of the cross-section.

After interpolation using the developed techniques the following temperature distributions were obtained, see Fig. 4.

To determine the load capacity a technique considering deformation model [4] was applied. Material properties were taken into consideration with recommendations provided by the second part of Eurocode 2 [5]. The calculations found that reduction of bearing capacity for investigational crossbars No. 1 and No. 2 (using different methods of interpolation, shown in Fig. 5) is time-dependent. These calculations show that the limit state of bearing capacity loss for both beams has been achieved. The limit condition was determined with regard to rated bending moment applied to the beams under question, according to the technical specifications (about 0.3 of destroying load) N = 50 kNm.

 DOI: 10.1051/matecconf/201711602024

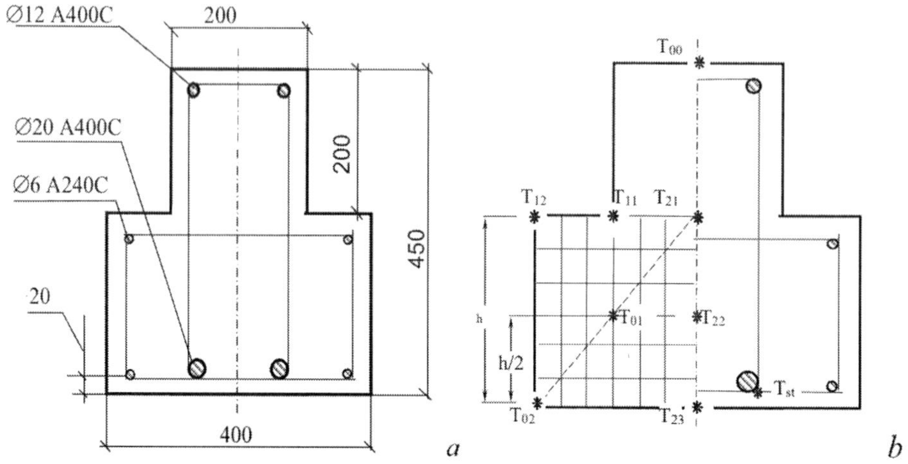

Fig. 3. Cross-sectional geometry (a) and layout of thermocouple sensing elements (b) of the beam for testing purposes.

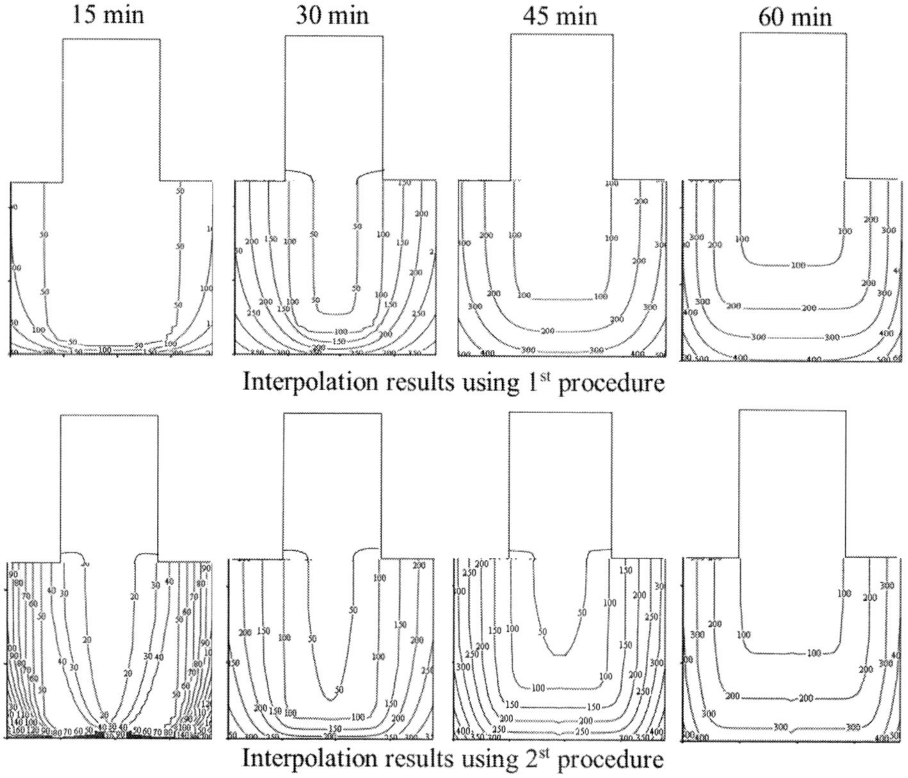

Fig. 4. Temperature distribution patterns in sample reinforced concrete beams, determined using interpolation method.

Based on the study the fire resistance calculation method for reinforced concrete beams using experimental and calculation methods was developed. Its application is schematically shown in Fig. 6.

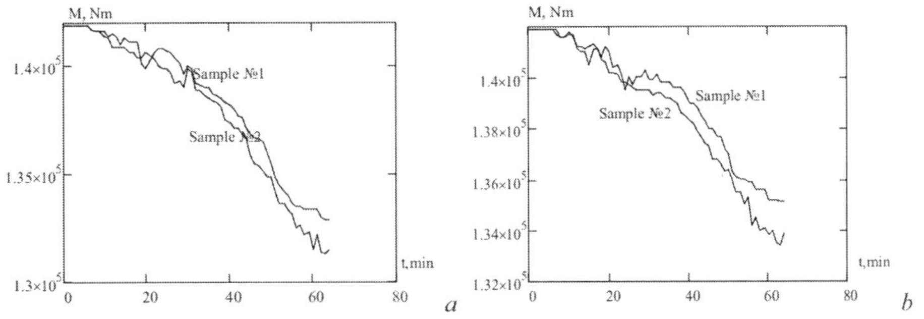

Fig. 5. Figure of the bearing capacity reduction for reinforced concrete beams according to the first interpolation method (a) and the second interpolation method (b).

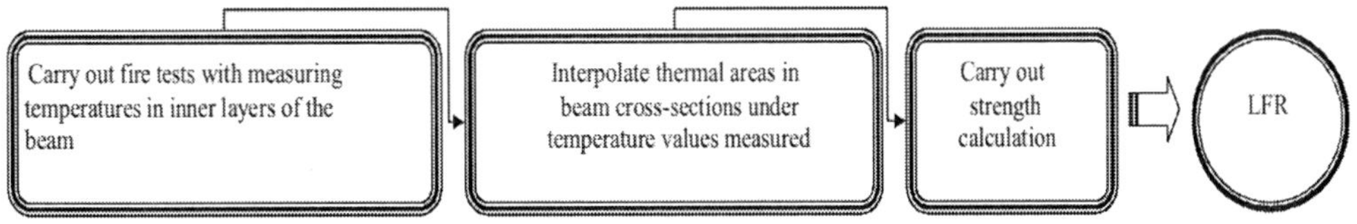

Fig. 6. Algorithm of operation life forecasting for reinforced concrete beams exposed to high temperatures using interpretation of the results of high-temperature tests.

As a result of the research the following conclusions were drawn.

1. The fire test of reinforced concrete beams was carried out and a temperature distribution in cross-section of beams was defined based on it by application of interpolation techniques developed.
2. The sequence of procedures of the strength calculation method based on temperature measuring in the inner layers of reinforced concrete beams was demonstrated.
3. Fire resistance of the reinforced concrete beams under question was evaluated based on the interpretation of the data obtained during the fire tests and using the proposed method of strength calculating.

References

1. *State Building of Ukraine, Fire protection. Building structures fire safety, DBN V.1.1-7-2002*, Kiev (2003)
2. *DSTU B V.1.1-13:2007, Fire protection. Beams. Fire resistance testing method*, Ukrarchbudinform, Kiev, 12 (2008)
3. *DSTU B V.1.1-4-98, Fire protection. Building structures. Fire resistance test methods. General requirements*, Ukrarchbudinform, Kiev (1998)
4. A. Omelchenko, S. Pozdeyev, Yu. Otrosh, [and others] Industrial building and engineering structures. Scientific and manufacturing magazine, **2/2014**, 14 (2014)
5. *EN 1992-1-2:2004 Eurocode 2: Design of concrete structures Part 1-2: General rules — Structural fire design*, Brussels (2004).

To the rationalization of the constructive solutions of the bridge supports

Olena Petrova[1*], *Olena* Lugchenko[1], *Mohammad* Hammud[1], and *Ali* Nazhem[1]

[1]Department of Building Structures, O.M. Beketov National University of Urban Economy in Kharkiv, 61002, Kharkiv, Ukraine

Abstract. Reduction of material consumption and labor intensity of building structures is a topical task for both theory and practice of building production. Today there are many ways to achieve proper economic effect while reconstruction or erection of building and structures. One of them is to rationalize well-known constructive solutions using as criteria the value of density of potential energy of deformation of structures. In this connection, the new constructive solution for short massive supports with outer reinforcement made of expanded metal sheet and frames (reinforcing rings) which are used in construction of road and railway bridges and high-rise buildings is proposed in this paper. The theoretical approach for the definition of its rational parameters is given. Numerical analysis of the highlighted method is also carried out. As a result of study the increase in the bearing capacity in concrete pillars made in rational mesh clips with and without frames was obtained.

Introduction

During the last decades, the reduction of the material consumption of building structures, while maintaining their high rigidity and strength characteristics has been the major task for both theoretical research and experimental verification. At present, this objective can be achieved either by introducing absolutely new effective constructive solutions, or by improving and rationalizing the already known ones. One of the promising ways of this direction, in particular, is the use of a high-tech expanded metal sheet as an external reinforcement of structures.

The use of external sheet reinforcement in the form of a mesh outer ring can simultaneously perform protective, technological and power functions. For example, if we consider steel bridge supports or heavily loaded supporting columns of high-rise buildings and structures, the use of such a clip allows us to limit the displacement of concrete inside it in the transverse direction. In turn, the three-axially compressed concrete located inside the clip increases the bearing capacity of the support itself. This fact is especially important for supports experiencing a high level of both static and dynamic loading. In addition, the use of expanded metal sheet allows achieving of about 30-60% reduction of metal consumption, which, in turn, leads to a reduction of the weight of the whole structure.

* Corresponding author: petrova.bcd@gmail.com

Formulation and solution of the problem

Currently, the works by V.Shmukler and O.Sinkovskaya [1] are known, in which short reinforced concrete supports with external reinforcement made of expanded metal sheet are considered. The effectiveness of the constructive solution proposed by the authors is confirmed experimentally. Based on the established rational tilt angle of the generatrix of the cell of expanded metal sheet, in this work we propose a new design solution of the outer ring with reinforcing frames as well as a procedure for selecting of rational parameters for such an element.

In this connection, the structure of a circular cylindrical short quasi-centrally loaded support (Fig. 1), which is made of concrete encased in a steel mesh clip, is considered. This clip is manufactured by non-waste technology from expanded metal sheet. The connection of the edges of the clip is carried out by welding. To increase the crimping effect of the clip special rings, including reinforcing steel called "the frames", are used. They are installed with a constant step l and have a cross-sectional area A_f. Since the column is considered to be loaded with a quasi-central load, the problem is assumed to be axial symmetric one.

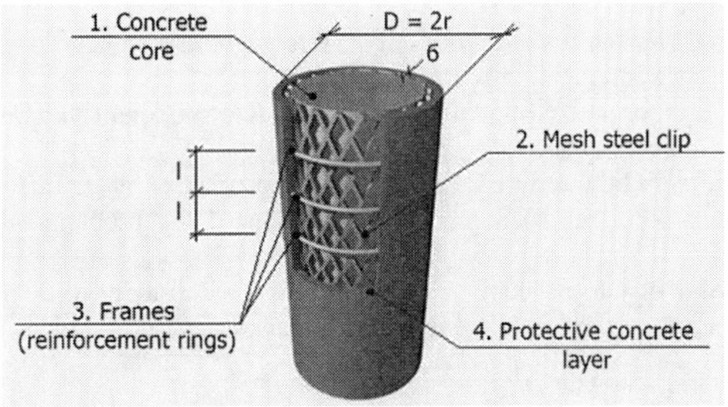

Fig. 1. Cylindrical support with a mesh clip and reinforcing frames.

The geometric dimensions of the column (first of all, the *H/D* ratio, where *H* is the height of the column, and *D* is its diameter) are chosen in such a way that the possibility of loss of stability of equilibrium is excluded. When considering the design model of such support, it is considered that longitudinal compressive forces are applied to the ends of the concrete core (Fig. 2).

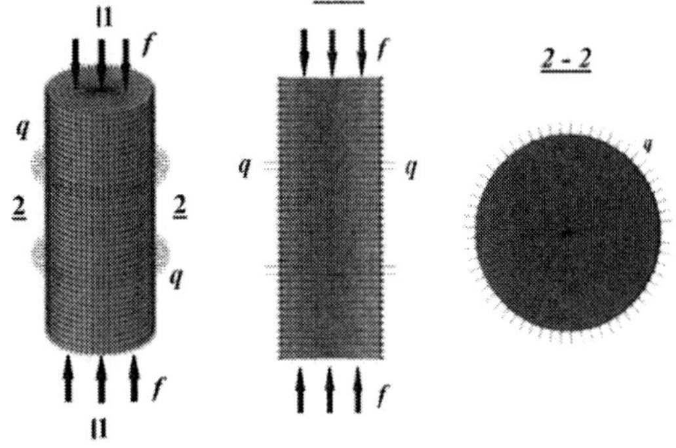

Fig. 2. 3D finite-element calculation model of the support.

As is known, the efficiency of reinforcing of the concrete pillar via frame sharply decreases in the case of appearance of the accidental (calculated) eccentricities. However, it is somewhat mitigating that the design we propose here is intended to be used for erecting pillars of road and railway bridges, as well as columns of high-rise buildings, which are usually projected in a form of inflexible, loaded, mainly with longitudinal forces applied with low eccentricity supports.

While solving the problem of searching for a rational constructive solution of the proposed support, the following simplifying hypotheses and assumptions are accepted:

- massive short rods (with flexibility $\lambda \leq 10$) are considered;
- material properties are specified by the "tension – deformation" indicator diagrams;
- materials are considered conditionally nonlinearly elastic;
- the volume deformation is approximately taken to be equal to $\varepsilon_0 = 0$. In this case, the third hypothesis of the theory of small elastic plastic deformations about the coincidence of the indicator diagram of the material with the dependence of "tension intensity - deformation intensity" is valid;
- the stress-strain state of the element is considered quasi-homogeneous; that means the assumption of $e = const$, where e is the density of the potential energy of the deformations (DPED);
- the compressive stresses f are assumed to be uniformly distributed along the ends of the element and are applied only to its concrete part;
- it is considered (for the reserve) that only the radial stresses occur at the point of contact of the concrete core and metal clip;
- the value of the normalized potential energy of deformations (NPED) is adopted as the point of conditional resilience for a concrete element;
- a linear dependence between PED and material consumption is assumed.

Further, it is necessary to identify the criterion of the problem and introduce some constraints.

As a criterion of the problem, the condition for maximizing the value of the contact pressure between the frame and the metal clip is adopted here:

$$P \to sup \tag{1}$$

where P is the circumferential contact pressure between the frame and the mesh.

Restrictions are accepted as following:

- boundary conditions - the ends of the element are free from fastenings [2];
- isoperimetric condition $e = const;$ (2)
- condition of compatibility of deformation on the whole range of loading

$$\varepsilon_c = \varepsilon_s, \tag{3}$$

where ε_c, ε_s is the deformation of the concrete and the meshclip in their contact zone.

Further, it is conventionally considered (because of the reduced thickness of the mesh in comparison with the radius of the concrete core) that the pressure P is completely transferred to the concrete.

In accordance with the accepted hypothesis, the radius of the element is chosen from the condition:

$$\lambda = \frac{4H}{D} \leq 10 \tag{4}$$

Basing on specified height of the element and the radius determined from (4), the vector of the variable (controlling) parameters will have the form:

$$\{\bar{x}\}^T = \{h_{red}; l; d\} \tag{5}$$

where h_{red} is the reduced thickness of the mesh clip, l is the step of the reinforcing rings (frames) and d is the diameter of the cross-section of the frames.

Let the reduced thickness be

 DOI: 10.1051/matecconf/201711602025

$$h_{red} = \alpha \cdot r, \tag{6}$$

where r is the radius of the concrete core, $D = 2r$; $\alpha \in (\alpha_{min}; \alpha_{max})$ is the coefficient, and $\alpha_{\min}, \alpha_{max}$ are known constants, and

$$A_f = \frac{\pi d^2}{4}, \text{ where}$$

A_f is a cross-sectional area of the frame ring, $d \in (d_{\min}; d_{max})$, $d_{\min}, d_{max}$ are known constants, and, finally,

$$l = \gamma \cdot r \tag{7}$$

$\gamma \in (\gamma_{min}; \gamma_{max})$ is the coefficient, $\gamma_{min}, \gamma_{max}$ are known constants.
Considering (6) and (7) we can set:

$$\{\overline{\mathrm{x}}\}^{\mathrm{T}} = \{\alpha; \gamma; d\} \tag{8}$$

under the next constraints

$$\begin{aligned} \alpha_{min} \leq \alpha \leq \alpha_{max}; \\ \gamma_{min} \leq \gamma \leq \gamma_{max}; \\ d_{min} \leq d \leq d_{max}. \end{aligned} \tag{9}$$

To achieve stated goal, it is also necessary to solve a number of subsidiary tasks.
Thus, using the solution of the Lame problem, we can write

$$q = b \cdot f, \tag{10}$$

where q is acontact radial stresses between the concrete core and the mesh clip;

$$b = \frac{\mu_c n \omega}{(1-\mu_c) n \omega + \mu_s}, \tag{11}$$

where μ_c, μ_s are the Poisson coefficients of concrete and steel, respectively;

$$n = \frac{E_s}{E_c}, \omega = \frac{h_{red}}{r} = \frac{\alpha r}{r} = \alpha$$

where E_c, E_s are the secant modulus of deformation of the first kind of concrete and steel.

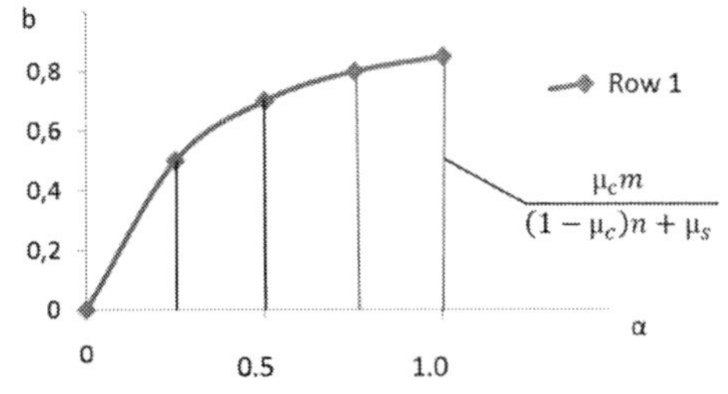

Fig. 3. Dependence $b_{(\alpha)}$.

Having the expression for the contact distributed stresses (10) and using the solution of S.P. Tymoshenko, who determines the radial contact annular stresses that arise between the frames, located with the same constant step, and mesh clip, after the transformations, we obtain [3]:

$$P = b \cdot \psi \cdot f \tag{12}$$

$$\psi = \frac{2}{\beta} \cdot \frac{1}{\frac{2h_{red}}{\beta A_f} + \frac{sh\beta l + sin\beta l}{ch\beta l - cos\beta l}}; \tag{13}$$

$$\beta = 4\sqrt{\frac{3(1-\mu_s)}{r^2 \cdot h_{red}^2}}.$$

Taking into account designations accepted previously, and the fact that $\mu_s = 0{,}3$

$$\beta = \frac{1{,}2}{r\sqrt{\alpha}}$$

The expression for ψ (13) will take the form:

$$\psi = 1{,}7 \cdot r\sqrt{\alpha} \cdot \frac{1}{2.12\alpha\sqrt{\alpha} \cdot \frac{r^2}{d^2} + \frac{sh\bar{\beta} + sin\bar{\beta}}{ch\bar{\beta} - cos\bar{\beta}}}, \tag{14}$$

where $\bar{\beta} = \beta l = \frac{1{,}2\gamma}{l\sqrt{\alpha}}$.

Let's set the following designations:

$$c_1 = 1{,}7\sqrt{\alpha} \tag{15}$$

$$c_2 = 2{,}12\frac{\alpha\sqrt{\alpha}}{d^2} \tag{16}$$

$$c_3 = \frac{sh\bar{\beta} + sin\bar{\beta}}{ch\bar{\beta} - cos\bar{\beta}} \tag{17}$$

Then the expression (14) will take next form:

$$\psi = \frac{c_1 r}{c_2 r^2 + c_3}$$

Calculating the derivative $\frac{d\psi}{dr}$ and using Fermat's theorem, we obtain

$$r_{opt} = \sqrt{\frac{c_3}{c_2}} \tag{18}$$

$$\psi_{ekc} = \frac{c_1}{2c_3} \cdot \sqrt{\frac{c_3}{c_2}}. \tag{19}$$

It is easy to see that (19) defines an extremum of the maximum type (Fig. 4).

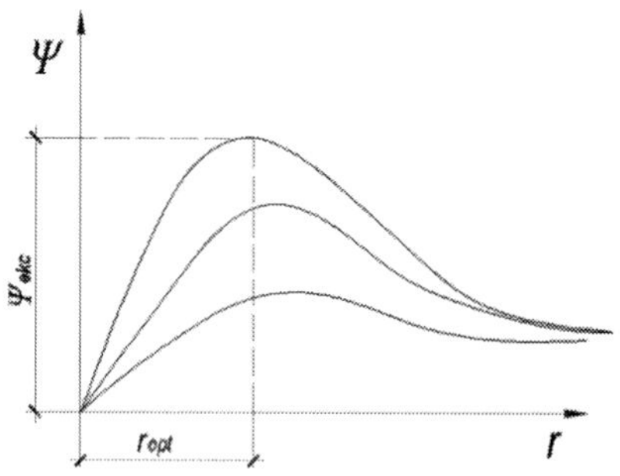

Fig. 4. Dependence $\psi_{(r)}$.

To evaluate the influence of the distance between the frames on the value of ψ, a numerical study was also carried out. Its results are demonstrated by the graphs shown in Fig. 5.

As follows from the graphs, for fixed values of the reduced thickness of the mesh clip h_{red}, the diameter of the frame d and the radius of the concrete core r the value of $\psi = const$.

Nevertheless, the complete solution of problem (1) ÷ (10) is of interest.

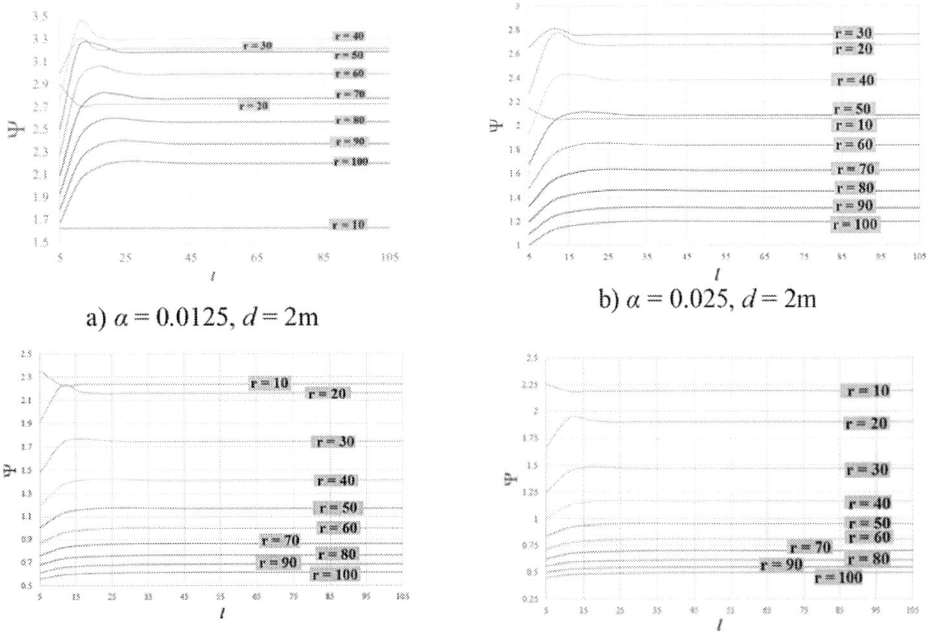

a) $\alpha = 0.0125$, $d = 2$m

b) $\alpha = 0.025$, $d = 2$m

c) $\alpha = 0.05$, $d = 2$m

d) $\alpha = 0.0625$, $d = 2$m

Fig. 5. Dependence $\psi_{(L)}$.

And, in particular, it is convenient to search for the maximum value and the corresponding values of the components of the vector (8) using the method of integral gradients [4]. In accordance with it, we can write:

$$sup\psi = \lim_{p\to\infty}\left[\int_{\alpha_{min}}^{\alpha_{max}}\int_{\gamma_{min}}^{\gamma_{max}}\int_{d_{min}}^{d_{max}}\psi[\alpha,\gamma,d]^p d\alpha d\gamma dd\right]^{1/p} \quad (20)$$

$$x_j^* = \lim_{p\to\infty}\frac{\iiint\limits_{\alpha\,\gamma\,d}\psi[\alpha,\gamma,d]^p\chi_j\cdot d\alpha\, d\gamma\, dd}{\iiint\limits_{\alpha\,\gamma\,d}\psi[\alpha,\gamma,d]^p d\alpha\, d\gamma\, dd} \quad (21)$$

where x_j^* is a coordinate of the global maximum point, $j = 1, 2, 3$, $x_1^* = \alpha;\ x_2^* = \gamma;\ x_3^* = d$.

The numerical solution of (20), (21) for constraints

$$0.004 \le \alpha \le 0.01$$
$$0.25 \le \gamma \le 2.0$$
$$0.020 \le d \le 0.036$$
$$0.25\text{m} \le r \le 1.0\text{m}$$

gives:

$$\psi_{max} = 0.77347\text{m},\ \alpha_{opt} = 0.06,\ \gamma_{opt} = 0.25,\ d_{opt} = 0.0366\text{m}. \quad (22)$$

As shown by the analysis of the numerical solution (20), (21), a significant increase in the maximum value of ψ can be obtained, basically, with increasing of the cross-sectional area of the frame A_f. In particular, for $A_f = 0.00785\text{m}^2$ (d = 0.1 m) solution of (20), (21) gives the following:

$$\psi_{max} = 0.150345\text{m},\ \alpha_{opt} = 0.1,\ \gamma_{opt} = 0.25,\ d_{opt} = 0.1\text{m}.$$

Having the expressions (10) and (22), it is possible to estimate the bearing capacity of the concrete core (actually considered support). In this case, the design model of the core should be adopted in accordance with the scheme shown in Figure 2 (sections "1 – 1" and "2 – 2").

As criterion for the limiting state, the criterion [5,6,7,8,9] is taken:

$$e_i \leq e_n, \tag{23}$$

where e_i is a DPED in i-s finite element; e_n the normalized (limiting) value of the DPED, which is determined by the formula:

$$e_n = 0{,}5\chi_\varepsilon[(\chi_\varepsilon + 1)e_{nc} + (\chi_\varepsilon + 1)e_{nt}] + (1 + W_\varepsilon)e_{nsh} \tag{24}$$

$\chi_\varepsilon = \frac{2\varepsilon_2-\varepsilon_1-\varepsilon_3}{\varepsilon_1-\varepsilon_3}$, $\varepsilon_1 \geq \varepsilon_2 \geq \varepsilon_3$ - the main linear deformations;

χ_ε - Lode-Nadai parameter for strained state

$e_{nc} = \int_0^{\varepsilon_{nc}} \sigma_c d\varepsilon_c$, $e_{nc} = \int_0^{\varepsilon_{nt}} \sigma_t d\varepsilon_t$, $e_{nsh} = \int_0^{\gamma_{nsh}} \tau d\gamma$.

$\varepsilon_{nc}; \varepsilon_{nt}; \gamma_{nsh}$ - normalized (limiting) deformations;

$\sigma_c = \sigma_c(\varepsilon_c)$; $\sigma_t = b_t(\varepsilon_t)$; $\tau = \tau(\gamma)$ - predetermined (known) indicator diagrams of concrete.

Each of the indicator diagrams is represented in the form [10]:

$$b = a_1\varepsilon + a_3\varepsilon^3 + a_5\varepsilon^5 + a_7\varepsilon^7 \tag{25}$$

where $\{\bar{\alpha}\} = \{E_0, R, \varepsilon_R, \varepsilon_u, \beta\}$

E_0 is the initial modulus of deformations;

R is the ultimate strength;

ε_R deformation, corresponding to the ultimate strength;

ε_u is the limiting deformation;

β is the stress, corresponding to the limiting deformation.

And the coefficients α_j (j = 1,3,5,7) are given by Table 1.

Table 1. Coefficients of a polynomial.

Coefficient	Value
$a_3 \cdot E_0^{-1}$	$\frac{\varepsilon_u^4}{2(\varepsilon_u^2-\varepsilon_R^2)^2}\left[\frac{1}{\varepsilon_R^2}\left(\frac{R}{E_0\varepsilon_R}-1\right)\left(5-7\frac{\varepsilon_R^2}{\varepsilon_u^2}\right)-\frac{1}{\varepsilon_R^2}\left(\frac{\varepsilon_R^2}{\varepsilon_u^2}-1\right)+\frac{2\varepsilon_R^4}{\varepsilon_u^6}\left(\frac{\beta R}{E_0\varepsilon_u}-1\right)\right]$
$a_5 \cdot E_0^{-1}$	$\frac{\varepsilon_u^4}{2(\varepsilon_u^2-\varepsilon_R^2)^2}\left[\frac{1}{\varepsilon_R^4}\left(7\frac{\varepsilon_R^4}{\varepsilon_u^4}-3\right)\left(\frac{R}{E_0\varepsilon_R}-1\right)-\frac{1}{\varepsilon_R^4}\left(1-\frac{\varepsilon_R^4}{\varepsilon_u^4}\right)-\frac{4\varepsilon_R^2}{\varepsilon_u^6}\left(\frac{\beta R}{E_0\varepsilon_u}-1\right)\right]$
$a_7 \cdot E_0^{-1}$	$\frac{\varepsilon_u^4}{2(\varepsilon_u^2-\varepsilon_R^2)^2}\left[\frac{1}{\varepsilon_R^4\varepsilon_u^2}\left(\frac{R}{E_0\varepsilon_R}-1\right)\left(3-5\frac{\varepsilon_R^2}{\varepsilon_u^2}\right)-\frac{1}{\varepsilon_R^4\varepsilon_u^2}\left(\frac{\varepsilon_R^2}{\varepsilon_u^2}-1\right)+\frac{2}{\varepsilon_u^6}\left(\frac{\beta R}{E_0\varepsilon_u}-1\right)\right]$

While calculating (comparing) (23), it is assumed that the solution of the problem is constructed using finite element method [4]. Here as the environment for the realization of calculation the "Lira" Software Package was used, which allows determining the value of e_i.

Conclusion

Analysis of these calculations shows that the increase in the bearing capacity of a short pillar due to the use of a rational mesh clip without frames is 20%, with frames - 35%.

As a consequence, it can be noted that the theoretical results which were obtained are fully justify the effectiveness of the proposed solution.

References

1. V.S. Shmukler, E.A. Petrova, E.V. Sinkovskaya. Scientific Bulletin of Building, Kharkiv, HNUBA, HOTW ABU, **74,** 148-156 (2013)
2. V.G Rekach. *Guidance for solving problems of the theory of elasticity* (Moscow: Higher School, 1966)
3. S.P. Timoshenko. *Resistance of materials*, v.**2**. *More complex theoretical questions and problems* (Moscow. "Science" Publishing House, 1965)
4. A.S. Gorodetsky, V.S. Shmukler, A.V. Bondarev. *Information technology of calculation and design of building structures* (Kharkov, NTU "KhPI", 2003)
5. G.V. Vasilkov. *Evolutionary problems of structural mechanics. Synergetic paradigm* (Rostov-on-Don, Infoservis, 2003)
6. G.V. Vasilkov. *Evolutionary theory of the life cycle of mechanical systems. Theory of Structures* (Moscow, LKI, 2008)
7. V.S. Shmukler, Y.A. Klimov, N.P. Burak. *Frame systems of a light type* (Kharkov, Golden Pages, 2008)
8. V.S. Shmukler. Proceedings of the International conference, Scotland, 183-190 (2008)
9. V.S. Shmukler. Third International Structural Engineering and construction Conference, Shunan, Japan, 539-545 (2005)
10. V.S. Shmukler. Collected Scientific Works of Ukrainian State University of Railway, UkrDUZT, **167**, 54-69 (2016)

The design of concrete elements strength under local compression based on the variational method in the plasticity theory

Volodymyr Pohribnyi[1*], *Oksana* Dovzhenko[1], *Leonid* Karabash[1], and *Iryna* Usenko[1]

[1]Poltava National Technical Yuri Kondratyuk University, Building Department, Pershotravneviy sq., 24, Poltava, 36011, Ukraine

Abstract. The two-dimensional concrete elements under the action of one-sided central crushing often occurs. For them, the strength problems are solved on the basis of the concrete ideal plasticity theory variational method with application of discontinuous velocities functions. The method takes into account the influence on concrete elements strength the compressive strength and tensile strength of concrete. It also takes the ratio of a sample height to the crushing platform length and friction coefficient between a surface of element and a load punch if it's necessary. The results of experimental researches were confirmed the assumed in the theoretical solution the kinematical schemes of elements failure. Higher convergence of theoretical strength with tested is received. The calculation sequence of the concrete element under one-sided central crushing is described. The joint calculation of all factors that determine the strength opens the possibility of more accurate assessment and improvement on its basis of structural solutions of concrete elements under the local compressive load.

1 Introduction

The concrete and reinforced concrete constructions and their elements under the action of a local compression are widely used (for example the joints of columns, the end faces of prestressed structures, the horizontal joints of panels of outside walls and floors, walls and supports of bridges in places of beams bearing, girders, lintels).

According to Eurocode 2, the crushing elements design is made by Baushinger's empirical relationship in which the ratio of the so-called "rated" area A_{c1} to the crushing area A_{c0} is assumed as the determining factor. However the value A_{c1} is conditional as there is no precise definition to find it.

The two-dimensional elements failure character under one-sided central crushing [1] confirms the influence on their strength of concrete axial tensile strength f_{ct} and ratio h/l_{loc} (h – a sample height, l_{loc} – the crushing platform length), not taken into account in Eurocode 2.

---

* Corresponding author: V.V.Pohribnyi@gmail.com

The reference strength design is not based on consideration of elements limiting stress-strain state (SSS) which is the cause of the above lacks.

Significant experimental material about concrete elements strength at local compression and various factors influence for strength now is saved, the numerous empirical dependences for determination of ultimate load (Table 1) are offered [2].

Table 1. The design dependence for determination of elements strength under one-sided central crushing.

№	The formula for determination φ_c	Authors
1	$\sqrt[3]{A_{c1} / A_{c0}}$	J. Baushinger (1876)
2	$4-3\sqrt{A_{c1}/A_{c0}}$	V. Chervonobaba (1959)
3	$1.2(2.3+5l_{loc}/l)/(\chi l_{loc}/h),$	B. Gladyishev (1968)
4	$(2.75+6l_{loc}/l)/(\chi l_{loc}/h)$	B. Gladyishev, M. Dyuzhenko (1970)
5	$2.7\sqrt{A_{c1}/A_{c0}} / \sqrt{\chi}$	G. Tsyskreli (1970)
6	$(A_{c1}/A_{c0})^{-0.3424}$	I. Rohlin (1972)
7	$0.42\left(\frac{1}{l_{loc}/l}+2\right)-0.29\sqrt{\left(\frac{1}{l_{loc}/l-1}\right)^2+5.06}$	S. Niyogi (1973)
8	$0.2+0.8\sqrt{A_{c1}/A_{c0}}$	B. Jensen (1977)
9	$6.92(A_{c1}/A_{c0})^{-0.47} / \chi$	A. Williams (1979)
10	$\varphi_c = 12.5\sqrt{40/f_{c,prizm}}, \quad if\ A_{c1}/A_{c0} \leq 320$ $\varphi_c = 0.7\sqrt{A_{c1}/A_{c0}}, \quad if\ A_{c1}/A_{c0} > 320$	K. Lieberum, H Reinhardt (1989)
11	$1/\left[1-0.55(1-A_{c1}/A_{c0})\right] \leq 1.7$	A. Griezic (1998)
12	$1+0.064\left(\sqrt{A_{c1}/A_{c0}}-1\right)$	N. Rak (2002)

It is difficult to choice one method of all for practical use, not only due to different quantitative results of calculation (Fig. 1) and difference of the basic preconditions.

In our opinion it is necessary to prefer a design method, which would be based on a general basis. A variational method of the concrete ideal plasticity theory is developed in Poltava National Technical Yuriy Kondratyuk University [3] may be such basis.

2 Design assumptions and design sequence

For development of a method the following preconditions were used:

– for concrete in a destruction stage the assumption about ideal plasticity is applied. Determining for applicability of the specified precondition is the condition of simultaneous existence of a limiting condition in more fragile (tensed) and more plastic (compressed) of concrete destruction zones;

– G. Geniyev's concrete strength condition [4] is assumed and considered as a plastic potential;

– relationship of deformation velocities to stresses is found from the associated low of plastic yielding;

– the model of a rigid-plastic body and problems solution in discontinuous velocities functions are used.

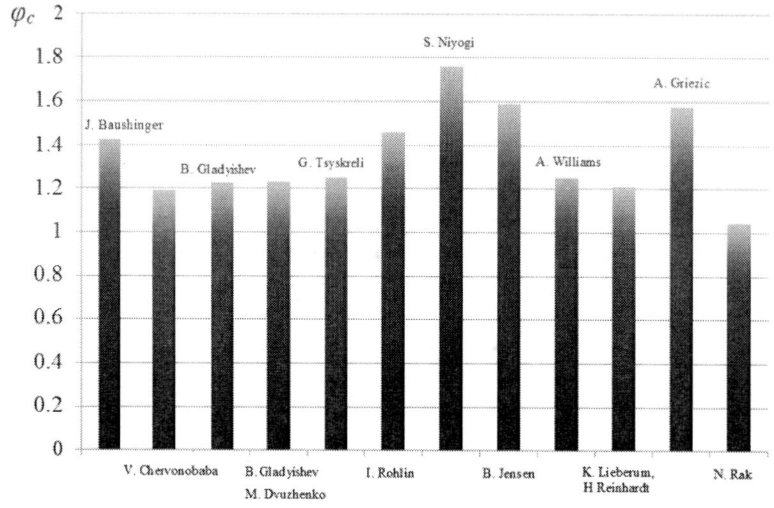

Fig. 1. The value of coefficient φ_c in accordance with formulas of Table 1 (for the ratio $h/l_{loc} = 4$ and concrete C20/25).

The calculation is carried out in the following sequence:

– for the being considered case of element crushing the kinematical possible element destruction scheme is assumed, that is the outline of the destruction surface (gaps of velocities) dividing before save element into being considered absolutely rigid parts, making mutual movement with some velocities in a destruction stage, is set. Thus geometrical parameters g_i determining a destruction surface and movement velocities of an element parts V_j are introduced; assumed cinematically possible scheme of the element destruction reflects the specificity of SSS in a destruction stage of the being considered element or construction;

– on a destruction surface S_l there are gaps of a normal ΔV_n and tangent ΔV_t to S_l velocity components expressed by parameters g_i, V_j;

– enters the functional of a principle of possible velocities speeds and stress [5], which on the valid SSS reaches to a minimum;

– limit load F_u expression is found through the parameters g_i and velocities ratio $k_j = V_j / V_F$, where V_F – velocity of a load application point;

– unknown parameters g_i and k_j are determined out of the condition of a limit load minimum. Then F_u is calculated.

Let's consider the calculation of the concrete element under one-sided central crushing (Fig. 2).

Cinematically possible destruction scheme includes disks 1 and disk 2, which move with velocities accordingly V_1 and - V_1, V_2. Disks are divided between themselves by the destruction surface (or surface of velocities gaps), which consists of parts A-B and B-C. On parts A-B the limit normal σ_u and tangent τ_u stresses act, which are determined according to the G. Geniyev's concrete strength condition. The area B-C is principal area with tensile stress $\sigma_u = f_{ct}$.

The being considered problem unknown quantities are the limit load F_u or resistance $f_{c,loc} = F_u / A_{c0}$, the angle γ (inclination angle of the destruction area A-B to a vertical) and velocities ratio $k = V_1 / V_2$.

Gaps of velocities normal and tangent components on the part A-B

$$\Delta V_n = V_1 \cos\gamma - V_2 \sin\gamma ,\ \Delta V_t = V_1 \sin\gamma + V_2 \cos\gamma . \tag{1}$$

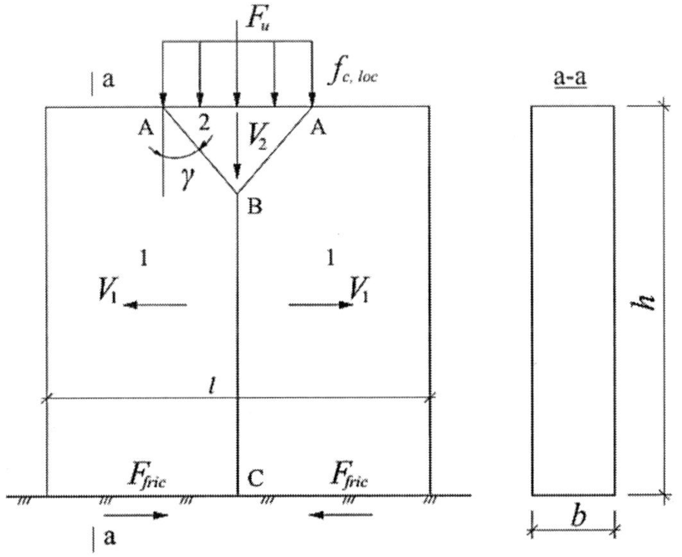

Fig. 2. Cinematically possible destruction scheme of element loaded with strip bilateral central load.

The part's area A-B length is

$$S = l_{loc} / 2\sin\gamma . \tag{2}$$

Accordingly on part B-C:

$$\Delta V_n = 2V_1,\ \Delta V_t = 0 . \tag{3}$$

Equal to zero the functional for the two-dimensional stressed state

$$I = m\left[\int_{S_t} 2B\left(1 + 0.25\left(\left(V_1 \sin\gamma + V_2 \cos\gamma\right) / \left(V_1 \cos\gamma - V_2 \sin\gamma\right)\right)^2\right)^{0.5} - 1\right] \times$$
$$\times\left(V_1 \cos\gamma - V_2 \sin\gamma\right) l_{loc} / \sin\gamma + f_{ct} V_1 \left(h - l_{loc} / 2tg\gamma\right) - f_{c,loc} b l_{loc} V_2 = 0. \tag{4}$$

The formula for determining the resistance of local compression was obtained after simple mathematical transformations

$$f_{c,loc} / m = \left\{\left[2B\left(\left(k - tg\gamma\right)^2 + 0.25\left(ktg\gamma + 1\right)^2\right)^{0.5} - k + tg\gamma\right] / tg\gamma + \right.$$
$$\left. + f_{ct} k\left(2\alpha tg\gamma - 1\right) \ / tg\gamma m\right\} / \left(1 - k f_{fric}\right), \tag{5}$$

where $B^2 = \left(1+\chi/(1-\chi)^2\right)/3$, $\chi = f_{ct} / f_{c,prizm}$, $m = f_{c,prizm} - f_{ct}$, $\alpha = l_{loc}/h$, f_{fric} – the friction coefficient between a surface of element and a load surface.

To simplify the calculation, the charts are offered. They make it possible to determine ultimate load for definite α and χ (Fig. 3).

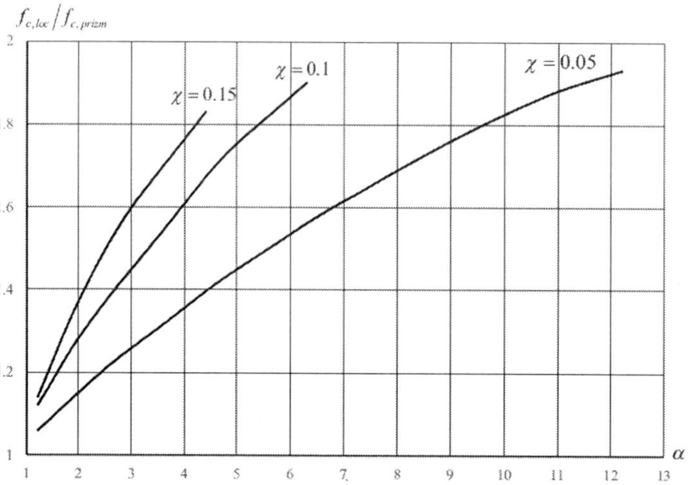

Fig. 3. The charts for determination of ultimate load for definite α and χ .

3 Experimental researches and convergence of theoretical strength with experimental

In a fig. 4 the comparison of experimental and theoretical strength is given at absence of friction forces between a surface of a sample and basic plates press [6-9].

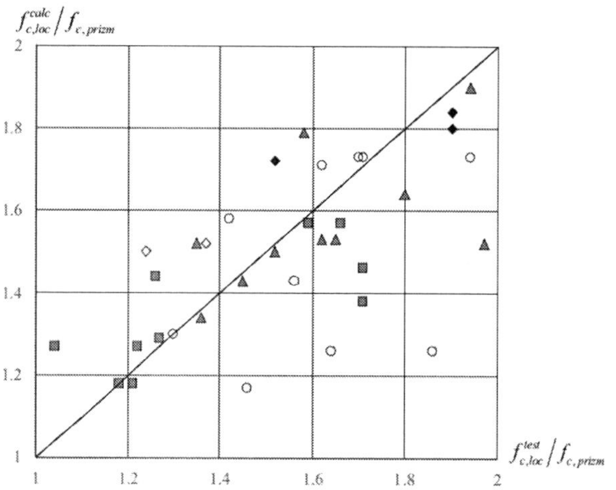

Fig. 4. Comparison of theoretical and experimental strength at one-sided central crushing of a concrete plate: ▪ – B. Gladyishev; ○ – TsNIISK (Russia); ◆ – G. Shapiro; ◇ – S. Kryilov, L. Zaytsev, I. Ulbieva; ▲ – A. Piradov.

The experimental researches includes tests 44 samples - cubes with the size of an edge 150 and 200 mm and samples of prisms 200x200x800 mm from heavy concrete. The tests were made on press PG-125. The load from a plate press was transferred through metal stamp.

The samples failure character, observable in experiments, has confirmed the kinematical possible failure schemes assumed in the theoretical solutions. At test of samples for one-sided crushing-splitting at stamps a shear cracks allocated wedges of condensation, on which ends a vertical tension crack is formed later.

The convergence analysis of theoretical strength F^{calc} with experimental F^{test} on size of the relation F^{calc} / F^{test} was carried out on the basis of data processing 36 samples, for which $h / l_{loc} = 1.5 - 6.7$, $f_{c,prism} = 1 - 30.0$ MPa and $f_{ct} = 0.4 - 2.39$ MPa. It has shown $\overline{x} = F^{calc} / F^{test} = 0.985$ with factor of a variation $v = 11.9\%$, that testifies to sufficient affinity of theoretical strength to experimental and serves a substantiation of applicability of a variational method of the ideal plasticity theory to the decision of the considered types of tasks.

4 Conclusion

The two-dimensional concrete elements strength problems under the action of crushing are solved on the basis of a variational method of the concrete theory ideal plasticity with application of discontinuous velocities functions. The received relationship is exacter in comparison with normative at the expense of the account to the stress-strain station in the case of one-sided central crushing. This account is carried out by means of kinematical scheme of failure displaying specificity of concrete task and introduction in account of a number of the factors: ratio h / l_{loc}, friction forces between the surface of a sample and weight surface etc.

The results of experimental researches confirmed the kinematical possible failure schemes assumed in the theoretical solution, and also influence of the factors, determining strength.

The theoretical strength of concrete elements at crushing, found on an offered method, well enough converges with tested.

References

1. D.H. Hladyshev, Resursoekonomni materialy, konstruktsiyi, budivli ta sporudy, Rovno, **29**, 94-101, (2014).
2. V.V. Tur, N.A. Rak, *Prochnost i deformatsii betona v raschetah konstruktsiy*, Brest, (2003).
3. V.P. Mitrofanov, Stroitelnaya mehanika i raschet sooruzheniy, Moskov, **6**, 23-28, (1990).
4. G.A. Geniev, V.N. Kissyuk, G.A. Tyupin, *Teoriya plastichnosti betona i zhelezobetona,* Moscow, (1974).
5. V.L. Kolmogorov, *Mehanika obrabotki metallov davleniem,* Moscow, (1986).
6. B.M. Gladyishev, *Mehanicheskoe vzaimodeystvie elementov strukturyi i prochnost betonov,* Kharkov, (1987).
7. S.M. Kryilov, L.N. Zaytsev, I.S. Ulbieva, Novyie konstruktsii selskohozyaystvennyih zdaniy industrialnogo izgotovleniya i metodyi ih raschyota, Moscow, **3**, 66-75, (1984).
8. B.P. Filipov, O.P. Vinokurov, Beton i zhelezobeton, Moscow, **1**, 42-43, (1983).
9. A.B. Piradov, *Konstruktivnyie svoystva legkogo betona i zhelezobetona,* Moscow, (1973).

Computational study of bearing walls fire resistance tests efficiency using different combustion furnaces configurations

Serhii Pozdieiev[1*], *Oleksandr* Nuianzin[1], *Stanislav* Sidnei[1], and *Stanislav* Shchipets[1]

[1]Cherkassy Institute of Fire Safety named after Chernobyl Heroes National University of Civil Protection of Ukraine, 18000 Cherkassy, Ukraine

Abstract. Imperfection of structure, metrological instruments and fuel and nozzle system control methods the conditions are created ensuring non-homogeneous distribution of temperatures throughout heated surfaces of tested structures and during their fire-resistance tests in combustion furnaces. The article contains results of numerical simulation of bearing walls fire resistance test performed using different configurations of combustion furnaces. Using computer-assisted gas and liquid flow simulating software temperature gradient for heated surfaces of bearing walls was created and temperature distribution for every minute of the computational experiment for each configuration was calculated. In temperature gradients 6000 to 7500 cells were located (depending on the particular configuration design), evenly distributed throughout the structure surfaces, containing temperature data in any moment during the computational experiment. As a result of processing of these data the value of temperature dispersion value was calculated. Difference between maximum and minimum temperatures on the surface of a reinforced concrete was also determined. Based on the curves representing temperature dispersion values at the surface of each of the simulated structures of the furnace chamber for every minute of the computational experiment the configuration with the most homogeneous temperature distribution throughout the heated surfaces of the bearing wall was defined, what allows reducing an error occurring due to temperature distribution non-homogeneity by heated surface structures during the fire resistance test.

Problem statement

Whereas testing in combustion furnaces [1, 2] are carried out in the "standard" temperature mode, there arises a question concerning homogeneousness of a reinforced concrete structure heating depending on design and configuration of the combustion furnace used, since internal design of its chamber, layout of burners and combustion products exhausting holes affects homogeneousness of temperature distribution throughout the heated surface of vertical structures, in particular this refers to wall structures.

* Corresponding author: svp_chipbbk@ukr.net

Analysis of recent achievements and publications

In works [3-5] a CFD FlowVision 2.5 software-assisted combustion furnace computer model creation and computational test process was described for a furnace used for real tests; during the process efficiency of heating process simulation for further use in studying of structural characteristics of combustion furnaces affect to their metrological characteristics is shown.

In accordance with the previous studies [3-5], imperfection of structure, metrological instruments and fuel and nozzle system control methods the conditions are created ensuring non-homogeneous distribution of temperatures throughout heated surfaces of tested structures and during their fire-resistance tests in combustion furnaces.

Purpose

The existing scientific works contain no researches with regard to effect of non-homogeneous temperature distribution throughout heated surfaces of bearing walls on fire resistance test results adequacy.

Based on the aforesaid, it is supposed to use existing structures of vertical combustion furnaces and an experience in designing of such quipment [6] taking into consideration general requirements [1, 2] in order to carry out the research.

Method

Using mathematical practice and models described in [3-4], and grounding on their adequacy, confirmed by [5], using computer-assisted modeling the article deals with a number of geometric configurations of vertical combustion furnaces (number and layout of burners and fume gases exhausting holes etc. were varied) and the way structural features of the stand can influence homogeneousness of the temperature distribution throughout heated surfaces of walls is shown. As a result, a configuration providing the most homogeneous temperature distribution throughout heated surfaces of a vertical structure was determined in course of the test.

Consideration on methods and results

During the research several configurations and stands for combustion furnaces used in testing of vertical building structures were studied. Their structural features were described in previous works [3-4]. In order to analyze their effectiveness, were studied temperature distribution throughout heated surfaces of a structure at the 60th minute of the test, as well as temperature value dispersion at a surface of each simulated configuration of a furnace for every minute of the computational experiment and curves for their change over time were studied.

Every configuration was assigned a unique symbol. Figure 1 illustrates temperature distribution throughout the structure heated surfaces for every configuration at the 60th minute of the test.

We believe that main disadvantage of the configuration named "A" (Fig. 1-a) is, compared to other configurations, small volume of furnace chamber, inefficient layout of the sole hole for fume gases exhausting. Due to that the heated reinforced concrete structure is heated non-homogeneously. The structure is heated less in central part of the furnace and above the hole for combustion gases exhausting.

After alternating of design of the furnace inner combustion chamber (Fig. 1-b) temperature distribution throughout the heated surfaces of the structure became even less homogeneous. However, more characteristic vertical distribution of temperatures is observed.

In "C" configuration (Fig. 1-c) more homogenous temperature distribution was achieved compared to "A" and "B" configurations, but due to location of combustion products exhaustion holes closely to burners, convection flows were directed towards these holes, too, and, as a result, this zone had higher temperature.

As it is shown in Fig. 1-d the hole in central part of the furnace leads to higher degree of temperature homogeneousness during heating of the structure compared to "C" configuration. That's why it was decided to continue research with regard to determination of the most homogeneous heating of a structure.

The following configurations of furnace in our temperature distribution research were "F" (Fig. 1-f) and "E" (Fig. 1-e) configurations. Advantage of the "F" configuration is the fact that heating is carried out using 4 burners, but their layout is altered compared to the initial configuration (Fig. 1).

The following stage of the research was determination of temperature dispersion value at surfaces of all simulated furnace chamber design for every minute of the computational experiment followed by building of time-dependent graph for change of this value.

Using computer-assisted gas and liquid flow simulating software, CFD FlowVision 2.5, temperature gradient for heated surfaces of bearing walls was created and temperature distribution for every minute of the computational experiment for each configuration was calculated [7]. In temperature gradients 6000 to 7500 cells were located (depending on the particular configuration design), evenly distributed throughout the structure surfaces, containing temperature data in any moment during the computational experiment. As a result of processing of these data the value of temperature dispersion value was calculated (Fig. 2).

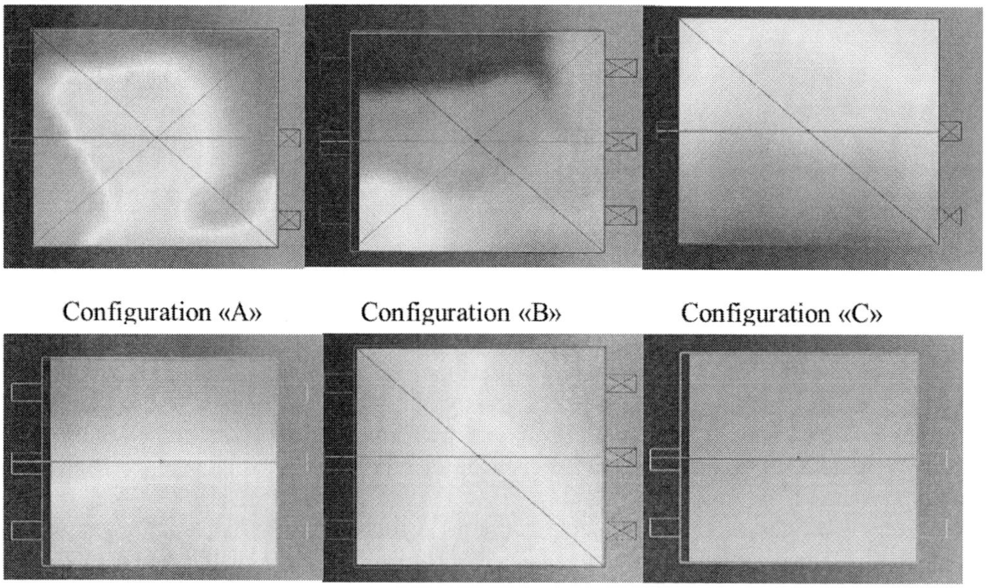

Configuration «A» Configuration «B» Configuration «C»

Configuration «D» Configuration «E» Configuration «F»

Fig. 1. Temperature gradient at the heated surface of the simulated structures configurations from A to F.

The graphs have one particular characteristic feature: most of curves, representing numerical value of temperature dispersion, have an extreme point. After the initial phase, increase of dispersion value begins to decrease continuously. It can be explained by reviewing of standard temperature curve of a fire event [1]. It is characterized by a drastic increase of temperature inside combustion chamber of the furnace during first minutes of an experiment and further decrease of difference between extreme points of maximum and minimum. Therefore, the experiment was limited by calculation of the 60th minute (Fig. 2).

Another discovered characteristic feature is the fact that the less such extreme point value is, the earlier it is reached referred to a temporal curve.

After thorough reviewing of the created configurations it becomes obvious that the least dispersion values during the whole time period of interest are observed in "E" and "F" configurations. The highest value of temperature dispersion in "F" configuration is observed at the 40th minute of the experiment, and for "E" configuration maximum point of this parameter is reached at 41st minute. At the same time, configurations "A" and "B" have reached the maximum almost at the 60th minute.

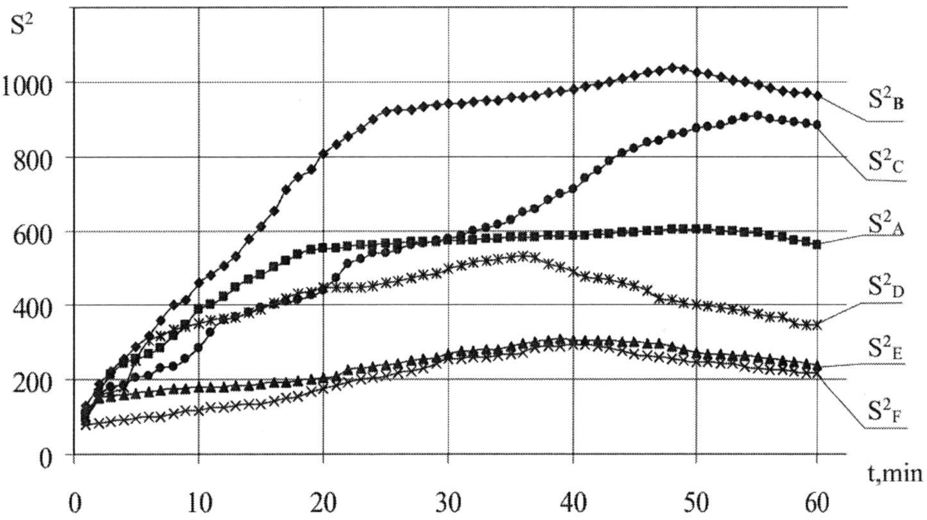

Fig. 2. Dispersion of the temperature of the heated surface of the bearing wall in the computational experiment. $S^2_{A\text{-}F}$ - The dispersion of the temperature of the heated surface of the bearing wall in the computational experiment in the configuration of the fire furnace, which corresponds to this index ("A-F" configurations).

Having reviewed all created configurations, one can say that average time value of the extreme point reach is 45th to 50th minute of the experiment, when allowable differential between maximum and minimum temperature in the furnace chamber decreases [1].

Thus, a conclusion can be drawn based on the obtained results.

Conclusions

This work shows results of the computational simulation of a range of computer configurations of stands for testing of bearing walls. Based on the curves representing temperature dispersion values at the surface of each of the simulated structures of the furnace chamber for every minute of the computational experiment (Fig. 2) the configuration with the most homogeneous temperature distribution throughout the heated surfaces of the bearing wall was defined, what allows reducing an error occurring due to temperature distribution non-homogeneity by heated surface structures during the fire resistance test.

References

1. *DSTU B V.1.1-4-98, Fire protection. Building structures. Fire resistance test methods. General requirements*, Ukrarchbudinform, Kiev (1998)
2. *DBN V.1.1-7-2002 Fire protection. Building structures fire safety*, Ukrarchbudinform, Kiev (2003)
3. A. Nuyanzin, S. Pozdeyev, S. Sidney [and others], Fire safety: theory and practice: collection of scientific works, Cherkasy, **18**, 91 (2015)
4. A. Nuyanzin, S. Pozdeyev, S. Sidney, Emergency situations: Education and science: international scientific and application magazine. Gomel, Republic of Belarus Emergencies Ministry, **15/1**, 158 (2015)
5. A. Nuyanzin, S. Pozdeyev, S. Sidney, Ukrainian fire safety scientific and research institute reporter: collection of scientific works, Kyiv, UkrNDITsZ, **31/1**, 4 (2015)
6. S. Novak, L. Nefedchenko, A. Abramov, *Structures and articles fire resistance testing methods*, Pozhimformtechnika, (2010)
7. *Liquid and gas simulating system. FlowVision Version 2.5.4.*, User manual TESIS, (2008)

The construction features of the deformation and force model of concrete and reinforced concrete resistance

Vasyl Romashko[1*], and *Olena* Romashko[1]

[1]Department of Architectural Designing Bases, Constructing and Graphics, National University of Water Management and Nature Recourses Use, Soborna Street 11, Rivne 33000, Ukraine

Abstract. The main features of the deformation and force model of deformation of reinforced concrete elements and structures based on generalized diagrams of their state are considered in the article. Particular attention is focused on the basic methodological problems and shortcomings of modern "deformation" models. It is shown that in the most cases these problems can be solved by the generalized diagrams of reinforced concrete elements and structures real state. Thanks to these diagrams, the developed method: provides a single methodological approach to the calculation of reinforced concrete elements and structures normal sections for limit states; allows to reveal the internal static indeterminacy of heterogeneously deformable elements and structures in their ultimate limit state calculation; justifies the application of the basic and derived criteria of reinforced concrete elements and structures bearing capacity exhaustion; retains the essence of the physical processes of concrete and reinforced concrete structures deformation. The defining positions of the generalized (universal) methodology for calculating reinforced concrete elements and structures are stated.

1 Introduction

In general, the modern theory of concrete and reinforced concrete continues to move towards its further improvement. However, until now it has not developed an unambiguous attitude to its main object - reinforced concrete. Within the framework of the above theory, some researchers recognize the existence of two objects: reinforced concrete-material and reinforced concrete-construction. Others consider reinforced concrete only as a composite material, which has a non-linear relationship between stresses and deformations, creep, micro- and macroformation of cracks, etc. The existing situation, although it is explainable, cannot remain unchanged, since any truly scientific theory has an unique attitude towards the research object.

2 Analysis of recent studies and publications

* Corresponding author: romashkovasyl@gmail.com

Already now, the basis for modern models of concrete and reinforced concrete resistance begins to be laid down for the sampling of calculation schemes and their representation as a set of elements of certain structural levels. In particular, Borovskikh A.V. distinguishes three hierarchical levels of research objects [1]: materials, elements of structures, structures and buildings. At the same time, Bondarenko V.M. [2] offers to classify objects of research for four structural levels: material, section (element), construction, constructive system of a building. However, one cannot agree with the identification of two structural levels: the section and the element. Therefore, it seems to be correct to classify the research objects according to the following five interrelated structural levels of hierarchical subordination. They are: material (a set of properties); cross-section (an aggregate or a combination of materials); an element (an aggregate or a combination of sections); structure (an aggregate or a combination of elements); constructive system of a building or construction (an aggregate or a combination of elements and structures).

Naturally, one of the main tasks in the modern concrete and reinforced concrete theory will continue to be the deep study and theoretical generalization of the specific properties of the above-mentioned composite material. And since it is always an integral part of the structure, then the theory of such material deformation should be only a part of a more general theory of concrete and reinforced concrete elements and structures deformation. Thus, a theory whose single-valued object will be an element or a structure will require clear, consistent and logical operations from researchers and will allow them to work purposefully on the generalized model of reinforced concrete elements and structures deformation.

In the existing deformation models [3, 4, 5], the actual stress-strain state of reinforced concrete elements and structures is reproduced by transferring the deformation diagrams of materials to their sections. Such a method, according to many researchers [6, 7, 8], is not very convincing and makes a number of reservations on their part. Therefore, in the proposed deformation and force model [9, 10], the entire process of concrete and reinforced concrete elements and structures deformation is reproduced using diagrams of their real state. In this case, there is no need for any transfer of the above diagrams to the cross-sections of these elements, because under certain facts and boundary conditions they are transformed into each other.

3 Issues not solved within the common problem

Ensuring the full methodological unity of concrete and reinforced concrete elements and structures calculations on the ultimate states of the first and the second group (on strength, rigidity and crack resistance) has always been one of the most important problems in the concrete and reinforced concrete theory.

This problem could not be solved in principle in the framework of the force model using the "plastic hinge" with a rectangular stress diagram in the compressed concrete. After all, because of the different initial assumptions, the calculations of reinforced concrete elements and structures on the ultimate states of the first and the second group generally remained completely autonomous.

Special hopes in solving this problem were assigned to the so-called "deformation" model. However, despite some successes in its construction, it still did not fully solve the main problem. Moreover, within its framework, the accompanying tasks remain unresolved:

• in most cases, the internal static uncertainty of heterogeneously deforming elements is revealed through numerous iterations;

• there are no clear criteria for the exhaustion (loss) of the bearing strength of reinforced-concrete elements and structures;

• for today, for reinforced elements and structures, not only exact, but also approximate analytical solutions of integral dependences $M = f(1/r)$ for the extreme criterion $dM/d(1/r) = 0$ are practically absent;

• in the calculations of the inhomogeneously deformed elements over the ultimate states of the second group, there are no recommendations or suggestions for disclosing their internal static indeterminacy, which leads to the inevitable use in calculations not only numerous iterative operations but also various kinds of empirical dependences.

4 Purpose and objectives of research

The research given are directed on working-out of the generalized universal calculation technique of reinforced concrete elements and structures on ultimate conditions which would:
– fully ensured the methodological unity of their implementation;
– was maximally devoid of empiricism;
– contributed to the disclosure of the internal static uncertainty of reinforced concrete elements and structures and to the significant reduction in iterative calculations;
– maximally preserved the physical essence of the deformation processes of reinforced concrete elements and structures;
– could be implemented by simple compact algorithms;
– preserved the methodological unity of precise and simplified calculation methods.

5 Results and discussion

To achieve this goal and cope with the above tasks, the entire process of reinforced concrete elements and structures deformation, and consequently the diagrams of their real state, in the deformation and force model is proposed to be reproduced using the rigidity function D. It is the latter that connects the main force (M) and deformation ($1/r$) parameters of the stress strain-state of these elements and structures. In this case, experiments show that [11], up to the exhaustion of the bearing strength, the rigidity of the elements depends on these parameters nonlinearly. Therefore, thanks to the hypothesis of "non-linearity" of rigidity [9, 10], the pronounced dependence

$$M/(1/r) = A - B \cdot (1/r)/(1/r_u) - C \cdot M/M_u, \quad (1)$$

the state diagram connecting the forces in a reinforced concrete element M with its curvature $1/r$ takes the following form (Fig. 1)

$$M = \frac{D_0 \cdot 1/r - M_u \cdot ((1/r)/(1/r_u))^2}{1 + (D_0/M_u - 2/(1/r_u)) \cdot (1/r)}, \quad (2)$$

where D_0 – is the initial value of the full ($\sum E_i I_i$) or reduced ($E_{c0} I_{red,0}$) rigidity of the reinforced concrete element; M_u – the bearing strength of a reinforced concrete rod (the maximum possible force in it when the limit state is reached); $1/r_u$ – the ultimate value of the reinforced concrete element curvature when its bearing strength is exhausted.

The universality of the generalized diagram of the element state (2) is confirmed by the fact that under certain initial conditions it is transformed into a well-known diagram of the concrete state laid down in Eurocode 2 [3]

$$\sigma_c = f_{ck} \cdot \frac{E_{c0} \cdot \varepsilon_c / (E_{cu} \cdot \varepsilon_{cu}) - (\varepsilon_c / \varepsilon_{cu})^2}{1 + (E_{c0} / E_{cu} - 2) \cdot (\varepsilon_c / \varepsilon_{cu})}, \tag{3}$$

where E_{cu} – the ultimate value of the secant modulus of concrete deformation, with axial compression $E_{cu} = E_{cf} = f_{ck} / \varepsilon_{cu}$; ε_{cu} – ultimate concrete deformation, with axial compression equal to critical $\varepsilon_{cu} = \varepsilon_{c1}$.

The curvature of the reinforced concrete element, which is the root of the square equation (2), is determined by the formula

$$\frac{1}{r} = \frac{1/r_u}{2M_u}\left[(1 - \frac{M}{M_u})\frac{D_0}{r_u} + 2M - \sqrt{((1 - \frac{M}{M_u})\frac{D_0}{r_u} + 2M)^2 - 4M \cdot M_u}\right]. \tag{4}$$

In the presence of cracks in the stretched zone of the reinforced concrete element, its curvature is corrected for expression $1/r^* = (1/r) \cdot \psi_p$ using a polynomial function of type

$$\psi_p = 1 + \rho_l / \alpha_s \cdot (1 - M / M_u) \cdot M / M_u, \tag{5}$$

where: α_s – the relative value of the armature elasticity modulus, $\alpha_s = E_s / 200000$; ρ_l – the reinforcement percentage of the element cross-section by longitudinal rods,%.

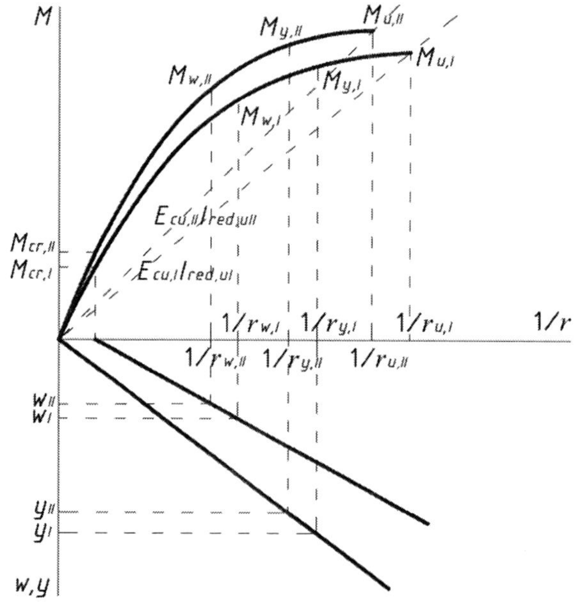

Fig. 1. Generalized characteristic and calculated state diagrams of the reinforced concrete elements and structures.

At the same time, using the hypothesis of flat sections, the current values of curvature are associated to the materials relative deformations in the most stressed cross-section or in the averaged cross-section of the most strained part of the element by the dependences

$$1/r = (\varepsilon_{c2} + \varepsilon_s)/d \text{ either } 1/r = (\varepsilon_{c2} + \varepsilon_{c0})/h_c, \tag{6}$$

where: ε_s – deformations of the most stretched armature in the element cross-section; ε_{c2}

and ε_{c0} – concrete deformation of the most and the least compressed fiber of cross-section; d and h_c – the working and total height of the reinforced concrete element cross-section.

Curvature, which depends on the basic strength and deformation characteristics of reinforced concrete elements, is one of the main parameters of the whole process of their deformation. Exactly the use of state diagrams, for which curvature is the defining parameter, provides a single methodological approach to calculating the reinforced concrete elements deflections regardless of the presence or absence of cracks in them.

Bearing strength exhaustion of reinforced concrete elements and structures in their normal cross-sections is characterized by a violation of one of the two well-known equilibrium equations $\sum N = 0$ and $\sum M = 0$. Usually, the second equation is more stringent, from which the defining condition of ultimate equilibrium follows $M_{Ed} \le M_u$, where M_{Ed} - the calculated value of the bending moment from the external load.

At all stages of concrete and reinforced concrete elements and structures deformation, the above equations have analytical and graphic solutions. Using these solutions dependence for the level of ultimate deformations of compressed concrete was obtained, using the notion of an extreme criterion of bearing strength $dM/d(1/r) = 0$ [10, 12]

$$\eta_{\varepsilon u} = \varepsilon_{cu} / \varepsilon_{c1} = 1 + 5^3 \cdot \alpha_s \cdot \frac{(k-1)}{(6-k)} \cdot \sum_{i=1}^{n} \rho_{li} \cdot \left(\frac{x_{si}}{x}\right)^2 + \beta_F \times$$
$$\times \left[\frac{(0{,}43 - 0{,}2(0{,}4 - m_h)^2)(1 - m_h^{3/2})(1 - m_h / k)\sqrt{(\ln k)^{(1 - m_h^{3/2})}}}{1 + ((1/6 - 0{,}1 \cdot m_h^2)(k-2)\ln(6/k - 2(0{,}1 - m_h)))^2} + \right. \tag{7}$$
$$\left. + \frac{(0{,}43 - 0{,}2(0{,}4 - m_b)^2)(1 - m_b^{3/2})(1 - m_b / k)\sqrt{(\ln k)^{(1 - m_b^{3/2})}}}{1 + ((1/6 - 0{,}1 \cdot m_b^2)(k-2)\ln(6/k - 2(0{,}1 - m_b)))^2} \right],$$

where k – characteristic of compressed concrete deformability (elastoplasticity), $k = E_{c0} \cdot \varepsilon_{c1} / f_{ck}$; x_{si} – the distance from the neutral line to the gravity center of compressed rods, the stresses in which, in the ultimate state, do not reach the yield point; ρ_{li} – the reinforcement coefficient of the element cross-section by the same rods, $\rho_{li} = A_{si} / (b_c \cdot h_c)$; β_F – the coefficient, depending on the type of stress-strain state of the element, is assumed to be equal to: for compressed elements – 1.0; for bending – 0.81; m_h – the inhomogeneity parameter of deformation of compressed concrete along the side h, $m_h = \varepsilon_{c0,h} / \varepsilon_{c2}$; m_b – the inhomogeneity parameter of deformation of the compressed concrete along the side b, $m_b = \varepsilon_{c0,b} / \varepsilon_{c2}$ (Fig. 2).

Exactly the dependencies (4) to (6), together with (7), make it possible to avoid or substantially reduce the number of iterations, thus contributing to the disclosure of the internal static uncertainty of reinforced concrete elements and structures cross-sections.

If we resort to a semantic analysis of the diagrams of element state (2) and the diagrams of material state (3), then it should be said that they can be characteristic (normative) or calculated (Fig. 1). It is known that the first of them can be obtained only by using the actual properties of the raw materials: concrete and armature. Therefore, they should be used to evaluate the real stress-strain state of elements and structures by the results of experimental studies (full-scale tests). According to these diagrams, the basic (characteristic) parameters of the stress-strain state of reinforced concrete element can be determined: the bearing strength $M_{u,II} = M_0$, the width of the cracks opening $w_{II} = w_0$, and the deflection $y_{II} = y_0$.

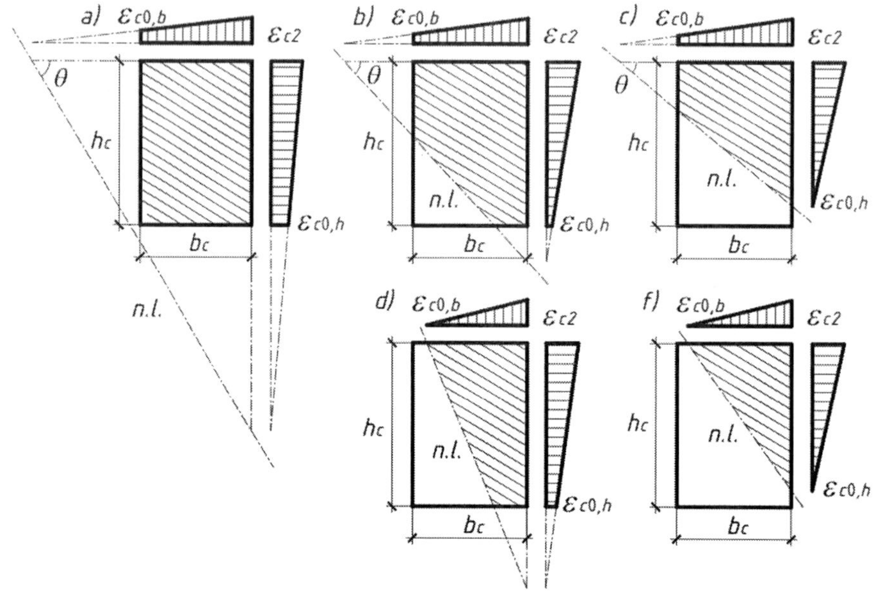

Fig.2. Possible cases of the position of the neutral line in the skew-deformed reinforced concrete element.

However, in the practice of reinforced concrete elements and structures designing, one should nevertheless rely on the calculated diagrams of their state. After all the real characteristics of the stress-strain state of reinforced concrete elements and structures $M_{u,I} = M_u$, $w_I = w_k$, and $y_I = y_k$ can differ significantly from the basic quantities M_0, w_0, and y_0, as a result of the possible variability of the strength and deformation properties due to the influence of various technological, structural and operational factors.

Consequently, the calculated state diagrams of reinforced concrete elements and structures as a whole provide possible weakening of concrete with a simultaneous increase in the parameters of its deformation. In such circumstances, there is no need to use various empirical coefficients that increase the basic parameters of reinforced concrete elements deformation ($w_0, y_0 = f_0$) to the level of calculated (w_k, $y_k = f_k$).

Therefore, taking into account the generally recognized principles of the mechanics of deformed solid body (MDSB) and dependences (2) to (7), the calculation of the reinforced concrete element in the most stressed cross-section (or in the average cross-section of the most stressed area) with its nonlinear deformation is expedient to perform with the help of computer or even "manual" solution of the following closed system of MDSB equations:

• static ratios with the hypothesis of ultimate equilibrium $M = f(\varepsilon_c, \varepsilon_{ct}, \varepsilon_s)$, $N = f(\varepsilon_c, \varepsilon_{ct}, \varepsilon_s)$;
• geometric ratios $1/r = f(\varepsilon_c, \varepsilon_{ct}, \varepsilon_s)$;
• physical ratios: the elements state $M = f(1/r)$ with an extreme criterion of bearing strength $dM/d(1/r) = 0$; materials state $\sigma_c = f(\varepsilon_c)$, $\sigma_{ct} = f(\varepsilon_{ct})$, $\sigma_s = f(\varepsilon_s)$ with an extreme strength criteria $d\sigma_c / d\varepsilon_c = 0$, $d\sigma_{ct} / d\varepsilon_{ct} = 0$. (8)

For approbation of the above method, calculations of various reinforced concrete elements tested by domestic and foreign researchers were made. The deflections of the bent elements were calculated according to the well-known Simpson's formula using the curvature dependence (4), while the deflections of the compressed elements in the ultimate state were calculated mainly by the classical expression

$$f_{\max} = s \cdot l^2 \cdot (1/r_x)_{\max}, \quad (9)$$

where s – coefficient, depending on the loading and fastening of the element; $(1/r_x)_{\max}$ – the average element curvature in the section with the greatest bending moment (the least stiffness).

The comparison results of the theoretical and experimental values of the investigated parameters, given in the table, indicate the legitimacy of developed methodology using in practical calculations.

Table 1. Comparison of theoretical and experimental parameters of reinforced concrete elements

The experiments authors, the number of samples	Deviation of the experimental data from the calculated data obtained by the method					
	Eurocode 2 [3]			Romashko V.		
	Δ	σ	v,%	Δ	σ	v,%
The bearing strength of eccentrically compressed, obliquely compressed and obliquely bended reinforced concrete elements						
Tal, K.E., Chistyakov, E.A. [13], 68 samples	1.023	0.0813	7.95	1.009	0.0763	7.56
Babich, S.V. [14], 15 samples	1.05	0.1269	12.09	1.026	0.1146	11.17
Rizak, V.V. [15], 15 samples	1.025	0.0979	9.55	1.016	0.0908	8.94
Babich, V.I. [16], 30 samples	1.004	0.0426	4.24	1.014	0.043	4.24
Total, 128 samples	1.022	0.0795	7.75	**1.013**	**0.0747**	**7.37**
Deflections of bent reinforced concrete elements						
Burns, N.H., Siess C.P. [17], 18 samples	1.022	0.1585	15.51	0.978	0.1104	11.29
McCollister, H.M., Siess, C.P. [18], 18 samples	0.978	0.065	6.64	0.997	0.063	6.32
Ernst G.C. [19], 20 samples	1.009	0.1881	18.64	1.018	0.1421	13.96
Rashid, M.A., Mansur, M.A. [20], 13 samples	1.137	0.2489	21.89	1.046	0.177	16.92
Total, 69 samples (234 experiments)	1.051	0.1807	16.94	**1.015**	**0.133**	**13.03**
Notes: $\Delta = M_{th} / M_{ex}$ (f_{th} / f_{ex}) and σ - arithmetic mean and standard deviations between experimental and calculated values, v - variance coefficient of deviations						

6 Conclusion

Thus, due to the generalized actual state diagram of concrete and reinforced concrete elements and structures $M - 1/r$, the proposed deformation-force model:

- provides a unified methodological approach to the calculation of reinforced concrete elements on the ultimate states of the first and the second group;
- allows to disclose the internal static indeterminacy of heterogeneously deformed elements in their calculations on ultimate states;
- justifies the use of the basic ($M_{Ed} \le M_u$) and derived ($dM/d(1/r) = 0$) criteria for the bearing strength exhaustion of reinforced concrete elements and constructions;

- does not require the use of complex software and allows to perform most calculations even "manually";
- to a certain extent, localizes the influence of empirical parameters on the methodology for calculating reinforced concrete elements and structures.

References

1. V.M. Bondarenko, V.I. Kolchunov, *Calculated models of power resistance of reinforced concrete* (St. Petersburg , Moscow, ASV, 2004)
2. A.V. Borovskikh, *Calculations of reinforced concrete structures on limiting states and limit equilibrium* (Moscow, DIA Publishing House, 2004)
3. EN 1992-1-1, Eurocode 2: *Design of Concrete Structures (Brussels*, CEN, 2004)
4. SNB 5.03.01-02, *Concrete and reinforced concrete structures* (Minsk, Stroytechnorm, 2003)
5. DSTU B V.2.6-156: 2010, *Construction of buildings and structures* (Kiev, Minregionbud Ukraine, 2011)
6. N.I. Karpenko, T.A. Mukhamediev, A.N. Petrov, Stress-strain state of concrete and reinforced concrete structures (Sci. Works), **1**, 7-25 (Moscow, NIIGB, 1986)
7. I.A. Uzun, IZV VUZ CONSTR ARCHIT, **11**, 7-12 (1991)
8. V.S. Dorofeev, V.Yu. Bardanov, *Calculation of bent elements taking into account the complete diagram of concrete deformation* (Odessa, OGASA, 2003)
9. V.M. Romashko, INT SCI J, **3**, 84-86 (2016)
10. V.M. Romashko, *The deformation and power model of concrete and reinforced concrete resistance* (Rivne, O. Zen, 2016)
11. B.Ya. Ryskind, G.I. Shornikova, Reinforced Concrete Structures: Sci. Works, **VI**, 42-71, (Chelyabinsk, UralNIIstromproekt, 1972)
12. V.M. Romashko, CONSTR OPT ENERG POTENT, **2(16)**, 75-83 (2015)
13. K.E. Tal, E.A. Chistyakov, Calculation of reinforced concrete structures. Experimental and theoretical studies to improve the calculation: works NIIGB, **23**, 127-195 (1961)
14. S.V. Babich, *Work and load-bearing capacity of compressed reinforced concrete elements with different eccentricity at the ends* (Thesis of Doct. PhD Tech. Sciences, Rivne, 1998)
15. V.V. Rizak, *Work and load-bearing capacity of compressed-bent ferro-concrete elements under short-term loading* (Thesis of Doct. PhD Tech. Sciences, Rivne, 2000)
16. V.I. Babich, *Experimental-theoretical studies of the operation of flexible reinforced concrete elements of rectangular cross-section on oblique eccentric compression* (Thesis of Doct. PhD Tech. Sciences, Lviv, 1972)
17. N.H. Burns, C.P. Siess, *Load-deformation characteristics of beam-column connections in reinforced concrete: A report of a research project* (Urbana, University of Illinois, 1962)
18. H.M. McCollister, C.P. Siess, N.M. Newmark, *Load-deformation characteristics of simulated beam-column connections in reinforced concrete. A technical report of a research project* (Urbana, University of Illinois, 1954)
19. G.C. Ernst, ACI J, **28 (12)**, 1119-1144 (1957)
20. M.A. Rashid, M. A. Mansur, ACI STRUCT J, **102 (3)**, 462- 471 (2005)

The application of penetrating compounds for repairing concrete, reinforced concrete and stone structures

Medeia Saliia[1,*], *Alexander* Rachkovskyi[2], *Allayar* Gasanov[2], *Ruslan* Shemet[1], and *Vladimir* Zemlyakov[1]

[1]Kharkiv National University of Civil Engineering and Architecture, Department of Reinforced Concrete and Stone Structures, 61002 Sumska st, 40, Kharkiv, Ukraine
[2]Kharkiv National University of Civil Engineering and Architecture, Department of Building Materials and Products, 61002 Sumska st, 40, Kharkiv, Ukraine

Abstract. Repair and protection of structural elements remain a relevant issue throughout the whole period of building operation and maintenance. Housing and public facilities regularly require rehabilitation and major repair. One of the main reasons behind a shorter lifespan of structural elements and units in a building is the impact on them of aggressive liquid and gaseous environments. The application of newly-developed cementitious products with penetrating properties is suggested in order to maintain and restore the protective properties of the concrete and reinforced concrete structures.

Buildings and structures in operation are subject to regular surveys aimed at assessing physical conditions of their structural elements. On the basis of visual observation and instrumental measurements a technical report is prepared which suggests further rehabilitation, repair and restoration measures to ensure safety, durability, overall performance and longevity of structural elements.

Deterioration and distress of concrete, reinforced concrete and stone structures are mainly caused by the integrated effect of various types of loads and exposure to aggressive environments. The most common are nonuniform shrinkage, temperature action, weathering effect and exposure to chemicals. The porous structure of building elements is vulnerable to external forces such as carbon dioxide, oxygen, and moisture, which trigger chemical reactions resulting in cracking and disintegration.

Over time, in the process of building operation, the pH value of the porous concrete structure in contact areas gradually approaches the point when it becomes aggressive for steel reinforcement. The porous structure of concrete itself results in electrochemical corrosion caused by exposure to vagrant currents. These adverse processes lead to corrosion of steel reinforcement with the subsequent deterioration of concrete and premature wear of their structural elements [1].

The surfaces of concrete and reinforced concrete structures can be effectively protected by means of applying special coatings. The most important property of protective

* Corresponding author: medeasalia@gmail.com

penetrating products is their ability to regenerate old concrete, which makes them useful for renovation and waterproofing works as long as there is no more need for prior cleaning and drying of concrete and stone backings.

The action principle behind penetrating waterproofing products is the following. Chemically active ingredients penetrate into the pores and capillary tracks of the concrete where they react with free lime and sediment on the surface of the pores. The pores (or micro cracks) are filled out reducing permeability and helping to waterproof the concrete. In addition, there is also a strong, 2-3-mm-thick coating of plaster on top of the concrete, which protects the concrete and prevents active ingredients from being washed away under direct water flow. Should the structure come in contact with water molecules, the reaction resumes and the sealing process advances deeper into the concrete. The same process lies behind the "self-healing" of micro cracks. Manufactured in the form of dry mix with penetrating properties, the products based on mineral bonding agents are used in foundations and backings with developed capillary porosity (concrete, reinforced concrete, brickwork, etc.) These are the following protective products: Penetron, Xypex, Kalmatron, Kalmaflex, Aquatron, Hydrotex, Viatron. Coatings up to 5 mm thick effectively protect concrete structures from external aggressive forces [2,3].

In order to ensure durability, strength and longevity of building structures, their surfaces require protection from external effects. Compounds manufactured in the form of dry mixes have received wide application.

Cementitious compounds with penetrating properties that are used to protect concrete surfaces have the following effect:

- restore deteriorated surfaces of concrete and masonry structures;
- improve the overall performance of the material, reduce permeability, protect from freeze-thaw cycles;
- enhance resistance to cracking under dynamic and static loads;
- protect structures and steel reinforcement from corrosion;
- protect structures in aggressive environments;
- ensure waterproofing [4].

The overall performance and longevity of restored structures is determined by the mineral composition of dry building mixes with cement binders, as well as by their proportions, micro- and macrostructure.

Regression analysis [5] was applied in the course of the experiment which helped to establish the composition of dry waterproofing mix (Table 1).

Table 1. Dry waterproofing mix

Component	Content, %
Portland cement	30
Quartz sand	65
Carbonate additive	3.8
Chopped glass fiber	0.1
Composite chemical additive	1.1

The method of mathematical modeling helped to establish the optimal content of fine-grained additives in a dry building mix with cement binders, the resulting quantities being 0.356 % of glass fiber and 13 % of carbonate aggregate. The experiment showed that the content of fiber glass additive ranging within 0.02 % - 0.5 % of the mass of the cement does not affect the strength of the samples.

The designed compositions include ingredients made from available local materials. The physical and mechanical properties, as well as the overall performance parameters of the composition designed in the course of laboratory experiments are shown in Table 2.

Table 2. Physical and mechanical properties and performance parameters of the samples after 28 days of standard curing.

Property	Unit of measure	Actual value
Compressive strength	MPa	46.5
Rupture strength	MPa	13.1
Resistance to direct pull from concrete backing	MPa	2.6
Water tightness	atm.	*W*12
Freeze-thaw resistance	cycles	more than *F*200
Resistance to cracking (observed visually)	–	absence of cracks
Water adsorption by mass	%	0.56

These compounds [6] are widely used to protect concrete surfaces and restore performance properties of various buildings and structures, water tanks, underground parking, sewage systems, water treatment facilities, foundations, as well as hydraulic structures such as piers, dams, docks, etc.

Upon the completion of the experimental part of the research, a trial batch of the newly-devised dry building mix was produced and used for renovation and construction works in a number of facilities in Kharkiv (Fig. 1-3).

a b c

Fig. 1. Examples of using dry building mix for waterproofing coating: a - plastered surface treatment; b - brickwork treatment; c - waterproofing of floor surfaces.

Fig. 2. Protective wall coating in a basement.

Cementitious capillary waterproofing compounds prove to be the most effective protective coatings for repairing concrete, reinforced concrete and stone structures. These protecting and repairing products can be applied both internally and externally in the form of plaster mix and external protective coatings. The research and practical application of dry mixes confirm all the properties and performance parameters of the devised compounds.

Fig. 3. Waterproofing of foundations and blind areas.

References

1. L. Bertolini, J. Wiley & Sons. Corrosion of steel in concrete: prevention, diagnosis, repair, 434 (2013)
2. T.A. Kostyuk. *Directional formation of the structure of cement compositions for waterproofing.* Dr. sci. diss., 289 (2015)
3. A.I. Voytov, V.L. Kozachuk, V.V. Lakin, A.A. Shkuratovskiy. *Modern waterproofing materials*, 204 (2002)
4. T. Andreychuk. Survey of capital construction, **2**, 16-20 (2004)
5. Saliia M.G. *Waterproofing coating on a cement base with increased crack resistance. PhD sci. diss.*, 153 (2012)
6. T.A. Kostyuk, A.G. Vandolovskiy, M.G. Saliia, D.A. Bondarenko. Dry mixes for repairs. Patent of Ukraine 57543 UA MPK S04V 24/00, (2011)

Thermophysical aspects of reconstruction of cold roof spaces

Oleksandr Semko[1*], *Oleg* Yurin[1], *Yurii* Avramenko[1], and *Serhii* Skliarenko[1]

[1]Architecture and Urban Construction Department, Poltava National Technical Yuri Kondratyuk University, Pershotravnevyi avenue, 24, 36011 Poltava, Ukraine

Abstract. The results of theoretical studies of thermal modernization of cold roof floor. The designs of overlap of long-life roofs of public buildings were considered. Overlap of these buildings are made by wooden or concrete beams and made of concrete slabs with upward edges. The problems of thermal modernization of these designs are consideration of the impact of heat-conducting inclusions on the resistance to heat. The heat-conducting inclusions of constructions are beams and vertical edges of the slabs. The results of calculations of the temperature fields were used for the analysis. The research was carried for the two temperature zones which Ukraine's territory is divided into. The analysis of thermal protection level of structures overlap was made. The magnitude of the impact of heat-conducting inclusions on resistance to heat after thermal modernization of cold roof floor was observed. The analysis showed that the thermal protection properties of structures less than standard, therefore the insulation of areas with heat-conducting inclusions is required. The options of insulation of these areas were considered. The optimal insulation options with the least amount of additional insulation over the heat conductive areas were selected.

One of the main tasks of external walling designing is to provide a normalized level of thermal and moisture conditions. However, when determining the thickness of the heat-insulation material of the walling the availability of the heat conductive inclusions with the thermal protection level smaller than in areas without it is ignored. In the roof floor buildings with long-life heat-conducting inclusions related to the vertical edges of concrete slabs, wooden beams and reinforced concrete. Ignoring of the properties of these areas leads to higher overall heat loss through external building walling. In addition to high heat in these areas there are such negative phenomena as lowering the temperature of the inner surface and increasing the amplitude of temperature variations on this surface. Increasing of the thermal-resisting properties is possible by additional insulation of heat conductive areas inclusions.

There are works of various authors dedicated to the issue of increasing the thermal-resisting properties of heat conductive areas inclusions of the external walling [1, 2]. Such as [3, 4] performed a series of tests of thermal external wall panels of residential and civil buildings. During the study the impact of "cold bridges" on the thermal quality of external

* Corresponding author: al.vl.semko@gmail.com

wall panels, developed technical solutions to their elimination [5, 6]. The approbation of theoretical methods for calculating the thermal characteristics of prefabricated buildings were performed. The results of thermal performance of prefabricated building walls were obtained [7]. Scientists [8] analyzed the thermal field forming of the exterior wall of frontal concrete panel with mounting hooks. Scientists [9, 10] researched the thermal qualities of sandwich panels with flexible links and reinforced concrete dowels.

The authors have concentrated on sections of wall constructions and walling complex configurations in the publications devoted to research the ways to improve the thermal barrier properties of the enclosure with heat conductive inclusions. The design of long-life roof floor buildings, requiring the thermal modernization and bringing them to the thermal regulations were not considered.

The work objective is to determine the optimal size of additional insulation sections with heat conductive inclusions for roof floor construction of long-life public buildings with their thermal modernization.

There are often could be the design with conductive areas of the cold roof spaces of buildings with significant lifetime. These structures include spaces on wooden or concrete beams and slabs of reinforced concrete with the upward edges (fig. 1).

a) b)

Fig. 1. The spaces of roof of buildings: a) construction of concrete slabs with upwards edges; b) - the wooden beams.

Drawing of these structures are shown in Figure 2.

The areas with heat conductive inclusions include the vertical edges of concrete slabs, wooden beams and reinforced concrete. When thermal modernization of such structures the availability of these sites and providing the additional insulation to ensure their compliance with the rules of thermal protection requirements listed in [6] shall be taken into account.

Analysis of the state of the existing structures was implemented by calculating the temperature fields. The thermal conductivity of the materials taken for conditions B. The research was carried out for two temperature zones. The estimated internal air temperature was taken as T_{in}=20 °C, and relative humidity as φ_{in}=55 %. The temperature fields of cold roof spaces is in 2 °C less then the ambient temperature estimated. The temperature fields of cold roof spaces in first temperature field is shown in Figure 3.

The results of calculation are shown in Table 1.

According to [6] for external walling buildings the following requirements of thermal insulation shall be carried out: $R_{\sum red} \geq R_{q\ min}$, $\Delta T_{red} \leq \Delta T_{sh}$, $T_{in,\ min} > T_d$.

As shown in Table 1, the first and the second thermal requirement does not apply in all the structures of cold roof overlap. So that all the structures require the thermal modernization. For the cold roof overlap structures thermal modernization there was the SUPERROCK insulation utility with the density of ρ_o=38 kg/m^3 and thermal conductivity λ=0,035 W/(m·K) and thermal. Typically, when the thermal modernization of constructions is applied the insulation existing (slag in the considered design) shall be removed.

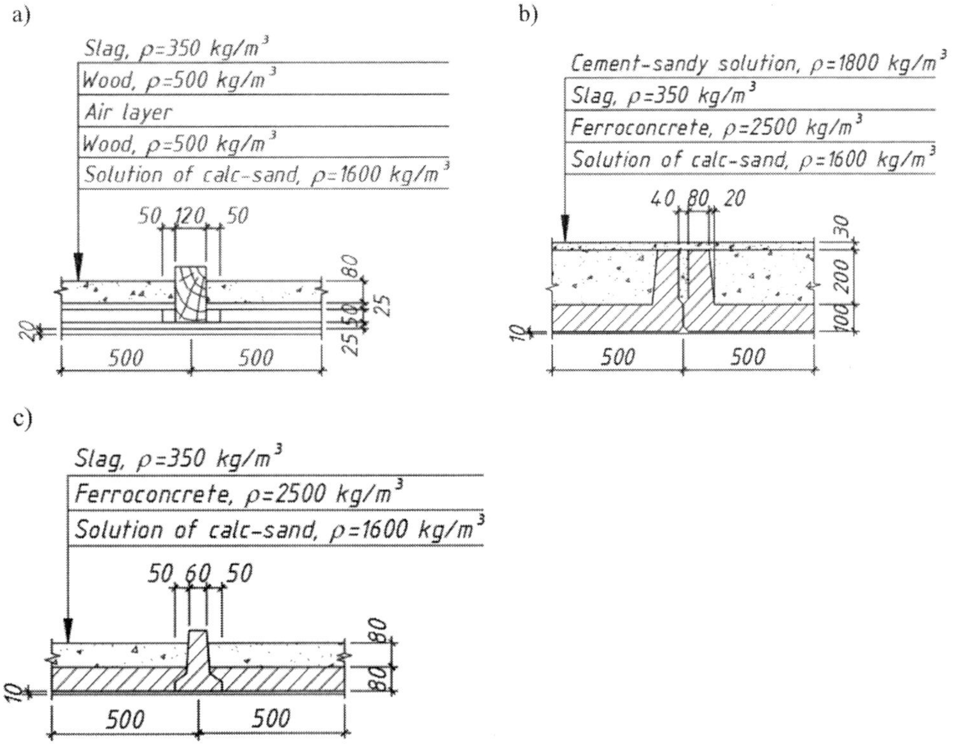

Fig. 2. The spaces of roof of buildings: a) - on the wooden beams; b) - on concrete slabs with upwards edges; c) - on reinforced concrete beams.

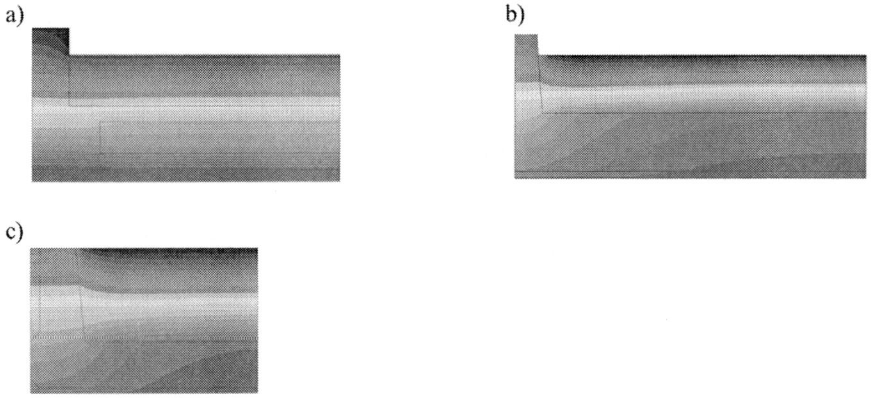

Fig. 3. The temperature fields of cold roof spaces in first temperature field: a) - on the wooden beams; b) - on concrete slabs with upwards edges; c) - on reinforced concrete beams.

According to the calculation of heating-performance without areas with heat conductive inclusions the thickness of insulation is (see Table 2).

Typically, the insulation utility is placed between the construction beams or ribs, and other sides could be open.

As shown in Table 3 with this version of insulation, the first requirement for heat protection is not fulfilled for all the constructions considered. In addition, when applying plates with upwards ribs, the second requirement is not fulfilled. That is, the insulation of areas with heat-conducting inclusions is mandatory.

Table 1. Thermal characteristic of the cold roof spaces.

Thermal field	Indicator	The design of the cold roof spaces		
		For wooden beams	For reinforced concrete beams	Made of concrete slabs with upward edges
1	$R_{\sum red}$, m^2·K/W	1,1	0,558	0,756
	$R_{q\ min}$, m^2·K/W	4,95		
	ΔT_{red}, °C	4,2	8,2	6,1
	ΔT_{sh}, °C	4		
	$T_{in,\ min}$, °C	15,7	12,8	15,6
	T_d, °C	10,7		
2	$R_{\sum red}$, m^2·K/W	1,1	0,558	0,756
	$R_{q\ min}$, m^2·K/W	4,5		
	ΔT_{red}, °C	3,9	7,6	5,6
	ΔT_{sh}, °C	4		
	$T_{in,\ min}$, °C	16	13,4	15,9
	T_d, °C	10,7		

Table 2. The thickness of insulation according to thermal calculations and excluding the areas with heat conductive inclusions as follows.

Thermal field	Indicator	The design of the cold roof spaces		
		By wooden beams	By reinforced concrete beams	Made of concrete slabs with upward edges
1	h_{ins}, m	0,16	0,17	0,17
2		0,14	0,15	0,15
1	R_{Σ}, m^2·K/W	5,209	5,105	5,115
2		4,638	4,534	4,544

As shown in Table 3 with this version of insulation, the first requirement for heat protection is not fulfilled for all the constructions considered. In addition, when applying plates with upwards ribs, the second requirement is not fulfilled. That is, the insulation of areas with heat-conducting inclusions is mandatory.

The results of calculation of temperature fields of structures are shown in Table 3.

The considered options for insulation of areas with heat-conducting inclusions were considered the next. Various thicknesses of the additional layer were taken over these sections and their width was determined, ensuring the fulfillment of the requirements for thermal insulation. Optimal option was considered with a smaller volume of additional insulation. The thickness of the additional layer was taken from top of the main heater. The volume of insulation for wooden and reinforced concrete beams was determined taking into account the heater located above the top of the beam to the level of the main heater.

The results of determining the optimum dimensions of the additional insulation are shown in Table 4.

Figure 4 shows the optimal options for the insulation of areas with heat-conducting inclusions, ensuring the fulfillment of the requirements of thermal insulation.

Table 3. Heat spaces properties of the cold roof spaces after thermal modernization.

Thermal field	Indicator	The design of the cold roof spaces		
		By wooden beams	By reinforced concrete beams	Made of concrete slabs with upward edges
1	$R_{\sum red}$, m²·K/W	3,812	1,787	0,997
	$R_{q\ min}$, m²·K/W	4,95		
	ΔT_{red}, °C	1,2	2,6	4,6
	ΔT_{sh}, °C	4		
	$T_{in,\ min}$, °C	19,1	18,7	17,7
	T_d, °C	10,7		
2	$R_{\sum red}$, m²·K/W	3,563	1,723	0,935
	$R_{q\ min}$, m²·K/W	4,5		
	ΔT_{red}, °C	1,2	2,5	4,6
	ΔT_{sh}, °C	4		
	$T_{in,\ min}$, °C	19,1	18,7	17,7
	T_d, °C	10,7		

Table 4. Heat spaces properties of the cold roof spaces after thermal modernization.

Thermal field	The design of the cold roof spaces	Dimensions of an additional layer of insulation over the heat-conducting sections		Insulation utility volume per 1 m of length, m³	$R_{\sum red}$, m²·K/W
		Thickness, m	Width, m		
1	By wooden beams	0,03	0,62	0,0216	4,959
		0,04	0,44	0,0206	4,955
		0,05	0,36	0,021	4,96
	For reinforced concrete beams	0,04	0,76	0,0334	4,954
		0,05	0,6	0,033	4,965
		0,06	0,5	0,033	4,963
		0,07	0,44	0,0338	4,963
	Made of concrete slabs with upward edges	0,14	0,82	0,10872	4,952
		0,15	0,76	0,10792	4,96
		0,16	0,72	0,10912	4,979
2	By wooden beams	0,05	0,46	0,0236	4,517
		0,06	0,38	0,0234	4,507
		0,07	0,34	0,0244	4,509
	by reinforced concrete beams	0,05	0,74	0,0388	4,542
		0,06	0,58	0,0366	4,508
		0,07	0,5	0,0368	4,508
	Made of concrete slabs with upward edges	0,15	0,8	0,10976	4,519
		0,16	0,74	0,10816	4,514
		0,17	0,7	0,10876	4,522

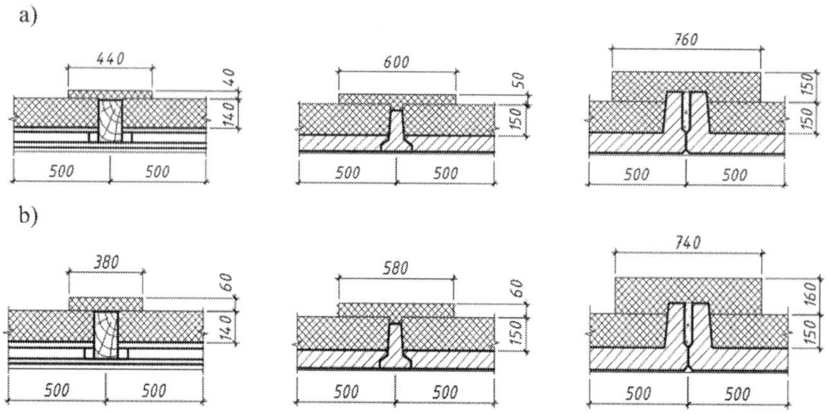

Fig. 4. Optimal insulation options for areas with heat conductive inclusions:
a) – I-st temperature field; b) – II-nd temperature field

Evaluation of the cold roofs structures spaces humidity mode after the thermal modernization performed in the climatic conditions in Poltava. Figure 5 shows the schedules of distribution for crossing the spaces of steam *e* resiliency and saturated steam *E* elasticity.

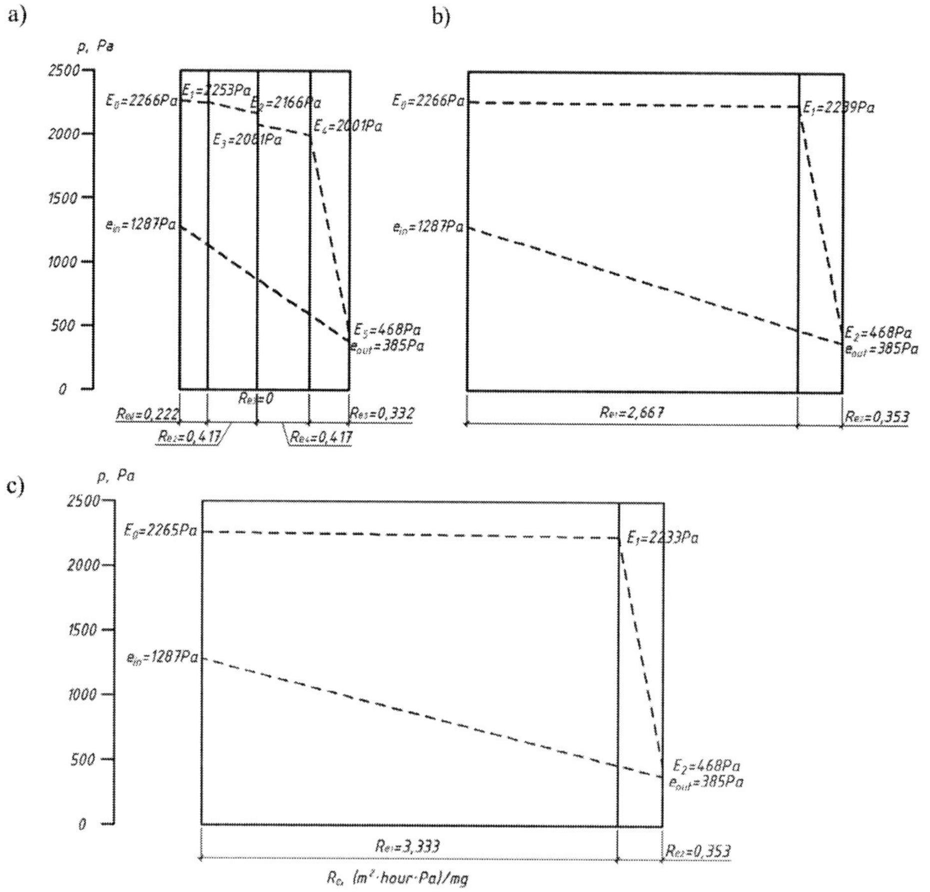

Fig. 5. The schedules of distribution for crossing the spaces of steam *e* resiliency and saturated steam elasticity *E*: a) – by wooden beams; b) – by reinforced concrete beams c) – made of concrete slabs with upward edges.

The schedules show that water vapour condensation inside capping as of January does not occurs. Since January is the most unfavourable month in terms of water saturation then the water saturation is not occur during other months of the year.

Summary:

1. When cold roof spaces structures thermal modernization the additional insulation of heat conductive areas inclusions shall be performed.
2. Ignoring of these requirements reduces resistance to heat reduction from 21% to 80% and increases the heat loss through the structure.
3. The optimal circuits of additional insulation of the areas with heat conductive inclusions allow to bring the constructions considered to thermal standards.
4. Additional insulation of heat-conducting sections depends on beams step or slabs of ribs, the thermal conductivity of insulation and temperature zones.
5. In perspective it is necessary to obtain the dependence size ratio of additional insulation of the following values for the constructions considered.

References

1. R. Sztanyi, Annals of the oradea university Fascicle of Management and Technological Engineering ISSUE, **1**, 126 (2015).
2. E. Megyesi, M. Brumaru, Advanced Materials Research, **899**, 24 (2014).
3. M. Citterio, M. Cocco, H. Erhorn-Kluttig, EPBD Buildings Platform, **28**, 4 (2008).
4. N. Norris, P. Ropell, M. Lawton, Building enclosure science & technology conference, Atlanta, **3** (2012).
5. Z. Ma, P. Cooper, D. Daly, L. Ledo, Energy Build, **55**, 889 (2012).
6. V. Murgul, MATEC Web of Conferences, **53**, 01046 (2016).
7. C. Diakaki, E. Grigoroudis, N. Kabelis, D. Kolokotsa, K. Kalaitzakis, G. Stavrakakis, Energy **35**, 5483 (2010).
8. N.D. Danylov, A.A. Sbakin, E.G. Slobodnikov, Housing construction, **11**, 46 (2013).
9. L. Tronchin, K. Fabbri, Energy Build. **40**, 1176 (2008).
10. Foucquier, F. Suard, S. Robert, L. Stephan, A. Jay, Renewable and Sustainable Energy Reviews, 23, 272 (2013).

Peculiarities of carrying capacity evaluations of cylindrical CFST columns with new type casing

Olena Sinkovskaya[1*], and *Andrey* Ignatenko[1]

[1]Kharkov National Automobile and Highway University, Department of bridges, constructions and building mechanics Yaroslava Mudrogo st. 25, 61002 Kharkov, Ukraine, Kharkov

Abstract. This paper deals with a cylindrical steel-concrete load-bearing unit. The feature of the unit is a steel mesh casing that is made from the expanded metal mesh. The optimal parameters of the suggested footing are determined taking into account anisotropy, i. e. different strength and rigidity in different directions that coincide and do not coincide with the direction of reinforcement. The optimal cell angle, when a casing deformation brings about maximum compression of a concrete core, is determined. Details for theoretical calculation of the bearing capacity of a cylindrical steel-concrete load-bearing unit are provided taking into consideration a complex three-dimensional structure of the suggested mesh casing. The validity of the calculations is confirmed by comparison them with the results of the Lira software package. The results of the calculations are provided in the form of the table showing the dependence of the adduced thickness of the plate on the longitudinal movement of the strip of the steel mesh casing. The dependencies to determine the adduced thickness of the mesh casing using the graphical method are provided for all the existing types of the mesh using concrete wedges or excluding them.

1 Introduction

The constructive decision of the intermediate column supports of a bridge has several routine solutions [1, 2] where reinforced concrete has been used for many years. However, it is only lately that steel-concrete bearing units have been used to build the footing of buildings and structures. The category of steel-concrete units includes tube-concrete ones as well as various systems with confinement-shear reinforcement in the form of radial rings, hoops, clamps, spiral windings etc. This category of the steel-concrete load-bearing units of the footing also includes the elements whose casing is made with a mesh [3] – a mesh casing. It differs from traditional solutions in the way that the casing is made from a mesh rather than a solid plate (Fig. 1).

The expanded metal mesh is a metal plate having a grooved surface with uniformly spaced cells. It has a number of advantages [4, 5].

The features and advantages of the operation of a mesh casing, as well as research by V.S. Shmukler [5, 6], substantiate the theoretical research into cylindrical steel-concrete structures having a mesh case where cells are distributed both along and across the generatrix.

* Corresponding author: iglema@meta.ua

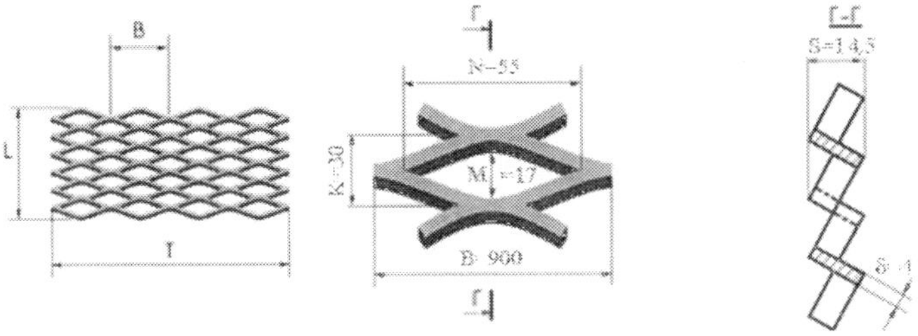

Fig. 1. Main geometric characteristics of the expanded metal mesh.

2 Objectives and tasks

The aim of this theoretical research is to investigate the features of the steel casing of a new type of cylindrical steel-concrete load-bearing units in the footing of buildings and structures. The tasks of this research are to choose the optimal parameters of the expanded metal mesh for the rational functioning of the steel-concrete structure having the mesh casing as well as to develop the method of the determination of the reduced thickness of the steel casing of a new type.

3 Determination of optimal parameters of the casing of a new type

The proposed casing of a new type is a composite where reinforcement is made from the expanded metal mesh (Fig. 2), which is made from a solid steel plate as a workpiece.

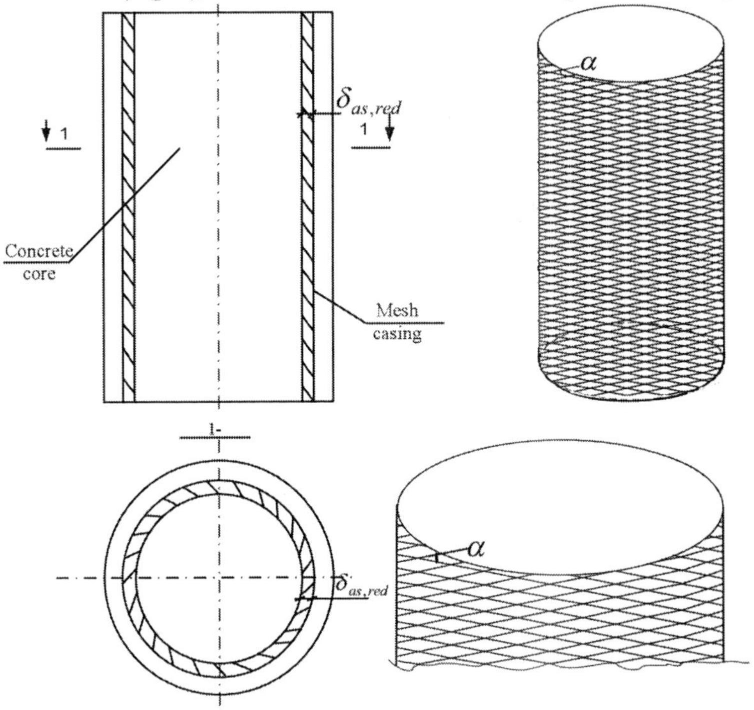

Fig. 2. Determination of an optimal angle for the direction of the mesh casing cells relative to the generatrix of a structure.

The feature of this material is anisotropy, i.e. different strength and rigidity in various directions that coincide and do not coincide with the direction of reinforcement [7]. It dramatically differs from superficially similar structures that have been researched by V. V. Vasiliev, V.L. Biderman, V.Eh. Vildeman, A. N. Elpatevskii, M. S. Saiidi, Y. C. Ou, Z. C. Girgin et al. These structures are made by means of layer-by-layer steel wire winding in the form of a net or spiral. Therefore, when forming the model, the vector parameters of expanded metal mesh are as follows:

$$\{\overline{x}\}^T = \{\alpha, R_r, \delta, S, B, M, N, K, G\}, \quad (1)$$

where α - is the inclination angle between the generatrixes of the cells of expanded metal mesh; R_r – the mean radius of the casing of a new type; δ– the thickness of a workpiece plate; S – the overall thickness of a plate; B – cell spacing used for height; M – cell width; N – cell height; K – cell spacing used for width; G – the width of cell feeding (GOST 8708-06).

The most representative parameter that integrates all the parameters is the inclination angle of expanded metal mesh cells. Therefore, taking into account the complex structure of the mesh casing of a cylindrical steel-concrete load-bearing unit, first of all, the optimum inclination angle of cells must be determined when the deformation of the casing brings about the maximum compression of a concrete core [8]. Depending on angle α, three types of deformation [9] of a cylindrical steel-concrete load-carrying unit can take place in the course of loading (Fig. 3). Figure 3 shows that the second type of deformation is the most optimum to gain maximum compression by the casing of a new type (Fig. 3 b).

a) b) c)

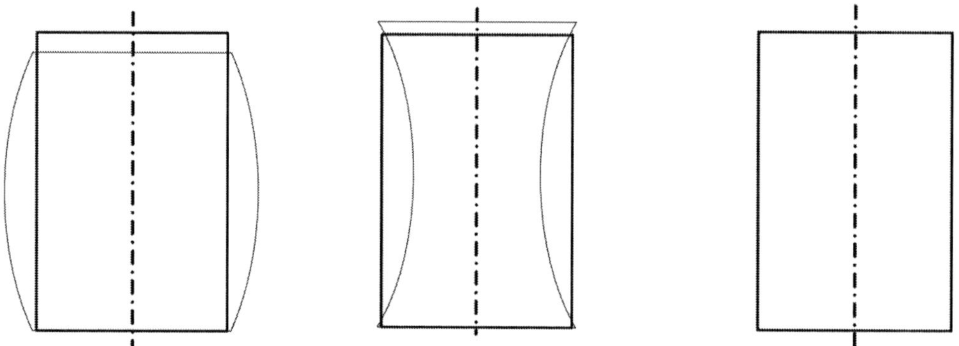

Fig. 3. The scheme of the deformation of the casing: a – the first type of deformation; b – the second type of deformation; c – the third type of deformation.

Table 1. Determination of inclination angles of cells of expanded metal mesh.

Marking of expanded metal mesh	Thickness of work piece plate, mm	Feeding Б, mm	Cell spacing B, mm	Overall thickness S, mm	Dimension of cell aperture $M \times N$	$\alpha = 2arctg\frac{M}{N}$	$\gamma = (180 - \alpha)^\circ = 2arctg\frac{N}{M}$

406	4	6	90	12.7	17 × 53	34°21′	145°39′
506	5	6	110	13.0	23 × 58	43°15′	136°44′
508	5	8	110	16.8	20 × 50	43°36′	136°24′
510	5	10	110	20.5	16 × 67	26°51′	153°08′
606	6	6	125	13.4	17 × 64	29°45′	150°15′
608	6	8	125	17.1	25 × 70	39°18′	140°41′
610	6	10	125	20.8	15 × 64	26°23′	153°37′

So, under the inclination angle of cells $\alpha = 2\beta < 70°32'$, the second type of deformation, which predetermines the horizontal orientation of cells, is rational for the compression of the concrete core.

The results of various inclination angles of cells in longitudinal and lateral directions for expanded metal mesh are in table 1. They are available in GOST [3].

Table 1 shows that an optimum direction for the maximum compression of the concrete core of the mesh casing is the direction of cells across the generatrix of a cylindrical steel-concrete load-carrying unit. As a result, this direction of cells relative to the generatrix of the structure is used by default during further theoretical research.

4 Methodology of the determination of the reduced thickness of the new type casing of cylindrical steel-concrete load-bearing units in the footing of buildings and structures

The current methods to calculate the load bearing capacity of a cylindrical steel concrete unit take into account the thickness of the steel casing. This fact considerably complicates the determination of the bearing capacity of the steel-concrete unit of a new type, which has the mesh casing. In view of it, we have developed several methods to determine the reduced thickness of the mesh casing. It enables us to verify their validity and representativeness.

Theoretical calculations are made using the superposition principle, i. e. the load bearing capacity of a concrete core and steel casing are considered as two independent components, which operate independently without any cooperation. In that case, to take into account the increase of the bearing capacity due to the compression of the concrete core by the casing, we can accept some points:

- massive (inflexible) structures are considered;
- the concrete core, which is encased in the steel casing, experiences a volumetric stressed state, i.e. triaxial compression. Radial stresses in the concrete core are equal to stresses in the places of contact between the casing and concrete core under the impact of the axial compression force;
- maximum hoop tension stress in the steel casing does not exceed the characteristic value of steel strength on the tensile yield limit;
- the value of the contact stress of the steel casing is equal to the value of the lateral resistance of the steel casing and the radial deformations of the concrete core resulting from the axial compression force; it depends on the ratio of the casing wall thickness (δ) to the mean radius of the steel-concrete unit (R_r);
- the value of the contact stress of the casing also depends on the reinforcement ratio of the lateral section of the steel-concrete load-bearing unit.

The geometric characteristics of the lateral section of the cylindrical steel-concrete load-bearing unit are determined, taking into consideration the features of the lateral section of the new type casing that contains the actual area of the steel mesh casing as well as concrete wedges, which fill the cells of the mesh casing.

The actual area of the cylindrical mesh casing is calculated as the product of the real perimeter of the steel mesh casing, taking into account the void of cells, and the thickness of the workpiece plate:

$$A_{as}^{actual} = S_{as} \cdot \delta_z = \pi \cdot k \cdot \delta_z (D_{ou} - \delta_{as}), \quad (2)$$

where S_{as} is the perimeter of the cylindrical mesh casing; δ_z – the thickness of the workpiece plate of the steel mesh casing; δ_{as} – the thickness of the mesh casing that is equal to the overall thickness of the mesh (S), according to GOST [3]; D_{ou} – the outer diameter of the steel mesh casing of a steel-concrete load-bearing unit; K – a coefficient, taking into account the geometric dimensions of mesh cells (Fig. 1), namely, G – feeding width, B – cell spacing used for height, A – an elongation dimension after every stroke of the press [3], which can be calculated as follows:

$$k = \frac{M}{K} = \frac{N}{B}$$

So the structure under review is a composite that has both the expanded metal mesh to reinforce the unit and concrete, which fills the cells void of the expanded metal mesh to bind material. During the calculation of the bearing capacity of the cylindrical steel-concrete load-bearing unit, the mass concrete in the form of wedges is included into the section of the mesh casing by means of the reduction of the area of concrete wedges to the steel area, using the equation:

$$A_{c,red} = \frac{A_{a,k} \cdot E_c}{E_a}, \tag{3}$$

where E_a, E_c are the deformation modules of the 1st type of metal and concrete correspondingly; $A_{a,k}$ is the area of the concrete wedges of the mesh casing, which is calculated as follows:

$$A_{c,k} = A_{as} - A_{as}^{actual}, \tag{4}$$

where A_{as} is the initial area of the lateral section of the mesh casing, which has the thickness of δ_{as}, nonregistering the complex structure of the mesh casing. It is as follows:

$$A_{as} = \frac{\pi \cdot D_{н}^2}{4} - \frac{\pi \cdot (D_{н} - \delta_{as})}{2} = \pi \cdot \delta_{as}(D_{н} - \delta_{as}). \tag{5}$$

Then, formula 2 looks as follows:

$$A_{c,red} = \frac{E_c}{E_a}[\pi \cdot \delta_{as}(D_{н} - \delta_{as}) - \pi \cdot k \cdot \delta_{as}(D_{н} - \delta_{as})] =$$

$$= \frac{\pi \cdot E_c}{E_a}(D_{н} - \delta_{as})[\delta_{as} - k \cdot \delta_z]. \tag{6}$$

Hence the mesh casing cross-section reduced area $A_{as,reed}$, taking into consideration the concrete wedges reduced cross-section, is as follows:

$$A_{as,red} = A_{c,red} + A_{as}^{actual} = \pi \cdot (D_{н} - \delta_{as})\left[\frac{E_c}{E_a}(\delta_{as} - k \cdot \delta_z) + k \cdot \delta_z\right] = \tag{7}$$

$$= \pi \cdot (D_{н} - \delta_{as})\left[\frac{E_c}{E_a}\delta_{as} + k \cdot \delta_z\left(1 - \frac{E_c}{E_a}\right)\right]$$

At the same time, concrete wedges, in addition to partly performing the function of the casing, due to their integrity with the concrete core, make a natural bond between the concrete core and the steel casing. As a result, it ensures the combined action of the constituents of the steel-concrete unit of a new type.

Taking into consideration the above-mentioned transformations, the reduced thickness of the mesh casing $\delta_{as,red}$ is determined:

$$\delta_{as,red} = \frac{A_{as,red}}{S_{as}} = \frac{\pi \cdot (D_{н} - \delta_{as})}{\pi \cdot (D_{н} - \delta_{as})}\left[\frac{E_c}{E_a}\delta_{as} + k \cdot \delta_z\left(1 - \frac{E_c}{E_a}\right)\right] = \tag{8}$$

$$= \left[\frac{E_c}{E_a}\delta_{as} + k \cdot \delta_z\left(1 - \frac{E_c}{E_a}\right)\right]$$

To test formula 8, a number of the problems to determine the reduced thickness of the new type casing are solved. The problems use various types of the concrete and the parameters of the expanded metal mesh (Fig. 1). Computing is made by the Lira software package.

Table 2. Results of the computation of the reduced thickness of the new type casing.

Grade of PVL	Computation method	Type of concrete	C12/15	C16/20	C20/25	C25/30	C30/35	C32/40
		$E_c(10^4), MP$	2.3	2.7	3.0	3.25	3.45	3.6
406	formula 8	$\delta_{as,red}$,mm	2.85	3.06	3.22	3.35	3.46	3.54
		ε, mm	0.0571	0.0535	0.0508	0.0487	0.0473	0.0462
	software package Lira	$\delta_{as,red}$,mm	2.91	3.02	3.09	3.15	3.21	3.26
		ε, mm	0.056	0.0542	0.0529	0.052	0.0509	0.0504
	discrepancy, %		2.1	1.32	4.21	6.35	7.69	8.6
506	formula 8	$\delta_{as,red}$,mm	3.54	3.72	3.87	4.00	4.10	4.18
		ε, mm	0.0462	0.044	0.0423	0.041	0.0401	0.0393
	software package Lira	$\delta_{as,red}$,mm	3.44	3.6	3.74	3.84	3.93	4.0
		ε, mm	0.0475	0,. 454	0.0439	0.0426	0.0417	0.0411
	discrepancy, %		2.91	3.33	3.48	4.17	4.33	4.5
508	formula 8	$\delta_{as,red}$,mm	4.27	4.53	4.74	4.90	5.04	5.14
		ε, mm	0.0385	0.0361	0.0347	0.0334	0.0325	0.0319
	software package Lira	$\delta_{as,red}$,mm	4.34	4.48	4.58	4.68	4.74	4.78
		ε, mm	0.0379	0.0366	0.0358	0.0351	0.0346	0.0343
	discrepancy, %		1.64	1.23	3.41	4.7	6.27	7.53
510	formula 8	$\delta_{as,red}$,mm	4.22	4.34	4.6	4.82	5.00	5.13
		ε, mm	0.03901	0.0379	0.0357	0.0340	0.0327	0.0319
	software package Lira	$\delta_{as,red}$,mm	4.59	4.66	4.70	4.73	4.76	4.78
		ε, mm	0.0357	0.0352	0.0349	0.0347	0.0345	0.0343
	discrepancy, %		8.78	7.5	2.17	1.9	5.0	7.32
606	formula 8	$\delta_{as,red}$,mm	4.08	4.27	4.42	4.55	4.65	4.72
		ε, mm	0.0403	0.0385	0.0371	0.0360	0.0353	0.0347
	software package Lira	$\delta_{as,red}$,,mm	3.9	4.01	4.10	4.16	4.23	4.26
		ε, mm	0.042	0.0409	0.0401	0.0395	0.0389	0.0386
	discrepancy, %		4.62	6.56	7.9	9.35	9.88	10.87
608	formula 8	$\delta_{as,red}$,mm	4.22	4.50	4.71	4.88	5.02	5.11
		ε, mm	0.0390	0.0364	0.0349	0.0336	0.0326	0.0320
	software package Lira	$\delta_{as,red}$,mm	4.19	4.32	4.41	4.49	4.55	4.61
		ε, mm	0.0393	0.0381	0.0372	0.0365	0.036	0.0356
	discrepancy, %		0.7	4.17	6.71	8.64	10.24	11.04
610	formula 8	$\delta_{as,red}$,mm	5.03	5.23	5.48	5.69	5.86	5.99
		ε, mm	0.0325	0.0313	0.0299	0.0288	0.0280	0.0273
	software package Lira	$\delta_{as,red}$,mm	5.81	5.85	5.94	5.98	6.0	6.03
		ε, mm	0.0282	0.0278	0.0276	0.0274	0.0273	0.0272
	discrepancy, %		15.5	11.94	8.38	5.02	2.32	0.6

The results of the computation are provided in table 2 and the diagrams of the dependence of the reduced mesh thickness and the longitudinal movement of a steel case strip (Fig. 4, Fig. 5, Fig. 6). The strip dimensions have the same number of cells in longitudinal and lateral directions as the new type casing when determining its optimum parameters [8].

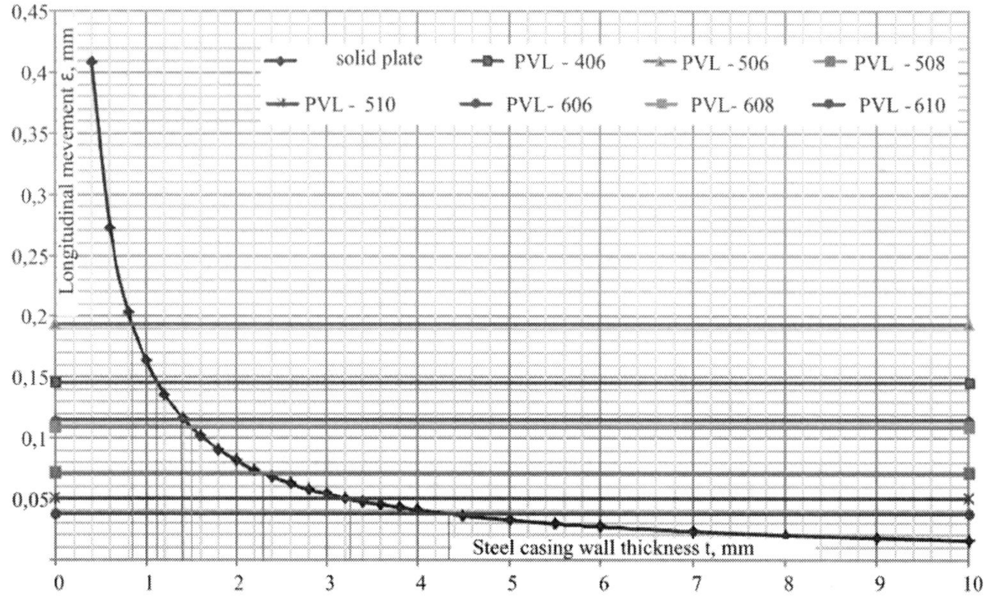

Fig. 4. Determination of the reduced thickness of the expanded metal mesh without concrete wedges.

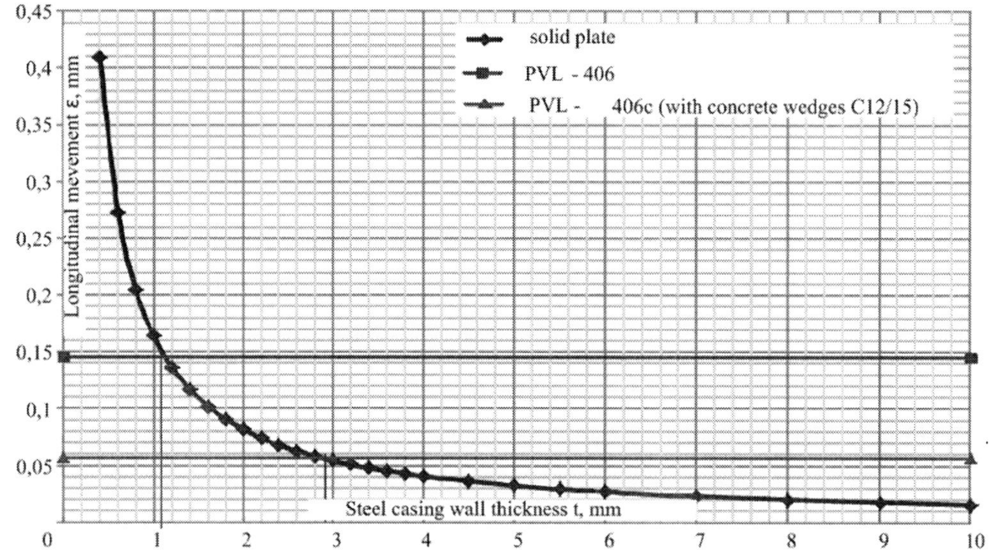

Fig. 5. Impact of concrete wedges in the cells of the expanded metal mesh upon the reduced thickness of the new type casing.

5 Conclusions

Thus, the diagrams (Fig. 6–8) enable us to determine the reduced thickness of the steel casing using a graphical method. The comparison of the theoretical and graphical calculations, which are provided in Table 2, shows that the discrepancies between them are 0.6–15.5%. It testifies that the formula is valid.

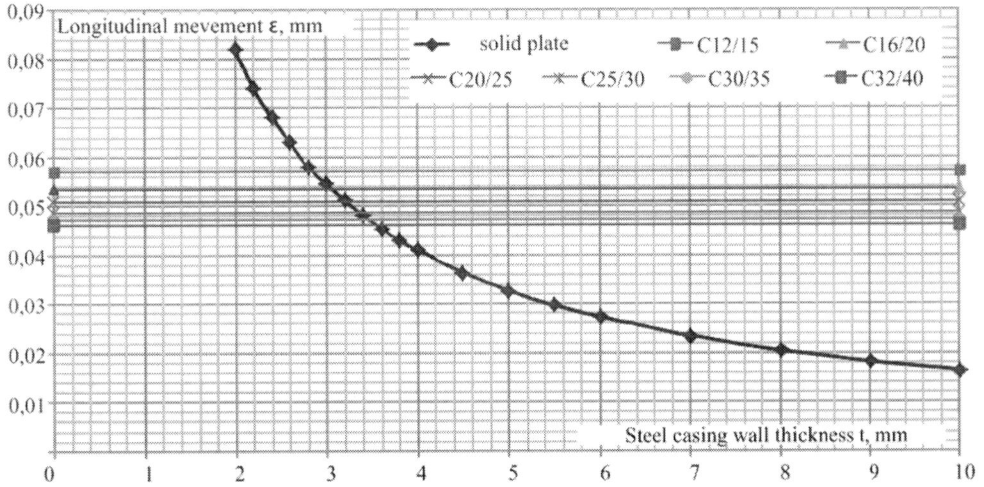

Fig. 6. Change of the reduced thickness of the casing depending on the type of the concrete to fill cells.

References

1. P.M.Salamahin, L.V.Makovskiy, V.I. Popov i dr., *Inzhenernye sooruzhenia v transportnom stroitelstve*, Moskva, (2007)
2. *Design of concrete structures – Part 1: General Rules and Rules for Building. EN 1992-1: 2001. Eurocode 2*, Brussels, (2002)
3. *Listy stalnye prosechno-vytiazhnye. Tehnicheskie uslovia. GOST 8706-78*, Moskva, (1978)
4. E.V. Sinkovskaya, Naukovyi visnyk budivnyctva, Kharkov, **68**, 96–100 (2012)
5. V.S. Shmukler, Yu.A. Klimov, N.P. Burak, *Karkasnye sistemy oblegchennogo tipa* Kharkov, (2008)
6. V.S. Shmukler, E.A. Petrova, E.V. Sinkovskaya, Naukovyi visnyk budivnyctva, Kharkov, **74**, 148–156 (2013)
7. N.A. Alfutov, P.A. Zinovev, B.G. Popov, *Rascet mnogosloinyh plastin i obolocek iz kompozitnyh materialov,* Moskva, (1984)
8. E.V. Sinkovskaya, E.A. Surzhan, Zbirnyk naukovyh prac Ukrainskoi derzhavnoi akademii zaliznychnogo transport, Kharkov, **149**, 169–178 (2014)
9. V.I. Feodocev, *Izbrannye zadaci i voprosy po soprotivleniu materialov* Moskva, (1967)

New design decisions of prefabricated girderless floors of multi-storeyed buildings

Leonid Storozhenko[1], *Dmytro* Yermolenko[2,*], *Alexander* Nyzhnyk[3], and *Ivan* Tegza[1]

[1] Poltava National Technical Yuri Kondratyuk University, Department of Structures from Metal, Wood and Plastics, Pershotravnevyi avenue, 24, Poltava, 36011, Ukraine
[2] Poltava National Technical Yuri Kondratyuk University, Department of Highway, Geodesy and Farm Building Architecture, Pershotravnevyi avenue, 24, Poltava, 36011, Ukraine
[3] Poltava National Technical Yuri Kondratyuk University, Department of Enginneering Management and Technology and Occupation Safety, Pershotravnevyi avenue, 24, Poltava, 36011, Ukraine

Abstract. Nowadays new trends are clearly manifested in building structures for buildings of various appointment and specific characteristics of their long-term operation. The priority in the construction industry takes place construction of high-rise residential and office buildings. Numerous studies in Ukraine and abroad proved that the construction of the frame with girderless floor provides resistance to both vertical and horizontal loads. It makes it possible to improve the traditional methods of frame building design. However, in the monolithic building of girderless floors there are a number of unresolved problems of structural, technological and organizational character. The authors have developed the construction of prefabricate reinforced concrete girderless floor that includes column drops, intercolumn slabs, span slabs. Column drops have bevelled lateral edges around the perimeter, forming a platform to support intercolumn slabs. Mounting of slabs occurs in the definite order. In the capacity of the columns it can be used as traditional prefabricated reinforced concrete structures as concrete filled steel tube ones. The advantage is that, comparing to modern reinforced concrete girderless floor, additional supporting equipment is not used during mounting slabs and scaffolding. The proposed girderless floors can be recommended for use in the construction of residential and public buildings for various purposes.

1 Introduction

Recently, approach to choice of design solutions, methods, technology and construction organization. Radically changed in our country. At present, new trends are clearly manifested in building structures for buildings of various appointment and specific characteristics of their long-term operation. The priority in the construction industry takes place construction of high-rise residential and office buildings. Reinforced concrete buildings with girderless frame represent a significant part of housing. It can be explained that this solution provides the possibility of building construction of any configuration with different space-planning decisions. Numerous studies in Ukraine and abroad proved that the

* Corresponding author: yermolenko-da@ukr.net

construction of the frame with girderless floor provides resistance to both vertical and horizontal loads. It makes it possible to improve the traditional methods of frame building design [1, 2, 3]. At the same time the current level of technological progress can create designs with high reliability, practicality of use and efficiency

It is known that the design girderless floors have been patented in the US in 1902. The first building with such floor was built in Moscow in 1908 under the direction of A.F. Loleyt. The first feasibility study was presented by A.A. Gvozdev and V.I. Murashev. in 1933. The results of the theory and calculating practice of the floor were entered the regulations in the 70s. Nowadays these documents are the main to regulate calculation of girderless floors.

2 Analyses of existing constructive solutions

Among the proposals of the prefabricated girderless floor construction the most interesting may be called the system "UFG" (unified framework girderless). A significant lack of these constructions is that the success of assembling depends on accuracy of column drop setting. It requires an unconventional technology and organization of work using additional tools, devices and supporting scaffolding to install the plates. This leads to some technical difficulties and a significant increasing in labor intensity to make a construction.

In the former USSR girderless floor were mainly implemented only in the industrial buildings and monolithic technology of their setting was used. The development of new opportunities in the design and calculation of structures using computer systems and software, technique and mechanisms has led to increasing the role of monolithic girderless floors in frame buildings. However, in the monolithic building of girderless floors there are a number of unresolved problems of structural, technological and organizational character. [4]. Structural problems should be considered:

- a great weight of monolithic floors comparing to running load,
- the complexity of the construction joint of columns and floor,
- unstudied impact of shrinkage deformation of concrete on stressed state of floor,
- creep and crack of concrete and deflections of floor under load.

Technological problems are

- time-consuming formwork, reinforcing and concrete work,
- technologically imperfect design of columns and floor slabs joints,
- lack of auxiliary equipment and recommendations for formwork with previous rise.

Organizational disadvantages include:

- lack of required documentation to organize and control a construction of reinforced concrete, taking into account regional and local conditions,
- not enough reliable quality control of monolithic construction,
- outdated form of certification of a new construction.

At the same time, a construction of shuttering for monolith girderless floors increases costs of the construction with concrete frame greatly. Some factors limits the use of monolithic floors in modern construction. They are:

- great amount of time of hardening concrete;
- significant labor costs at the construction site to formwork and scaffolding.

Wellknown prefabricated girderless floor is a system of modular panels supported directly with capitals of columns (Fig. 1). Capitals are mainly used to provide a hard joints of floors with columns, reducing the calculating span length of panels and create support for the panels. A grid of columns at this scheme is usually a square of 6×6 m

A system of prefabricated girderless floor consists of three main elements: the capitals, column drop and span slabs. A capital is based on the expansion of the column and takes up load of column drops, which are located in two mutually perpendicular directions, acting as beams. Prefabricated girderless floor acts as a ribbed floor with slabs, supported along four sides, where column drops perform a function of broad beams. In this case, the prefabricated elements division is made of the same type and size of slabs and slab connections are located where the value of bending moment is close to zero. Thus, the floor consists of panels that are different with reinforcement and embedded parts for setting up. Construction of prefabricated reinforced concrete girderless floor allows to perform work at a growing rate, with factory quality control and at minimum loss of time. But it requires more material and energy resources.

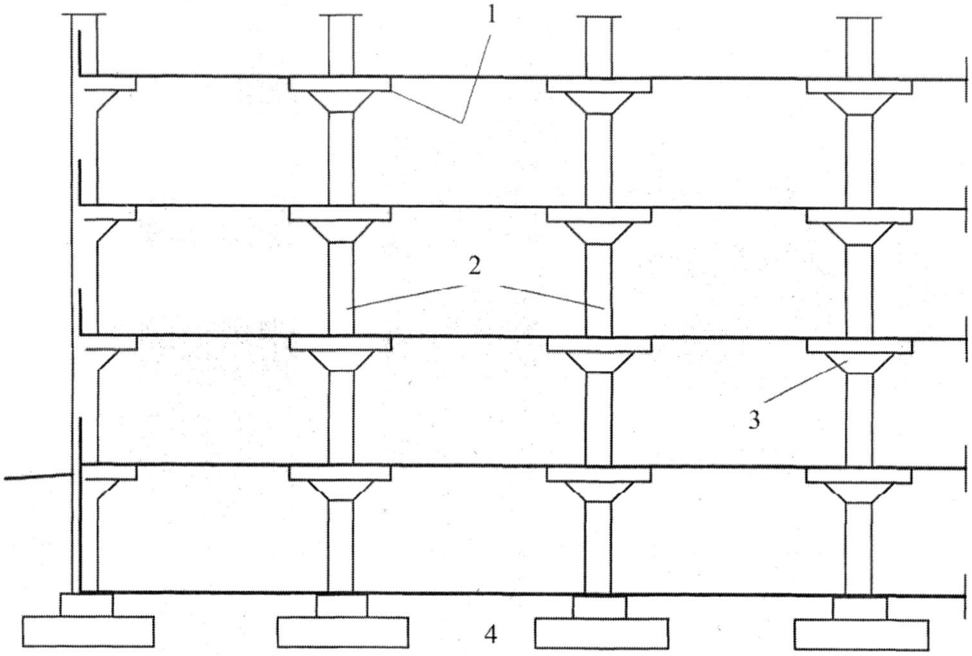

Fig. 1. Reinforced concrete girderless floor: 1 – cap plate; 2 – column; 3 – capital; 4 – foundation.

This fact determines the importance of choosing rational options of floor structures used in buildings according to constructability, strength and stiffness characteristics, efficiency of projects.

3 Purpose and objectives of the study

Based on the above, the department of structures from metal, wood and plastics Poltava national technical Yuri Kondratyuk university department led by Professor L.I. Storozhenko set a task to create and research girderless floor with sufficient bearing capacity by improving cross-section, its optimization and changing of manufacturing technique.

4 Main text

The problem is carried out in the following manner. Column drops, intercolumn slabs, span slabs are used in prefabricate reinforced concrete girderless floor. Column drops have bevelled lateral edges around the perimeter, forming a platform to support intercolumn slabs. The proposed constructive solution uses two types of connections: " column – slab";"slab – slab". All connections are relatively simple, understandable and are not labour intensive. Mounting of slabs occurs in the following order: column drops, intercolumn slabs and span slabs are installed after setting up columns in the design position. In the capacity of the columns it can be used as traditional prefabricated reinforced concrete structures (Fig. 2) as concrete filled steel tube ones (Fig. 3), with their numerous benefits [5].

If concrete filled steel tube is used (Fig. 3) the column drops are installed on the consoles that are welded to the columns. The joints between slabs are filled with mortar and total hardness is achieved by welding of embedded parts in all slabs.

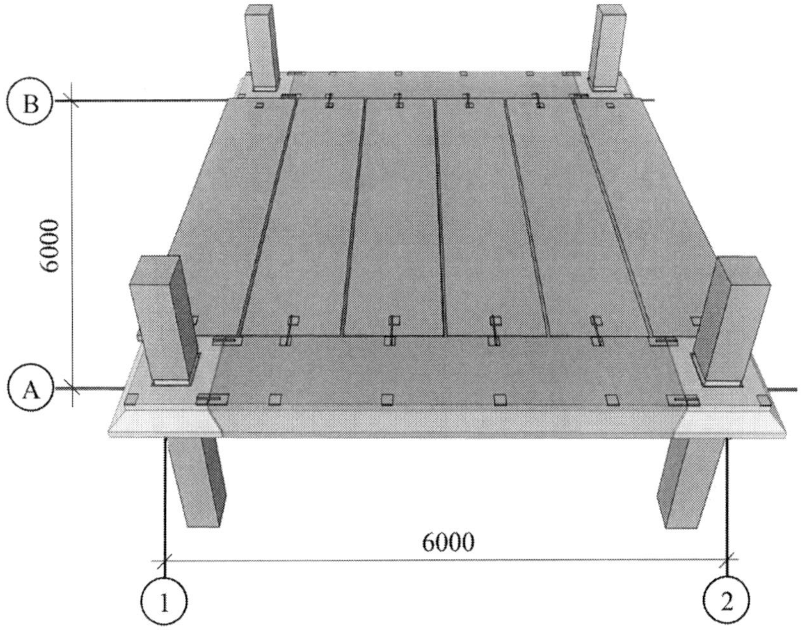

Fig. 2. General view of the prefabricated girderless floor with reinforced concrete columns.

Fig. 3 shows a lay-out plan of slabs in floor slab according to the closest analogue and its section (b).

Mounted slabs (2) of given frame, installed on concrete filled steel tube columns (1), have through- circular holes. Column drops have bevelled lateral edges around the perimeter, forming a platform to support intercolumn slabs (3). In a turn, span slabs (4) are supported at bevelled lateral edges of intercolumn slabs, forming a solid floor slab. The joints between slabs are filled with mortar and then embedded parts are welded.

Fabrication of columns can occur on the construction site. In case of need, it is necessary to use modern superplasticizers for concrete mix, which fill the tube to prevent concrete disintegration. Mounting steel tubular elements with embedded parts for column jointing and floor slab supporting are delivered to a construction site in the form of combined elements with optimum length for one floor.

Because of design features of concrete filled steel tube elements it is necessary to provide collaboration of connecting elements as steel shell, as concrete core at the design of the joints. Taking it into account, erection of structural elements of concrete filled steel tube columns is performed using electric welding. Elements are connecting end to end.

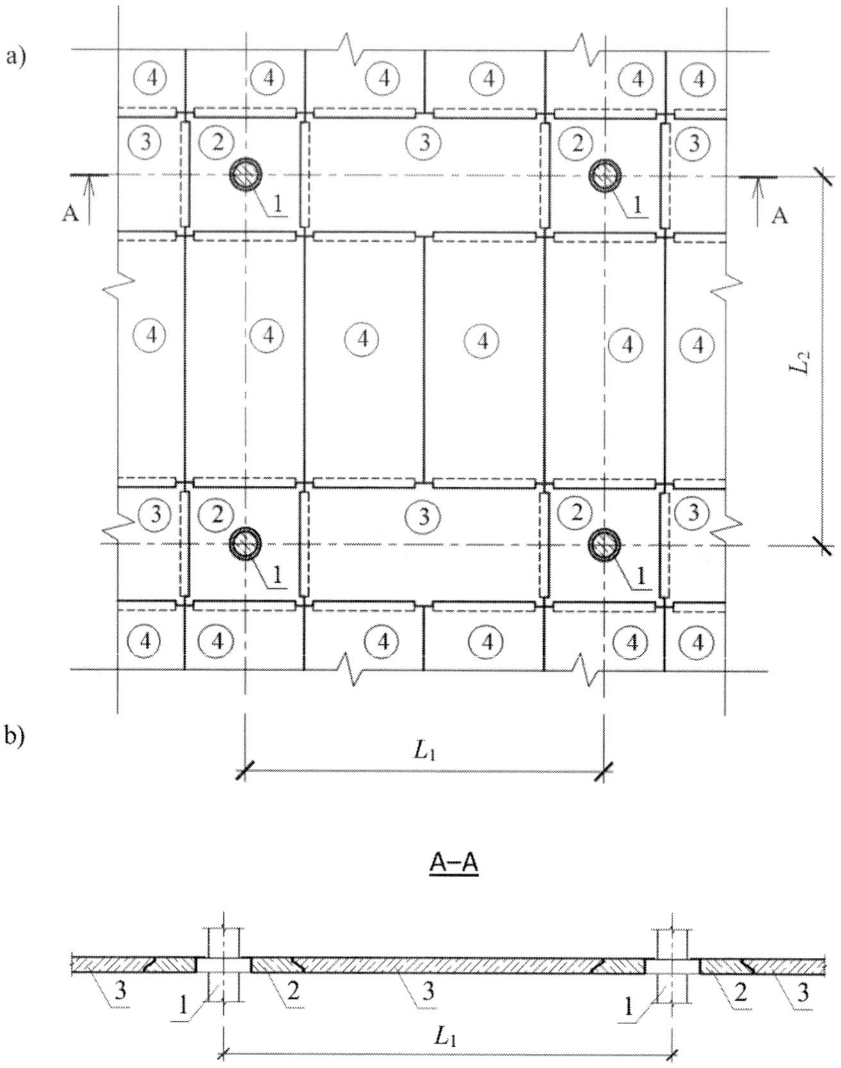

Fig. 3. Scheme of the prefabricated reinforced concrete girderless floor: 1 – concrete filled steel tube column; 2 – the column drop; 3 – intercolumn slab; 4 – slab spans.

5 Conclusions

The advantage is that, comparing to modern reinforced concrete girderless floor, additional supporting equipment is not used during mounting slabs and scaffolding.

Have compared the proposed construction with the closest analogues, the results show that they are different in the following:

a) a simple formwork is used to fabricate slabs of girderless floor system;

b) a simplified installation process of slabs at its high accuracy due to the lack of supporting equipment and scaffolding;

c) joints of slabs are relatively small and takes small amount of mortar without additional formwork.

Summing up, it should be noted that the proposed girderless floors can be recommended for use in the construction of residential and public buildings for various purposes.

References

1. S. N. Abovskaya, *Novye prostranstvennye stalezhelezobetonnye konstructsii.* (Stroyizdat, Krasnoyarsk, 1992)
2. N. I. Vatin, A. D. Ivanov, *Sopryazenie kolonny I bezrebristoy bezkapitelnoy plity perekrytiya monolitnogo zhelezobetonnogo karkasnogo zdaniya.* (S-PgODZPP, Sunk-Petersburg, 2006)
3. D. A. Pekin, *Plitnaya stalezhelezobetonnaya konstructsiya.* (ASV, Moskva, 2010)
4. A. E. Dorfman, L. N. Levontin, *Proektirovanie bezbalochnyh bezkapitelnyh perekrytiy.* (Stroyizdat, Moskva, 1975)
5. L. I. Storozhenko, D. A. Yermolenko, O. I. Lapenko, *Trubobeton*, (ASMI, Poltava, 2010)

Parameters rationalization for flat double-layer grid spatial structure (STiSK system)

Yevhen Surzhan[1*], *Kostyantin* Rapina[1], and *Tetyana* Rapina[2]

[1]Building Designs Department, O.M. Beketov National University of Urban Economy, 91000, Kharkiv, Ukraine
[2]Building Materials Department, O.M. Beketov National University of Urban Economy, 91000, Kharkiv, Ukraine

Abstract. It is the initial paper in rationalization of steel consumption for flat double-layer grid spatial structure of STiSK system. Brief historical review of metal spatial structures was given. Main economic and production advantages of the system were described Optimal geometry parameters that are satisfied to both ultimate limit and service ability states are offered. Some vital aspects related to the structure BIM design are clarified as well. Results of this research can be used by design engineers, as a basis for geometry determination of STiSk system. Outcomes of the paper can helpful for manufacturers of metal structures as well, and can help them to build up smart life-cycle designs (from engineer's idea to maintenance). BIM process that was presented in this paper will enhance rationalization process of STISK system and opens way for new designs based on this structure. The process can serve as a theoretical basis for full scale tests.

1 Introduction

Recently, in civilian multistory housing, there has been a significant increase in the use of composite reinforced concrete and steel structural slabs used to construct floors and ceilings [1]. This phenomenon can be explained, apparently by a number of advantages appropriate for such constructions:

- relatively small construction depth of the structure in relation to the span;
- reasonable use of the material (rods work on centric load);
- the possibility of laying engineering networks within the flooring;
- spatial system operations, which makes it possible to span largely without interior supports;
- increased reliability from sudden and progressive damage [2, 3];
- facilitation of roofing enclosure structures due to the frequent grid of nodes;
- maximum unification of nodes and rod elements, that creates conditions for the transition to the continuous manufacture of steel constructions;
- the possibility of using the absolute methods of assembly on the ground and surface lifting in large blocks or in a completely finished form;
- precast-mobility;

* Corresponding author: EvgenijSurzhan@gmail.com

- architectonic expressiveness and multipurposeness of application for buildings of various purposes.

In the world practice of steel structures constructing, a certain number of classical solutions are known, including: V.A. Kucherenko Central Research Institute of Building Constructions design, "Kislovodsk" and others [4]. Each of mentioned systems has a number of advantages and disadvantages, in connection with which, at the moment there is no well-formed optimal construction solution of the structural design. To the distinctive

complications that arise in the structural constructions design, the following ones can be referred: the complexity of the spatial contacting of the system rod elements; labour intensity of structural construction; ensuring of the joint work of all system elements.

Over the past decade of domestic construction practice, the structural system Structure and Folding (translation of the abbreviation STiSK) [4], developed at Kharkov National University of Municipal Economy, under the supervision of Professor V.S. Shmukler has approved itself. This system has a number of advantages in comparison with the known analogs: small relative height of the structure; relatively low elementary discharge of steel (kg/m^2); low labour intensity in production and installation; powerful application capabilities (overlapping span up to 60m). These positive aspects are achieved due to the constructive feature of the system, a detailed description of which can be found in the work [5].

The described system has approved itself well in practice in the construction of objects for various purposes in the city of Kharkov (Figure 1).

On the basis of the above mentioned information, it comes that the issue of the described system research for possible minimization of its own weight, as well as searching for possible constructive improvements in the folded structure, is very relevant in the view of the positive implementation of the Structure and Folding (STiSK) system in design and construction practice.

Fig. 1. Examples of application of the Structure (STiSK system):a- Misto swimming basin; b- confectionery shop; c- metro station Tsentralnyi Rynok entrance canopy;d- children's entertainment center.

The purpose of this study is to find possible ways of steel consumption minimization (kg/m^2) in the fold.

The object of the study is the coating of the structural Structure and Folding (STiSK) system.

The subject of the study is the parameters that affect the steel consumption of the chords of the research object.

The intended purpose of the research can be achieved by a numerical analysis of the system operation in two groups of limit states. The most common system spans in the design practice (from 18m to 40m) were accepted for analysis, with relative construction heights from 1/10 to 1/30. As a result, the dependence of the steel consumption on the relative heights of the structure cross section should be obtained for each span. At the same time, the criterion for a minimum of the system's own weight is the analytical point of steel consumption intercrossing in the ultimate and service limit states, that theoretically, makes it possible to use the construction material most effectively.

As simplifying conditions, the following are accepted:

- The model for work analyzing of the fold is accepted as a basic structural component, formed by two trusses (a modular element).
- It is assumed that the chords of the truss work together (the compliance of the braces is not taken into account).
- The angle of braces slope in the modular element is conventionally adopted as close to the optimal value [5].
- The angle of truss slope to each other is constantly 45 degrees (due to the design features of the system).
- The external load distributed uniformly defaults to the node load.
- When calculating in the ultimate limit state, the chords section is taken from a square tube with a constant wall thickness (4 mm) and a variable section height.
- When calculating the service limit state, the section of the chords is taken in the form of continuous squares (Figure 2).

2 Calculating in the service limit state

The calculation of the limit deformations is made on the assumption of the normatively limited permitted values of the deflections of such structures. In this case, the permitted deflection is accepted:

$$f_{max} = \frac{1}{200} * l \tag{1}$$

Substituting the restriction *(1)* into the expression for the maximum deflection of a pin-ended beam, we will obtain:

$$\frac{1000}{384}\frac{ql^3}{EI_x} = 1 \tag{2}$$

Further, from *(2)* we will express the fold moment of inertia, at which the system deflection volume will be equal to *(1)*:

$$I_x = \frac{1000ql^3}{384E} \tag{3}$$

Taking into account the accepted simplifications, we will write the expression of the moment of inertia of a section of the fold (Fig. 2):

$$I_x = 4\left[\frac{b^4}{12} + b^2\left(\frac{h-b}{2}\right)^2\right] = b^4 - 2b^3h + b^2h^2 \tag{4}$$

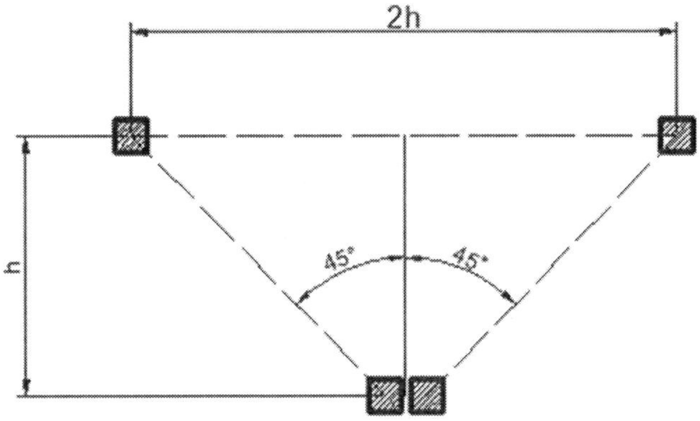

Fig. 2. Cross section of the fold.

Substituting (*3*) into (*4*) we will obtain:

$$b^4 - 2hb^3 + h^2b^2 - \frac{1000ql^3}{384E} = 0 \tag{5}$$

Solving (*5*) in general form, it can be noted that this equation has two real roots, one of which is suitable in the physical sense of the task [5]:

$$b = \frac{1}{2}\left(h + \sqrt{h^2 - 4\sqrt{\frac{1000ql^3}{384E}}}\right) \tag{6}$$

The obtained solution (*5*) allows us to determine the minimum permitted (according to the service limit state) required cross-sectional area of chords of the Structure and Folding (STiSK) system in relation to the span of the structure (*l*), the external load (*q*) and the adopted system altitude (*h*). For visual illustration of the working efficiency of this approach, a numerical solution analysis (*6*) was performed. Having set the external load distributed uniformly (q = 500 kg/m²), the steel consumption for the structure for spans of 18, 24, 30, 36, 40 meters is determined, depending on the height of the structural system.

3 Calculating in the ultimate limit state

When calculating in the ultimate limit state, the chords of the trusses were calculated as centrally compressed rods in accordance with [7]:

$$\frac{N*\gamma_n}{\varphi*A*R_y*\gamma_c} \leq 1, \tag{7}$$

where φ is the stability coefficient under central compression, which value at $\lambda \geq 0.4$ is to be calculated according to the formula:

$$\varphi = \frac{0.5}{\bar{\lambda}^2}\left(\delta - \sqrt{\delta^2 - 39.48\bar{\lambda}^2}\right), \tag{8}$$

where $\bar{\lambda}$ is conventional flexibility.

The value of the coefficient δ should be calculated according to the formula:

$$\delta = 9.87\left(1 - \propto + \beta * \bar{\lambda}\right) + \bar{\lambda}^2, \tag{9}$$

where α and β are the coefficients characterizing the initial imperfections and residual intensions and are determined in accordance with [7] (for the considered example $\alpha = 0.03$, $\beta = 0.06$).

Having been given with the initial data, analogous to the calculation in the service limit state, the steel consumption for the system under research is determined, depending on the height of the structure and the span, based on the limitations of the material strength. The calculation was carried out by successively iterating of the buckling coefficient φ, while the convergence was determined from the condition of the difference between the steps of iteration $\leq 1\%$.

Taking into account the adopted simplifications, in this calculation the length of the rods of the compression chord of the fold depended only on the altitude of the structure, and the radius of cross of the rods at each step was determined from the required design altitude of the square tube in the previous step.

On the basis of the obtained data, diagrams of the dependence of material consumption on the structural Structure and Folding (STiSK) system depending on the height of the structure for spans of 18 ... 40 m are shown Fig. 3, 4. From the data in the diagrams, it comes that without significant influence from the span, the material consumption functions on the system chords along different limit states intersect in the system height range l/20 - l/21.

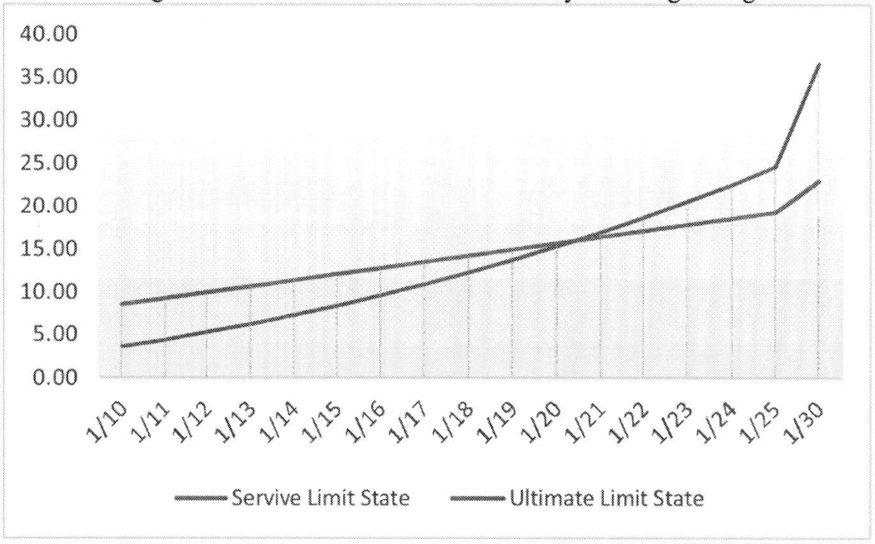

Fig. 3. Steel consumption kg/m² for the span 18m.

4 Verification

To confirm the obtained theoretical results, it is necessary to perform verification calculation of structures using FEM. The difficulty in this case is the creation of numerous computational schemes when varying geometric parameters.

The geometry of the structure depends on a number of parameters. To determine them, it is needed to understand the logic of structural construction. Thus, in the Structure and Folding (STiSK) system, the modular element (truss) (Fig. 5) forms the basic structural element (Fig. 6), by connecting the former at an angle of 45°. In its turn, the structural plate body is enlarged from the structural elements in the transverse direction.

Among which the defining ones are the span of the structure, its construction height, as well as the angle of slope of the rods in the modular elements. (Fig. 5.) The structure is formed from the basic structural components, which in their turn form a complete structure with the necessary dimensions.

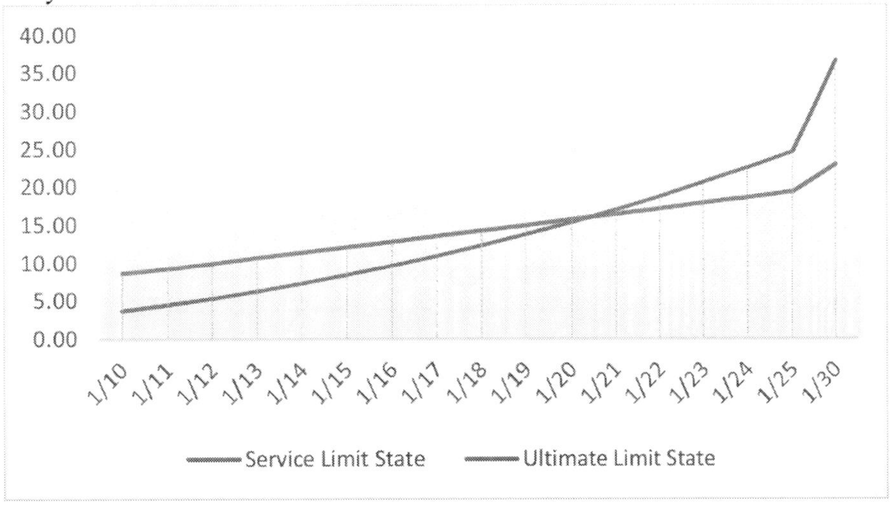

Fig. 4. Steel consumption kg/m² for the span 40m.

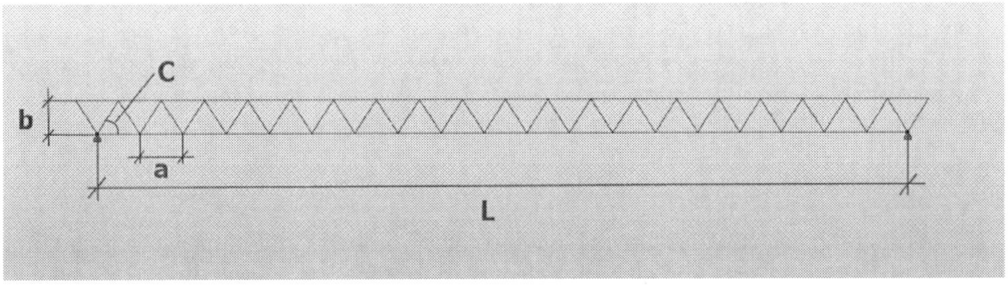

Fig.5. Truss, modular element. *a*– spacing of nodes, *b*– truss height on the support, *c*– the angle of slope of the rods in the truss and *L*– truss span.

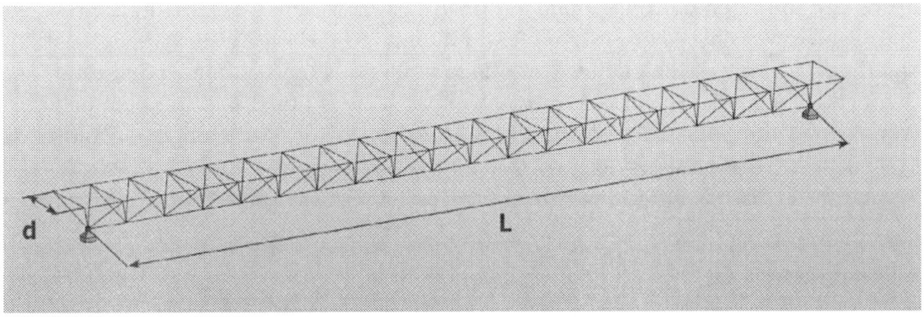

Fig. 6. General view of the basic structural component. *d*– width of the modular element.

Thus, the main benchmark data determining the outer look of the structure is the parameters of the truss. This is primarily the span of the structure (*L*), depending on which the construction height of the future structure is assigned. It lies in the range from 1/10-1/30 of the main span.

It should be noted that the reducing of the height of the structure, that is not used directly, reduces the cost of building operation. Therefore, structures with the minimum possible height are of particular interest [8].

The angle of slope of the rods in the trusses is taken on the basis of the optimal material consumption [5], and lies within the range. The datum angle is assumed to be 65°. Then, the spacing of nodes in the truss (a) is calculated, as well as the number of steps in the truss. The obtained number is rounded up to a nearest whole number and the angle of slope of the rods is recalculated. The angle is checked for compliance with the criterion of optimal material consumption. If it was, then the calculation is continued, otherwise the number of panels changes to match the criterion of the angle of slope of the rods.

Concerning the width of the basic structural components "d" in the transverse direction, it depends on the angle of the trusses slope in this direction. Inasmuch as the angle is constant (45°) and is determined by design considerations, the width is always equal to the height of the truss on the support defining the geometry, according to the initial structure parameters (d and L). This algorithm was used to write the program in Dynamo 1.3.0 and for creating the calculation schemes for Autodesk Robot Structural Analysis. Also, the possibility of creating a BIM model for Autodesk Revit was carried out for the purpose of node detailing and connections in Autodesk Advance Steel or Teckla Structure to simplify the process of real design of structures.

Further, the obtained calculated schemes were calculated with the parameters obtained during the research. Noteworthy, that the initial assumptions and the obtained results can be correctly compared only for a number of parameters, such as the cross-sectional area of the chords and lattice profiles, the ultimate deflections of the structure. For each span, the construction height was chosen, the value of which was obtained as an analytical intersection of the two steel consumption curves for the service and ultimate limit states.
The sections of the chords were assigned from paired angles, and the lattices were assigned from the square tubes. At the same time, the area of the section of the chords corresponded to calculated ones as close as possible. Noteworthy, that those rolled sections were applied that are rollable outside Ukraine.

5 Conclusion

The results of the calculation showed that the results of the calculations given in the first part of the article are confirmed with a high degree of probability. Especially constructions with construction depths, defined earlier, satisfy both limit states and are approximated to the minimum material consumption. But it should also be noted that further study of this issue is necessary, taking into account the rods of the lattice of the trusses, and the runs along the upper and lower chords of Structure and Folding (STiSK). It is also worth mentioning that significant change of the obtained results is possible when binding to domestic rolled steel products, which also requires additional research. It is noted that the obtained data in the work shows the possibility of work simplification of the Structure and Folding (STiSK) system, which for today is seemed to be a very priority and topical direction, taking into account wide interest in this construction in the field of design and building.

References

1. S.S. De Silva, D.P. Thambiratnam, Computers & Structures, **87(17),** 1067-1076 (2009)
2. Y. Ding, W. LIN, Z. LI, Engineering Mechanics, **3,** 97-101, (2007)
3. Y.Alashker, S. El-Tawil, F. Sadek, Journal of Structural Engineering, **10**, 1187-1196 (2010)
4. V.S. Shmukler, N.P. Burak, A.I. Duker, I. Vassim, patent of Ukraine 84560
5. V.S. Shmukler, E.V. Berezhnaya, V.V. Gerasimenko, O.A. Kalmikov, V. Ismail, Vestnik KhNADU, **49,** 75-83 (2010)
6. *DSTU B V.1.2-3:2006. Progini ta peremishennya. Vimogi proektuvannya*, Kiev (2006)
7. *DBN B V 2.6-198:2014. Stalevi konstruktsii. Normi proektuvannya*, Kiev (2014)
8. J. Kneifel, Energ Buildings, **3**, 333-340 (2010)

Urban renovation of the hot water supply system

Oleksandr Taraday[1], *Oleksandr* Gvozdetskiy[1*], and *Sergii* Fomich[1]

[1]Kharkov National University of Civil Engineering and Architecture, Department of Heat and Gas supply, Ventilation and Use of Thermal Secondary Energy Resources, Sumska str., 40, Kharkiv, Ukraine, 61002.

Abstract: Issues related to improving the reliability of hot water supply systems are considered. Currently, centralized hot water supply systems are in an emergency condition due to the fact that the external networks are made of black pipes without anticorrosive coatings. A fundamental difference between hot water supply systems and heating systems is the nature of black metal corrosion. The purpose of this article is to conduct a research of the state of hot water supply systems and consider options for their recovery. The options suggested for restoration of hot water supply systems in cities and settlements of Ukraine are the following: complete replacement of failed networks with new pre-insulated plastic pipelines; refusal from external networks of centralized hot water supply systems and a central heat point with the installation of individual heat points in each house; decentralization of hot water supply systems by installing electric water heaters. A technical and economic comparison of these options is carried out, and the advantages and disadvantages of each option are considered. The arrangement of a set of individual heat points instead of one central point cannot be considered as the recommended method, because of the need for large additional costs for the reconstruction of heating, water supply and electricity supply systems. The most technically and economically justified method from the considered renovation methods of hot water supply systems is the complete replacement of failed steel pipelines with new nonmetallic pre-insulated structures.

1 Introduction

At present, most of centralized hot water supply systems of cities and settlements of Ukraine are in the state of failure due to the fact that the external pipework is made of black pipes, without any internal anticorrosive coatings. The main difference between hot water supply systems (HWSS) and heating systems is the nature of the black metal corrosion. In heating systems where chemically purified de-aerated water is circulating [1], mainly, external corrosion occurs due to various causes: stray currents, leaks from other systems and chemical composition of the soil [2-3]. More than 90% of the defects in the HWSS are resulted from internal corrosion [1-3]. Internal corrosion is caused by the chemical

* Corresponding author: npp-tghv@ukr.net

composition of the heated under-aerated tap water and by air-blasting and, most importantly by the total lack of any protection of the internal pipe surface.

2 The objective

The objective of this article is to carry out the study of the state of hot water supply systems and consider options for their recovery.

3 Matters of the article

Table 1 is an example of the state of hot water supply piping. It shows the number of breakthroughs in the hot water pipelines in the city of Zaporozhye where vacuum deaeration is used to protect the inner pipelines surface [4-6].

Table 1. Failure rate of systems in 2005-2014.

Boiler house	2005	2006	2007	2008	2009	2010	2011	2012	2013	2014	2005-2014
double-pipeline system											
Sv.Nikolayastr., 79a + Shkolnayastr., 2a	42	20	34	29	49	47	53	39	54	31	398
Nakhimova str., 4	78	37	94	85	91	134	111	122	113	88	953
four-pipeline system											
Tovarischeskayastr., 47	15	12	16	20	19	20	21	19	17	22	181
Zadneprovskayastr., 7	18	13	13	16	19	32	29	37	39	28	244
Chubanovastr., 3d	28	27	19	24	22	12	18	22	46	25	243

It should be emphasized that the data given in Table 1 characterize the state of the hot water supply system of a large regional center, the only one in Ukraine where protection against internal corrosion is established. In other cities of Ukraine the number of "failures" is an order of magnitude greater. Despite excessive corrosion in some cities, including Kiev, centralized hot water supply continues to be implemented in Kharkov [7-9]. At the same time heating supply companies expend resources at times exceeding regulatory ones. This in turn leads to a tariffs increase. As the world experience shows centralized hot water supply systems can function normally only if they are made of galvanized, brass or non-metallic pipes [9-11].

An important issue of reliable and economical operation of the hot water supply system is restoring circulation in the pipework. However, life requires the restoration of hot water supply in all cities and towns of Ukraine. We believe that the restoration of hot water supply is possible in several options.

The first option is a complete replacement of failed pipe networks with new pre-insulated plastic pipelines.

The second option is the refusal of the existing external networks of a hot water supply system and central heat point (CHP) with installation of individual heat points (IHPs) in every house.

The third option is an installation of capacitive electric heaters in each apartment.

In the first and second options funding is provided by municipalities and the state. In the third option funding is provided, mainly, by the apartment owners themselves.

When reasoned the second option it should be emphasized that the installation of several new IHPs for the preparation of hot water instead of one operating CHP is not a reasonable technical and economic decision. Calculation and practice show that hot water is more advantageous and easier to heat at a single location and fed through pipes to a group of buildings.

Some experts propose to replace one CHP by many IHPs. They justify their positions allegedly there is a need to install IHP everywhere in each house for heating needs. Due to this they as if "extend" the heating IHP giving it the function of preparing hot water. This presentation of the problem is not justified and correct. The installation of IHP, equipped with a variety of equipment, is not a necessity today neither from a technical point of view nor even from an economic point of view.

The question should be formulated clearly and unequivocally: how many installations for the preparation of hot water do we construct, one or several? The answer to this question is not unambiguous since it requires techno-economic justification.

Today we don’t have to explore the matter of the rational number of CHPs for a micro district. The works of several authors gave an answer to this question in the eighties of the last century that it is more advantageous to have one CHP. Semicentennial practice of exploitation in Ukraine and other countries fully confirmed the theoretical development correctness of the past years.

Three options of heat supply systems for a micro district with twenty buildings with a total thermal load of 11.95 Gcal/h, including 5.45 Gcal/h for hot water supply, are considered. Further we will carry out a techno-economic comparison of three options of renovating hot water systems.

The first option is a restoration of hot water supply through the complete replacement of all hot water pipelines and circulation pipelines, and external four-pipe heat networks of the micro district without sacrificing maintenance of the existing network configuration while maintaining the operating CHP.

Modern pre-insulated pipelines for hot water supply are ready-made for external non-canal laying structure in the form of straight or rolled up construction like "Isoproflex".

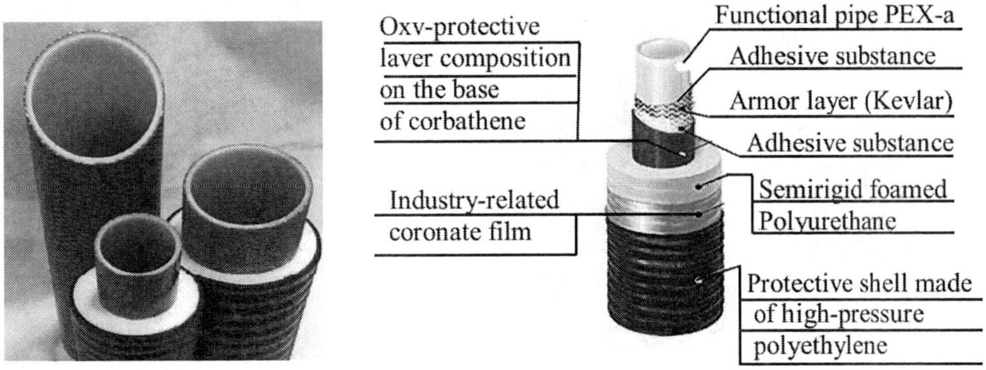

Fig.1. Pre-insulated pipe for hot water supply systems.

The laying of new pipelines is done on a non-canal principle using conduits in which the old networks of the hot water supply system were previously laid. In this case the conduits are filled with sand after dismantling the old pipes, and new pre-insulated pipelines are laid on a sand base, naturally, without any sliding supports. The matter of restoring conduits covering with slabs is decided in each individual case depending on the laying depth or the presence of a traffic way above the heating main.

Advantages of the first option:

- replacing only the external hot water supply networks we restore the efficiency of the entire system as heat exchange and pumping equipment at the CHP (TRS), as a rule, has kept its operability anywhere and does not require any replacement;

- preserving thermal load of hot water supply on the operating source of thermal energy, boiler house or CHPP.

Disadvantages of the first option:

- the need to break existing heating ducts; breaking and restoring road surfaces and lawns;

- dependence on hydraulic control and schedules of the heat supply company.

The second option. Instead of a single CHP preparation of hot water is provided at IHP in the basement of each building. At the same time the central heat point is virtually eliminated as an installation for the preparation of hot water.

The source of thermal energy for heating up hot water stays the same (CHPPor boiler house). Fig. 2 is a photograph of such IHP installed in the basement of a 16-storeyed residential building with a thermal load on heating and ventilation - 362,000 kcal/h, on hot water supply - 297,000 kcal/h.

Fig. 2. Photograph of individual heat point in residential building.

Advantages of the second option:

-the possibility of preparing hot water for your own home regardless of other buildings in the micro district;

- cost reduction of the heat losses component by external heat networks in the price of hot water supply;

- thermal load conservation of hot water supply at the operating heat source: a boiler house or CHPP.

Disadvantages of the second option:

- the need for capital input to construct a new individual heat point. For the 16-storeyed house considered above the costs are 430 000 UAH;

- cost increase of the operating costs component in the price of hot water supply (additional personnel for servicing an individual heat point, and additional power costs for newly installed equipment);

-dependence on hydraulic control and schedules of a heat supply company;

- additional capital input for increasing the diameters of the existing submain heating networks due to a significant load increase on them related to the addition of a load on the HWS system. For example, prior to reconstruction thermal load delivered by the inlet and return pipeline at the entrance to the 16-storeyed house was 362,000 kcal/h. After reconstruction it is needed to start another 297,000 kcal/h along the same pipes, i.e. almost twice as much. At the same time, the load on all sections of the submain "heating" heating network increases to 84%. For the main section the output load from the disconnected central heat point, which provides heat supply to 30 residential houses will be 11.95 Gcal/h. instead of 6.5 Gcal/h;

- diameters increase of the external cold water supply system;

- revision and partial reconstruction of the micro district power supply system due to the transfer of the location for heating up hot water from CHP to IHP installed in the basement of each house.

The third option. Preparation of water in each apartment by means of a mini electrical boiler.

Advantages of the third option:

- independence: hot water is always at the consumer's request;
- savings: consumer has full cost control of hot water supply;
- simplicity of installation and operation of the hot water preparation plant;
- apartment owners' self-solution about capital investments
- significant reduction of all operating costs for hot water in comparison with the first and second options.

Average costs per person per month we have calculated (with hot water allowance of 3 m3 per month) to pay for hot water supply at current tariffs in 2016 will be UAH 90.

In the third option an advantage for the power supply company is that it acquires an additional consumer. Especially valuable and important for the power supply company is that the new consumer strives to expend electrical energy to heat up hot water at night, i.e. in the period of power consumption "valley value".

If to consider consumers changeover to independent apartment hot water supply as a trend in the development of the electric power industry, one can view this as a positive phenomenon in the future. Because the changeover to the direct electric water heating in the apartment gives a performance factor increase in the entire energy system by reducing losses when heating and transporting water heat-transfer fluids.

To drawbacks, and more correctly, to the expenses of the third option should be attributed:

- theelectricity supply company with the number of electric water heaters increasing should increase the build-up or redistribution of its capacities as well as to reconstruct domestic and outdoor networks and transformer substation;

- additional difficulties arise for water supply companies as they must ensure an almost twofold increase in cold water running supply to the apartment. Like in the second option they should perform the re-laying of external running water supply pipelines;

-in the case when the source of heat supply is a Central Heating and Power Plant (CHPP) there is mass disconnection of hot water consumers

– this is a decrease in its technical and economic indicators of the heat supply system as a whole.

The main indicators for the three options are shown in Table 2.

Table 2. Reduced costs for 1 GCal and the cost of 1 m^3 of hot water.

Number of option	Title	Reduced costs, UAH	Cost of 1m^3 of hot water per person	Source of funding
I	Complete replacement of existing hot water supply networks while maintaining the CHP	2560	120	Municipality
II	Refusal of existing hot water supply networks and CHPs replacing them with IHPs in each house	3600	120	Municipality; Electricity supply company
III	Installation of electric water heater in an apartment	2820	90	Apartment owner; Municipality; Electricity supply company

4 Conclusions

-The use of black steel pipes for hot water supply, which is a gross violation of normative documents, led to massive damage to the pipelines of hot water systems caused by increased internal corrosion. The result of this approach is the almost complete cessation of centralized hot water supply in most cities and settlements of Ukraine.

-The only major regional city of Ukraine, where the centralized hot water supply system continues to operate normally enough, is the city of Zaporozhye. As everywhere it is made from black steel pipes without internal covering. Sufficient level of the central hot water supply system efficiency is ensured by using vacuum-aeration treatment of hot water to remove oxygen from it.

- Complete replacement of corroded pipelines with new non-metallic pre-insulated structures is a method which is the most technically and economically proved of the renovation methods of hot water supply systems considered above.

- Installation of a number of IHPs instead of a single CHP cannot be considered as a recommended method of mass renovation because of the need for heavy additional costs for the reconstruction of the entire heat supply system, water supply, and electricity supply partially.

- Changeover to apartment capacitive electric heaters is the most real and cost-efficient from the consumer's point of view. This method allows reanimating the hot water supply system in the shortest time, which is already being done everywhere in Ukraine.

-The use of the third method creates prerequisites for increasing the efficiency of power supply in the system, but leads to a decrease in technical and economic indicators of heat supply systems with a source of thermal energy – CHPP.

References

1. Yu.V. Balaban-Irmenin, V.I. Sharapov, A.M. Rubashov, Elektricheskie stantsii, **6**, 42-46 (1993)
2. A.A. Ionin, *Reliability of the systems of thermal networks,* M.: Stroyizdat, (1989)
3. V.M. Borovkov, V.G. Kiselev, Izvestiya RAN. «Energetik», **5**, 110-119 (2006)
4. O.M. Taraday, S.V. Fomich, P.M. Glamazdin, Ventilyatsiya, osvitlennya ta teplogazopostachannya KNUBA, **19**, 117-124 (2016)

5. O.M. Taraday, M.O. Yaremenko, E.S. Esin, S.V. Fomich, Naukoviy visnik budivnitstva HNUBA, **83**, 158-161 (2016)
6. O.M. Taraday, E.S. Esin, S.V. Fomich, N.S. Bolotskih, Naukoviy visnik budivnitstva HNUBA, **87**, 188-191 (2017)
7. A.Ya. Naymanov, N.G. Nasonkina, V.N. Maslyak, N.I. Zolotov, *Bases of reliability of the engineering systems of communal economy*, Donetsk IEP NAN Ukraine (2001)
8. B.A. Permyakov, Yu.V. Kananovich, *Economic and ecological comparison centralized and decentralized systems of teplosnabzheniya*, M.: MGSU, (2002)
9. E.M. Malaya, S.A. Sergeeva, Novosti teplosnabzheniya, **4**, 41-44 (2006).
10. O.F. Redko, O.M. Taraday, M.O. Yaremenko, I.G. Kirilenko, Naukoviy visnik budivnitstva HNUBA, **45**, 233-236 (2008).
11. O.M. Taraday, O.F. Redko, I.G. Kirilenko, Naukoviy visnik budivnitstva HNUBA, **45**, 128-132 (2008).

Study of an effective structural system with rational parameters based on main energy principles

Katerina Toporkova[1,*]

[1]Building Designs Department, O.M. Beketov National University of Urban Economy, 91000, Kharkiv, Ukraine

Abstract. The effective architectural and construction system with rational parameters based on main energy principles applied to construction of buildings and constructions with a small-step and large-step frames, representing the metal frame of full factory production collected in spatial system on high-strength bolts and previously strained combined prestressed concrete slabs which formed without timbering is proposed in this paper. The main constructive and technological features of the proposed frame, which allows reducing construction period, increasing working efficiency, and reducing labor intensity by using factory-made materials, quick erection of all process elements through the use of high-strength bolts is considered. The advantages of this constructive system in comparison with alternative systems are shown. The basic concepts of "rational decisions" to the design, namely, the objective of the optimal management of the structure parameters, which can not only improve its basic performance indicators, but also, and most importantly, improve operational reliability, is presented.

1 Introduction

In today's world, the construction of steel and reinforced concrete frames, both for single-storey and multistorey buildings and structures, is very popular because of constructive and technological advantages of as compared to other systems.

The intensive growth of new design and technological developments opens up unlimited possibilities for the erection of objects implemented by one of the modern building methods, but each of these methods has a number of significant drawbacks.

Thus, for example, most of the shortcomings in the construction of a monolithic building are related to the characteristics of concrete which, depending on the building season or adverse weather conditions, require additional human and financial resources, which in turn has a significant impact on the cost and period of work execution [1]. The above shows one of the basic concepts of R. Descartes, represented by the thesis that any saving is ultimately a saving of time. In addition, the concrete casting process itself is

* Corresponding authors: tes290692@gmail.com

highly labour-intensive and cost-based, and it requires expensive equipment for continuous supply of concrete mix, its consolidation and heating.

This was the reason for the creation of a new and effective form-factor that lies in the merging of the metal frame and the precast floor slabs with the use of high-strength bolts into a single system "S&F" (the abbreviation - slabs and frame).

Objective of the study: Creation of a new structural system with rational parameters based on main energy principles.

Object of the study: Metal dimensional frame with prestressed concrete slabs which formed without timbering.

Subject of the study: Control parameters for this design.

2 Construction and technological features of a design

The proposed design found its application in two implementation alternatives, a large-step frame and a small-step frame (Fig. 1).

Fig. 1. Offered construction of a large-step frame.

In order to determine the qualitative indicators of the system, it is necessary to become familiar with the characteristics of its elements in particular and the integrity of the system as a whole, namely:

- Factory-made metal columns and girders, which are assembled into a dimensional system on high-strength bolts (usually from the rolled and, if necessary, perforated wasteless elements);
- Prestressed concrete slabs, which formed without timbering is considered [2];
- The small weight of the elements;
- Absence of protruding elements from the floor and ceiling planes;
- The possibility of using modern wall and roof cladding (any facade and roofing systems);

- Rapid erection under the conditions of minimized labour-intensiveness of all technological operations (depending on the nomenclature of large and small mechanization);
- Almost unlimited combinatorics of architectural and planning concepts (owing to an arbitrary choice of steps and bays);
- Simple renovation of premises, if necessary;
- Year-round possibility of carrying out the construction works.

The most distinctive features are those of a large-step metal dimensional frame, with possible variation of cells 7,2 x 7,2 m; 9 x 9 m; 12 x 12 m.

The skeleton frame is formed by means of conjunction using highly-strength bolts of bridle wire columns and girders located in the same plane in two (not necessarily perpendicular) directions (Fig. 2).

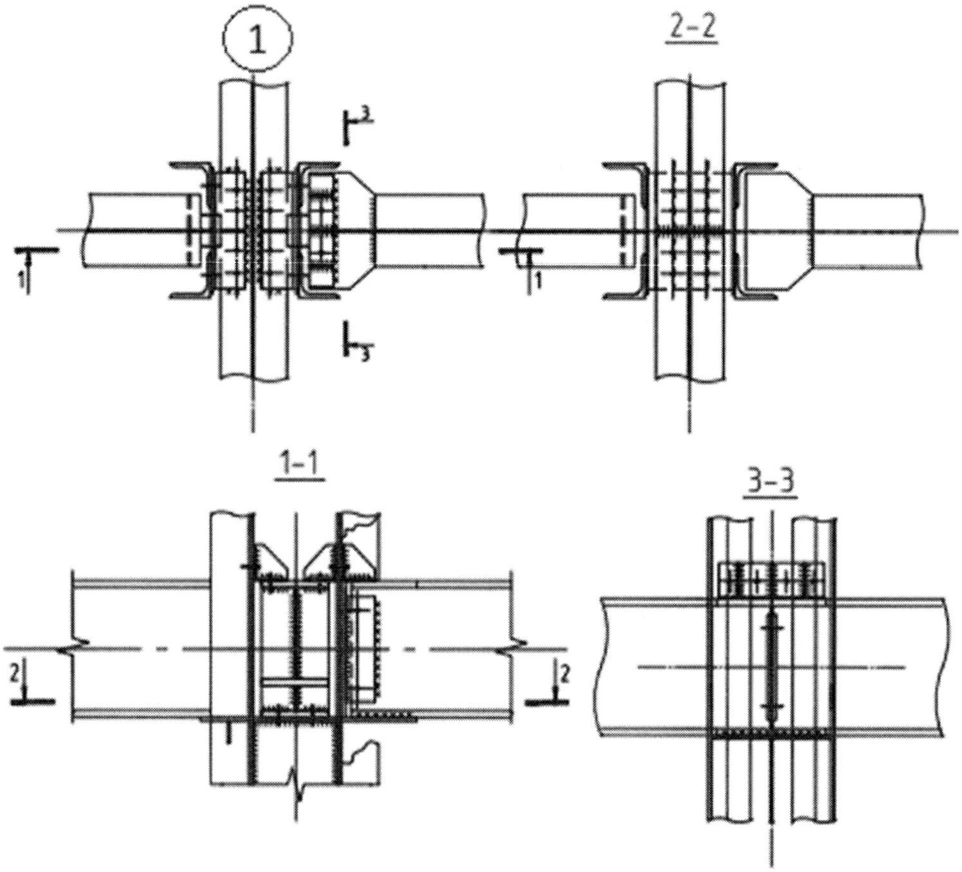

Fig. 2. The column and the girder junction on the high-strength bolts.

Moreover, girders and columns can be equipped with perforated walls, which are manufactured on a non-waste technology from a rolled section, resulting in the reduction of material consumption and structure cost [3]. An overlap is formed by means of combining prestressed concrete slabs which formed without timbering is considered with the rotation in adjacent cells by 90 °, which in turn reduces the load on the girder two times (Fig.3). This property is reflected in the reduction of the size of the bolt section (the degree of perforation), that is, it forms the weight of the element. The proposed solution is fully consistent with the famous aphorism by B. Fuller: "If you want to measure the degree of a building perfection, weigh it."

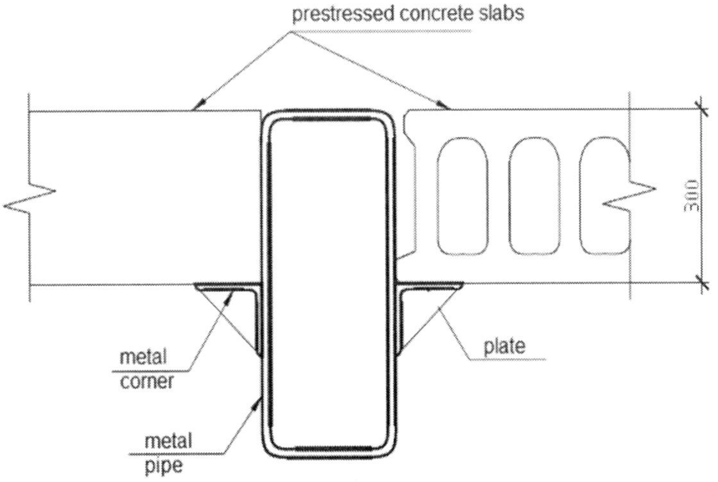

Fig.3. Slabs leaning on the girder.

In addition, the proposed slabs may have different lengths without loss of material. This decision is remarkable for the developed combination of space planning decisions, and an ability to create consoles in a designed building (Fig. 4).

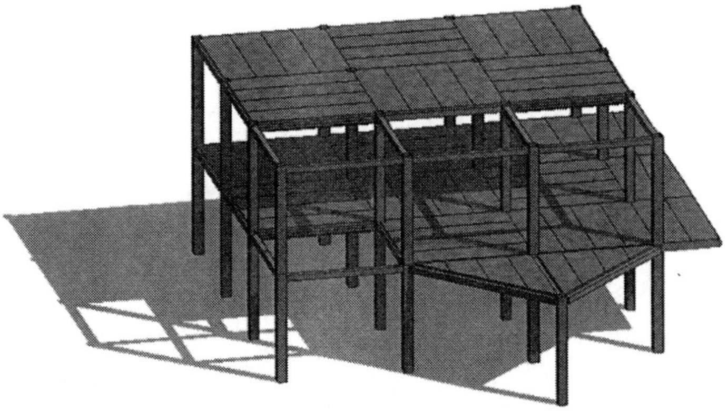

Fig. 4. Combinatorics of space planning decisions of metal frame.

Moreover, the use of heavy-duty concrete (C30/35 and above) avoids the use of the working transverse reinforcement and all types of structural reinforcement, which significantly reduces the price of the slabs.

In comparison with the monolithic reinforced concrete applied in recent years the pre-tension, system is less expensive and can be controlled at all stages of technological processes.

A construction arrangement of the frame can be braced and frame-braced. Number of storeys is calculated.

The stability and stiffness of the frame is ensured by the floor slabs and stiffening diaphragm (double girder, steel ties - portal or four-way the subsequent laying of foam concrete block or brick).

No less used for skeleton-type buildings is the small-step metal frame (the size of the cell does not exceed 6 m).

 DOI: 10.1051/matecconf/201711602035

Small-step and large-step frames require creation of floor slabs operating as quasimonolithic system and consisting of pre-tensioned slabs and reinforcement frames with using fine-grained concrete on expanded cement. At that frames are fixed to cross-beams and create rigid floor slab where re-bars serve as connecting system to avoid sliding of slabs (Fig. 5).

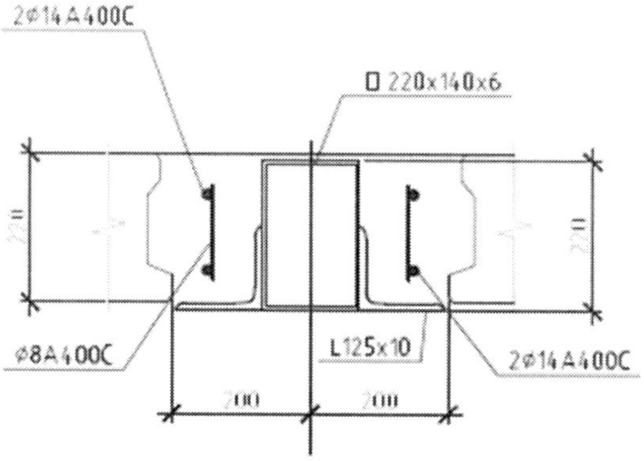

Fig. 5. Formation of a floor slab by means of reinforcement frames.

The main idea of the "S&F" system is to create buildings with a small-step or large-step frame, depending on its purpose, which will allow the realization of a light transformation, including replanning and reshaping, organization of parking spaces, inbuilt or separate, commercial areas, if necessary (Fig. 6). At the same time, the use of high operational compatibility materials using bolted connection of elements reduces labour-intensiveness, resulting in the reduction of frame-construction time.

Fig. 6. Proposed frame under construction conditions.

3 Rational decisions of construction

An analysis of the available information has made it possible to define the basic concepts of "rational decisions" to the design, namely, the objectives of optimal management of structure parameters, which can not only significantly improve its basic performance indicators, but also, and most importantly, improve operational reliability.

Management of the settings of rational systems consists of the creation of iterative procedures that gradually improve the geometric and/or physical-mechanical parameters of the design.

The frame system is characterized by some external and internal parameters. External parameters describe the resistibility of the system while the total volume of the material remains unchanged. For example, these parameters define the dimensions of the elements, the coordinates of the reference links, the coordinates of the external loads locations, their types and so on. The internal parameters describe the topology, volume and properties of the specified materials [4].

The provided ideology found its display in two directions [5]:

- the adoption of a single target function as a criterion;
- adoption of an isoperimetric criterion that determines the regulation of the stressed and strained state (SSS) of the system.

Under these conditions, it may be possible to achieve a significant simplification in defining the rationalization of the design parameters problem. It is also assumed that a uniform record of the target function could be replaced by the traditional criteria (cost, weight, volume, etc.) by one, and related to the nature and characteristics of SSS.

This requirement also applies to the restrictions governing the SSS management.

In turn, the virtual diversity of limitations that form the scope of permissible solutions (SPS) has led to the creation of a two-stage procedure:

- search for an ideal decision;
- formation of a final decision.

In order to create an ideal decision, new provisions are applied which are based on energy principles, namely:

- the provisions that for managed systems with a fixed amount of material, a number of external and internal links under a static external load - self-weight, the potential energy of deformation (PED) after conversion reaches the lower bottom on a rational combination of geometric values [5]:

$$U = \inf_{a} U\,(a)^k \quad k = 1{,}2 \ldots \infty, \tag{1}$$

where U - PED; k - number of the comparison option; $a \epsilon \mathrm{M}$; M - multitude of the allowed values of external geometric parameters.

The condition (1) can be distributed if there is a dominant among multiple virtual loads. For example, you can minimize the torque while creating a thin-walled element that has a bending strain with torsion by passing a load vector through the center of the bend, which in turn corresponds to the position (1) [5];

- the conditions of the isoenergetic state of the system (structure), i.e. when

$$e[\{\overline{x}\}] = const, \tag{2}$$

where e - is the density of the potential energy of deformation (DPED);

- $\{x\}$ - a vector of internal parameters.

Solving of practical task per condition (1) proves its operating efficiency, and also possibility to control geometric and physical & mechanical properties of structural element.

In the process the above information analyzing the calculation of potential energy of deformation of the beam with perforated wall for the frame with a pitch of 9x9 m and load of 4,5 t/m in LIRA software complex was performed. Beam size was calculated based on condition that maximum rolled I-beam 60 does not fit as per two ultimate conditions [6]. At that the calculation of potential energy of deformation of the same beam with continuous wall was performed (Fig. 7).

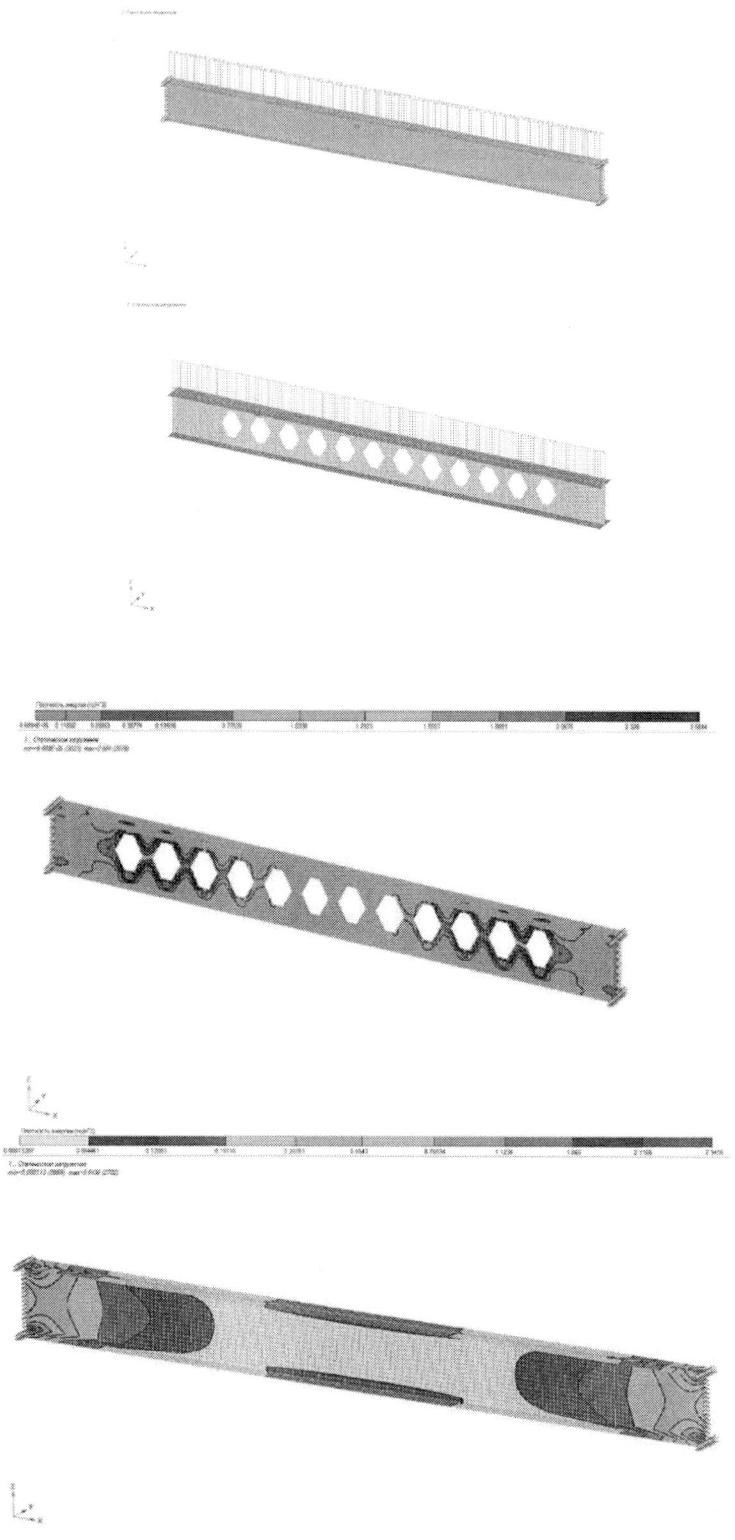

Fig. 7. Potential energy of deformation of the beam with perforated wall.

Calculation results have showed that PED for the beam with perforated wall is less than for the beam with continuous wall; moreover, the weight of perforated beam is much less than the weight of continuous beam resulting in cost-cutting of structure and system as a whole.

4 Conclusion

Application of steel and reinforced concrete frame with prestressed concrete hollow slabs which formed without timbering for construction of buildings and structures allows reducing construction period, increasing working efficiency, and reducing labor intensity by using factory-made materials, quick erection of all process elements, and also selection of optimum structural elements with rational parameters based on energy principles.

References

1. S.M. Anpilov, *Tehnologiya vozvedeniya zdaniy i sooruzheniy iz monolitnogo zhelezobetona. Uchebnoe posobie* (Izdatelstvo Assotsiatsii stroitelnyih vuzov, Moscow, 2010)
2. V.S. Shmukler, E.G. Stoyanov, O.M. Pustovoytova, E.N. Tertyichnyiy, I.G. Lvovskiy, R.V. Korenev, Komunalne gospodarstvo mist, **114,** 22-31 (2014)
3. B.S. Taranath, *Structural analysis and design of tall buildings* (CRC press, NW, 2016)
4. V.S. Shmukler, Yu.A. Klimov, N.P. Burak, *Karkasnyie sistemyi oblegchennogo tipa* (Zolotyie stranitsyi, Kharkiv, 2008)
5. V.S. Shmukler, Collective scientific works of UkrSURT, Kharkiv, **167,** 54-69 (2017)
6. *DSTU B V.1.2-3:2006. Progini ta peremishennya. Vimogi proektuvannya*, Kiev (2006)

Simulation of performance of circular CFST columns under short-time and long-time load

Glib Vatulia[1*], *Alexey* Lobiak[1], and *Yevhen* Orel[1]

[1] Ukrainian State University of Railway Transport, Structural Mechanics and Hydraulics Department, Feuerbach sq., 7, Kharkiv, 61050, Ukraine

Abstract. The method of calculation of concrete-filled steel tubular (CFST) columns with consideration of physical nonlinearity of materials, geometric nonlinearity of the confinement and the effect of the gain in strength of the core is considered. The method uses a step iteration algorithm, which involves analytical dependencies and the ultimate element simulation method. Allowance for creep of concrete is based on using the generalized kinetic long-term deformation curve and phenomenological deformation development equations. Creep of concrete is controlled through new structural factors that determine the structure of cement rock layers between sand and mortar grains between chip grains. The method is validated by comparing experimental results and theoretical data. The suggested method allowed to study the stress-strain and limit state of circular concrete-filled steel tubular columns, as well as to evaluate their effectiveness with account for the time factor.

1 Problem statement

Today, many analytical calculation methods for compressed concrete pipe elements give quite close values of bearing capacity [1]. Both general approach to the calculation of concrete structures with rigid or confinement reinforcement and such methods as bringing concrete to steel, numerical calculation methods based on the nonlinear deformation model, as well as approaches recommended by Eurocode 4 are applicable for steel concrete [2]. In general, the existing methods of taking into account the rheological properties of the concrete core involve complicated calculations and considerable error. Despite the wide choice, elaboration of a fully functional technique to calculate the stress-strain state under short-term and long-term loads, taking into account the deformation diagrams of the concrete core under conditions of volumetric stress, contact interaction between the confinement and the core and the geometric nonlinearity has not been completed yet.

2 Calculation for short-term loads

2.1 Theoretical principles

* Corresponding author: glebvatulya@gmail.com

The set task is proposed to be solved based on the nonlinear model taking into consideration the specifics of core and confinement deformation under non-homogeneous stress. The main problem in calculation of the pipe concrete under such conditions is absence of core deformation diagrams $\sigma_{c3} - \varepsilon_{c3}$, the appearance of which will be determined by the value of the lateral pressure of the steel concrete on the concrete σ_{cr} which is unknown in advance (Fig. 1).

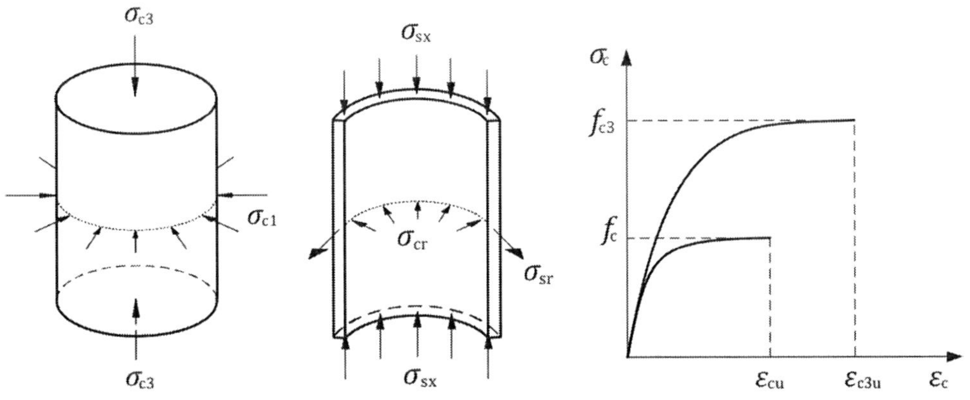

Fig. 1. Stressed state of the concrete core and steel confinement reinforcement.

In view of the state-of-the-art computer technologies, this solution can be implemented mathematically using a computer complex as the main simulation instrument [3]. The suggested calculation method uses a step-iteration algorithm. It is accepted at the first iteration of the first load, step lateral pressure $\sigma_{cr} = 0$ and the strength of the core is determined exclusively by the axial compression strength of the cylinder f_{cc}. Further, iteration search for actual core strength values f_{c3} in accordance with Karpenko criterion [4] under triaxial compression of the element is performed:

$$f_{c3} = f_{cc} + k\sigma_{cr} \tag{1}$$

Lateral pressure ratio k is determined depending on the relative edging level $m = \sigma_{c1}/\sigma_{c3}$:

$$k = 1/(0.1 + 0.9m). \tag{2}$$

The last iteration will determine the actual components of the stress-strain state of the first load step: axial σ_{c3} and transverse σ_{c1} stresses in the concrete, principal compressive σ_{sx} and radial σ_{sr} stresses in the confinement, as well as respective relative deformations ε_{c3}, ε_{c1}, ε_{sx}, ε_{sr}. Consequently, the second and further load steps are calculated up to the exhaust of the carrying capacity. It is thus assumed that the limit state for pressure with little eccentricities occurs when one of three conditions is fulfilled: either the strength of the concrete core $\sigma_{c3} = f_{c3}$, or of the confinement $\sigma_{sr} = f_y$ is exhausted or as a result of reaching ultimate deformations $\varepsilon_{c3} = \varepsilon_{cu}$.

The ultimate element model for calculation of the stress-strain state consists of two blocs (the confinement and the concrete core) united into combined action using unilateral connection accepting compression and shear (Fig. 2, a).

The confinement is simulated taking into account physical and geometrical nonlinearity, while the core material is given by variable deformation laws. Load is transferred through an auxiliary part providing distribution of forces along the loading plane. The triangulation circuits were obtained as a result of solving a separate problem of finding a compromise between time and accuracy of calculation.

DOI: 10.1051/matecconf/201711602036

2.2 Calculation data

The method was validated by comparing the calculation data with the experimental data. The adopted concrete pipe elements were 500 mm long, with diameter 102 mm, confinement wall thickness 3 mm, and the core made of concrete with prism strength 16 MPa. Load transfer is accepted for the combined cross-section, i.e. simultaneously to the concrete and the pipe.

The results for the ultimate condition were visualized on the stress fields which are shown separately for the confinement (Fig. 2, a, b) and the core (Fig. 2, d, e).

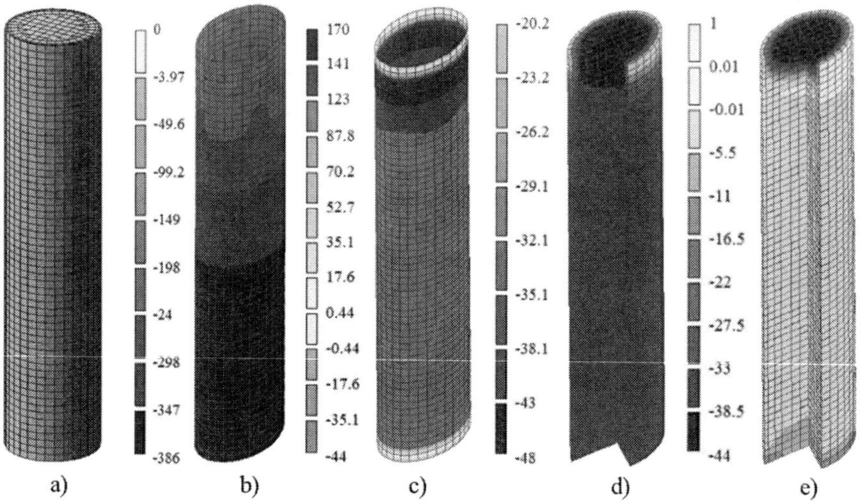

Fig. 2. Stress state of the concrete pipe element (a – general view of the model, b and c – stresses in the confinement σ_{sx} and σ_{sr}, d and e – stresses in the core σ_{c3} and σ_{c1}).

The calculation data were analyzed for dependencies of stress changes (Fig. 3, a) and deformations (Fig. 4).

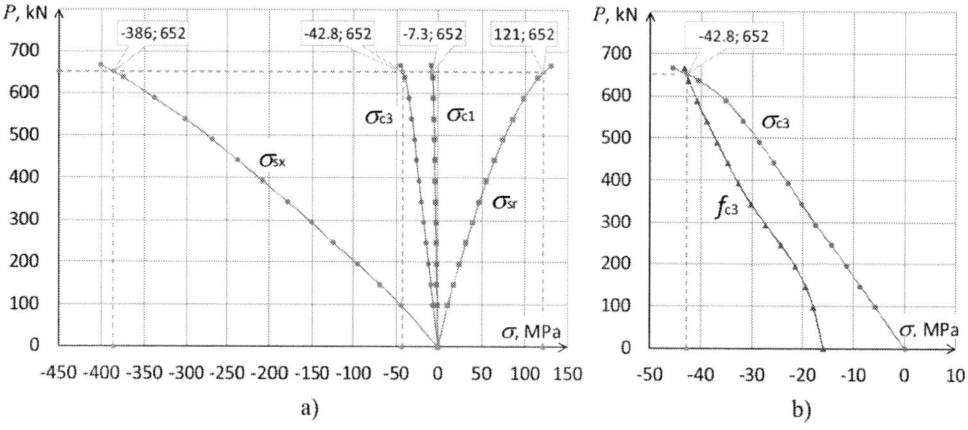

Fig. 3. Diagrams of stress change in the concrete pipe element (a – stress diagrams, b – core strength diagram f_{c3} together with stress diagram in the concrete σ_{c3}).

The core strength diagram f_{c3} (Fig. 3, b) has the common solution with the axial stress development curve in the concrete σ_{c3}, which determines the strength criterion of the stress state ($N_{plf} = 652$ kN). Load-carrying capacity of the concrete pipe element was obtained based on the value of limit relative concrete deformations ε_{cu} ($N_{pl\varepsilon} = 648$ kN).

Analysis of the diagrams demonstrates approximately similar type of dependencies of deformation changes in the core and in the confinement that indicates their joint performance.

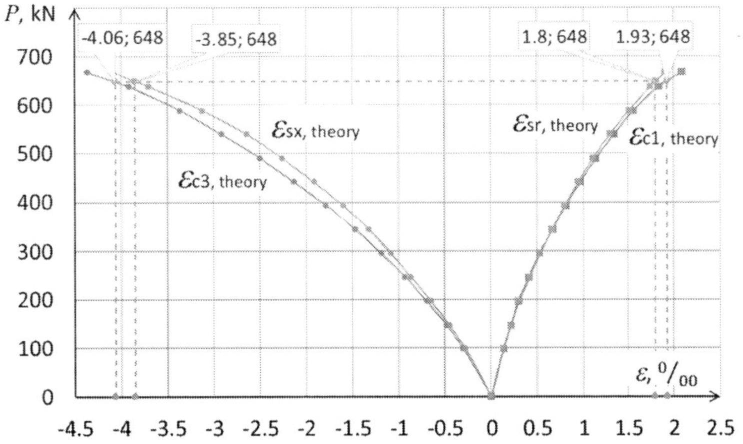

Fig. 4. Diagrams of deformation change in the concrete pipe element.

The factor of numerical adjustment of the effect of the confinement is determined by the relation of the core strengths in the limit and initial state. For the reference relation D/t=34, strength of the concrete core increase 2.6 times.

Test specimens for experimental studies were concreted in the laboratory of the Department of structural mechanics of the Ukrainian State University of Railway Transport (Fig. 5). Fine grain concrete c W/C=0.55, portlandcement M500 and broken stones of 2.5-7 mm fraction were used to prepare the concrete mixture. Deformations in the concrete were measured in the geometric centre of the core using low-depth strain gauge load cells [5]. Deformations of the confinement were measured on the surface in the middle of the element length. The mean error obtained for relative deformations in the core and the confinement was found to be 10.5%. In general, the error for the load-carrying capacity is not more than 2.5%.

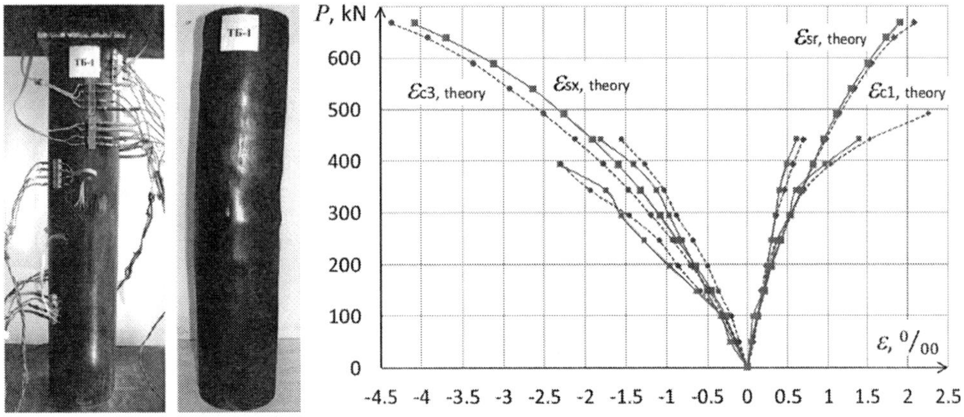

Fig. 5. Experimental data.

Assuming satisfactory convergence of theoretical and experimental data for short-term loads, the next step considers the influence of the time factor on deformation in the concrete core and, in general, on the stress-strain state of the concrete pipe element.

3 Calculation for short-term loads

3.1 Theoretical principles

Numerous experimental studies have demonstrated that inelastic deformations occur in constructions in which concrete is used exposed to long-term load action, which can exceed several times the initial, conditionally elastic deformations. The issue of creep is critical for the structures made of steel-reinforced concrete and in particular, for concrete slabs [6, 7] and pipe concrete. Therefore, the issues of forecasting the long-term deformation of the concrete core in concrete pipe elements and synchronous redistribution of efforts remain relevant.

The national building codes of Ukraine harmonized with Eurocode 2 [8] are based on the simplest – phenomenological – approach which, in turn, is based on experimental data and establishing dependency between deformations and time through the creep coefficient $\varphi(t, t_0)$. In accordance with this method, the ultimate creep deformations are determined by a coefficient that depends on the average strength of the concrete, its age and relative humidity, while the development of creep in time t depends on relative humidity and cross-section.

$$\varepsilon_{c,t} = \sigma_c/E_c\,[1 + \varphi_0\beta(t, t_0)], \tag{3}$$

where: $\beta(t, t_0)$ – coefficient describing the development of creep over time; t_0 – age of the concrete at the time of the first load.

Despite its advantages, this method cannot reveal the actual mechanism of long-term deformation of the concrete, and thus does not allow to control this mechanism and influence the ultimate values of creep deformations. Since alternative methods of creep accounting can be applied using special software, we propose a method based on the colloidal chemical representation of the mechanism of long-term deformation of concrete in accordance with the Plugin theory [9].

The theory is based on the more accurate representation of the kinetic curve of concrete deformation (Fig. 6). Based on the findings of the study, the equations of deformation development were obtained depending on 4 stages of concrete performance: compression in the conditionally elastic setting, the stage of high-rate creep (1), ordinary creep, consisting of a nonlinear (2) and a linear (3) part, as well as long-term creep (4). The data are confirmed by experimental studies with continuous loading of 10x10x40cm cement prisms.

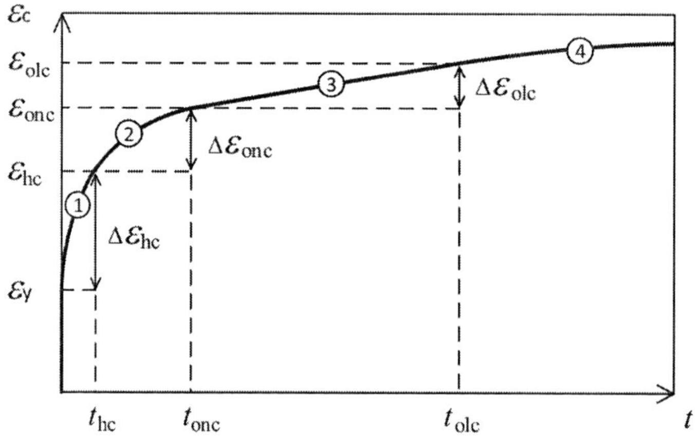

Fig. 6. Generalized kinetic nomogram of creep of concrete.

In accordance with Fig. 1, the pattern of the long-term deformation of the concrete under compression and the general concrete deformation equation becomes the form:

$$\varepsilon_c = \varepsilon_y + \Delta\varepsilon_{hc} + \Delta\varepsilon_{onc} + \Delta\varepsilon_{olc} + \Delta\varepsilon_{lc}. \tag{4}$$

Specific equations simulating deformation kinetics at each of the above stages:

$$\begin{aligned} \varepsilon_{c,hc}(t) &= \varepsilon_y + \Delta\varepsilon_{hc}\left(1 - e^{-t/T_r^{hc}}\right), \\ \varepsilon_{c,onc}(t) &= \varepsilon_{hc} + \Delta\varepsilon_{onc}\left(1 - e^{-t/T_r^{onc}}\right), \\ \varepsilon_{c,olc}(t) &= \varepsilon_{onc} + k_{olc}(t - t_{onc}), \\ \varepsilon_{c,lc}(t) &= \varepsilon_{olc} + k_l(t - t_{olc}), \end{aligned} \tag{5}$$

where: T_r^{hc}, T_r^{onc} – relaxation time according to the Tertsagi's dispersion phase model on the exponential deformation stages;
k_{olc}, k_l – kinetic coefficients reflecting rate of general (on the linear section) and long-term creep deformation.

In view of the latter, expression for determination of the creep coefficient at any concrete deformation stage becomes the form:

$$\varphi(t) = \left[\varepsilon_c(t) - \varepsilon_{c,y}\right]/\varepsilon_{c,y} \tag{6}$$

Total relative concrete deformations, corresponding to completion of the high-rate (ε_{hc}) and ordinary (ε_{onc}, ε_{olc}) creep stage, are determined based on the hypothesis about the behaviour mechanism of submicroscopic structure of the concrete consisting of electrically charged particles (globules) of hydrosilicate gel when it is exposed to compressing load. It is assumed that ultimate creep deformations are determined by gel deformations due to compression of double electric layer, water filtration from the compressed areas of the construction into those stretched according to the pressureless water penetration law, while long-term creep kinetics will depend on the scale factor during filtration water squeezing from the gel.

Creep deformations can be controlled through new structural coefficients determining the structure of cement stone layers between the sand and mortar grains between chip grains [9]:

$$\begin{aligned} \alpha &= V_{csm}/V_{emp}^{bs} = \left(\frac{S}{\rho^S} + \frac{C}{\rho^C} + \frac{W}{\rho^W}\right)\Big/\left(\frac{BS}{\rho_{pour}^{bs}} \cdot S_{emp}^{bs}\right); \\ \mu &= V_{cs}/V_{emp}^{S} = \left(\frac{C}{\rho^C} + \frac{W}{\rho^W}\right)\Big/\left(\frac{S}{\rho_{pour}^{S}} \cdot S_{emp}^{bs}\right). \end{aligned} \tag{7}$$

The best coefficients α_{opt} and μ_{opt} were obtained from the condition of the fullest contact which decreases disjoining pressure and, respectively, compressive deformations and hair cracks.

A calculation algorithm for the long-term load action was suggested (Fig. 7), which in combination with Lira-SAPR software allows to perform adjusted calculation of constructions for long-term exposure and control creep deformations through the structural characteristics α and μ.

PCC – specialized software developed at the Department of building materials, constructions and structures of the Ukrainian State University of Railway Transport – is used to determine dosing of concrete components according to the specified data.

 DOI: 10.1051/matecconf/201711602036

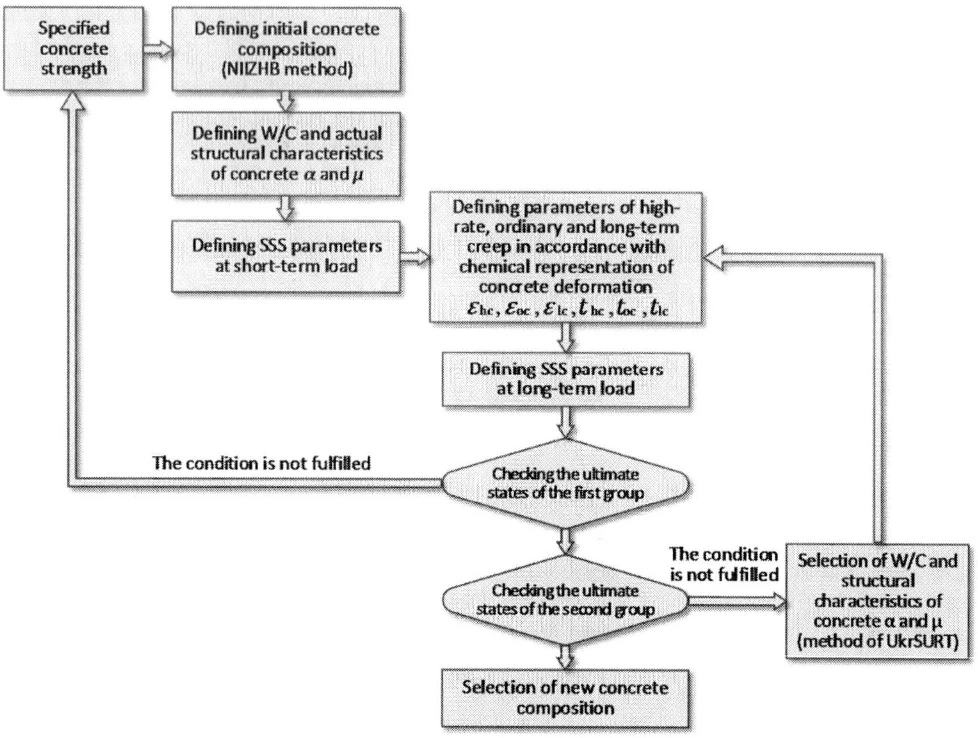

Fig. 7. Algorithm of minimization of limit creep deformations.

3.2 Calculation data

For the considered pipe concrete elements, calculation taking into account long-term load exposure determined significant stress redistribution towards the confinement (Fig. 8). For instance, in 730 days under stress of 50% of the carrying capacity, axial σ_{sx} and transverse σ_{sr} stresses in the confinement increase by 60% and 70%, respectively. Axial stresses in the core σ_{c3} decrease by 40%, while transverse stresses and σ_{c1}, on the contrary, increase by 50%. Strength of the concrete core f_{c3}, during the long-term loading, increases additionally by 20%. Axial deformations in the core ε_{c3} increase by 70%.

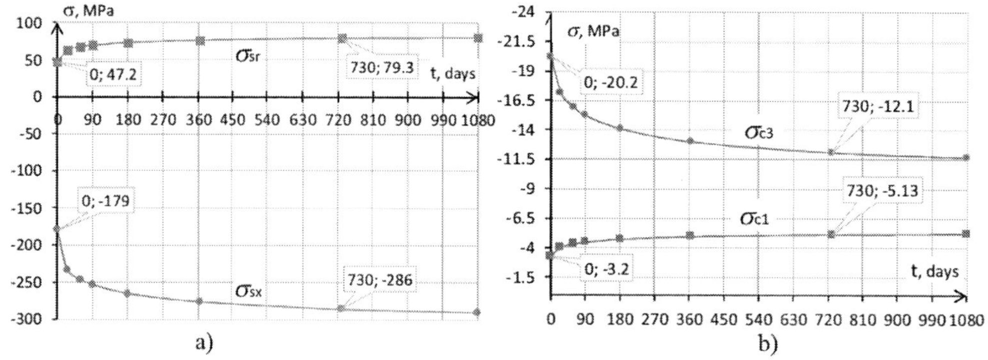

Fig. 8. Diagrams of stress changes in a pipe concrete element (a – stress changes σ_{sx} and σ_{sr} in the confinement, b – stress changes σ_{c3} and σ_{c1} in the core).

The performed studies resulted in obtaining a method for calculating the stress-strain state and bearing capacity of pipe concrete elements taking into account the short-term and long-term load action, physical nonlinearity of materials, geometric nonlinearity of the confinement, compressing effects of concrete, and increase of the strength of the core. Validity of the method is provided by comparing experimental and theoretical results.

The adopted long-term load accounting method includes an algorithm for creep controlling through the structural characteristics of concrete. It is also established that the long-term processes that occur in the concrete core affect significantly the stress-strain state. The performance of pipe concrete does not decrease with time and will depend on the reasonable concrete composition and the optimal proportions of design parameters.

References

1. L.I. Storozhenko, A.V. Semko, Urban economy, Kharkiv, **63**, 59-70 (2005)
2. *Design of composite steel and concrete structures – Part 1-1: General Rules and Rules for Bridges. EN 1994-1: 2001. Eurocode 4*, Brussels (2006)
3. A.S. Gorodetsky, *Computer models of structures* (Fact, Kyiv, 2007)
4. N.I. Karpenko, *General models of reinforce concrete mechanics* (Stojizdat, Moscow, 1996)
5. G.L. Vatulia, E.I. Galaguria, D.G. Petrenko, Transport, Transport Facilities and Ecology, Perm, **2**, 48-56 (2014)
6. A.V. Lobiak, D.V. Golovko, Steel concrete structures: design, construction and maintenance, Poltava, **11**, 156-162 (2014)
7. A.V. Lobiak, A.A. Plugin, D.V. Golovko, News of ODABA, Odesa, **63**, 79-83 (2016)
8. *Design of Concrete Structures– Part 1-1: General Rules and Rules for buildings EN 1992-1-1. Eurocode 2,* Brussels (2004)
9. A.N. Plugin, A.A. Plugin, O.A. Kalinin et al. *Osnovy teorii tverdeniya, prochnosti, razrusheniya i dolgovechnosti portlandtsementa, betona i konstruktsiy iz nikh* (Naukova dumka, Kyiv, 2012)

Study of technological features of tubular compressed concrete members in concreting

Olena Voskobiinyk[1*], and *Olga* Gukasian[1]

[1]Poltava National Technical Yuri Kondratyuk University, Department of Civil Engineering, Pershotravnevyi Avenue, 24, Poltava, 36011, Ukraine

Abstract. The technological features of core concreting were analyzed as the main factor in ensuring of strength and reliability of compressed concrete-filled steel tubular (CFST) members. We have conducted the analysis of existing concreting methods of CFST members. In this respect, the most dangerous types of possible technological defects of concrete core of CFST members are inhomogeneity along the height, voids, caverns, and concrete "weak spots". The authors considered the influence of such technological factors of concreting: placeability, time, concrete mixture compaction method, concreting height on the concrete core strength of CFST members. Based on the experimental studies conducted we suggested the regression correlations for determining the concrete strength of CFST members of different length depending on the movability of concrete mixture and a time for its compaction. The authors performed the correlation analysis of technological factors of concreting on the strength of the concrete core. We carried out the comparison of data on the concrete core strength of CFST members, that were determined by non-destructive methods (sclerometer test results, ultrasonic method) and direct compression strength tests. We experimentally proved that using movable mixtures with the slump of about 4 – 9 cm the overall variation coefficient of concrete core strength of CFST members along the height reaches nearly 13%. Based on the experimental studies conducted we suggested the guidelines on optimal regimes of concrete compaction during manufacturing CFST members at a construction site environment.

1 Introduction

CFST elements are commonly used in global civil engineering practice. Their prevalence is justified by high strength characteristics, technological and economic indicators [1]. The modern state of the civil industry requires bearing structures to be a rational combination of high strength and low material consumption. CFST members meet the requirements completely.

Use of compressed CFST members as bearing structures is particularly important for industrial, residential and civil buildings. However, the known studies of CFST members [2, 3] are mostly dedicated to the study of its operational features under the load and to its stress-strain state.

* Corresponding author: elenvosko@gmail.com

Nevertheless, relevant literature summary [4] proves that the studies are primary focused on the characteristics and features of CFST members. These structures have the relatively small cross section area, but they can bear large weights. In these structures, due to the volumetric stressed state, concrete bears longitudinal force which is far beyond its prism strength. In previous surveys CFST members were tested in steel shell-tube with the diameter of 102 to 108 mm and the wall thickness of 3 to 3.5 mm. Such specimens have a significant reinforcement percent. Steel shell-tube acts as external reinforcement and carries the load. A significant part of a load is born by steel part of a cross section of CFST members. It is hard to estimate the influence of concrete core defects (its size and location) on general strength of a specimen.

2 Analysis of possible technological defects of CFST members concreting

Known CFST members manufacturing methods divided into two groups: concreting of columns at the plant or directly at a construction site. From a technological point of view, when CFST members manufactured at a plant its concreting (filling in the shell-tube with concrete) could be carried out in vertical, horizontal and inclined positions.

When concreted inclined, the tubes placed at an angle of $25^0 - 30^0$ on the vibratory plate and filled with concrete mixture. Compaction carried out by vibration. The disadvantages here are the possibility of cavern and voids emerging which localized at the wall of shell-tube. Among the advantages of this method, there is a possibility of using the heavy concrete mixtures (V4, S1).

Vertical concreting applied for manufacturing the CFST members with the length up to 5 m. Concrete mixture supplied from the top, then it compacted by external vibrating. The disadvantage of this method is the possibility of concrete mixture separation when it supplied from a large height. It could cause the inhomogeneity of concrete structure along the height, as well as weak spots areas and incomplete compaction of a concrete mixture.

When horizontally concreted (centrifuging), the rotary form filled with concrete and compacted by rotation with the frequency of 900 rotations per second. Concrete mixture distributed across the form driven by centrifugal forces to the periphery of the form. Advantages of this method are high density and strength of concrete, low water consumption, permeability and increased resistance to aggressive influences. It is possible to concrete the tubes and other long items by centrifuge with low labor costs. The disadvantage of this method is the void with offset center and nonuniform concrete distribution on the walls (fluctuations in the thickness of the concrete layer).

CFST members concreting at a construction site carried out in three different ways: concreting the pack of tubes, concreting in design position and concreting through the perforation in a steel tube. Construction site concreting could cause the emerging of concrete core defects.

A pack of tubes fixed at an existing overlaying of a building and filled with movable concrete mixture out of the mixer. Specimens compacted by vibrating. One of the methods is a fixation of a depth-vibrator on the tube wall

When concreting the installed in design position tubes, concrete mixture supplied from the overlaying. Compaction of specimens made by external or deep vibrating. Tubes filled with concrete layer by layer, which could cause inhomogeneity of concrete features.

Concreting through a perforation in a tube carried out through the opening (with the diameter of 5 cm) in shell-tube under the pressure. The advantage of this concreting method is an easy quality control process (it's possible to drill small holes with the diameter of less than 5 mm for control of filling the tube). The disadvantage of this method is the possibility of damaging the steel shell-tube. Moreover, the way of development of formwork defects couldn't be foreseen.

Manufacturing method and materials for CFST members have a major influence on its strength. Careful attention must be paid to the concreting quality. The concrete mixture has a significant variability of its physical and mechanical properties. Nonuniform distribution of the properties along the height could result in decreasing of carrying capacity for an entire structure.

Concrete non-homogeneity may be caused by a set of systematic and accidental factors [5]. These factors cause primary (natural), and secondary (virtual) non-homogeneity of features. True dispersion of concrete features caused by its natural non-homogeneity: variation of concrete components features, different compaction conditions, and other technological factors. Secondary non-homogeneity is an entropy due to the unclear method of sampling.

Analyses of CFST members manufacturing defects shows that the most significant ones are concreting anomalies (defects). These anomalies caused by technological complications which connected to the peculiarities of placing and compaction of concrete mixture into shell-tube inner space. Among them the most widespread there is caverns and voids formation at the contact surface between a core and a shell. These defects vary by the size, configuration and propagation depth in concrete body.

Accidental defects should also be contributed to technological violations of concreting. Among them, there are strength difference along the height of a specimen, concrete mixture separation, strength variability, non-homogeneity of structure and physical and mechanical features along the height of a specimen. Numerous surveys [6, 7] show that separation could be avoided and homogeneity of a core could be reached even if the concrete mixture dropped from a significant (10 – 15 m) height during vertical concreting inside of a steel tube. Thereby, according to its genesis, the defects could be systemized in two groups: technological and accidental.

Concrete with a slump range from 6 to 16 cm recommended for filling in the CFST members (placeability marks S2, S3, and S4) [8]. Using lower values of concrete slump causes the defects of the concrete core. These defects are empty segments (voids and caverns at the tube wall) and inhomogeneity of strength along the height.

This research aimed at studying the influence of concreting technology on uniform distribution of strength (non-homogeneity of concrete strength along the height). As well as at studying the influence of vulnerable spots localization along the height of CFST members on its collapse character and strength.

3 The methods of experimental researchers (Material and methods)

We have conducted a set surveys in order to investigate the influence of technological factors on the core strength of CFST members. We used Box and Behnken experimental procedure of "3/1/15" type (table 1). It is equivalent to fractional factorial procedures 3**(k-p), but it also more economic due to applying the complex interaction effects. That allowed a minimum number of experiments to be conducted. According to the procedure, 15 experimental specimens were made of CFST members. To minimize the influence of force on resulting carrying capacity of CFST members we applied PVC tubes with the diameter of 109 mm, filled with heavy concrete with C 20/25 strength class ($f_{cm,cuf}$=24.18 MPa). We used polycarboxylates as a plasticizing additive for increasing the placeability of a concrete mixture.

During experiment planning, we investigated the influence of following factors: specimens height (X_1), concrete mixture compaction time (X_2) and placeability of a concrete mixture (X_3). Each factor varied on three levels: "–1", "0", "+1". Characteristics of specimens shown in table 1.

Table. 1. Procedure of experimental research of the concrete core strength of CFST members

Specimen cypher	Variable parameters and variation levels								
	Specimens height, mm			Compaction time, min			Placeability mark		
	X_1			X_2			X_3		
	-1	0	+1	-1	0	+1	-1	0	+1
CFST-1(s)	400			0			V4		
CFST-2(s)	400					4	V4		
CFST-3(m)		1200			2		V4		
CFST-4(l)			2000	0			V4		
CFST-5(l)			2000			4	V4		
CFST-6(s)	400				2			S1	
CFST-7(m)		1200		0				S1	
CFST-8(m)		1200			2			S1	
CFST-9(m)		1200				4		S1	
CFST-10(l)			2000		2			S1	
CFST-11(s)	400			0					S2
CFST-12(s)	400					4			S2
CFST-13(m)		1200			2				S2
CFST-14(l)			2000	0					S2
CFST-15(l)			2000			4			S2

During experiment planning, we investigated the influence of following factors: specimens height (X_1), concrete mixture compaction time (X_2) and placeability of a concrete mixture (X_3). Each factor varied on three levels: "–1", "0", "+1". Characteristics of specimens shown in table 1.

All experimental specimens of CFST members hardened in the same environment. The specimens were cut into equal parts with the height of $l = 2d$ before the testing. Two short specimens made of each specimen marked "s", 6 short specimens made of each specimen marked "m" and 10 short specimens made of each specimen marked "*l*". The overall quantity of specimens was 90 units. Each one was measured, weighted and its strength was tested by three methods: non-destructive (using sclerometer and ultrasonic method) and destructive.

4 Results and Discussion

Results of research (figures 1 – 3) grouped by qualitative factor – initial length of the specimen (concreting height). Experimental data on the strength of specimen's component divided into three groups: 1 – specimens with the height of 400 mm (figure 1); 2 – specimens with the height of 1200 mm (figure 2); 3 – specimens with the height of 2000 mm (figure 3). Characteristic of concrete strength change with the height presented for each group.

First group – specimens with a height of 400 mm (figure 1). Concrete strength varies by 7%. But the average strength of concrete of CFST -1(s) is the lowest ($f_{cm,cul}$=14.83 MPa), it is 61% of the design strength. Compression strength decreasing of concrete is due to empty voids inside of it (diameter is 15 mm, depth 15 mm, 8 mm, 7 mm, and4 mm). Theses defects are caused by poor compaction.

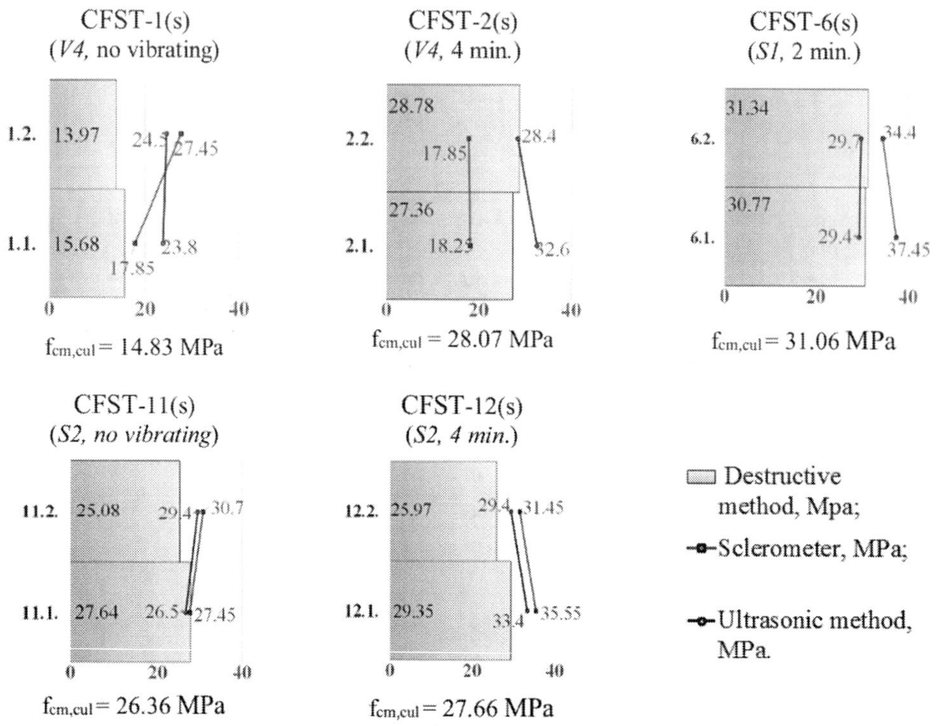

Fig. 1. Concrete strength variation along the height of CFST members with initial height of 400 mm.

The voids discovered during specimen cutting. The average value of the strength is 2 times higher ($f_{cm,cul}$ = 28,07 MPa) for the CFST-2(s) specimen. This one is of the same placeability mark but a compaction time was 4 minutes. That proves the necessity of using compaction for manufacturing the CFST from the stiff concrete mixtures. For all other specimens, the strength reserve varies from 9 to 23 percent. The average strength of specimens CFST-11(s) and CFST-12(s) is different by 5 % under the same compaction time and concrete placeability mark.

Compression strength testing results of CFST-7(m) specimen are the lowest ($f_{cm,cul}$=14.83 MPa). That proves that the absence of compaction decreased the concrete strength of CFST-7(m) specimen by 30 %. And CFST-8(m) specimen's average strength doesn't exceed 60 % ($f_{cm,cul}$=25.22 MPa) compared to CFST-7(m).

The only difference here is the way of compaction – CFST-8(m) compacted by vibrating. The average value of its strength is in design limits. The only peculiarity of this specimen is decreasing of its strength at the top part. The difference in the technology of concreting CFST-8(m) and CFST-9(m) samples is compaction time. CFST-9(m) specimen (compaction time is 4 minutes) possesses the same strength of concrete along the height. Thereby, concreting defect (inhomogeneity at the top part) of CFST-8(m) specimen is the consequence of the lack of compaction. Specimens CFST-3(m) ($f_{cm,cul}$=27.17 MPa) and CFST-9(m) ($f_{cm,cul}$=27.24 MPa) with almost the same average concrete strength are homogenous along the height. That proves a rational way of its manufacturing. Concrete strength along the height of CFST-13(m) specimen decreases two times from the bottom to the top. This caused by not rational compaction way. Comparing the results of testing of CFST members of medium height, there is an obvious stratification into 2-3 strength areas which emerge when the manufacturing methods are not rational

The second group is the specimens with the height of 1200 mm cut into 6 research fragments (figure 2).

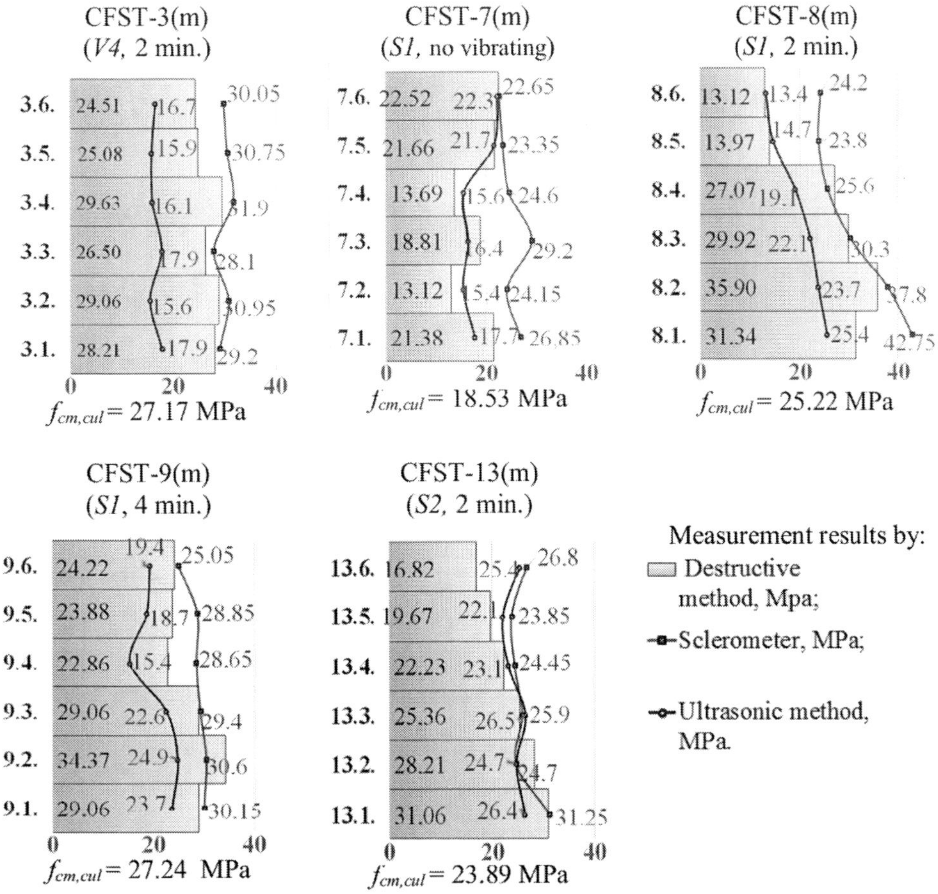

Fig. 2. Concrete strength variation along the height of CFST members with initial height of 1200 mm.

The third group is the specimens with the height of 2000 mm cut into 10 fragments (figure 3).

The average concrete strength of CFST-4(*l*) and CFST-5(*l*) decreased (63 % and 81 % of design strength). CFST-5(*l*) specimen characterized by nonhomogeneous structure and concrete features along the height (strength along the height reduce in 3 times, a lot of concreting defects). Manufacturing conditions of CFST-4(*l*) and CFST-5(*l*) specimens are different in compaction way only. Concrete strength homogeneity along the height of CFST-4(*l*) specimen varies within the limits of 50 %. During concreting a long CFST members with V4 strength mark, defects couldn't be avoided irrespective of compaction. The average concrete strength of CFST member CFST-10(*l*) is 18 % higher than calculated ($f_{cm,cul}$ = 28.71 MPa). However, at this concreting method, the strength of specimen along the height is homogeneous, except for one fragment. Such spot of a high strength is likely to emerge during compaction (at the place of vibrating device location). The average value and homogeneity of CFST-14(*l*) ($f_{cm,cul}$ = 24.75 MPa) and CFST-15(*l*) ($f_{cm,cul}$ = 25.25 MPa) specimens is almost the same when the placeability mark of concrete is the same. The movable concrete mixture (the slump is 50 – 90 mm) fills in even the small empty space completely without vibrating. There weren't any voids or caverns in CFST-14(*l*) and CFST-15(*l*) specimens.

Non-destructive control methods used to provide data for comparison with concrete compression tests results. They show the same dynamics of strength variation for homogeneous concrete structure. When there were some defects in concrete core, non-destructive control methods show increased strength values. Thereby, such control methods are not rational when applied to complete CFST members in case of possible defects of core concreting.

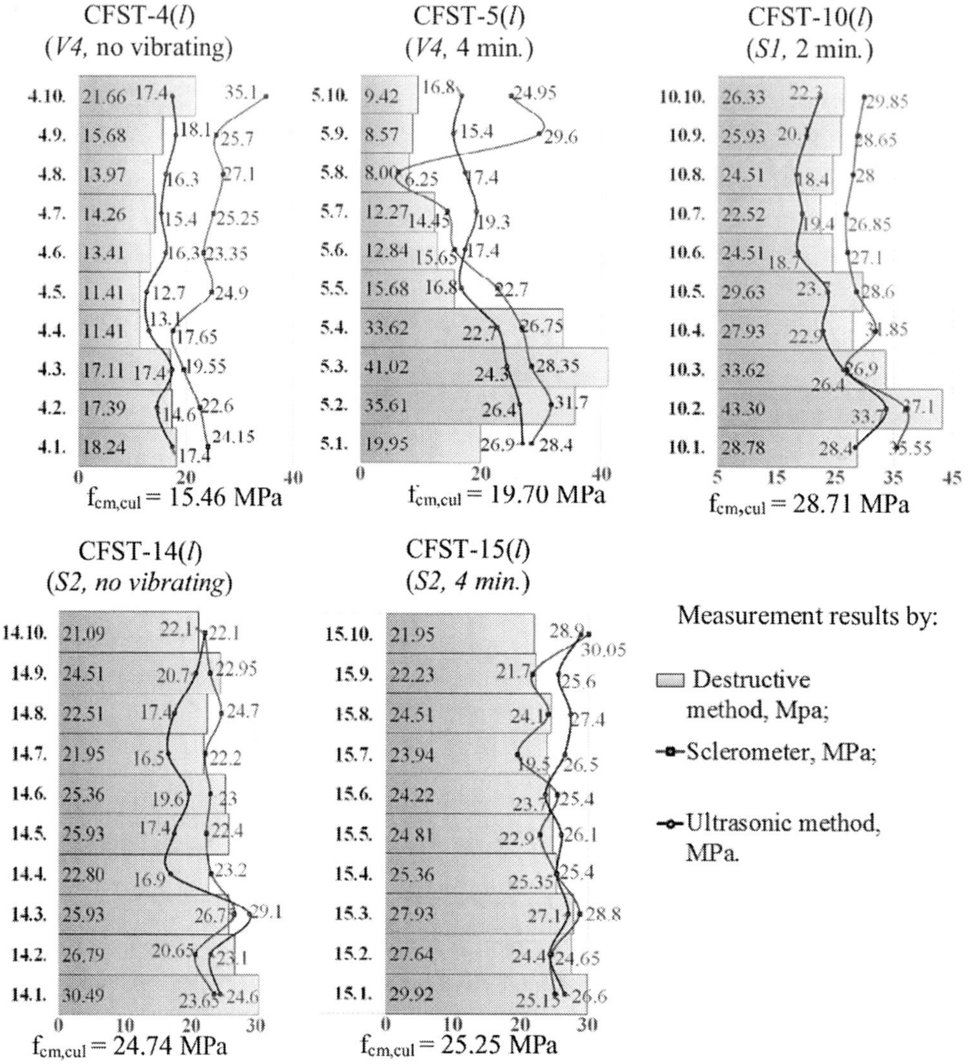

Fig. 3. Concrete strength variation along the height of CFST members with initial height of 2000 mm.

According to research procedure, for an approximation of the obtained research data the following polynomial expression used:

$$M_u = a_o + a_1 x_1 + a_2 x_2 + a_3 x_3 + a_{11} x_1^2 + a_{22} x_2^2 + a_{33} x_3^2, \quad (1)$$

Statistical data processing carried out using STATISTICA 10 software (figure 4). The obtained equations checked by Fisher criterion. Obtained regression equation allows estimation of the factors influence on the average compression strength of a specimen which calculated as an arithmetic average of all components (s – 2, m – 6 or l – 10 values) for one specimen.

According to the result of statistical analysis, we established the correlations between variable factors and the strength. The optimal compaction time for short CFST members of “s” series is 2 minutes (figure 4, a). S2 placeability mark recommended for short specimens. Optimal compaction time is 4 minutes.

a) the influence of compaction time and specimen height on the compression strength

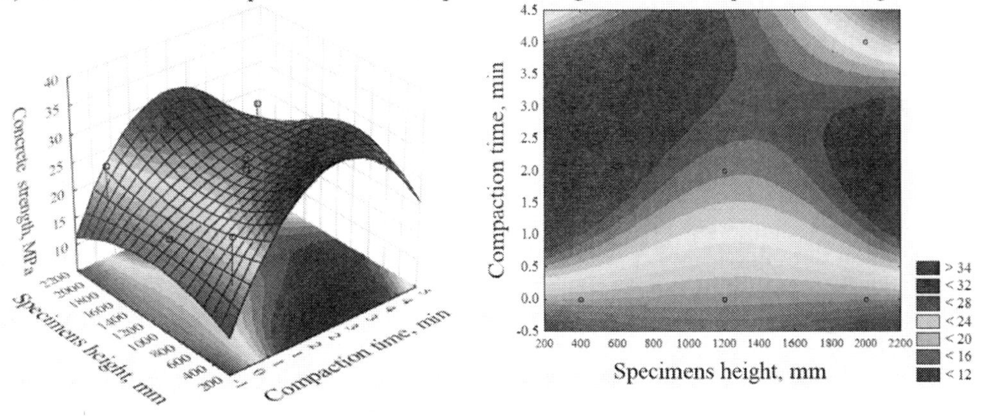

b) the influence of concrete placeability mark and the specimen height on the strength

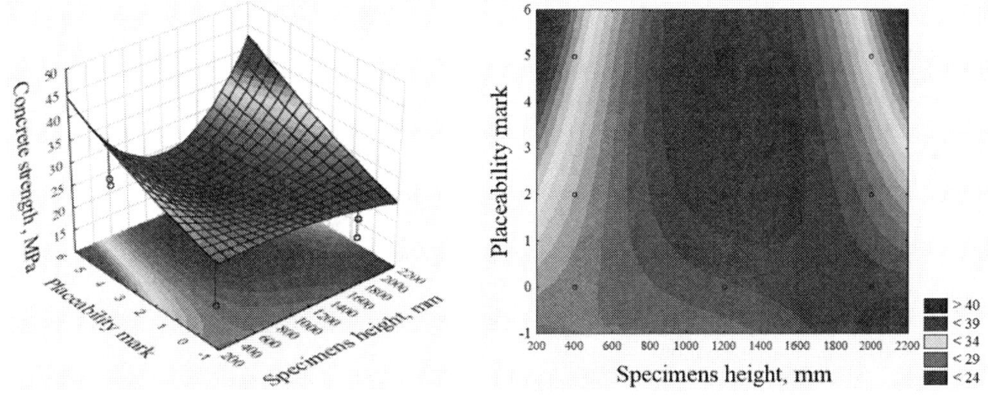

c) the influence of concrete placeability mark and compaction time on strength

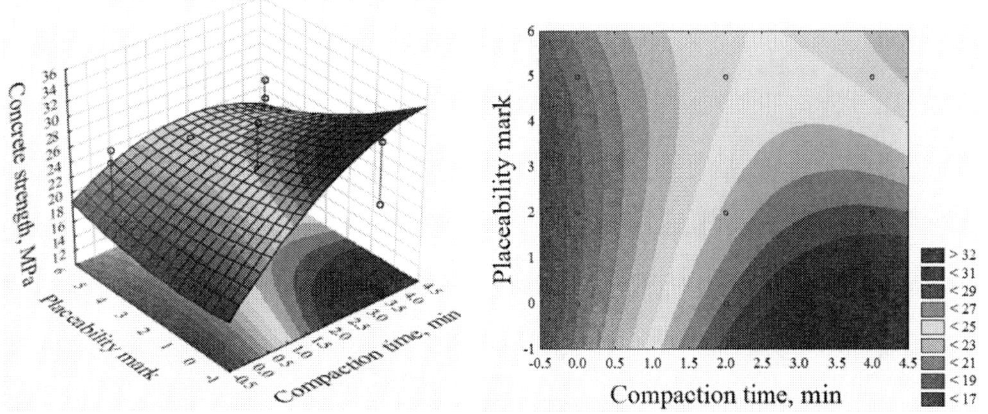

Fig. 4. Approximating surface of variation of the average value of concrete strength of CFST members in respect to experiment factors

According to the result of statistical analysis, we established the correlations between variable factors and the strength. The optimal compaction time for short CFST members of "s" series is 2 minutes (figure 4, a). S2 placeability mark recommended for short specimens. Optimal compaction time is 4 minutes.

5 Conclusion

According to the developed method, we investigated the influence of concreting technology on strength distribution along the height of concrete core of CFST members.
Results of the research show that inconstancy of concrete strength along the height depends on placeability mark of the concrete mixture. Constant of variation of concrete core strength along the height of CFST members when stiff concrete mixtures used (V4) is 1.5…3.5 times higher than that of movable ones (S1 and S2).

Statistically significant correlation between the compaction time and concrete strength discovered when stiff mixtures used (V4) and movable mixtures with slump ranging from1 to 4 cm (S1) 2 minutes of vibration time recommended as an optimal regime for concrete compaction. When movable mixtures with the slump of about 4 – 9 cm used, the total constant of variation of concrete strength along the height is about 13% irrespective of compaction time.

References

1. O.V. Semko, *Probabilistic aspects of calculating composite constructions,* Poltava (2004)
2. L.I. Storozhenko, *Tube confined concrete*, Kiev (1978)
3. L.I. Storozhenko, D.A. Ermolenko, O.I. Lapenko, *Tube confined concrete*, ASMI, Poltava, 306 (2010)
4. B. Uy, Z. Tao and L.H. Han, J CONSTR STEEL RES, **67**, 360-378 (2011)
5. Leschynskyy A.M. Budivelnik, Kiev, **3**, 47-49 (1986)
6. Kikin A.I., Sanzharovskii R.S. and Trul V.A., *Constructions made of steel tubes filled with concrete*, Stroyizdat, Moscow (1974)
7. Komar A.G., *Technology of concrete and reinforced-concrete products*, Stroyizdat, Moscow (1984)
8. *Design of concrete structures – Part 1-1: General Rules and Rules for Bridges. EN 1992-1-1:2004 + AC:2010 Eurocode 2,* Kiev (2012)

Probabilistic analysis of deformed mode of engineering constructions' soil-cement grounds

Yuriy Vynnykov[1]*, *Olena* Voskobiinyk[1], *Maksym* Kharchenko[1], and *Valentyn* Marchenko[1]

[1]Poltava National Technical Yuri Kondratyuk University, Department of Civil Engineering, Pershotravnevyi Avenue, 24, Poltava, 36011, Ukraine

Abstract. The results of the analysis of probabilistic methods that are used to assess the deformed state of the foundations of engineering structures are presented. A finite element analysis of the stress-strain state of the "man made soil ground – foundation – structure" system was carried out. A method for probabilistic calculation using the finite element method is proposed. On a real example, the level of reliability of a design decision based on a deterministic calculation is estimated by probabilistic calculation. On the basis of the statistic data obtained by imitational modeling, the probability of failure and no-failure operation of the structure regarding the absolute value of settlement and regarding the value of tilt against the reinforcement ratio of soft soil grounds settlements was determined. The probability of failure regarding the value of tilt against the reinforcement ratio was taken (15 to 25%), which is 0.03 – 0.05.

1 Introduction

The standard approaches to calculations of soil grounds of the structures and foundations are based on limit states principle. For such calculations, the average values of soil characteristics and load components are used. However, those parameters have the variable nature. That is why taking into account its variety is carried out by the instrumentality of safety factors. By such method one check, the criterions of reaching of one of the limits states: strength or serviceability of structure [1].

The up to date requirements of structure's design on deformable soil grounds provide calculations of the "soil ground – foundation – structure" system taking into account stiffness and strength of soil ground, a stiffness of a whole structure and its separate construction elements [2-4].

To consider the variable character of stated system's component it is necessary to consider the soil ground and the structure like a united (integrated) tridimensional stochastic system and development of methods of its calculation. In this case, the loads and influences (including geological processes), mechanical properties of structure's and ground's elements will become the main factors. The states factors have variable character. That is why for determination of the «soil ground – foundation – structure» system's reliability one need to use probabilistic approach considering the variable nature of values and processes, which influence on it.

* Corresponding author: vynnykov@hotmail.com

2 Analysis of resent research and publications

There is a line of scientific works devoted to probabilistic nature of loads and influences. It is known that one cannot determine exact values of loads on constructions due to their variable nature.

In addition, the stochastic nature of such characteristics as geometrical parameters of the structure, physical and mechanical properties of its materials was determined [1, 5, 6]. The statistic parameters of stochastic values of material properties one can find in scientific literature and building standards. The spreading of values of those parameters is influenced by construction works execution quality and construction exploitation quality.

The physical and mechanical properties of soils have stochastic nature due to the variety of its staff and composition, the influence on it geological processes, the polydispersity, the polyphasic and polyminerality the character of internal bonds, the value of loads etc. These properties typically change over time due to geological processes. Characteristics of man-made grounds of foundations have similar stochastic nature [7-16].

Due to the number of factors, which influence values of soil characteristics, the statistic parameters and distribution laws of those values were not studied sufficiently. In addition, the limited number of the samples for soil properties determination during investigations was a complication factor.

Thus, we saw that spreading of soil characteristics values is considerable. For reducing this irregularity uses engineering methods for soil improvement (soil compaction, reinforcement, cementation etc.) Due to the number of factors, which influence system serviceability, it is perspective to use finite element method (FEM) for probabilistic analysis of the “soil ground – foundation – structure” system.

3 Establishing of the objective and tasks of the study

The main objective of this study is the reliability evaluation of the “man-made ground – foundation – construction” system by FEM using the probabilistic approach.

The tasks for achievement of that objective include: 1) analysis of experimental statistic data of physical and mechanical properties of soils and man-made grounds; 2) analysis of world-wide progress in probabilistic calculations of foundation grounds; 3) FE analysis of deformed mode of man-made ground with the use of elastic-plastic soil model and imitational simulation.

4 The main part of the study

As statistic parameters of stochastic values of soil properties accepted mean value μ_x, standard deviation σ_x, a coefficient of variation v_x and distribution law (probability density function). Statistic parameters of properties of natural soils according to various authors are given in Table 1.

One of the effective methods for soil improvement is a construction of soil-cement grounds. By the researching, performed under the direction of prof. M. Zotsenko were determinate the statistic parameters of soil-cement properties distribution. The coefficient of variation is 10 to 15%.

Based on the experience of different researchers, it was determinate, that distribution law of permanent vertical loads and temporary loads is normal with coefficients of variation 10% and 50%, respectively [12], and horizontal variable loads described by Gumbel law with a coefficient of variation 25%. Findings of researches of stochastic nature of loads on foundations are given in table 2.

For reliability evaluation of constructions developed a row of methods such as: linearization, O.S. Lychov's enumerative technique, V.P. Chyrkov's method of numerical integration, method of statistical testing well known as Monte Carlo Simulation (MCS), Response Surface Method (RSM), Point Estimate Method (FO–PEM & A–PEM) and First Order Reliability Method (A–FORM) etc. [9, 11-15, 19-24].

Table 1. Coefficient of variation v_x (COV) values and distribution laws of natural soil characteristics.

Characteristic	Coefficient of variation υ_x (COV), % Distribution function (probability density function)			
	Sand	Sandy loam	Loam	Clay
Moisture content, w	30–50/4.4–49/- Normal	10–30/6.2–27.7/8–30 Normal	8–28/3.8–15/8–30 Normal	4–25/12.65/8–30 Normal
Porosity coefficient, e	3–13/1.1–6.7/- Normal	6–12/2.3–16.5/- Normal	6–25/3.5–14.2/- Normal	3-22/19.3/- Normal
Density, ρ	2–7.5/0.5–3.2/<10 Normal	2–4.5/0.5–2.5/<10 Normal	2.5–7.5/0.8–3.7/<10 Normal	2–6/4.3/<10 Normal
Density of soil particles, ρ_s	-/0–0.3/- Normal	-/0.2–0.65/- Normal	-/0.2–0.6/- Normal	-/0.8/- Normal
Plasticity index, I_p		25–50/-/10–40 Normal	5–35/-/10–40 Normal	7–30/-/10–40 Normal
Plasticity limit, w_p		6–17/-/6–30 Normal	5–25/-/6–30 Normal	7–27/-/6–30 Normal
Fluidity limit, w_L		5–16/-/6–30 Normal	5–20/-/6–30 Normal	5–20/-/6–30 Normal
Shearing resistance, τ	-/-	9–27/-/10–30 Lognormal	6–29/-/10–30 Lognormal	-/-/20–50 Lognormal
Friction angle, φ	-/-/5–15 Normal	-/-/5–15	-/-/12–56	-/-/12–56
Stiffness modulus, E	-/-/15–65 Lognormal	-/-	15–35/18.6–65.4 Lognormal	-/-

Note: over line data of Ermolaev M.M. and Mikheev V.V. [1] / data of Bukhrov O.K. and Shilin V.G. [1] / data of Phoon K.K and Kulhaw [23]; under line distribution law according to Fenton G.A. [8]

Table 2. Coefficient of variation (COV) values for loads on foundations.

Load type	COV
Dead load	0.1
Variable load (50 years maximum)	0.25
Snow load (50 years maximum)	0.26
Wind load (50 years maximum)	0.37
Earthquake load (50 years maximum, West and East of the USA)	1.38

Every of those methods are the convenient instrument for solving of particular problem. The main universal methods are MCS, RSM and PEM. They are used for solving of geotechnical problems by FEM. Such approach is implemented in Ansys (USA), Phase^2 (Canada) and in the special utility for Plaxis (under the direction of S. Baars).

The failure of foundation consist of such components: 1) failure (the ultimate limit state), the stresses under foundation is more than the limit soil resistance $p>p_u$; 2) absolute $S>S_u$ or different settlements $\Delta S/L>(\Delta S/L)_u$ of ground exceeded the limit settlement values (the serviceability limit state).

The foundation ground failure determination regarding the serviceability limit state is complicated the fact of settlement can be linear S_L or nonlinear S_p.

The probabilistic analysis can be performed by numerical simulation with the use of FEM. There is a row of probabilistic geotechnical problems, which cannot be solved by analytical methods.

In exemplification the allowable irregularity of loading of two silo-type grain-storages, which share one foundation plate on soft soil ground was determinate (Fig. 1). The allowable irregularity means that its value does not exceed the normative limit. For this reason, FE analysis of the deformed mode of soft soil ground of foundation plate of grain-storages, which was improved by soil-cement was performed. In Figure 2, the diagrams of settlements over time according to results of instrumental measurements and simulations are presented.

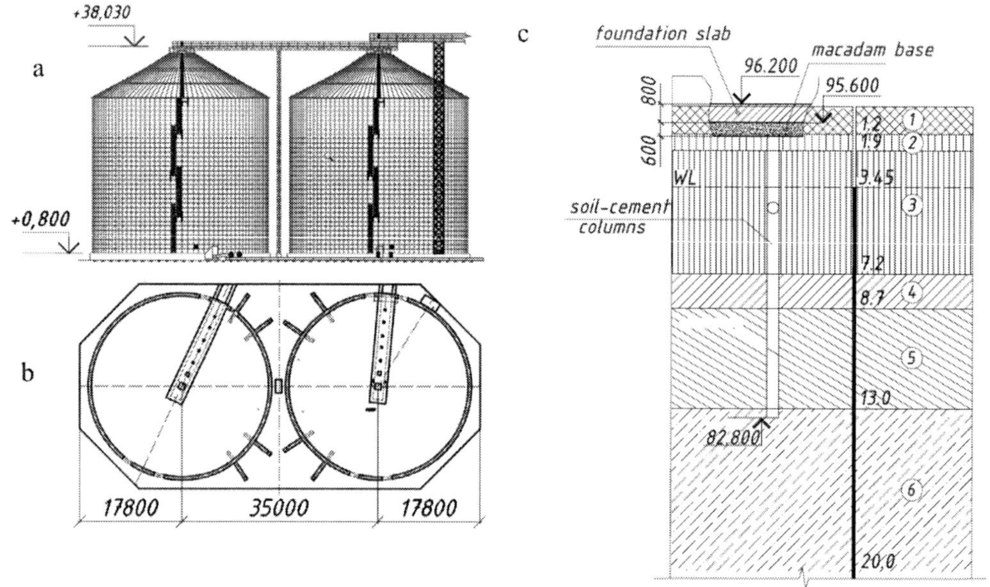

Fig. 1. Overall view of silo-type grain storages with the capacity of 24121 m³ each: a – front view; b – plan at the level of +0.800; c – ground profile.

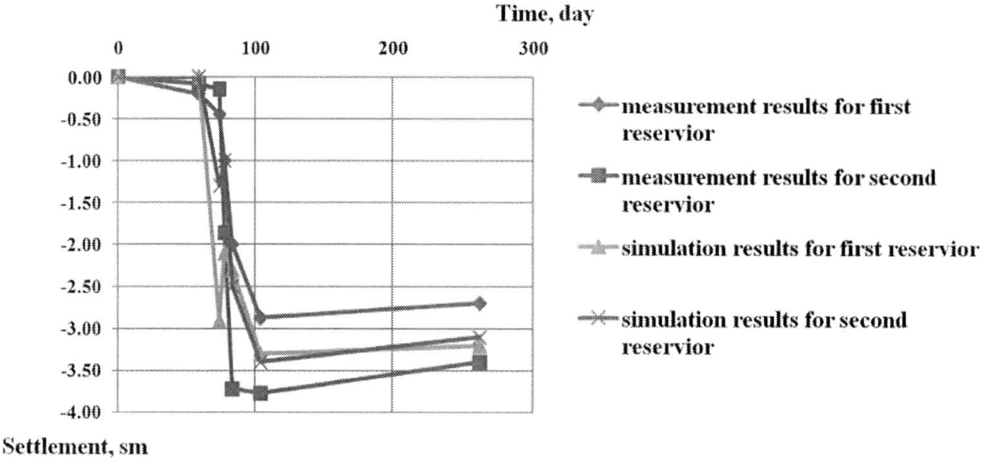

Fig. 2. Diagrams of settlements over time according to results of instrumental measurements and simulations.

Table 3. Input data for probabilistic analysis of foundation's improved grounds.

Characteristic	Distribution law	Statistic parameters
		COV, %
Stiffness modulus of soil, MPa	lognormal	15
Stiffness modulus of soil-cement, MPa	normal	15
Loading of reservoir	normal	30

It was accepted to improve soil ground under the foundation plate down to 12 m depth. To ensure required strength and stiffness properties, the ratio of reinforcement of ground was approximately 25%. The diameter of soil-cement columns was 0.5 m and spacing between them 0.9 and 0.95 m. Statistical parameters of stochastic values of input characteristics are given in table 3. PEM was used to avoid a large amount of iterations during imitational numerical simulation by FEM (Fig. 3). Parameters were changed simultaneously on two levels ($\mu \pm 3\sigma$, where μ – mean value, σ – standard deviation): 1) stiffness modulus of soil layers of reinforced ground E; 2) loading of reservoir #1; 3) loading of reservoir #2.

The finite element analysis was performed with different combinations of stochastic values and was determine settlements and tilt of foundation plate in the particular stages of silo-type storages exploitation. In the Fig. 4, a 3-D deformed FE mesh of analytical model on one of stages of imitational modeling by PEM is given. The distribution and statistic parameters of the stochastic value of reinforced ground settlements after initial loading and unloading are given in Fig. 5 according to results of an imitational simulation.

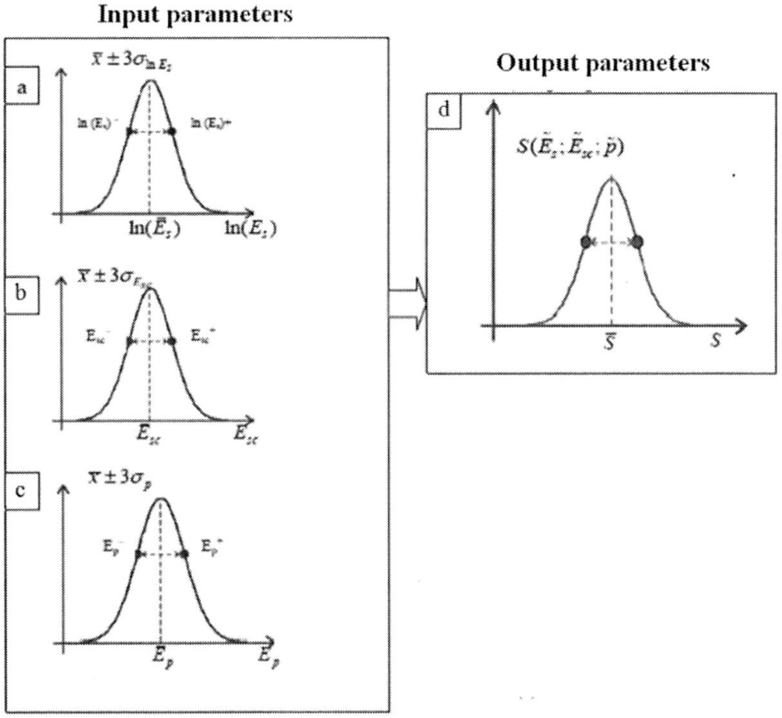

Fig. 3 Scheme of probabilistic analysis by PEM: a - density of distribution of stiffness modulus of soil E_s; b - density of distribution of stiffness modulus of soil-cement E_{sc}; c – density of distribution of the pressure to ground p; d - density of distribution of improved ground settlements.

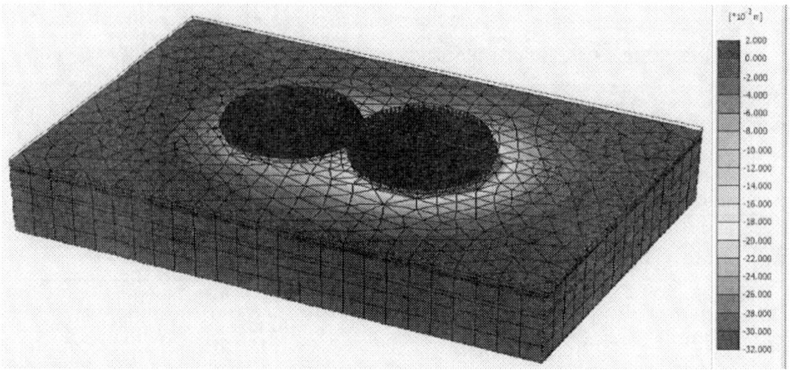

Fig. 4. Deformed 3-D FE mesh of the analytical model of imitational simulation regarding to PEM.

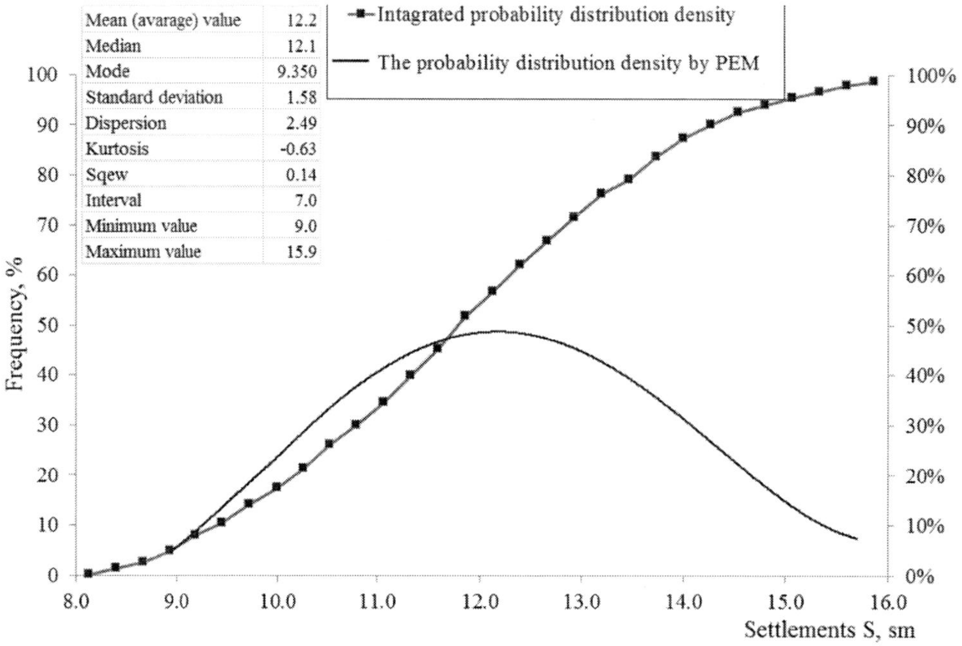

Fig. 5. Distribution of stochastic value of improved soil ground for foundation plate according to results of imitational simulation by FEM.

The distribution and statistic parameters of stochastic values of tilt of the foundation plate are given in the Fig. 6 (the value of tilt was multiplied by 104). During the subsequent loadings of reservoirs, the absolute values of settlements increase and have damped character. In that case, the probability of failure regarding of settlements is not more 0.001 and regarding of tilt – 0.07.

On the basis of the statistic data obtained by imitational modeling, the probability of failure and no-failure operation of the structure regarding the absolute value of settlement (Fig. 7) and regarding the value of tilt against the reinforcement ratio of soft soil grounds settlements was determined. The probability of failure regarding the value of tilt against the reinforcement ratio was taken (15 to 25%), which is 0.03 – 0.05.

According to the results of imitational simulation, the allowable irregularity of loading of reservoirs with shared foundation plate on the improved soil ground is 20%. The probability of structure failure regarding criterion of limit allowable tilt of foundation plate (i=0.002) is less than 0.01. On the base of imitational simulation by FEM using PEM for

 DOI: 10.1051/matecconf/201711602038

reliability rate of p=0.9 was matched the minimal required reinforced ratio of soil ground by soil-cement columns (it is 19 %).

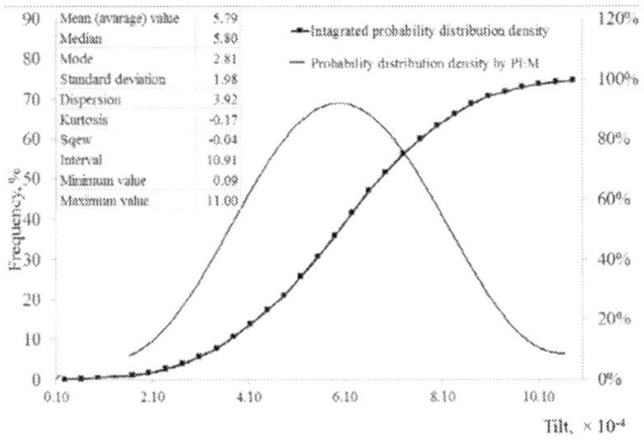

Fig. 6. Distribution of stochastic value of foundation plate tilt according to results of imitational simulation by FEM (the value of tilt was multiplied by 10^4).

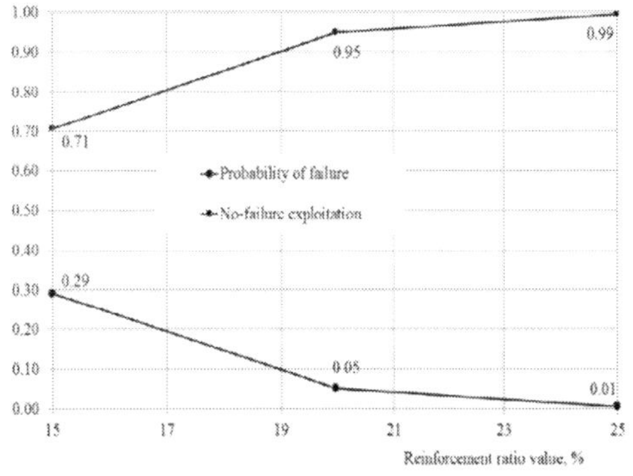

Fig. 7. Probability of failure and no-failure exploitation of structure regarding to criterion of absolute settlement against of reinforcement ratio value.

This problem shows that through the instrumentality of probabilistic approach the required reinforcement ratio of soft soil ground for the ensuring the no-failure exploitation of construction during its life was validated.

5 Conclusions

For probabilistic design calculations of the grounds of foundations, the statistic data of stochastic values of physical and mechanical properties of nature soils can be used according to Table 1, loads to soil ground – Table 2, improved grounds by soil-cement columns – Table 3.

Stochastic nature of linear and nonlinear stages of deformation process of soil grounds should be taken into account to determine the reliability of the grounds of foundations regarding criterions of its strength and stiffness.

It is reasonable to use MCS, RSM, and PEM that were approved enough by scientists including authors of this article to determine the reliability of nature and man-made soil grounds of foundations by numerical methods of calculation.

Decisions which allow reaching the required reliability rate should be taken for the benefit of the rational design of the grounds of foundations.

References

1. A.K. Bugrov, V.H. Shilin, *Opredelenie verojatnostnuh haracteristik aktivnogo davlenia grunta metodom Monte-Carlo,* Reconstruction of cities and geotechnical construction, **5**, 3 (2003)
2. A.V. Perelmuter, *Izbrannye problemy nadejnosty, bezopasnosty stroitelnyh construksiy*, Moscow (2007)
3. S.F. Pishugin, *Nadejnost stalnyh construksiy proizvodstvennyh zdaniy*, Poltava (2009)
4. J. Xue, D. Nag, *Reliability analysis of shallow foundations subjected to varied inclined loads*, Proc. of the 3rd Intern. Symposium on Geotechnical Safety and Risk, Munich, **7** (2011)
5. A.S. Lychev, *Nadejnost stroitelnyh construksiy,* Moscow (2008)

6. V.P. Chircov, *Pricladnye metody teorii nadejnosty v raschetah stroitelnyh construksiy,* Moscow (2006)
7. A.N. Trofimchuk, V.G. Chernyi, G.I. Chernyi, *Nadejnost sistem soorujenie – gruntovoe osnovanie v slojnyh injenerno-geologicheskih usloviah*, Kyiv (2006)
8. G.A. Fenton, *Probabilistic Methods in Geotechnical Engineering*, Utah, **96** (1997)
9. A.L. Rechenmacher, Z. Medina-Cetina, R.G. Ghanem, *Calibration of heterogeneous, probabilistic soil models,* Proc. 16th Intern. Conf. on Soil Mechanics and Geotechnical Engineering, Osaka, **3** (2005)
10. Y. Wang, B. Wang, *Study on autocorrelation model and reduction function of variance of soil random field,* Proc. of the 1st Intern. Symposium on Geotechnical Safety and Risk, China, Shanghai, **7** (2007)
11. G.B. Beacher, J.T. Cristian, *Reliability and statistics in geotechnical Engineering*, New York, 619 (2003)
12. A.J. Bond, *A procedure for determining the characteristic value of a geotechnical parameter*, Proc. of the 3rd Intern. Symposium on Geotechnical Safety and Risk, Germany, Munich, **7** (2011)
13. Y. Honjo, *Challenges in Geotechnical Reliability Based Design*, Proc. of the 3rd International Symposium on Geotechnical Safety and Risk, Germany, Munich, **6** (2011)
14. C. Pereira & L. Caldeira, *Shallow Foundation Design through Probabilistic and Deterministic*, Proc. of the 3rd Intern. Symposium on Geotechnical Safety and Risk, Germany, Munich, **8** (2011)
15. J.Y. Won *A probabilistic approach to estimate one dimensional consolidation settlements,* Proc. of the 17th Intern. Conf. on Soil Mechanics and Geotechnical Engineering, Egypt, Olexandria, **3** (2009)
16. S. Baars, M.A. Deptula, E. Dembicki, Ph. Gotteland, *Adaption of finite element models for probabilistic design*, Proc. of the 11rd Baltic Sea Geotechnical Conf. "Geotechnics in Maritime Engineering", Poland, Gdansk, **6** (2007)
17. M. Zotsenko, Y. Vynnykov, M. Kharchenko, *Evaluation of failure probability of soil cushions*, Proc. of the 3rd Intern. Symposium on Geotechnical Safety and Risk, Germany, Munich, **8** (2011)
18. G. Falsone, N. Impollania, *A new approach for the stochastic analysis of finite element modeled structures with uncertain parameters,* Computer Methods in Applied Mechanics and Engineering, **191(44),** 18 (2002)
19. M. Fredlund, *Finite elements stochastic analysis*, Proc. of 57th Canadian Geotechanical Conf. and 5th Joint IAH-CGS Conf., Canada, Quebec, **5** (2004)
20. A. Haldar, S. Mahadevan, *Reliability Assessment Using Stochastic Finite Elements Analysis*, New York, 220 (2000)
21. M. Manjuprasad, C.S. Manohar, *Adaptive Random Field Mesh Refinementsin Stochastic Finite Element Reliability Analysis of Structures*, CMES, **21** (2007)
22. G.M. Pechl, H.F. Schweiger, *Reliability analysis in geotechnics with deterministic finite elements – a comparison of two methods,* Proc. of 5th European Conf. on Numerical Methods in Geotechanical Engineering, Paris, **5** (2002)
23. K.K. Phoon, *Reliability-based design in geotechnical engineering. Computions and applications*, New York (2008)
24. B. Sudret, A. Kiureghian, *Stochastic Finite Elements Methods and Reliability. A state-of-the-Art Report,* Berkeley, Civil and Environment Engineering University of California (2000)

Study of heat shielding qualities of brick wall angle with additional insulation located on the outside fences

Oleg Yurin[1], and *Tatiana* Galinska[1*]

[1]Poltava National Technical Yuri Kondratyuk University, Department of Civil Engineering, Pershotravnevyi Avenue, 24, Poltava, 36011, Ukraine

Abstract. The results of theoretical studies of heat-shielding properties of the angle of external wall a residential building are presented. The brick wall thickness of 510 mm was considered. The studies were performed for the I-st temperature zone of Ukraine. Heat insulation was located on the external side of enclosure. Thermal protection problems of the angle are consideration of the impact of heat-conductive inclusion on adjusted heat transfer resistance of enclosure. Conductive inclusion on the external angle area is its configuration. An additional layer of insulation in the angle area is necessary to use for bringing the level of thermal protection of angle area to the standards. The results of calculations of two-dimensional temperature fields were used for the analysis of thermal protection. Two options for additional insulation were studied. The first option: continuous strip insulation covering the external wall angle. Another one provides division of additional insulation layer into two parts and displacement them from the angle within the design scheme. Studies were performed for different thickness of additional insulation. There was additionally taking into account the value of insulation displacement from the angle for the second option. The optimal location and size of additional insulation were determined. The minimum amount of additional insulation under which the thermal protection requirements are performed was accepted like optimality criterion.

1 Introduction

One of the main tasks of designing external fencing structures of residential buildings is to provide a normalized level of heat shield. Herewith in the determination of thickness of the insulation in external walls the seats with heat conductive inclusions, in which the level of heat shield usually less than in areas without them are ignore. Areas with heat conductive inclusions include the following: place of relying on the wall of overlap constructions or balcony plate, walls abutment to windows or balcony doors, external wall angle. Ignoring the properties of these areas leads to higher overall heat loss through external fencing structures. Besides high heat, losses in these places there are such negative phenomena as

* Corresponding author: galinska@i.ua

lowering the temperature of the inner wall surface and increase amplitude of oscillations the temperature on this surface. Increasing of heat shielding properties of external wall angle is possible through its additional insulation.

1.1 Review of the recent studies and publications

Work of many authors dedicated to issue of increasing heat shielding properties of angles to the requirements of norms [1-3]. So, in [4-6] described variants of additional insulation on the external side by increasing the thickness of the primary insulation in the area of building angle. The option of internal insulation is also considered in this article. The variants of additional insulation of frame-monolithic buildings angle are considered in [7]. The impact of the relative position of the external building angle and window aperture on the temperature of the internal surface angle was considered in [8].

1.2 Assignment aspects of the general problem unsolved earlier

In publications dealing with the study of ways to improve the heat shielding properties of angle to the heat transfer norms, the use continuous additional layer of insulation located on the external surface of the fence was proposed. The option of separation of additional insulation into two parts and displacement it from angle within the design scheme is not considered. No recommendation was given concerning the ratio of the thickness to length of the additional insulation layer under which material expenses will be minimal.

1.3 Setting objectives

The objective of this work was determining the optimal location and the possible minimum size of additional insulation in the external brick wall angle under which heat shielding conditions will be performed. Minimal required capacity of additional insulation adopted like criterion of optimality.

2 Basic material and results

The construction of a brick wall with external insulation is shown in Fig. 1. Fencing layers located on the external side insulation have no significant effect on the heat-shielding properties of the fence due to little thickness. That is why in the calculations they are changed to a layer of cement-sand mortar with the thickness of 5 mm. A leveling layer of cement-sand mortar between the insulation and brickwork the thickness of 5 mm is added.

Characteristics of materials of the fencing structure layers are shown in Table 1.

Table 1. Characteristics of materials of the fencing structure layers.

Fencing structure layers	Density, kg/m³	Thickness, m	Thermal conductivity, W/m·K
Sand-lime mortar	1800	0.02	0.93
Brick	1800	0.51	0.81
Cement-sand mortar	1600	0.005	0.81
SUPERROCK Basalt mineral wool	135	90	0.035
Cement-sand mortar	1600	0.005	0.81

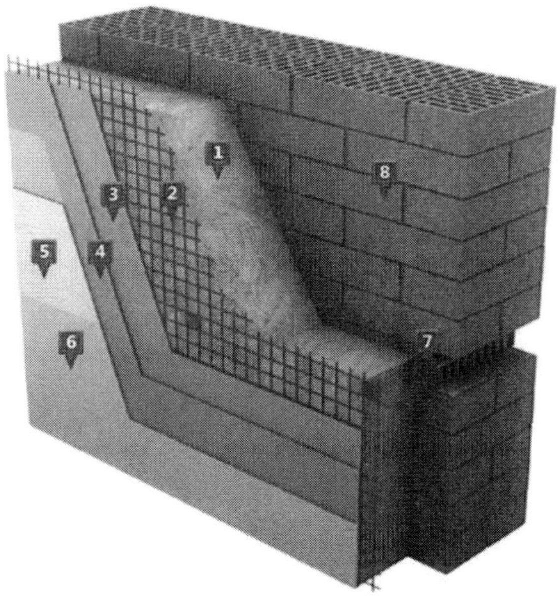

Fig. 1. The construction of a brick wall with external insulation. 1- insulation; 2 - steel net; 3 - primer coat; 4 - leveling coat; 5 - decorative plaster; 6 - facade paint; 7- steel rock bolting; 8 - brick wall.

The thickness of the base layer of insulation defined by the heat engineering calculation for residential building in the I-th temperature zone is 90 mm.

Adjusted heat transfer resistance of external angle is determined on the basis of temperature field calculations. Options for offered additional insulation angle are compared with the scheme provided by A. Prischenko in [4-5] (Fig. 2).

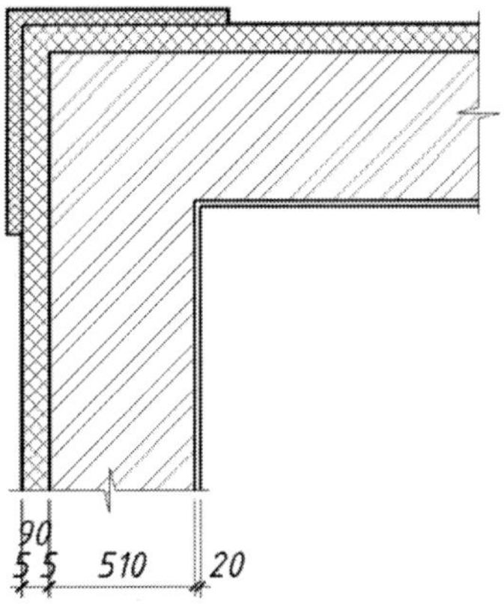

Fig. 2. Options for additional insulation angle according to the source [4-5].

Studies have shown that the minimum capacity of additional insulation (to meet the requirements of heat shield) will be determined for thickness of 40 mm and length (from the angle) of 1840 mm. In other words, additional insulation length is close to the length of the design scheme. The minimum capacity of additional insulation was 0.14 to 1 m high of the wall. The temperature field of angle for this option is shown in Figure 3.

For these dimensions of additional insulation:

- $R_{\Sigma\ red} = 3.305\ m^2 - W\ K > R_{q\ min} = 3.3\ m^2 - W\ K$;

- $\Delta t_{red} = 1.5\ ^oC < \Delta t_{s.h} = 4\ ^oC$;
- $\tau_{in.min} = 16.9\ ^oC > t_{dew} = 10.7\ ^oC$.

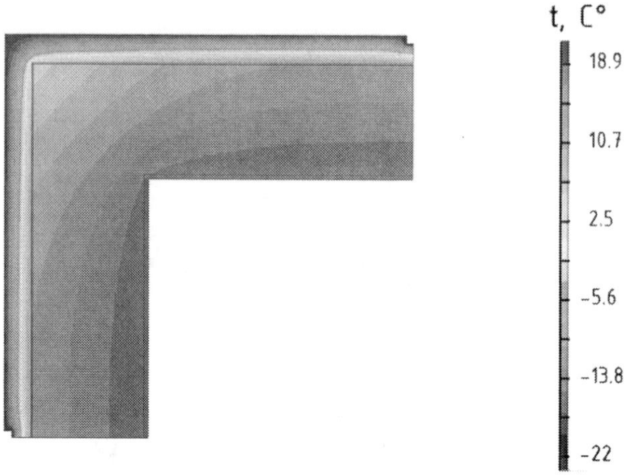

Fig. 3. The temperature field of angle with a solid layer of additional insulation.

The proposed option of additional insulation with a shift from the wall angle is shown in Figure 4.

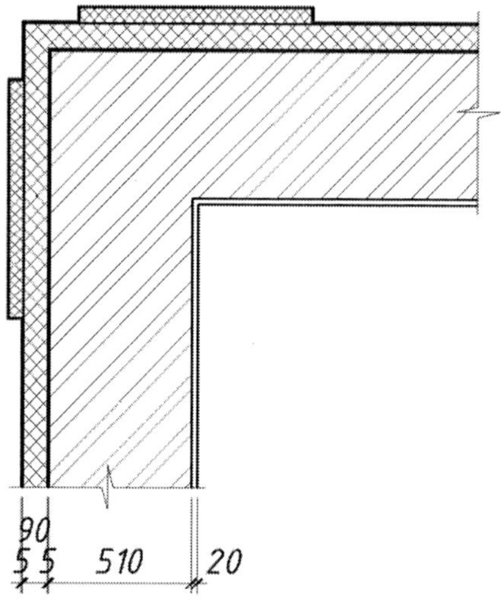

Fig. 4. The proposed option of additional insulation with a shift it from the wall angle.

The options of additional insulation layer with a thickness of 60 mm, 50 mm, 40 mm and 30 mm were considered.

The length of the design scheme is not enough to meet the requirements of heat shield at a thickness of additional insulation 30 mm.

The last option of the location of additional insulation, at each of its thickness, is when it coincides with the edge of design scheme.

As shown in Table 2, the optimal option (minimum capacity of insulation) was additional insulation angle with layer thickness of 40 mm, a length of 1700 mm at its location on the edge of design scheme (at a distance of 160 mm from the wall angle).

Table 2. The results of the study of heat-shielding properties of the angle by the proposed scheme, at a thickness of additional insulation 50 mm.

Scheme	Thickness of additional insulation layer	Distance from angle, mm	Length of additional insulation, mm	Adjusted heat transfer resistance	Capacity of additional insulation, m^3	Percentage of capacity of additional insulation by the scheme according to the source [4, 5], %
According to the source [4, 5]	40	0	1830	3.305	0.1448	100
Proposed	60	50	1450	3.308	0.174	120
		100	1410	3.311	0.1692	117
		150	1380	3.307	0.1656	114
		200	1360	3.308	0,1632	113
		250	1340	3.306	0.1608	111
		300	1320	3.304	0.1584	109
		350	1300	3.304	0.156	108
		400	1290	3.305	0.1548	107
		450	1270	3.3005	0.1524	105
		500	1260	3.305	0.1512	104
		550	1250	3.307	0.15	104
		600	1230	3.3007	0.1476	102
		660	1220	3,305	0,1464	101
	50	50	1580	3.3004	0.158	109
		100	1540	3.3007	0.154	106
		150	1510	3.301	0.151	104
		200	1490	3.303	0.149	103
		250	1470	3.303	0.147	102
		300	1450	3.301	0.145	100
		350	1440	3.307	0.144	99
		400	1420	3.304	0.142	98
		450	1400	3.302	0.14	97
		480	1390	3.303	0.139	96
	40	50	1780	3.303	0.1424	98
		100	1740	3.304	0.1392	96
		160	1700	3.302	0.136	94

The temperature field of option for the location of additional insulation shown in Figure 5.

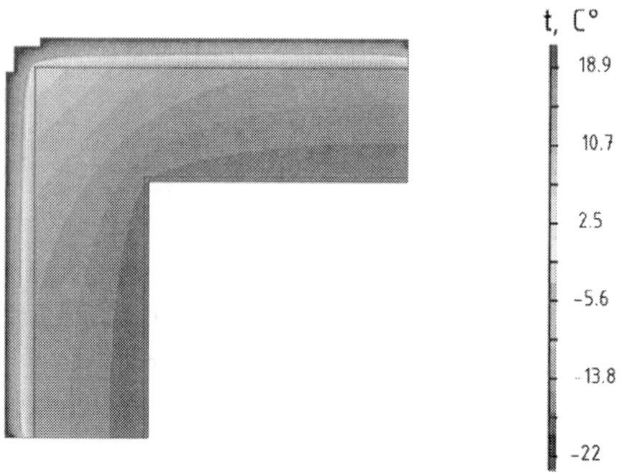

Fig. 5. The temperature field of wall area with additional insulation length of 1700 mm and thickness of 40 mm at a distance of 160 mm from the angle (on the edge of design scheme).

A graph of temperature distribution on internal surface of the wall area is shown in Figure 6.

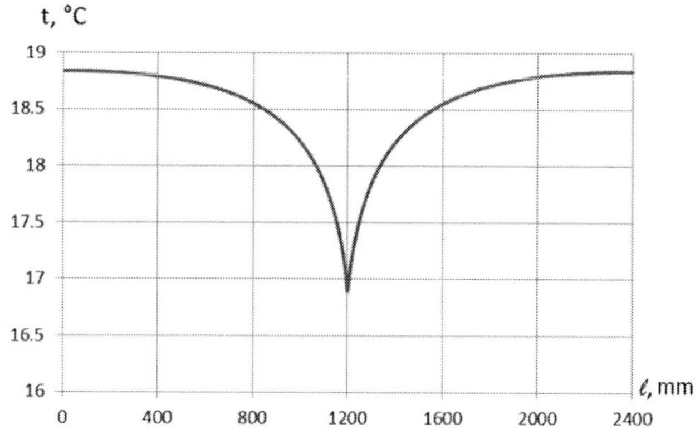

Fig. 6. A graph of temperature distribution on internal surface of the wall area with additional insulation length of 1700 mm and thickness of 40 mm at the distance of 160 mm from the angle (on the edge of design scheme).

At these sizes and location of additional insulation:

$R_{\Sigma\ red} = 3.302\ m^2 - W\ K > R_{q\ min} = 3.3\ m^2 - W\ K$;

$\Delta t_{red} = 1.5\ ^oC < \Delta t_{s.h} = 4\ ^oC$;

$\tau_{in.min} = 16.9\ ^oC > t_{dew} = 10.7\ ^oC$.

All heat shielding requirements are fulfilled.

3 Conclusions

With decreasing the thickness of additional insulation layer its capacity needed to meet the heat shielding requirements is decreases.

With displacement of additional insulation layer from the wall angle, its capacity necessary to meet the heat shielding requirements is decreases.

The lowest capacity of additional insulation needed to meet the heat shielding requirements is observed at its location on the edge of design scheme.

The lowest capacity of additional insulation will be at its least possible thickness and location on the edge of design scheme.

For the considered construction of external wall insulation and applied insulation according to the scheme when it consists of two parts shifted from the wall angle, decreasing of the capacity of additional insulation is 6% compared with the scheme when a continuous strip insulation covering angle.

References

1. *DBN V.2.6-31:2006. Konstruktsiyi budinkiv i sporud. Teplova izolyatsiya budivel /* Minbud Ukrayini. Kiev, (2006)
2. A.E Piir., O.A Kozak., V.B Kuntysh. Science & Technique. **16(2)** 931(2017)
3. A.E. Zakharevich. Science & Technique. **15(6),** 940, (2016)
4. A.M. Prishchenko, T.V. Zhmykhova, M.V. Tymofieiev. Budivelni konstruktsii. **79,** 16 (2013)
5. A.N. Pryshchenko Visnyk Donbaskoi natsionalnoi akademii budivnytstva i arkhitektury. **2,** 13, (2014)
6. G. Vasilyev, V. Leskov, V. Gornov, V. Lichman, M. Kolesova, A. Dmitriev. MATEC Web of Conf. **40,** 05004 (2016)
7. O.I. Yurin, T.A. Galinska, M.V. Levchuk *Doslidzhennia temperaturnykh poliv zovnishnikh kutiv karkasno-monolitnykh budivel*. Zbirnyk naukovykh prats studentiv budivelnoho fakultetu PoltNTU imeni Yuriia Kondratiuka 87 (2013)
8. O.V. Semko, O.I. Yurin Academic journal Series: Industrial Machine Building, Civil Engineering **4(2),** 182, (2013)

Revisiting the reasons for contact fatigue defects in rails

Alexander Darenskiy[1*], *Dmitry* Potapov[1], *Yuseph* Tuley[2], *Natalia* Bugaets[1] and *Alina* Malishevskaya[1]

[1]Department of Track and Track Facilities, Ukrainian State University of Railway Transport, Feierbakh Square 7, 61050, Kharkiv, Ukraine
[2]Track Department JSC "Ukrzaliznytsia", Tverska Street 5, 03150, Kyiv, Ukraine

Abstract. As it is known rail is one of the most significant elements of the whole railway construction. Operation under alternating loads from wheels of the rolling stock and different ambient temperatures lead to appearance and development of rail defects and damages. A great variety of operational factors (freight traffic density, axial loads, traffic speeds, track layout and profile) as well as special features of manufacturing and thermal treatment of rails create certain difficulties while identifying reasons for defects and damages. The article deals with an attempt to estimate influence of track layout and lateral forces on appearance of defects and damages in rails on the base of long-term observations of rail operation in Kharkiv Metro. On the basis of the vehicle/track mathematical model which considers structural features of both rolling stock and permanent way in underground systems, the level of lateral forces in curves was calculated. The coefficients of correlation between the track curvature, the level of forces and the amount of defected rails removed were later obtained, that made it possible to determine the dominant factor which may lead to appearance and development of contact fatigue defects in rails laid in curves.

Introduction

In order to provide the optimal level of aggregate costs for track maintenance it is necessary to solve the problem of longer life of all elements of the permanent way, first and foremost, the rail, being the most expensive and vital one.

The problem can only be solved if all reasons for failures in these elements are definitely established, that will make it possible to plan and implement measures aimed at effective solutions.

The life cycle of rails now is determined, mostly, by failures due to defects and damages in rails, first of all, of contact fatigue nature. The study attempts to establish reasons for appearance of contact fatigue defects in rails by the example of Kharkiv Metro. The research presented in [1-6] establishes that the principle reason for appearance of these defects is insufficient metal strength. However, the long-term observations, conducted by the authors of the article, have revealed that a large number of defects have appeared in

* Corresponding author: ppx_xiit@kart.edu.ua

curves on the track. But studies [1-6] do not provide an answer to the question about the reason for a significant number of defects in curves.

The issue of a life cycle of rails in underground systems in Ukraine is the theme of the above-mentioned studies. It should be noted that foreign studies [7-16], while paying attention to high intensity of defect appearance in curves, do not uniquely answer the question about the reason for the phenomena. Thus, investigation into the reasons for contact fatigue defects in rails has to be continued.

The objective of the study is to reveal reasons for contact fatigue defects in rails by the example of Kharkiv Metro. The results of long-term observations of rail operations, numerical research into the train/track dynamics in underground systems with application of the theory of probability methods for generalization of results are on the basis of this research.

The basic part

Nowadays Kharkiv Metro includes three lines. The track mileage of main lines is approximately 75 km (about 20 km of them are curves, the radii being from 350 to 3000 m).

The basic regulation in force for Ukraine's metro systems which establishes the type of a rail defect and also regulates the rail rolling stock operation on rails with defects and damages is *the Classification and Catalogue of Defects and Damages in Rails on Ukrainian Railways*. All defects and damages in *the Classification* are coded as three digital units.

Table 1. Rail defects by the Classification in force

Name of the defect and main reasons for appearance and development	Digital code of defect	Diagram
the layer separation and spalling of metal on the rolling surface due to breakdowns in the manufacturing technology	10.1-3	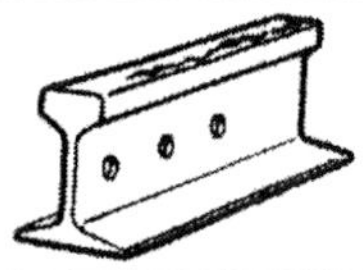
the metal spalling on the corners due to insufficient contact fatigue metal strength	11.1-3	
cross cracks in the railhead in form of bright and dark spots and fractures as a result of insufficient contact fatigue metal strength	21.1-3	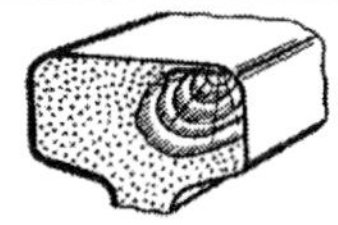
corrosion, scorching or local wear of rail foot and fatigue cracks leading to them	69.1-3	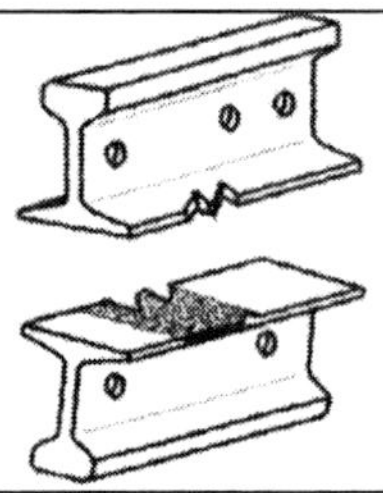

The structure of the code is as follows: the first digit stands for the type of a defect or damage, it also indicates the location in the cross-section (rail head, rail web, rail foot), the second digit stands for the kind of a defect or damage indicating reasons for the appearance and propagation, and the third digit determines the defect or damage location along the rail.

The first two digits of the code are separated from the third one by a point mark. The diagram and the description of the most typical defects and damages in rails for Kharkiv Metro are presented in Table 1.Results of observations of rail operation in Kharkiv Metro revealed that from 1986 to 2001 the main rail defects were ones by code 11.1-2-3. This state was the most characteristic for lines I and II, where the number of rails removed due to this defect was 60% of the total number of rails removed. The conclusion regarding line III was not made due to lack of the statistical information.

Besides, it has been established that the track layout also impacts the appearance of contact fatigue defects in rails, all other operational factors being constant. The less the radius of the curve is the more the damageability of rails in terms of this defect.

During 2002-2012 the total number of rails removed by code 11 was already 73.24% out of all rails removed (763 units). There exists a trend of a greater number of defected rails due to this defect if the radius of the track decreases. In curves the defected rails removed accounted for 48.05%, for straight sections it was 25.19%.

The final confirmation of the conclusions about the basic defect types in rails on main lines in Kharkiv Metro is the data on the rails removed from January 2012 till October 2014. The analysis conducted established the following:

- rail defects by code 11.1-3 was 55% (line I – 35.63%, line II – 13.75%, line III – 5.62%)
- rail defects by code 10.1-3 was 14.38% (line I – 8.75%, line II – 5.63%, line III – 3.12%)
- rail defects by code 69 was 6.25% (all these rails were removed from line I).

During this period with the help of the defectoscopy means the following was detected: two rails with defects by code 21.2 which accounted for 1.25% out of the total number.

The removal of rails in curves (Fig. 1) by the main defect (code 11) was 37.5%. And for straight sections the number of rails removed was only 17.5% which again confirmed the influence of the track layout on the appearance of defects in rails.

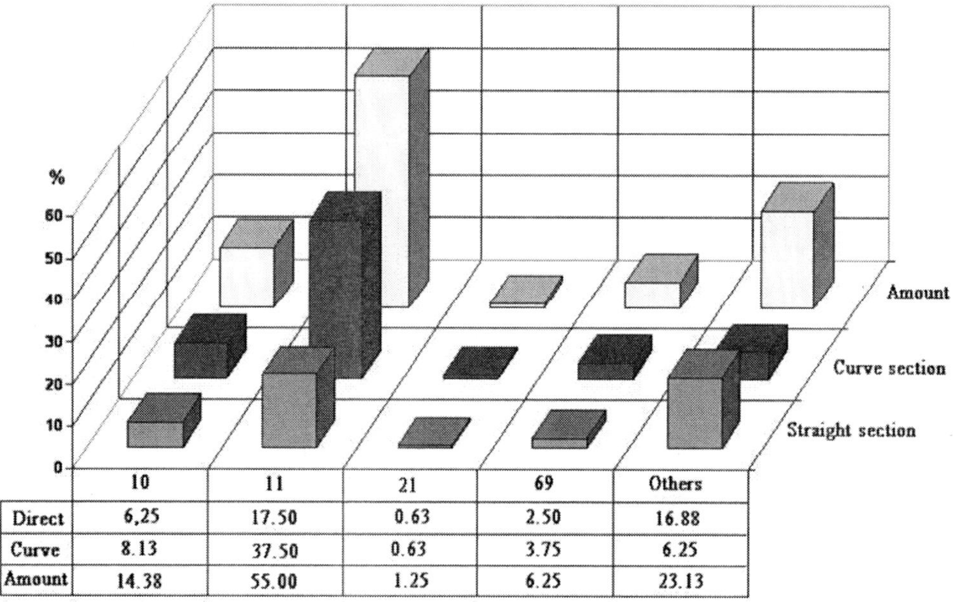

	10	11	21	69	Others
Direct	6,25	17.50	0.63	2.50	16.88
Curve	8.13	37.50	0.63	3.75	6.25
Amount	14.38	55.00	1.25	6.25	23.13

Fig. 1. The total number of rails removed in Kharkiv Metro in dependency on the track layout.

The analysis conducted on the basis of the information obtained brought to an assumption about the main reason for contact fatigue defects in rails, which is the impact of horizontal transverse forces.

It should be noted that experimental determination of these forces even for main line railways are rather costly and time consuming, besides it does not provide a definite answer to the question about the influence of rolling stock characteristics on the level of these forces as the input data (actual speed values, axial loads, stiffness of spring groups, etc.) cannot be virtually determined. For underground systems owing to narrow circumstances these work is practically impossible. Therefore the authors made a solution on conducting numerical research into the dynamic vertical and horizontal transverse forces for underground systems. Thus, a mathematical model of the spatial vehicle/track dynamic system was developed in order to solve the problem [20].

A large amount of research, which considers the problems of the train/track interaction on main line railways in order to determine the level of vertical, horizontal longitudinal and transverse forces influencing the track in train operation along straight lines and curves, applies only two-dimensional mathematical models of vehicles.

Besides, while determining forces of the train/track interaction, the vertical and horizontal stiffness of track is taken as constant. However, the analysis of features of arrangements and track operation under the influence of rolling stock in underground systems shows that such an assumption cannot be taken.

It should also be noted that presence of curves of small radii, structural features and operational modes of rolling stock in underground systems require consideration of significant nonlinear dependencies, including ones of the tearing type. Since the principle of superposition cannot be applied for nonlinear systems, the authors proposed a nonlinear spatial model of the vehicle.

In the mathematical model, the design scheme of a four-axle vehicle considers the basic structural characteristics of underground cars of the types Ezh3, Em-508T, 81-717, 81-714, 81-718, 81-719, namely:

- the two-stage spring suspension;
- all the wheel sets are leading;
- tractive and braking efforts from the wheel sets to the body frame transfer through the traction drive unit.

Considering all the above-mentioned the spatial vibrations of elements of the "vehicle" subsystem are described with a system of 44 differential second-order equations.

Unlike known mathematical models the "track" subsystem was presented as a spatial structure of rail bars rested on numerous elastic and dissipative supports of nonlinear characteristics. The forced vibrations of rails under the influence of loads from the rolling stock were described with a system of ten equations for rail vibrations in the vertical plane and ten equations for rail vibrations in the horizontal cross plane. And the bend of rails combined with torsion was also considered [23], as well as the distributed rail mass.

The spatial elastic and dissipative characteristics of rail supports were defined on the basis of the analytical analysis of operation of the "Metro" type fastenings and the experiments conducted [17-19]. A mathematical model of the spatial vehicle/track dynamic system was applied in Mathcad software.

The summarized results of actual loads from underground cars on track are presented in Table 2.

In order to determine the influence of the layout (track curvature) and horizontal lateral forces on appearance of contact fatigue defects in rails, the statistic treatment of the observation data in terms of the number of rails removed and the data of numerical research into horizontal lateral forces for establishing a level of the influence of these parameters on the damage rate were conducted.

The correlation coefficient was taken as a criterion for determination of an influence level.

Table 2. Design values of vertical and lateral forces influencing the rail head from the rolling stock in Kharkiv Metro.

	Radius, m	Value of elevation of the high rail, mm	Speed, km/h	Values of loads obtained, *N*			
				Vertical		Horizontal (lateral)	
				min	max	min	max
1	400	120	80	35746	83856	4630.3	9891.8
2			60	42033	75866	-1584	-722.24
3			40	51620	75293	-9707.2	-4574.8
4	500	120	80	35968	84559	2047	4367.7
5			60	42015	76297	-4620.6	-2201.3
6			40	50967	75941	-10891	-5154.7
10	600	120	80	36081	84792	297.41	868.84
11			60	42327	76219	-6544.7	-3151.2
12			40	50958	75462	-11867	-5645.7
16	800	90	80	35911	84917	283.32	691.78
17			60	42290	75927	-4983	-2355.6
18			40	50863	75763	-8792.8	-4153.5
22	1000	80	80	36106	84829	-505.78	-266.61
23			60	42200	76164	-5017.2	-2415
24			40	50641	75408	-8107.3	-3816.5
25	1500	70	80	36331	85007	-2591.3	-1203.7
26			60	41851	76427	-5590.5	-2698.2
27			40	51099	75670	-7633.6	-3609.2
31	straight	0	80	36026	84881		
32			60	42087	75770		
33			40	50804	76030		

This criterion was calculated according to [21, 22] from the expression:

$$r = \frac{\sum x_i \cdot y_i - k \cdot \bar{x} \cdot \bar{y}}{\sqrt{\sum x_i^2 - k \cdot \bar{y}^2} \cdot \sqrt{\sum y_i^2 - k \cdot \bar{x}^2}} \tag{1}$$

where x_i is the value of argument (track curvature or the value of lateral forces);

y_i is the value of function (damageability of rails by contact fatigue defects);

$\bar{x}$ is the average value of argument;

$\bar{y}$ is the average value of function;

k is the number of observation points.

The significance of the correlation coefficient was checked by Kolmogorov's criterion

$$r \cdot \sqrt{k-1} \geq T \tag{2}$$

where T is the table value of random deviations of products $r \cdot \sqrt{k-1}$ from zero [21, 22].

The correlation coefficients obtained for the values under investigation for all three lines of Kharkiv Metro are presented in Table 3. Table 4 presents the results of validation of the correlation coefficients obtained.

Table 3. Values of coefficients of correlation between the track curvature, lateral forces and the number or defected rails removed.

Factor	Values of correlation coefficients			
	Line I	Line II	Line III	throughout the Metro
Track curvature	0.302	0.200	0.262	0.323
Level of lateral forces	0.748	0.262	0.394	0.354

Table 4. Validation of correlation coefficients.

Line number	Dependency	Number of observations	Value of $r \cdot \sqrt{k-1}$	Table value of random deviations, T
Line I	track curvature	48	2.07	1.95
	lateral forces	48	5.12	1.95
Line II	track curvature	29	1.06	0.95
	lateral forces	29	1.39	0.95
Line III	track curvature	34	1.51	1.21
	lateral forces	34	2.26	1.21
throughout the Metro	track curvature	111	3.38	3.01
	lateral forces	111	3.71	3.01

The analysis of values in Tables 3 and Tables 4 shows that values of the coefficients of correlation between the level of lateral forces and the number of defected rails removed 10-148% higher than values of the coefficients of correlation between the track curvature and the number of rails removed. The test of significance for correlation coefficients showed that design values were considerably higher than table critical ones. The most obvious is the value of the coefficient of correlation between the level of lateral forces and the number of defected rails removed for line I of Kharkiv Metro, which is in operation for over 40 years.

Conclusions

1. The research into the correlation coefficient has shown that the level of lateral forces is the most significant reason for appearance of contact fatigue defects by code 11 for rails in Kharkiv Metro.

2. The results of the research make it possible to give grounded recommendations about the change in operational characteristics of curves for underground systems in Ukraine in order to lower the susceptibility to contact fatigue defects. It deals with admissible traffic speeds and elevation of the high rail in curves of different radii, as well as possible correction of these parameters.

3. The research conducted can serve as the basis for development of recommendations on frequency of track maintenance works.

References

1. R.M. Yosifovich, *Naukove obgruntuvannya i praktichne virishennya problemi pidvishchennya ekspluatatsiynogo resursu reyok Kiïvs'kogo metropolitenu. PhD Thesis, DETUT,* 24 (2015).
2. E.I. Danilenko, R.M. Yosifovich, O.A. Oliynik, O.O. Soroka, Zb. nauk. prats' DETUT. Seriya «Transportni sistemi i tekhnologii», **(22)**, 10-20 (2013).

3. V.V. Kosarchuk, O.V. Agarkov, Zb. nauk. prats' DETUT. Seriya «Transportni sistemi i tekhnologii», **(18)**, 61-72 (2011).
4. V.V. Kosarchuk, A. V. Agarkov, Zb. nauk. prats' DETUT. Seriya «Transportni sistemi i tekhnologii», **(20)**, 77-89, (2012).
5. O.V. Agarkov, R.M. Yosifovich, «Science and Transport Progress». Bulletin of Dnipropetrovsk National University of Railway Transport, **4(58)**, 71-86, (2015).
6. R.M. Yosifovich, «Science and Transport Progress». Bulletin of Dnipropetrovsk National University of Railway Transport, **6 (60)**, 36-41 (2015).
7. S.L. Grassie, J. Kalousek, *Proceedings of 6th International Heavy Haul Conference, The International Heavy Haul Association, Cape Town, South Africa,* 381-404 (1997).
8. J.W. Ringsberg, H. Bjarnehed, A. Johansson, B.L. Josefson, *Proceedings of the Institution of Mechanical Engineers, Part F: Journal of Rail and Rapid Transit*, **1(214)**, 7-19 (2000).
9. M. Taraf, E.H. Zahaf, O. Oussouaddi, A. Zeghloul, Tribology International, **3(43),** 585-593 (2010).
10. A. Johansson, B. Palsson, M. Ekh, J.C. Nielsen, M.K. Ander, J. Brouzoulis, E. Kassa, Wear, **1(271)**, 472-481 (2011).
11. V.V. Kuznetsov, V. S. Lysyuk, Vestnik VNIIZhTa. **6**, 69-94 (2000).
12. M.F. Verigo, Vestnik VNIIZhTa, **21 (5)**, 25-37 (2001).
13. A.Yu. Abdurashitov, Put' i putevoe khozyaystvo, **11**, 16-20 (2002).
14. M.Kh. Akhmetzyanov, Vestnik VNIIZhT, **2**, 36-43 (2003).
15. Beli, Ya., I. Nemet, Put' i putevoe khozyaystvo, **5**, 33-37 (2011).
16. A.Yu. Abdurashitov, Put' i putevoe khozyaystvo, **11**, 16-20 (2002).
17. A.S. Malishevskaya, D.A. Fast, Ivanovo: Nauchnyy mir, **2**, 48-55 (2015).
18. A.S. Malishevskaya, Zb. nauk. prats'. Ukraïns'kogo derzhavnogo universitetu zaliznichnogo transportu, **161**, 55 – 60 (2016).
19. O.M. Darenskiy, Yu.L. Tuley, D.O. Potapov, A.S. Malishevs'ka, «Science and Transport Progress». Bulletin of Dnipropetrovsk National University of Railway Transport, **6(66)**, 96-104 (2016).
20. O.M. Darenskiy, A.S. Malishevs'ka, Information and control systems at railway transport, **6**, 56-64 (2016).
21. M.N. Stepnov, *Statisticheskie metody obrabotki rezul'tatov mekhanicheskikh ispytaniy.* Moscow, 232 (1985).
22. K.A. Braunli, *Statisticheskaya teoriya i metodologiya v nauke i tekhnike.* Moscow, 408 (1977).
23. A.N. Darenskiy, Ya.S. Leybuk Information and control systems at railway transport, **2(123)**, 52-60 (2017).

A mathematical model of the rail track presented as a bar on elastic and dissipative supports under the influence of moving loads

Alexander Darenskiy[1]*, *Volodymyr* Vitolberg[1], *Denis* Fast[1], *Andrii* Klymenko[1] and *Yaroslav* Leibuk[1]

[1]Department of track and track facilities, Ukrainian State University of Railway Transport, Feierbakh Square 7, 61050, Kharkiv, Ukraine

Abstract. At present, the most common track model is the one in which rails are presented as bars of infinite length rested on continuous elastic foundation. However, some specialists consider the model to be rather ideal for railways in terms of track and the technical state of track. Calculation of track as a bar rested on numerous elastic supports with variable characteristics of stiffness under static loads has shown that application of methods of elastic foundation gives results understated by 17-24%. The study presents mathematic models of the vehicle/track dynamic system, and a design scheme of track presented as a bar on numerous elastic dissipative supports with non-linear characteristics, which is taken on the base of this system. The authors developed models and methods to define the reduced vertical stiffness of the track in the wheel/rail contact point, which considers rail elastic and geometric characteristics, stiffness of supports, distance between supports and distributed track mass. The value of stiffness is variable by time for each wheel at any time and various for the vehicle's wheels. The mathematical model proposed has been implemented in Matlab software and will make it possible to conduct numerical research into the track/vehicle dynamics.

Introduction

An investigation into rail track operation under loads from the rolling stock by applying numerical methods makes it possible to obtain information on the impact on track and features of the deflected mode of the whole structure in various operational conditions at much lower costs if compared with conducting experiments. Numerical methods make it possible to specify the input design data, which is virtually impossible in experiments.

However, numerical methods can only be implemented when mathematical models adequately reproduce the actual operation of both the vehicle/track dynamic system and track structure. It refers entirely to simulation of the track/vehicle interaction. Thus, simulation of the track/vehicle interaction is an urgent task.

* Corresponding author: ppx_xiit@kart.edu.ua

The analysis of studies and publications

On the basis of theoretical research conducted by E.M. Bromberg, M.F. Verigo, A.Ya. Kogan, G.M. Shahunjanc [1-4] and others mathematical models of the track/rolling stock interaction and the deflected mode of track based on that design scheme were developed. A large number of experiments have revealed that for main line railways the theoretical and experiment findings have very good convergence.

However, some specialists (K.D. Belykh [5], N.S. Nikerov [6] and others) consider this model to be rather ideal for railways in terms of track and the technical state of the track. V.V. Rybkin when calculating track as a bar rested on numerous elastic supports with variable characteristics of stiffness under static loads showed that methods of elastic foundation when applied gave results understated by 17-24% [7].

The study [8] gives mathematic models of the dynamic vehicle/track system on the base of which a design scheme of track as a bar on numerous elastic dissipative supports with non-linear characteristics was taken. Implementation of these models in Matlab software allows us to define spatial forces from rolling stock on track.

However, these models do not consider the inertial characteristics of track, which limits their application by cases when speeds do not exceed 40 km/h. The mentioned conditions exist, for example, on non-public railways.

The general statement of the problem

The study deals with solution to the problem of absence at present a mathematical model of track as a bar rested on numerous elastic and dissipative supports with consideration of the inertial track characteristics. The development of such a model will considerably increase the accuracy of calculation. The objective of the research is development of a mathematical model of track as a bar on elastic supports, which is inertial under moving loads from the rolling stock.

The basic part

The study [9] presents the differential equation of transverse vibrations of bars under the influence of transverse loads changing by the harmonic law:

$$y(x) = y_0 Asx + \frac{y_0'}{s} Bsx - \frac{M_0}{s^2 EI} Csx - \frac{Q_0}{s^3 EI} Dsx - \frac{q}{s^4 EI}\left(A_{sx} - 1\right), \quad (1)$$

where y_0, y_0', M_0, Q_0 are the initial parameters: bend, turning angle, bending moment, transverse force at the origin of coordinates, respectively;

Asx, Bsx, Csx, Dsx are Krylov functions;

E is the coefficient of elasticity;

I is the moment of inertia of the cross-section;

s is the characteristic number defined by the ratio:

$$s^4 = \frac{m\theta^2}{EI}, \quad (2)$$

where m is the distributed track mass involved in vibration;

θ is the vibration frequency.

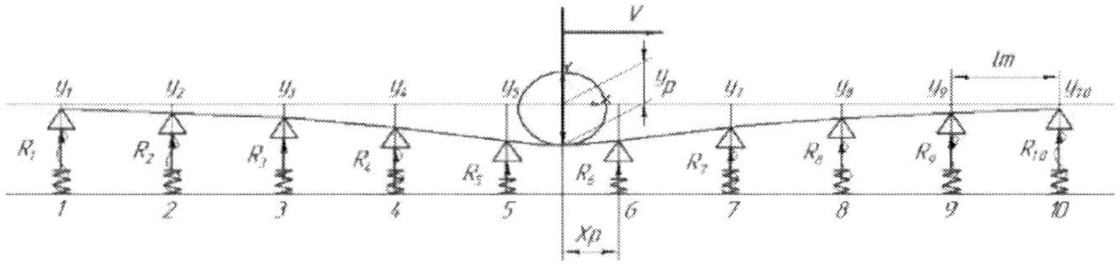

Fig. 1. The design scheme of track on elastic supports under the moving force

Let us consider track as a design scheme presented in Fig.1. A large amount of research, the results of which are presented, for example, in [8], has shown that a bending wave of the rail does not exceed 2.5m from the load application point. Therefore, let us consider the track as a bar rested on ten supports. Under the influence of concentrated force moving along the track equation (1) is generally presented as:

$$y(x)=\frac{1}{s^3EI}\big(-R_1Ds(x-x_1)-R_2Ds(x-x_2)-R_3Ds(x-x_3)-R_4Ds(x-x_4)-R_5Ds(x-x_5)- \qquad (3)$$
$$-R_6Ds(x-x_6)-R_7Ds(x-x_7)-R_8Ds(x-x_8)-R_9Ds(x-x_9)-R_{10}Ds(x-x_{10})+PDs(x-x_P)\big),$$

where R_i is the reaction of the supports;
P is the wheel load;
x_i is the distance from each support to the cross-section under consideration.

$$\left.\begin{aligned}
&y(1)=\frac{1}{s^3EI}(-R_1Ds(l-l)-R_2Ds(l-2l)-R_3Ds(l-3l_{ш})-R_4Ds(l-4l)-R_5Ds(l-5l)-\\
&-R_6Ds(l-6l)-R_7Ds(l-7l)-R_8Ds(l-8l)-R_9Ds(l-9l)--R_{10}Ds(l-10l)+PDs(l-x_P))\\
&y(2)=\frac{1}{s^3EI}(-R_1Ds(2l-l)-R_2Ds(2l-2l)-R_3Ds(2l-3l)-R_4Ds(2l-4l)-R_5Ds(2l-5l)-\\
&-R_6Ds(2l-6l)-R_7Ds(2l-7l)-R_8Ds(2l-8l)-R_9Ds(2l-9l)-R_{10}Ds(2l-10l)+PDs(2l-x_P))\\
&y(3)=\frac{1}{s^3EI}(-R_1Ds(3l-l)-R_2Ds(3l-2l)-R_3Ds(3l-3l)-R_4Ds(3l-4l)-R_5Ds(3l-5l)-\\
&-R_6Ds(3l-6l)-R_7Ds(3l-7l)-R_8Ds(3l-8l)-R_9Ds(3l-9l)-R_{10}Ds(3l-10l)+PDs(3l-x_P))\\
&y(4)=\frac{1}{s^3EI}(-R_1Ds(4l-l)-R_2Ds(4l-2l)-R_3Ds(4l-3l)-R_4Ds(4l-4l)-R_5Ds(4l-5l)-\\
&-R_6Ds(4l-6l)-R_7Ds(4l-7l)-R_8Ds(4l-8l)-R_9Ds(4l-9l)-R_{10}Ds(4l-10l)+PDs(4l-x_P))\\
&y(5)=\frac{1}{s^3EI}(-R_1Ds(5l-l)-R_2Ds(5l-2l)-R_3Ds(5l-3l)-R_4Ds(5l-4l)-R_5Ds(5l-5l)-\\
&-R_6Ds(5l-6l)-R_7Ds(5l-7l)-R_8Ds(5l-8l)-R_9Ds(5l-9l)-R_{10}Ds(5l-10l)+PDs(5l-x_P))\\
&y(6)=\frac{1}{s^3EI}(-R_1Ds(6l-l)-R_2Ds(6l-2l)-R_3Ds(6l-3l)-R_4Ds(6l-4l)-R_5Ds(6l-5l)-\\
&-R_6Ds(6l-6l)-R_7Ds(6l-7l)-R_8Ds(6l-8l)-R_9Ds(6l-9l)-R_{10}Ds(6l-10l)+PDs(6l-x_P))\\
&y(7)=\frac{1}{s^3EI}(-R_1Ds(7l-l)-R_2Ds(7l-2l)-R_3Ds(7l-3l)-R_4Ds(7l-4l)-R_5Ds(7l-5l)-\\
&-R_6Ds(7l-6l)-R_7Ds(7l-7l)-R_8Ds(7l-8l)-R_9Ds(7l-9l)-R_{10}Ds(7l-10l)+PDs(7l-x_P)\\
&y(8)=\frac{1}{s^3EI}(-R_1Ds(8l-l)-R_2Ds(8l-2l)-R_3Ds(8l-3l)-R_4Ds(8l-4l)-R_5Ds(8l-5l)-\\
&-R_6Ds(8l-6l)-R_7Ds(8l-7l)-R_8Ds(8l-8l)-R_9Ds(8l-9l)-R_{10}Ds(8l-10l)+PDs(8l-x_P)\\
&y(9)=\frac{1}{s^3EI}(-R_1Ds(9l-l)-R_2Ds(9l-2l)-R_3Ds(9l-3l)-R_4Ds(9l-4l)-R_5Ds(9l-5l)-\\
&-R_6Ds(9l-6l)-R_7Ds(9l-7l)-R_8Ds(9l-8l)-R_9Ds(9l-9l)-R_{10}Ds(9l-10l)+PDs(9l-x_P))\\
&y(10)=\frac{1}{s^3EI}(-R_1Ds(10l-l)-R_2Ds(10l-2l)-R_3Ds(10l-3l)-R_4Ds(10l-4l)-R_5Ds(10l-5l)-\\
&-R_6Ds(10l-6l)-R_7Ds(10l-7l)-R_8Ds(10l-8l)-R_9Ds(10l-9l)-R_{10}Ds(10l-10l)+PDs(10l-x_P)
\end{aligned}\right\} \qquad (4)$$

By using the basic equation (3), equations were compiled for each of the ten supports under the influence of the force P.

In these equations, the distance from the nearest support to the load application point is defined as:

$$x_P = 5l + \left(1 - \left\{\frac{x_{ni}}{l}\right\}\right) \cdot l\ , \tag{5}$$

where $\left\{\frac{x_{ni}}{l}\right\}$ is the fractional part of the ratio $\frac{x_{ni}}{l}$;

System (4) can be presented as a matrix:

$$A \cdot y = H\ , \tag{6}$$

where A is the matrix of elastic and inertial characters of the track;
y is the matrix of unknown deformations in the supports;
H is the external load matrix;

The matrix of elastic and inertial characteristics of the track is of the form.

$$\left\{\begin{matrix}
s^3EI & cD(l-2l) & cD(l-3l) & cD(l-4l) & cD(l-5l) \\
cD(2l-l) & s^3EI & cD(2l-3l) & cD(2l-4l) & cD(2l-5l) \\
cD(3l-l) & cD(3l-2l) & s^3EI & cD(3l-4l) & cD(3l-5l) \\
cD(4l-l) & cD(4l-2l) & cD(4l-3l) & s^3EI & cD(4l-5l) \\
cD(5l-l) & cD(5l-2l) & cD(5l-3l) & cD(5l-4l) & s^3EI \\
cD(6l-l) & cD(6l-2l) & cD(6l-3l) & cD(6l-4l) & cD(6l-5l) \\
cD(7l-l) & cD(7l-2l) & cD(7l-3l) & cD(7l-4l) & cD(7l-5l) \\
cD(8l-l) & cD(8l-2l) & cD(8l-3l) & cD(8l-4l) & cD(8l-5l) \\
cD(9l-l) & cD(9l-2l) & cD(9l-3l) & cD(9l-4l) & cD(9l-5l) \\
cD(10l-l) & cD(10l-2l) & cD(10l-3l) & cD(10l-4l) & cD(10l-5l)
\end{matrix}\right.$$

$$\left.\begin{matrix}
cD(l-6l) & cD(l-7l) & cD(l-8l) & cD(l-9l) & cD(l-10l) \\
cD(2l-6l) & cD(2l-7l) & cD(2l-8l) & cD(2l-9l) & cD(2l-10l) \\
cD(3l-6l) & cD(3l-7l) & cD(3l-8l) & cD(3l-9l) & cD(3l-10l) \\
cD(4l-6l) & cD(4l-7l) & cD(4l-8l) & cD(4l-9l) & cD(4l-10l) \\
cD(5l-6l) & cD(5l-7l) & cD(5l-8l) & cD(5l-9l) & cD(5l-10l) \\
s^3EI & cD(6l-7l) & cD(6l-8l) & cD(6l-9l) & cD(6l-10l) \\
cD(7l-6l) & s^3EI & cD(7l-8l) & cD(7l-9l) & cD(7l-10l) \\
cD(8l-6l) & cD(8l-7l) & s^3EI & cD(8l-9l) & cD(8l-10l) \\
cD(9l-6l) & cD(9l-7l) & cD(9l-8l) & s^3EI & cD(9l-10l) \\
cD(10l-6l) & cD(10l-7l) & cD(10l-8l) & cD(10l-9l) & s^3EI
\end{matrix}\right\} \tag{7}$$

The other matrices are of the form:

$$y = \begin{Bmatrix} y_1 \\ y_2 \\ y_3 \\ y_4 \\ y_5 \\ y_6 \\ y_7 \\ y_8 \\ y_9 \\ y_{10} \end{Bmatrix}, (8) \quad H = \begin{Bmatrix} P(t)Ds(l-x_p) \\ P(t)Ds(2l-x_p) \\ P(t)Ds(3l-x_p) \\ P(t)Ds(4l-x_p) \\ P(t)Ds(5l-x_p) \\ P(t)Ds(6l-x_p) \\ P(t)Ds(7l-x_p) \\ P(t)Ds(8l-x_p) \\ P(t)Ds(9l-x_p) \\ P(t)Ds(10l-x_p) \end{Bmatrix}, (9) \quad \{R_i\} = \begin{Bmatrix} R_1 = y_1 \cdot c \\ R_2 = y_2 \cdot c \\ R_3 = y_3 \cdot c \\ R_4 = y_4 \cdot c \\ R_5 = y_5 \cdot c \\ R_6 = y_6 \cdot c \\ R_7 = y_7 \cdot c \\ R_8 = y_8 \cdot c \\ R_9 = y_9 \cdot c \\ R_{10} = y_{10} \cdot c \end{Bmatrix}. \quad (10)$$

The number of unknown support bends in system (4) is equal to the number of equations; therefore matrix A of equation (6) is square with ten lines and columns. Let us indicate as Δ the determinant of matrix A (7), as Δ_j the determinant, which is obtained from Δ by deleting the jth column (j=1, j=2 ... j=10) and insert instead the column of absolute terms b_j of matrix H (9). By Cramer's rule, the elements y_j of matrix (8) can be defined as:

$$y_j = \frac{\Delta_j}{\Delta}. \quad (11)$$

This equation, as well as expressions (7), (8) and (9), allows calculating elastic depressions of supports and reactions in them (matrix 10).

The vertical stiffness of track in the wheel/rail contact point can be defined on the basis of the following considerations. By taking the external load as equal to 1, let us define elastic depressions of all supports R_i under the unit force of equation (11).

The value of rail depression in the wheel/rail contact point can be defined as:

$$y_e = y_5 \frac{x_p}{l} + y_6 \frac{(l-x_p)}{l} + \frac{(l-x_p)^2 x_p^2}{3EIl}. \quad (12)$$

So the stiffness required is equal to:

$$C_b = \frac{1}{y_e}. \quad (13)$$

Conclusions

The authors propose the methods to define the reduced vertical stiffness of track in the wheel/rail contact point, which allows consideration of elastic and geometric rail characteristics, stiffness of supports, distance between supports and distributed track mass. The stiffness value is variable by time for each wheel at any time and various for the vehicle's wheels. The mathematical model proposed was applied in Matlab software.

References

1. V.G. Al'breht, E.M. Bromberg, N.B. Zverev, *Besstykovoj put'*, 206 (1982).
2. E.M. Bromberg, M.F. Verigo, V.N. Danilov, *Vzaimodejstvie puti i podvizhnogo sostava*, 280 (1956).
3. G.M. Shahunjants, *Zheleznodorozhnyj put', Monografija*, 497 (1987).
4. M.F. Verigo, A.J. Kogan, *Vzaimodejstvie puti i podvizhnogo sostava*, 589 (1986).
5. K.D. Belyh, M.K. Umanov, G.N. Malyshko, Metallurgicheskaja promyshlennost', **5**, 78-79 (1976).
6. N.S. Nisirov, Trudy LIIZhT, **416,** 26-35 (1977).
7. V.I. Klimov, V.V. Rybkin, Trudy DIITa, **32(244)**, 72-84 (1987).
8. O.M. Darens'kij, *Teoretichni ta eksperimental'ni doslizhennja roboti zaliznichnih kolij promislovogo transportu*. Kharkiv, 204 (2011).
9. O.M Darens'kij, Y.S. Leibuk, Information and control systems at railway transport. (to be published).
10. K.Wang, C.Huang, W. Zhai, P. Liu, S. Wang, Journal of Traffic and Transportation Engineering (English Edition), **1**, 209–220 (2014).
11. S. Dailydka, Transport, **(3)23**, 236–239 (2008).
12. J. Otero, Journal of Rail and Rapid Transit, **1(226)**, 62–71 (2011).

Software analysis for modeling the parameters of shunting locomotives chassis

Anatoliy Falendysh[1*], *Mykyta* Volodarets[1], *Viktoria* Hatchenko[1], and *Ivan* Vykhopen[1]

[1]Ukrainian State University of Railway Transport, Operation and Repair of traction rolling stock Department, Feuerbach square 7, 61050, Kharkiv, Ukraine

Abstract. The article provides an overview of software designed to perform the simulation of structures, calculate their states, and respond to the effects of loads applied to any of the points in the model. In this case, we are interested in the possibility of modeling the locomotive chassis frames, with the possibility of determining the weakest points of their construction, determination of the remaining life of the structure. For this purpose, the article presents a developed model for calculating the frame of the diesel locomotive chassis, taking into account technical, economic and other parameters.

1 Introduction

Because of the high degree deterioration of the Ukrainian railways traction rolling stock (more than 90%) on the one hand, and the lack of free investment resources for modernization, on the other hand, and production capabilities of manufacturing plants, tasks to ensure the safe operation of traction rolling stock fleet cannot be solved by its instant updating [9]. Accordingly, searching for a better use of qualitative characteristics of the available operated machinery is necessary, including through the achievement of safe operation of the traction rolling stock beyond the established standard service life.

The condition of the operated fleet of traction rolling stock on industrial transport is hardly better, requiring no less expert studies to assess the remaining resource with the possibility of extending the service life. And every year, this need increases at times.

If now enterprises are prohibited to operate locomotives that have already outlasted the standard service life, it is unclear how they can continue their activity. The situation will be same for Ukrzaliznytsya, except for locomotives of M62 and ChME3 series, which make up a significant part of its operated fleet, the service lives of has already been successfully extended.

Despite this, an important area of research for scientists is further development and improvement of effective control of the traction rolling stock resource which will significantly increase the time of its safe operation. Especially, increasing the role of assessment of the technical state of the base parts of locomotives in order to determine the

*Corresponding author: fap_hiit@ukr.net

remaining resource and the possibility of its further operation, its further development and improvement of methods and means.

The possibility of further operation of diesel locomotives after expiration of their standard service life is due, firstly, to a sufficient margin of strength and reliability of the equipment, and secondly, to the fact that most of the mechanism components are suitable for repair with the possibility of restoring the resource.

Standard service life for each locomotive type and series is set by the manufacturer on the based on the reliability of its basic construction parts, provided it will be operated intensively and maintained in accordance with the standards.

However, the conditions of locomotive operation and repair basically differ from those offered.

The standard resource depends significantly on how the locomotive was operated and maintained throughout its "life", and whether or not it got into emergency situations.

Thus the important components of the examination in terms of technical diagnosis are:

—assessment of intensity and operating conditions;

—analysis of the faults accumulated during operation;

— assessment of the training level of the personnel engaged in operation;

—assessment of the conformity of the locomotive repair system to the industrial standards.

The combination of all these factors is individual for each locomotive, but only analyzing them in aggregate allows to predict most accurately the value of the remaining life of the locomotive.

One of the bases is the chassis of the locomotive. Creating the most complete calculation model that takes into account a multitude of factors is necessary to perform the qualitative and efficient calculation of its parameters. The task can be solved only using a computer.

2 Objective of work

Analysis of existing methods for simulating the chassis of rolling stock, and the technical means that are used for this purpose with the aim of constructing a model for calculating the chassis parameters of TGM4 and TEM2diesel locomotives, to perform work on modeling the strength and reliability of their chassis parts. The main requirement for the model is taking into account in the calculation such important parameters, as performance characteristics; technical specifications; economic and safety parameters [10].

3 Main part

The procedure for determining the residual life and service life of a locomotive is a large complex of measures, including:

a) analysis of technical documentation and information on operating conditions of locomotives;

b) determination of technical condition, collection and analysis of operational data on damage to the base parts of the locomotive series;

c) dynamic strength tests of the locomotive;

d) bench tests of the base parts of the locomotive, their fragments or samples of materials to assess their fatigue resistance;

e) computational studies of the stress-strain state, fatigue life and survivability of the base parts of the locomotive;

f) choosing criteria and making conclusions about the extent of the residual resource;

g) development of recommendations on design and technological measures to ensure the necessary resource;

h) deciding on the termination or continuation of the operation with the establishment of an additional service life.

Schematically, the algorithm for determining the residual life and service life of the traction rolling stock and the general rolling stock is shown in Figure 1.

The use of modern technologies makes it possible to simplify and speed up the process of performing certain procedures included in the complex for determining the residual life of rolling stock, by eliminating the need to perform dynamic strength tests of the locomotive, bench tests of the base parts of the locomotive, their fragments and material samples to assess their fatigue resistance. All this is achieved by modeling, and in addition to the above benefits, there is a significant reduction in the cost of implementation of these procedures.

Simplification, acceleration and increase of accuracy in the calculation of the study of stress-strain state, fatigue life and survivability of locomotive base parts, the selection of criteria and the formation of conclusions about the magnitude of the remaining resource using a computer.

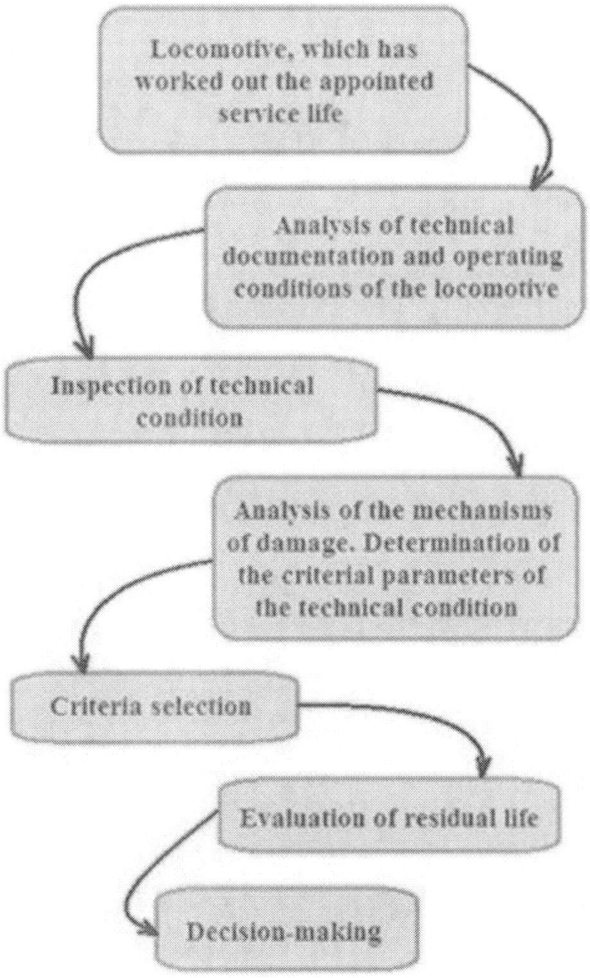

Fig. 1. Block diagram of the procedure for determining the remaining service life extension.

By this time, a number of software programs have been developed enabling to perform a multivariate analysis of calculations taking into account all the loads to obtain sufficiently accurate results. In most calculations, the frame of locomotive chassis is adopted in the form of a rod system. As the practice has shown, in the areas of the junction of the frame beams, in the areas of attachment of brackets, overlays and in other nodes of complex shape, the stresses cannot be determined with the help of a bar chart [1]. In this regard, it is advisable to use modern software and methods of calculation, primarily the finite element method, the application of which will improve the accuracy of calculations and determine the stresses in elements of complex shapes, primarily in the joins of the beams and in the areas where individual parts are attached to the bearing elements.

In work [2], the stressed strain state of the lateral frame of the freight car was investigated using the finite element method with the Siemens NX package-CAD software. For the calculation, the construction of the lateral frame of the 4th execution of the stamped TsNIIKhZO car was used. The method of geometric modeling was used, which was to specify the geometric limits of the model, the dimensions of the elements and to assign the necessary forms of control. In the form of a solid state finite element, a tetrahedron was used (figure 2), because the actual construction is a solid in three-dimensional space. The frame model with an attachment of a tetrahedral grid 7 mm long was divided into 336,416 elements. The number of grid nodes is 104,565. On Figure 3, simulating the calculation of the static space of the deformation theory was carried out in package-CAE software of the Siemens NX. The sections of stresses, displacements and deformations were considered and numerical results were obtained in the form of graphs, tables and diagrams. The result of the calculation showed the accumulation of high stresses in the area of technological hollow and the technological opening of the side frame of the trolley, which precisely determines the "dangerous zones" of the structures, the increase in rigidity of which will result in increasing the strength of the side frame of the freight car.

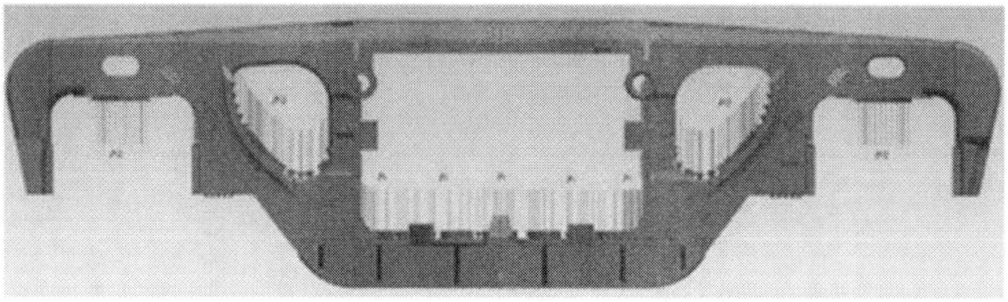

Fig. 2. Use of a tetrahedron as a solid-state finite element.

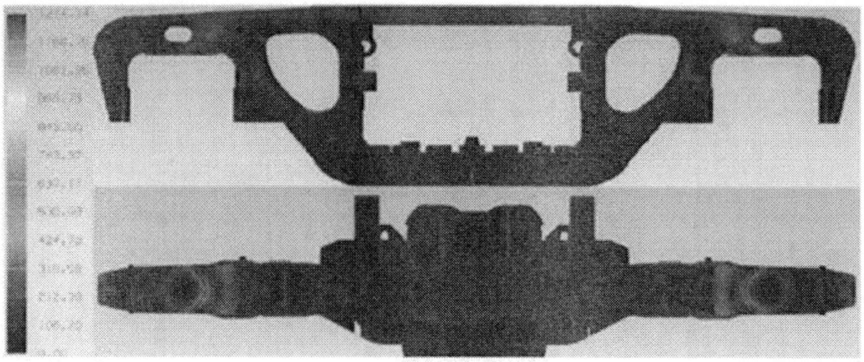

Fig.3. Modelling the calculation of the static space theory of deformation in the package-CAE software of Siemens NX.

In work [3], the task was to develop and verify the procedure for calculating the strength, cast side frames of freight car chassis, taking into account internal defects. The proposed methodology consists of the following main stages: development of initial data for constructing a solid model of a new side frame design; construction of a solid model of the side frame without casting defects and calculation for strength; development of technology and simulation of casting processes when pouring the side frame, determining areas and sizes of internal defects; construction of a solid model of the side frame with internal casting defects, and calculation of strength; development of drawings of the parts and casting of the side frame; manufacturing.

Modern MAGMA software complexes allow determining the areas of occurrence of internal casting defects [4], but their dimensions, distances from the casting surfaces to the defect can be estimated only conditionally, not according to the solid model, but from the JPEG image (fig. 4).

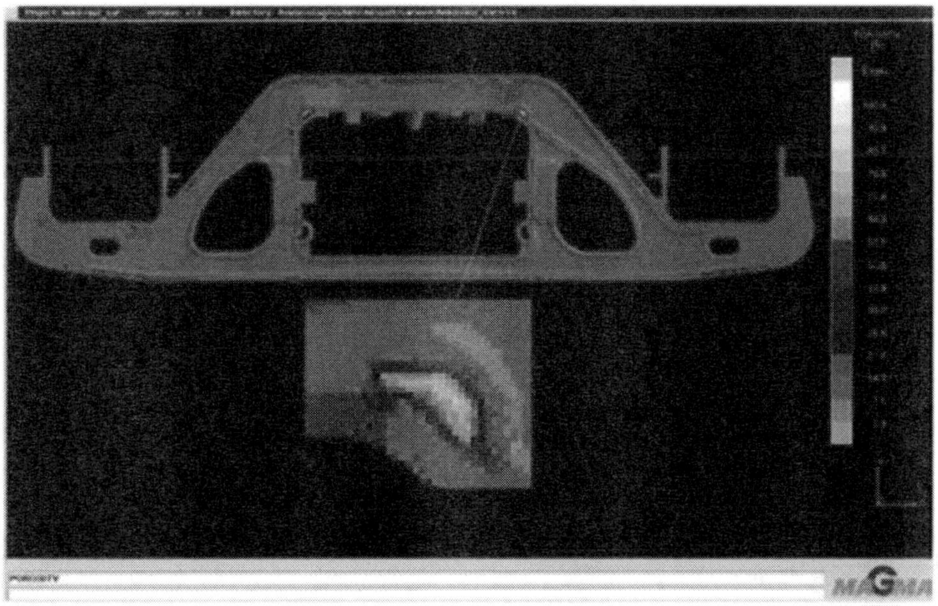

Fig. 4. The distribution of internal casting defects in the longitudinal section of chassis side frame 18-100, determined in MAGMA software.

The results of simulation of casting processes in MAGMA with subsequent strength calculations of the lateral frame with defects cannot be used directly, MAGMA does not picture defects in the solid model. However, the areas of possible location of defects, the type of defect, the nature of its boundaries (sharp edges, rounded forms) can be predicted.

The influence of internal casting defects with sharp peaks are estimated most correct by the methods of fracture mechanics. The technique [4] uses an algorithm for constructing an internal crack with a small radius at the vertex and determining the stress intensity factor (KIN) for KI separation implemented in the finite element ANSYS package version 14.5. Since this software allows to define a semi-elliptical crack only on the surface, an internal crack inside the side frame element directly cannot be constructed, then an alternative method for modeling the internal acute defect is used. In ANSYS 14.5, the lengths of the semi-axes and the radius at the vertex are specified. The finite element mesh for the side frame and for the created crack are automatically built in the Crack module, the boundary conditions are set and the calculation is started. The result of the calculation are the KIN KI values at the crack tip (fig. 5), the Mises stresses.

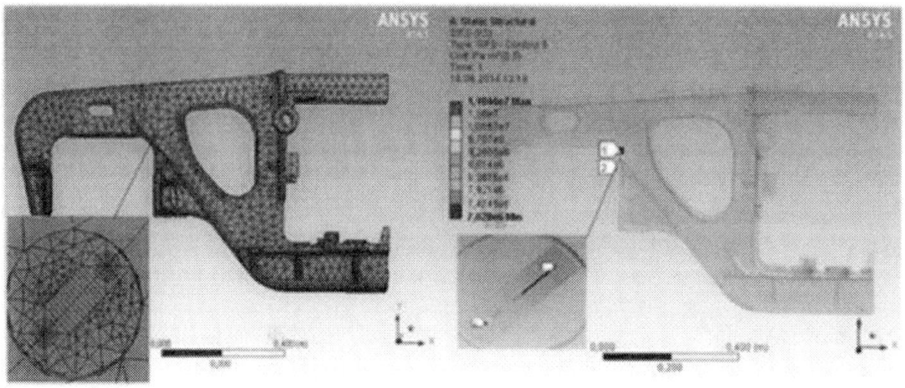

Fig. 5. The finite element grid at the top of the crack (a) and the results of calculating KI (b) from the effect of the vertical load on the side frame 18-100.

In work [5], an FEM was implemented in the SolidWorksSimulation software package for analysis of the stress-strain state, in the editor of which a calculation model was created. The application is focused on the preparation of a full-fledged finite element model with maximum capabilities, taking into account the geometric features, strength character and the performance of various types of calculations. Calculation of the force factors effecting the chassis was carried out according to [6]. At generating a grid of finite-element idealization of the object of investigation, volume finite elements in the form of a tetrahedron were used. The model consists of 46,155 elements and 90,898 nodes. The results of calculating the stress values in the frame of the diesel locomotive chassis under static loading are shown in Fig. 6.

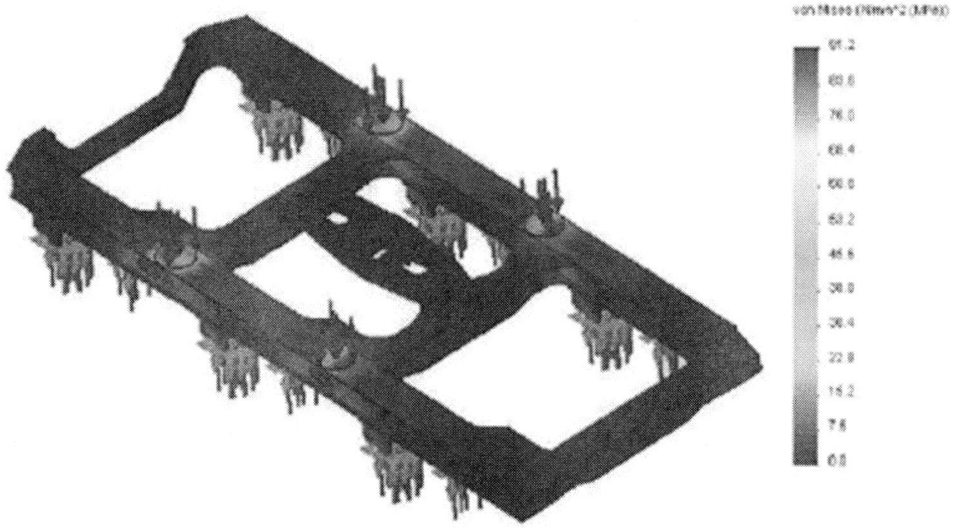

Fig. 6. Results of calculation of stress values in the frame of the diesel locomotive chassis under static loading in the SolidWorksSimulation software package.

Currently, in the software market, this direction is represented by a rather large number of universal software, such as SYMPACK, MSC. ADAMS, LMS. DADS and applications focused on specific objects (MEDYNA, NUCARS and Vampire are aimed at modeling the dynamics of rail vehicles) [7]. All applications of this type automate the process of forming the motion equations of a specific mechanical system based on the description of inertial, geometric and kinematic parameters, models of force interactions selected or specified by the user. To further investigate the dynamics of an object, numerical methods for analyzing the equations of motion are used, for example, numerical integration. The Universal Mechanism software complex (UM) [8], developed in the laboratory of computing mechanics of the Bryansk State Technical University has become quite widely used.

Currently, the complex includes a powerful universal core that meets all modern requirements and a number of specialized modules for simulating the dynamics of cars, railway chassis, caterpillar vehicles, optimization modules, calculating the longevity (figure 7), etc. At the same time, the UM complex has interfaces with other software products (the elastic body modelling module supports the import of data from ANSYS and MSC. NASTRAN, the UMControl module imports models from Matlab / Simulink, the module for importing 3D models from CAD programs now supports SolidWorks, KOMPAS, AutodeskInventor and Pro/ENGINEER).

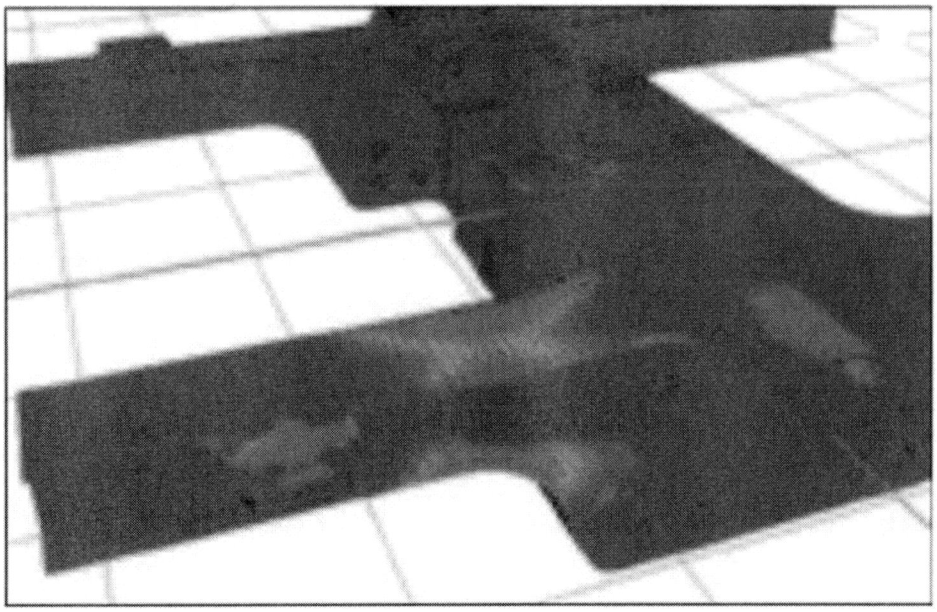

Fig. 7. Results of calculating the longevity of the locomotive chassis frame in the "Universal Mechanism" PC.

In the models considered in the article, during performing calculations, the complex of technical, operational; economic and security parameters is not taken into account.

Taking into account the above parameters, which must be allowed for in calculations, a more detailed model of traction rolling stock chassis technical condition and a separate calculation model have been developed. Mathematically, it has the form

$$P = P1 + P2 + P3 + P4, \qquad (1)$$

where*P1* — is the set of technical parameters;
P2 — is the set of operation parameters;
P3 — set of economic parameters;
P4 — is the set of security parameters.

For more detailed explanation of calculations structure in the developed model, Fig. 8 schematically shows the sequence of the procedure for calculating the parameters of locomotive chassis.

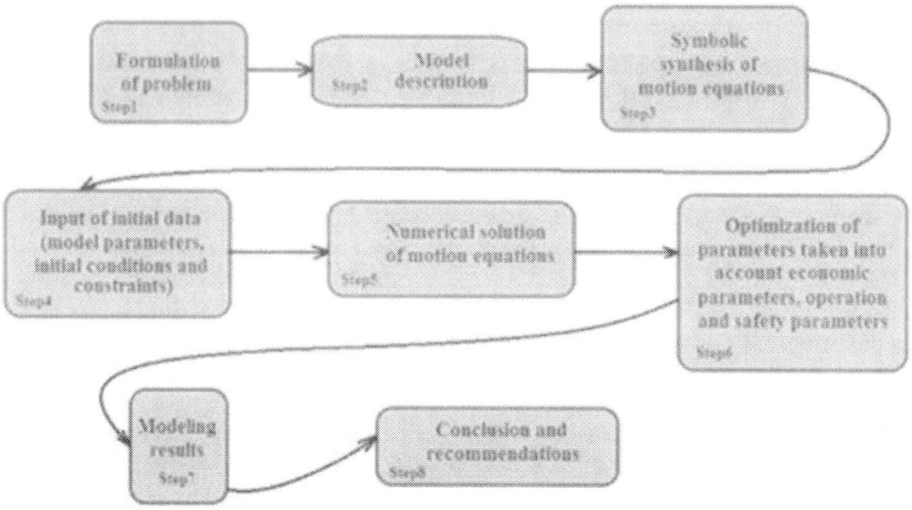

Fig.8. The procedure for calculating the calculation parameters for rolling stock chassis.

The initial step is defining the tasks, in our case this is determination of the residual resource of chassis of TGM4 and TEM2shunting locomotives, in particular, determination of the areas of heaviest impact able loads causing significant loss of strength, which leads to failure of the entire structure and its destruction.

The next step is the implementation of the model description, careful detailing of the design features of diesel locomotive chassis, in this case, TEM2 and TGM4shunting locomotives.

After the symbolic synthesis of the equations of locomotive chassis motion is implemented which implies a painstaking process of combining different parameters characterizing the process of locomotive chassis movement into a structured set. Here, all the features of the process are taken into account, such as acceleration, loads from unevenness of the track, loads from the interaction of the locomotive itself with the elements of the trolley and so on.

The next step is one of the most important in the model, when the input of initial data is performed, such as model parameters, its main dimensions, strength values of materials from which the chassis was manufactured, their strength characteristics etc., initial conditions and values of limitations exceeding. It is inadmissible for the elements of the chassis structure.

Next, is a numerical solution of the motion equations of the bogies is made based on all the initial data. The calculation is carried out automatically by the software.

After the optimization of intermediate results, the equations the locomotive chassis movement is calculated in accordance with economic parameters, operation parameters and, most importantly, safety of their operation as one of the most important requirements. The final output of the results simulation of locomotive chassis is carried out on the basis of which further conclusions and recommendations are made regarding the technical condition, the remaining life, the strength and so far of the simulated locomotive chassis.

4 Conclusions

1. The work analyzes the existing software for simulating the parameters of rolling stock chassis.

2. A calculation model for chassis parameters for TEM2 and TGM4 diesel locomotives was created allowing for technical, operation, economic and safety parameters.

3. Further, a model of TGM4 and TEM2 locomotive chassis should be compiled in order to carry out the corresponding calculations in ANSYS and SolidWorks software.

References

1. A.Beljaev, B.Bunin, S.Golubjatnikov. *Increasing the reliability of locomotive chassis part*, Moscow (1984)
2. G.Alushaeva, S.Duzel'baev, A.Omarbekova. SWorld magazine, **10**, 1-12 (2013)
3. J.Ruzmetov, A.Jakushev, S.Komichenko. Internet journal "NAUKOVODENIE" **3(5-6)**, 12-18(2014)
4. S.Glebov, A.Jakushev. Journal of Transport of the Russian Federation, **2**, 63– 5, (2011)
5. N.Zajniddinov. Izvestiya St. Petersburg State University of Communications, **3**, 12 (2010)
6. *Norms for calculating and assessing the strength of load-bearing elements, dynamic qualities and the impact on the path of the crew part of locomotives of the railways of the Russian Federation, gauge 1520 mm*, Moscow (1998)
7. JohnMcPhee. Multibody System Dynamics: Research Activities.
8. *Universal Mechanism: Software Lab.*

The influence of external field on the lubricity of mineral oil for railway transport

Serhii Voronin[1*], *Pavlo* Konovalov[1], *Ivan* Safoniuk[1], and *Oleksandr* Kebko[1]

[1]Ukrainian State University of Railway Transport, Department of Construction, track and handling machines, Feuerbach sq., 7, Kharkov, Ukraine, 61050

Abstract. The use of mineral oil is associated with its gradual operational degradation caused by its natural aging and contamination with various impurities. As the concentration of impurities increases, the number of active surface molecules which determine the operational properties of mineral oils decreases. A promising method of recovery of the operational properties of oils is the treatment with an electric field, which makes it possible to enhance the activity of surfactants in the tribo-contact area. This statement is proved through the improvement of the wettability of the bronze surface with mineral oils after their treatment with an electrostatic field. However, the method of electrical treatment is associated with the need to increase the requirements for the purity of liquids, especially to the presence of water, which requires creating an oil pre-treatment system. As an alternative, a method of electrical treatment with special field parameters is proposed enabling to accelerate the coalescence process. The major parameter that accelerates the coalescence process is the electric field oscillation frequency. The results of the study give grounds for choosing the optimal field parameters.

1 Introduction

The abrasion resistance of a friction pair depends on a number of performance characteristics of the lubricants with the universal decimal classification applied. One of the most important is their lubricity [1], which, in turn, is characterized by the ability of oils to form strong interfacial lubricating films on the surface of a friction pair [2, 3]. These properties of oils are estimated based on their ability to wet the friction surface. However, oil loses its wetting ability during operation as a result of the decrease in the concentration of polar molecules caused by the formation of molecular aggregates in the oil volume, the ingress of wear products, dust, and water. The existing cleaning methods, which include, among others, filtration, centrifugation, and sedimentation, are quite an energy- and time-consuming. Therefore, the search for new effective methods to support the operating properties and purity of mineral oils remains an urgent task.

1.1 Analysis of recent studies and publications

* Corresponding author: voronin.sergey@ukr.net

Mineral oil is a liquid isotropic dielectric, consisting of polar and nonpolar molecules, which arbitrarily move in volume. As they get into the external electric field, the oil is polarized [4, 5]. As a result, the oil molecules in the volume become someway ordered, which consists in the fact that the dipole moments of the molecules, under the action of the external field, pass from random to organized orientation, close to equally directed, and pass from the micellar to the single state. An increase in the number of single molecules in the oil volume leads to the formation of a stronger molecular layer on the working surfaces.

1.2 Statement of the objective and tasks of the study

The objective of the study is to establish the possibility of applying EST to mineral oils used in the operation of construction and track-laying machines.

To achieve the set objective, the following tasks were solved:
- to determine the influence of the electrostatic field on the process of the metal surface wetting as an indirect indicator of the tribological properties of mineral oils;
- to propose a method of purification of the mineral oil before EST;
- to determine the dependence of the water droplet coalescence rate on the electric field parameters.

2 The basic part of the study

One of the tasks of our paper is an experimental study of the wetting properties of axial and industrial oils with respect to the bronze surface. The experiment was conducted with the use of a part made of bronze (GOST 613-79). The surface roughness of the used part is similar to the roughness of the friction surface of the axial bearing liner R_z^b =7.5 μm [6].

Before each experiment, the surface of the part was degreasared with 96% alcohol solution and treated mechanically with coal dust powder, according to recommendation [7]. The surface of the part was maintained in a horizontal state.

The studies were conducted with axial oil of grade "C", and with industrial oil I-30A. Each of the experiments compared the parameters of the wetting properties of the oil before and after its treatment with an electrostatic field of strength E =0.5×10^6 V/m.

The dynamics of the wetting process was estimated using the wetting factor [8]

$$k_{var} = \frac{(S_t - S_0)}{S_0}, \quad (1)$$

where S_0 – the surface area occupied by an oil drop at the time of its application;
S_t – the surface area of the same drop 300 seconds after observation.

The experiments were carried out at an oil temperature of (18-19) °C. The diameter of the oil drops at the time of application onto the part surface was 6.5-8.5 mm. The drop sizes during the experiments were recorded with a digital camera and then measured in the graphical program ACAD. Video shooting interval: the first two minutes – after 15 seconds, and the rest period – after 30 seconds.

Diagrams of the relative variation in wetting factor for axial and industrial oils with or without electrostatic treatment (EST) are shown in Fig. 1:

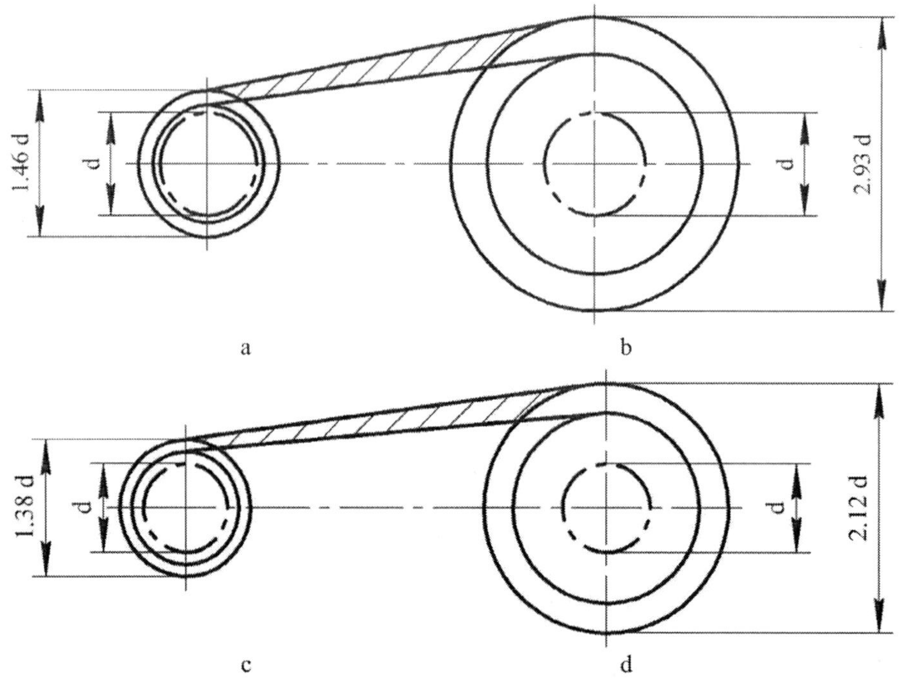

Fig. 1. The variation diagram of factor k_{var} during 300 seconds for: a – axial oil without EST; b – the same with EST; c – industrial oil without EST; d – the same with EST.

As can be seen from Fig. 1, EST of axial and industrial oils leads to activation of the process of the metal surface wetting. The wetting factor for axial oil after EST was 2.93 against 1.46 before treatment, and for the industrial oil – 2.12 against 1.38 before EST, respectively.

Following on from the conducted studies, we can conclude about the effectiveness of the proposed method of improvement of the anti-wear properties of mineral oils, resulting in a slower process of losing their lubricating properties. Activation of the wetting process by EST is explained by an increase in the adsorption potential of axial oil, which is an indirect indicator of the improvement of its anti-wear properties.

On the other hand, the implementation of electrical treatment requires a high electric field strength between the electrodes and, at the same time, prevention of electrical breakdown of the oil [5, 9]. It is known that during operation, the impurities of various nature enter the hydraulic oil, and significantly change its electrical conductivity, especially water. Such a problem makes it impossible to implement EST in many technical systems. Therefore, the implementation of EST in hydraulic drives of transport means should be accompanied by a preliminary removal of water, not settled in tanks, i.e., fine water.

The solution of this problem during the implementation of the on-board EST centers around of the study and creation of energy-saving (cold) methods for the removal of dispersed water. The main task is to remove droplets smaller than 5 μm. For such droplets, the action of the resistance force exceeds the action of gravitational forces, which makes their precipitation impossible. In this case, in order to accelerate the process of water precipitation in oil, it is necessary to create conditions for combining these droplets into larger ones. If a direct contact is necessary for combining the droplets, the main condition for their effective consolidation is as follows:

$$A \geq D - d \quad (2)$$

where A – drop oscillation amplitude;
D – distance between the random adjacent drops;
d – drop diameter.

The distance between the drops depends on the water concentration in the oil and the dispersion degree. The oscillation amplitude, in turn, depends on the geometric dimensions of the droplets, viscosity, density and parameters of the external electric field, namely, its intensity and frequency.

By changing the parameters of the external field we can affect the frequency and amplitude of the oscillations of microdroplets, however, in order to achieve maximum efficiency of the process of removing water from the oils, it is necessary to establish the rational values of the field parameters for each type of oil and the degree of initial dispersion of water.

We conducted a study of the process of water coalescence in a mineral oil environment. The study used an emulsion of industrial oil I-20 GOST 20799-88 (97%) and industrial water (3%). Since the coalescence of droplets is long and does not clearly time-restricted, we took the emulsion clarification time as the indicative time. The process of the emulsion clarification is shown in Fig. 2:

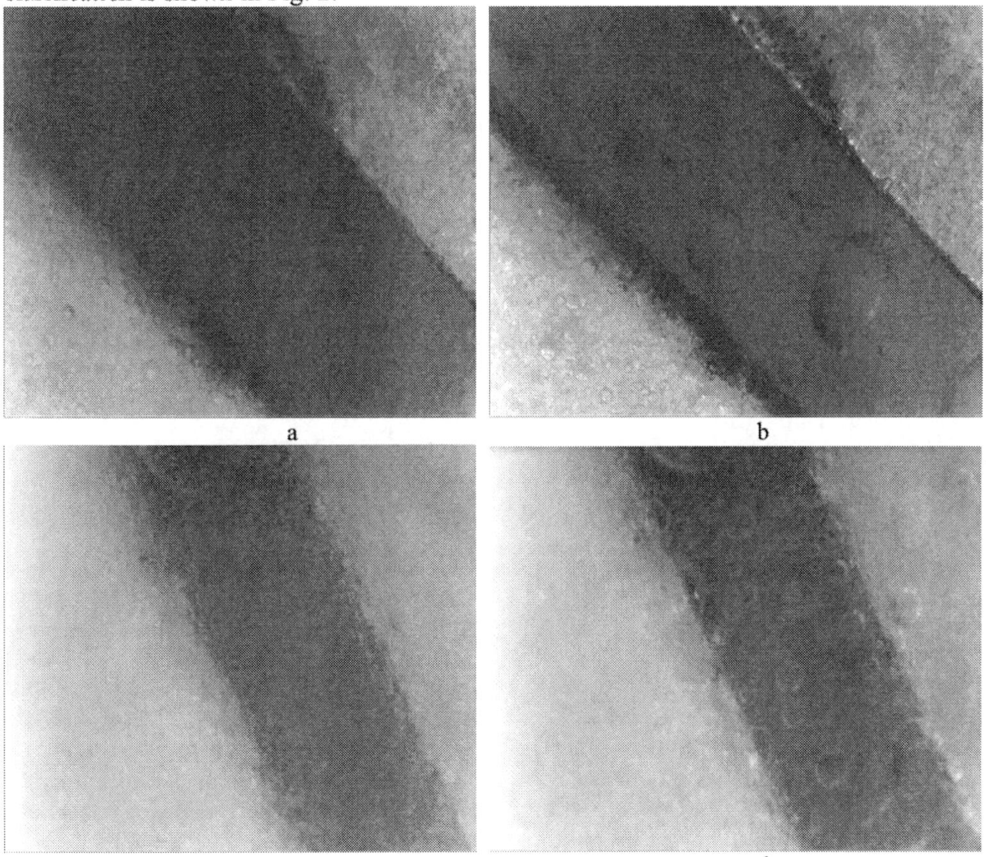

a b

c d

Fig. 2. The effect of electrical field on the emulsion: a – the initial state of the emulsion with non-insulated electrodes, b – clarified emulsion with non-insulated electrodes, c – the initial state of the emulsion with insulated electrodes, d – clarified emulsion with insulated electrodes.

The equipment for determining the emulsion clarification time consists of an optical microscope, a high voltage electrical signal generator, flat electrodes and a stopwatch. The method for determining the emulsion clarification time is as follows. The required oscillation frequency and voltage are set on the frequency generator. A drop of the prepared emulsion is placed on the electrodes of the laboratory unit. A de-oiled and dried 0.17 mm thick glass was used as a dielectric. The microscope eyepiece is focused on the middle of the drop. At the same time, the frequency generator and the stopwatch are turned on. The experiment ends with the clarification of the emulsion and the clarification time is recorded.

3 Results and discussion

The results of the studies show that the frequency of the external electric field has a significant effect on the coalescence of the oil. Based on the calculations, the time to field frequency curves was obtained, Fig. 3. The nonlinear relationship between the emulsion clarification time and frequency is established. At a frequency close to 300 Hz, emulsion clarification begins after 50 seconds and lasts long (about 5 minutes), and at 3 MHz, the clarification begins after 7 seconds and lasts for 9 seconds.

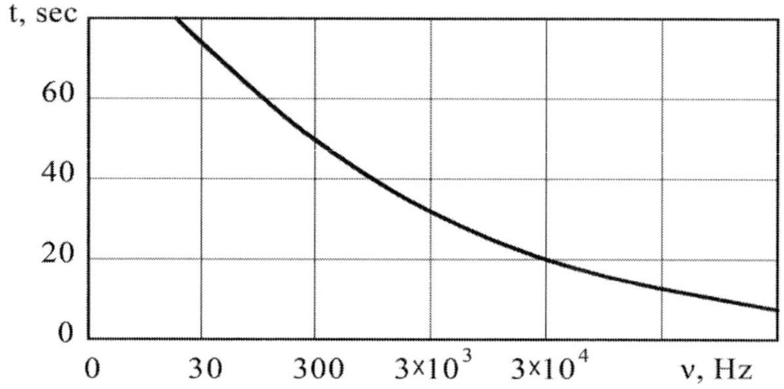

Fig. 3. The relationship between the emulsion clarification time and the oscillation frequency of the external electric field.

Such data allow determining the coalescence rate of water, as well as its variation depending on the frequency of the external field. For example, the relative coalescence rate at a frequency of 300 kHz is 0.02 s^{-1} seconds, and at a frequency of 3 MHz – 0.143 s^{-1}.

4. Conclusion

An essential reserve for the improvement of the anti-wear properties of axial and industrial oils, resulting in their slower loss of lubricity, is electrostatic treatment. The electrostatic treatment of oils intensifies the process of wetting of the metal surfaces. An increase in the intensity of the wetting process is characterized by an increase in the wetting factor for axial oil – by 2.93 times, and for industrial oil – by 2.12 times, respectively.

Activation of the wetting process is explained by the adsorption potential of mineral oil increasing under the influence of EST and the intensification of the process of formation of the interfacial lubricating layer on the friction surfaces, which is an indirect sign of improvement of its lubricating properties.

The possibility of cold removal of dispersed water through the coalescence of the droplets in an external electric field was proved. The results obtained can be used in the hydraulic systems of various transport means as a method for cleaning oil from water, and for preparing oil for EST.

References

1. I.I. Karasik, F.V. Savvin, *Trenie, iznashivanie i smazka. Terminy i opredeleniya* (Izdatyelstvo standartov, Moscow, 1988)
2. S.V. Voronin, O.O. Skoryk, Ye.M. Korostelov, Eastern-European Journal of Enterprise Technologies. **4** (**80**). 11-17 (2016)
3. S.V. Voronin, O.O. Skoryk, V.O. Stefanov, D.V. Onopreychuk, Ye.M. Korostelov MATEC Web of Conferences. (to be published)
4. A.S. Akhmatov, *Fizika. Elektrichestvo i stroenie atoma* (Nauka, Moscow, 1974)
5. E.N. Lysikov, V.B. Kosolapov, S.V. Voronin, *Nadmolekuliarnye struktury zhidkikh smazochnykh sred i ikh vliianie na iznos tekhnicheskikh sistem* (EDENA, Moscow, 2009)
6. V.A. Azarenko, *Povyshenie nadezhnosti motorno-osevykh podshipnikov skolzheniia magistralnykh lokomotivov* (Thesis by Ph.D. in Engineering, Moscow, 1984)
7. A.S. Akhmatov, *Molekuliarnaia fizika granichnogo treniia* (Publ. house fiz.-mat. lit., Moscow, 1963)
8. A.D. Zimon, *Chto takoe adgeziia* (Nauka, Moscow, 1983)
9. S.V. Voronin, A.V. Dunaev, Friction and wear. **1 (36)** 41-49 (2015)

Study of the predominant defect development in rails of underground systems after preventive grinding and lubrication

Serhii Voronin[1], *Oleksii* Skoryk[2], *Volodymyr* Stefanov[1], *Dmytro* Onopreychuk[1], and *Yevhen* Korostelov [2*]

[1]Ukrainian State University of Railway Transport, Department of Construction, track and handling machines, Feuerbach sq., 7, Kharkov, Ukraine, 61050
[2]Ukrainian State University of Railway Transport, Department of Track and track facilities, Feuerbach sq., 7, Kharkov, Ukraine, 61050

Abstract. In order to determine the method of contact surface treatment and to define the rational surface roughness in the wheel/rail contact for the minimal growth rate of contact fatigue defects the program of experiments in operational conditions was developed. On the basis of the results and calculations the coefficients of regression equations, which describe dependencies of the wear rate on the parameters of the contact surface treatment in real operational conditions were obtained. Besides, the diagrams of dependencies of the friction coefficient on the roughness of railhead contact surface of the high rails with and without lubrication, when laid in curves of underground systems, were built. The experiments conducted in operational conditions of Kharkiv Metro revealed the influence of roughness of the railhead side surface of the high rails in curves on the contact fatigue defect development after preventive grinding and lubrication. According to the results of the experiments it was established that the rail preventive grinding with simultaneous lubrication led to slower contact fatigue defect development in rails of underground systems. The rational roughness of the railhead side surface of the high rails in curves after preventive grinding, which led to the minimal rate of defect development, was defined.

1 Introduction

Development of the defect according to figure 11.1-2 [1, 2] as dominant for rails of Kharkiv Metro is a multistage process accompanied by the appearance of nano- and microcracks, propagation of cracks, layer separation and spalling the contact surface material. These effects are of contact fatigue nature. The rate of their development depends on the contact material properties. Taking it into consideration it is possible to slow the defect development in side surfaces of the high rails in curves of underground systems at initial stages.

* Corresponding author: dek_str@ukr.net

The primary inspection of rails in Kharkiv Metro demonstrated the presence of contact fatigue defects of the «head checking» type indicating, as the authors believe, the initial stage of the defect, according to figure 11.1-2 [2], which appeared due to lack of contact fatigue strength of the rail steel. This may lead to early removal of such rails.

The average life of rails laid in curves of Kharkiv Metro is 1.5-2 years [2], which is too long.

The measures for preventing the mentioned defects include, mostly, slower development of contact fatigue defects at initial stages [1].

1.1 Analysis of recent studies and publications

The issue of a low rate of the contact fatigue defect development in the wheel/rail system with consideration of all contact features is not solved so far [3]. Considering differences in contact conditions the issue can be solved for certain contact types [4-6]. The research was mostly conducted with consideration of some contact features with many assumptions.

The analysis of existing ways to decrease the rate of the contact fatigue defect development on the railhead side surface, as the most vulnerable for contact conditions of the wheel/rail system in curves of underground tracks, is presented in [3].

The analysis of previous research has shown that the grinding of the railhead during operation on Ukraine's railways is mostly aimed at shaping the railhead [7]. And the issue of achieving certain characteristics of the microgeometry of contact surfaces and the friction control for slow contact fatigue defect development has not been considered at all. The friction control, however, is the most effective method to decrease the rate of defect progress in rails on tracks of underground systems [3].

Studies [8] define the influence of the traffic speed and roughness of contact surfaces on the stress rate in the wheel/rail contact for main line railways. The research of this kind has not been conducted for underground systems so far.

Article [9] proposes rail grinding for achieving rational characteristics of the microgeometry of contact surfaces. And there is a need to consider the features of contact conditions and the material of contact elements in grinding. This is possible for underground rail systems owing to uniformity of the rolling stock and the track elements.

In [9] for the quarry rail transport in order to increase the friction coefficient in the wheel/rail contact and the running-in speed of contract surfaces, the value of rational roughness of the railhead after grinding was defined. The research was conducted in order to increase the friction coefficient by providing contact surfaces with a certain roughness in difficult conditions of the quarry rail transportation. They cannot be applied for determination of the rational friction coefficient in the wheel/rail contact in curves of underground tracks, though they can indicate the correlation between the roughness of contact surfaces and the friction coefficient.

The research into the influence of parameters of the microgeometry of the wheel/rail contact surfaces by example of the underground system is considered to be promising [1]. In [10] on the friction device for such research the wheel/rail contact along the railhead side surface in a curve in the underground system was simulated. The number of experiments [2] of the influence of the roughness of railhead surfaces of the high rails in curves of underground tracks on the friction force in a simulated wheel/rail contact was conducted. Thus, the rational roughness values of the railhead side surfaces of the high rails in curves of underground tracks were obtained. The results of the bench tests [2] are confirmed by the research in operational conditions on tracks in Kharkiv Metro.

The authors in [11] developed an analytical method of defining the rail life of underground systems with preventive grinding and lubrication on the base of previously conducted bench tests. However the method is applicable only for a dry contact. The application of an analytical method for lubrication conditions is not appropriate considering the lubrication film properties. The study presents the research which allows developing the method to define the rail life for underground systems after preventive grinding and lubrication.

2 The basic part of the study

The study program in operation conditions of the tracks in Kharkiv Metro included the study on the section Maidan Konstytutsii-Prospekt Haharina. However, considering a large number of contact fatigue defects on this section the authors chose an additional section of similar characteristics on the line Zakhysnykiv Ukrainy-Arkhitektora Beketova, its rail service time being two months.

According to the theoretical and bench tests [2], the range of values of the surface roughness in the wheel/rail contact in curves of underground systems was established and accounted for R_a=4-0* µm (* the value of roughness R_a=0.008µm corresponds to the 14th class of roughness by the FEPA standards.) And the contact surfaces were treated with grinding paper of the P150, P320 and P500 granulation.

The values of roughness R_a=0*µm was obtained by grinding the contact surfaces. Lubrication of the railhead side surfaces was conducted manually with a lubricant designed for these tests in the laboratory of the Chair «Construction, track and handling machines» Department of Ukrainian State University of Railway Transport [2].

During the experiments in operational conditions, the pictures taken with a digital USB-microscope were later analyzed in terms of presence of contact fatigue defects and their average size. An average size of contact fatigue defects on the underground section was defined by analyzing the pictures (Fig. 1).

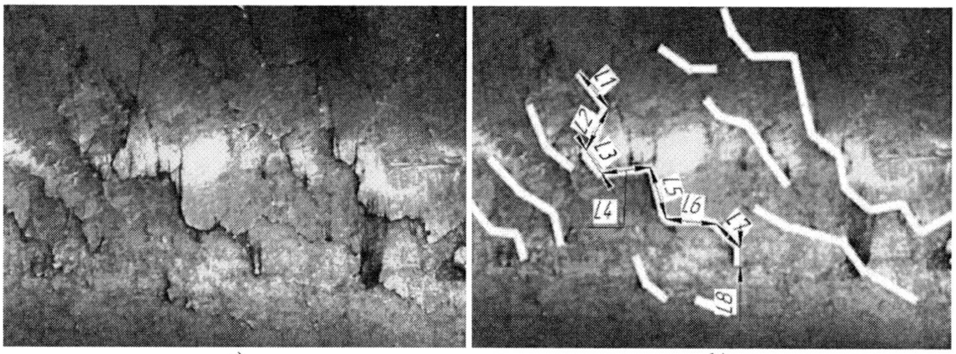

a) b)

Fig. 1. Determination of the size of contact fatigue defects by analyzing pictures taken by a digital microscope: a) – the picture to be analyzed; b) – the analysis of the picture regarding the average size of the contact fatigue defects; L_1-L_n – the lengths of the defects

Determination of an average size of the contact fatigue defects on the test sections was made by the formula:

$$\bar{l} = \frac{\sum_{i=1}^{n} L_i}{n} \quad (1)$$

The research program included two double-factor experiments where the rate of defect development was the response function (Table 1). In order to determine the influence of the rate of the contact fatigue development on the factor of the section tonnage, two double-factor experiments were conducted, each included ten tests with the frequency equal to two.

Table 1. Research program in operational conditions

Number of experiment	Lubrication	Roughness	Operation frequency *T*, MT.gross
1	-	R_a=4μm	1.3
		R_a=3μm	
		R_a=2μm	
		R_a=1μm	
		R_a=0*μm	
2	+	R_a=4μm	
		R_a=3μm	
		R_a=2μm	
		R_a=1μm	
		R_a=0*μm	

3 Results and discussions

The results of the experiments in operational conditions in Kharkiv Metro are given in Table 2:

Table 2. The results of operational tests on the defect development for the test sections

№	*T*, MT. gross	Without lubrication						With lubrication				
		Without grinding	R_a=4μm	R_a=3μm	R_a=2μm	R_a=1μm	R_a=0*μm	R_a=4μm	R_a=3μm	R_a=2μm	R_a=1μm	R_a=0*μm
An average size of the defect on the section Maidan Konstytutsii-Prospekt Haharina $\bar{l}$,mm												
1	23.4	5.500	5.500	5.500	5.500	5.500	5.500	5.500	5.500	5.500	5.500	5.500
2	24.7	5.963	5.954	5.948	5.940	5.956	5.959	5.910	5.903	5.894	5.913	5.915
3	26.0	5.651	5.620	5.723	6.574	6.630	6.633	6.502	6.497	6.458	6.518	6.525
4	27.3	7.462	7.392	7.405	7.309	7.421	7.428	7.190	7.154	7.110	7.227	7.230
5	28.6	8.364	8.245	8.193	8.120	8.299	8.312	7.949	7.855	7.827	8.012	8.223
6	29.9	9.340	9.164	9.003	8.992	9.248	9.264	8.765	8.612	8.596	8.859	8.870
7	31.2	10.380	10.138	10.062	9.915	10.257	10.295	9.630	9.621	9.410	9.760	9.893
8	32.5	11.474	11.161	11.013	10.883	11.320	11.403	10.538	10.293	10.263	10.707	10.801
9	33.8	12.622	12.228	12.173	11.893	12.431	12.442	11.485	11.164	11.151	11.696	11.706
An average size of the defect on the section Zakhysnykiv Ukrainy-Arkhitektora Beketova $\bar{l}$,mm												
1	2.6	0.540	0.540	0.540	0.540	0.540	0.540	0.540	0.540	0.540	0.540	0.540
2	3.9	0.560	0.563	0.571	0.564	0.563	0.569	0.566	0.568	0.570	0.564	0.572
3	5.2	0.619	0.625	0.624	0.527	0.626	0.632	0.628	0.632	0.637	0.625	0.632
4	6.5	0.719	0.723	0.726	0.723	0.727	0.734	0.720	0.721	0.732	0.717	0.721
5	7.8	0.861	0.855	0.854	0.850	0.865	0.873	0.840	0.842	0.853	0.838	0.845
6	9.1	1.043	1.021	1.015	1.008	1.038	1.044	0.986	0.984	0.996	0.986	0.994
7	10.4	1.266	1.218	1.202	1.194	1.245	1.251	1.157	1.164	1.161	1.161	1.172
8	11.7	1.533	1.447	1.416	1.409	1.486	1.493	1.352	1.349	1.347	1.362	1.370
9	13.0	1.838	1.708	1.692	1.650	1.762	1.784	1.569	1.562	1.551	1.588	1.592

On the basis of the results presented in Table 2, the regression equations describing the development of contact fatigue defects in railhead side surfaces on the test sections of the underground railway in dependency on the tonnage carried at various treatments of the contact surface were obtained [11] (Table 3).

Table 3. The regression equations presenting the dependency of the rate of contact fatigue defect development on the surface roughness in the wheel/rail contact in curves of underground systems

Treatment characteristics		Test section on the line Maidan Konstytutsii-Prospekt Haharina	Test section on the line Zakhysnykiv Ukrainy-Arkhitektora Beketova
Grinding to	Lubrication		
-	-	$\bar{I} = 5.5 + 0.328 \cdot T^{1.315}$	$\bar{I} = 0.54 + 0.011 \cdot T^{2.017}$
R_a=4µm	-	$\bar{I} = 5.5 + 0.326 \cdot T^{1.293}$	$\bar{I} = 0.54 + 0.141 \cdot T^{1.889}$
R_a=4µm	+	$\bar{I} = 5.5 + 0.292 \cdot T^{1.289}$	$\bar{I} = 0.54 + 0.016 \cdot T^{1.776}$
R_a=3µm	-	$\bar{I} = 5.5 + 0.319 \cdot T^{1.294}$	$\bar{I} = 0.54 + 0.015 \cdot T^{1.841}$
R_a=3µm	+	$\bar{I} = 5.5 + 0.292 \cdot T^{1.286}$	$\bar{I} = 0.54 + 0.017 \cdot T^{1.741}$
R_a=2µm	-	$\bar{I} = 5.5 + 0.314 \cdot T^{1.287}$	$\bar{I} = 0.54 + 0.014 \cdot T^{1.838}$
R_a=2µm	+	$\bar{I} = 5.5 + 0.282 \cdot T^{1.281}$	$\bar{I} = 0.54 + 0.019 \cdot T^{1.693}$
R_a=1µm	-	$\bar{I} = 5.5 + 0.322 \cdot T^{1.308}$	$\bar{I} = 0.54 + 0.014 \cdot T^{1.911}$
R_a=1µm	+	$\bar{I} = 5.5 + 0.293 \cdot T^{1.301}$	$\bar{I} = 0.54 + 0.014 \cdot T^{1.815}$
R_a=0*µm	-	$\bar{I} = 5.5 + 0.333 \cdot T^{1.304}$	$\bar{I} = 0.54 + 0.013 \cdot T^{1.932}$
R_a=0*µm	+	$\bar{I} = 5.5 + 0.301 \cdot T^{1.318}$	$\bar{I} = 0.54 + 0.014 \cdot T^{1.838}$

The results of the research in operational conditions can be reproduced in accordance with Cochran's C test and the models obtained of the rate of contact fatigue defect development in high rail in curves of underground systems proved to be adequate according to the F-test.

By the data given in Table 3 the diagrams of dependency of defect development for the test sections on the tonnage handled are built (Fig. 2-3):

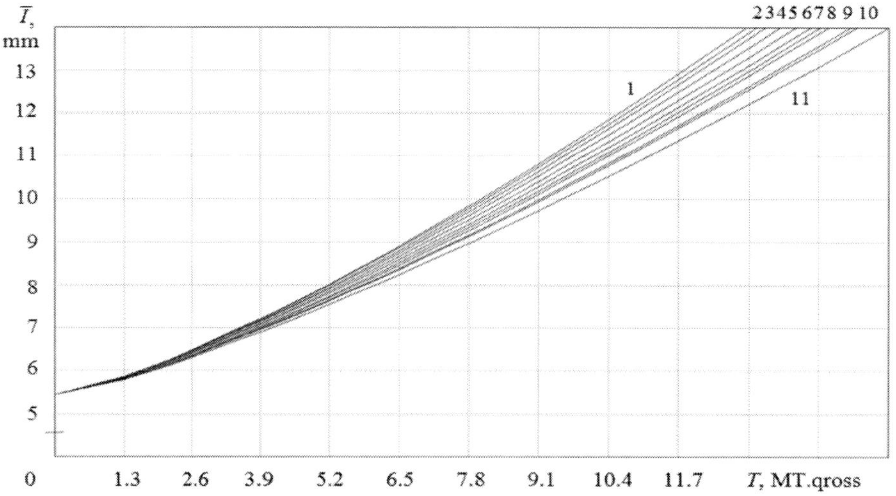

Fig. 2. The dependency of the contact fatigue defect development of the railhead surface of the high rails in curves of underground systems on the tonnage handled for the test section on the line Maidan Konstytutsi-Prospekt Haharina: 1 – without grinding and lubrication; 4, 5, 6, 3, 2 – grinding to R_a=4, 3, 2, 1, 0* µm without lubrication, respectively; 9, 10, 11, 8, 7 – grinding to R_a=4, 3, 2, 1, 0* µm with lubrication

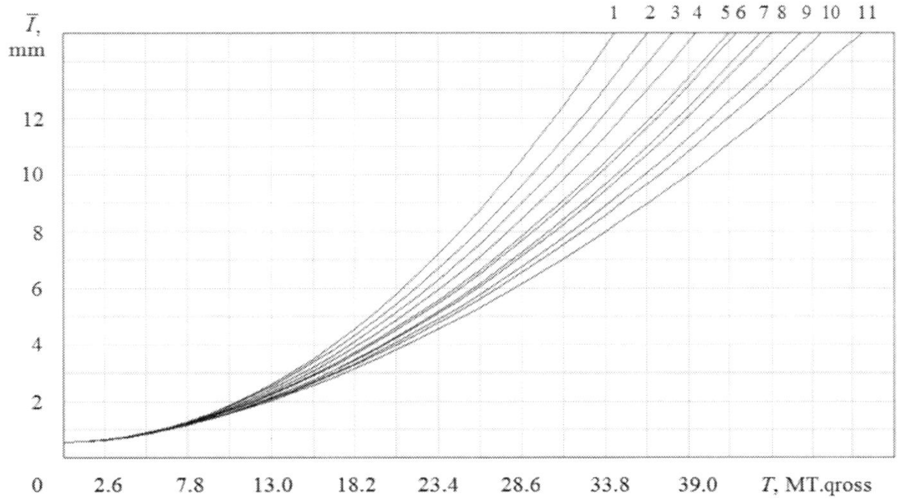

Fig. 3. The dependency of the contact fatigue defect development of the railhead surface of the high rails in curves of underground systems on the tonnage handled for the test section on the line Zakhysnykiv Ukrainy-Arkhitektora Beketova: 1 – without grinding and lubrication; 4, 5, 6, 3, 2 – grinding to R_a=4, 3, 2, 1, 0* µm without lubrication, respectively; 9, 10, 11, 8, 7 – grinding to R_a=4, 3, 2, 1, 0* µm with lubrication

The experiments conducted in operational conditions established that preventive grinding with lubrication led to slow development of contact fatigue defects on railhead side surface of the high rails in curves of underground systems. The values of the rational roughness of railhead side surfaces of the high rails in curves of underground systems after preventive rail grinding of rails with R_a=2µm, when contact fatigue defects propagate with the minimal speed, have been established. Besides, the analysis of dependencies presented in Fig. 2-3 demonstrates that the rational solution regarding retardation of the contract fatigue defect development is preventive grinding of railhead side surfaces in curves of underground rail systems with lubrication of the treated surfaces since the beginning of their service.

4 Conclusion

In order to determine the method of a contact surface treatment and to define the rational surface roughness in the wheel/rail contact for minimal growth rate of contact fatigue defects the program of experiments in operational conditions was developed.

The research conducted in operational conditions in curves of Kharkiv Metro revealed the influence of the roughness of railhead side surfaces of the high rails in curves after preventive grinding and lubrication on the contact fatigue defect development.

The results of the experiments carried out in the operational conditions made it possible to establish that the rail preventive grinding with simultaneous lubrication led to slower contact fatigue defect development in rails of underground systems. The rational roughness of the railhead side surface of the high rails in curves after preventive rail grinding at R_a=2µm with minimal rate of defect development was defined. And the preventive grinding with consequent lubrication of treated rail surfaces should begin with laying down rails in curves of underground systems.

References

1. Ye. Korostelov. Nauchnye trudy SWorld. **3 (44)**, 25-30 (2016).

2. S. Voronin, O. Skoryk, Ye. Korostelov. Eastern-European Journal of Enterprise Technologies. **4** (**80**). 11-17 (2016).
3. S. Voronyn, Ye. Korostelov. Informatsiyno-keruyuchi systemy na zaliznychnomu transporti, **3**, 22-27 (2014).
4. A. Alonso, J. Gimenez. TRIBOL INT, **8**, 755-768 (2008).
5. P. Remington, J. Webb. J. SOUND VIB. **5-6**, 83-102 (1996).
6. A. Rovira, A. Roda, M. Marshall, H. Brunskill, R. Lewis. WEAR, **5-6**, 911-924 (2011).
7. D. Potapov, V. Vitolberh. Informatsiyno-keruyuchi systemy na zaliznychnomu transporti, **5-6**, 70-73 (2006).
8. W. Wang, R. Lewis, B. Yang, L. Guo, Q. Liu, M. Zhu. WEAR, **15**, 146-152 (2016).
9. A. Keropyan. Procedia Engineering, **150**, 406-410 (2016).
10. S. Voronyn, S. Karpenko, Ye. Korostelov. Zbirnyk nauk. prats UkrDAZT, **148**, 201-204 (2014).
11. S. Voronin, O. Skoryk, Ye. Korostelov. Zbirnyk nauk. prats UkrDUZT, **167**, 156-162 (2017).

AUTHOR INDEX

AUTHOR INDEX